Lecture Notes in Computer Science 10225

Commenced Publication in 1973
Founding and Former Series Editors:
Gerhard Goos, Juris Hartmanis, and Jan van Leeuwen

More information about this series at http://www.springer.com/series/7412

Jesús Angulo · Santiago Velasco-Forero
Fernand Meyer (Eds.)

Mathematical Morphology and Its Applications to Signal and Image Processing

13th International Symposium, ISMM 2017
Fontainebleau, France, May 15–17, 2017
Proceedings

Springer

Editors
Jesús Angulo
Center for Mathematical Morphology
MINES ParisTech, PSL-Research University
Fontainebleau
France

Fernand Meyer
Center for Mathematical Morphology
MINES ParisTech, PSL-Research University
Fontainebleau
France

Santiago Velasco-Forero
Center for Mathematical Morphology
MINES ParisTech, PSL-Research University
Fontainebleau
France

ISSN 0302-9743   ISSN 1611-3349   (electronic)
Lecture Notes in Computer Science
ISBN 978-3-319-57239-0   ISBN 978-3-319-57240-6   (eBook)
DOI 10.1007/978-3-319-57240-6

Library of Congress Control Number: 2017937501

LNCS Sublibrary: SL6 – Image Processing, Computer Vision, Pattern Recognition, and Graphics

Printed on acid-free paper

This Springer imprint is published by Springer Nature
The registered company is Springer International Publishing AG
The registered company address is: Gewerbestrasse 11, 6330 Cham, Switzerland

Preface

This volume contains the articles accepted for presentation at the 13th International Symposium on Mathematical Morphology (ISMM 2017), held in Fontainebleau, France, during May 15–17, 2017. The 12 previous editions of this conference were very successful, and the ISMM has established itself as the main scientific event in the field. We were delighted to celebrate the 50th anniversary of the foundation of the Center for Mathematical Morphology (CMM) and connect it with the second and third generations of worldwide mathematical morphology researchers by hosting ISMM 2017 at the CMM, on the campus of MINES ParisTech, in Fontainebleau.

We received 53 high-quality papers, each of which was sent to at least three Program Committee members for review. Based on 161 detailed reviews, we accepted 40 papers. A few of these papers were accepted after a substantial revision by the authors in response to reviewer comments. The authors of these 40 articles are from 15 different countries: Australia, Austria, Brazil, France, Germany, Greece, India, Japan, The Netherlands, Poland, Serbia, Sweden, Turkey, USA, and UK.

The ISMM 2017 papers highlight the current trends and advances in mathematical morphology, be they purely theoretical contributions, algorithmic developments, or novel applications, where real-life imaging and computer vision problems are tackled with morphological tools.

We wish to thank the members of the Program Committee for their efforts in reviewing all submissions on time and giving extensive feedback. We would like to take this opportunity to thank all the other people involved in this conference: firstly, the Steering Committee for giving us the chance to organize ISMM2017; secondly, the three invited speakers, Dan Claudiu Cireşan, Stephane Gaubert, and Pierre Vandergheynst, for accepting to share their recognized expertise; and finally, the most important component of any scientific conference, the authors for producing the high-quality and original contributions. We acknowledge the EasyChair conference management system that was invaluable in handling the paper submission and review process and putting this volume together. Last but not least, we would like to thank the financial and logistic support of our institution, MINES ParisTech, and the members of the local Organizing Committee, especially Anne-Marie De Castro.

March 2017

Jesús Angulo
Santiago Velasco-Forero
Fernand Meyer

Organization

ISMM 2017 was organized by the Center for Mathematical Morphology, MINES ParisTech, PSL-Research University, France.

Organizing Committee

General Chair, Program Chair

Jesús Angulo   MINES ParisTech, France

Program Co-chairs

Santiago Velasco-Forero   MINES ParisTech, France
Fernand Meyer   MINES ParisTech, France

Local Organization Chair

Michel Bilodeau   MINES ParisTech, France

Local Organization Co-chair

Beatriz Marcotegui   MINES ParisTech, France

Local Organization Assistant

Anne-Marie De Castro   MINES ParisTech, France

DGCI Liaison and Advisory

Gunilla Borgefors   Uppsala University, Sweden

Steering Committee

Jesús Angulo   MINES ParisTech, France
Junior Barrera   University of Sao Paulo, Brazil
Jón Atli Benediktsson   University of Iceland
Isabelle Bloch   Telecom ParisTech, France
Gunilla Borgefors   Uppsala University, Sweden
Jocelyn Chanussot   Grenoble Institute of Technology, France
Renato Keshet   Hewlett Packard Laboratories, Israel
Ron Kimmel   Technion, Israel Institute of Technology, Israel
Cris Luengo Hendricks   Flagship Biosciences Inc., Colorado, USA
Petros Maragos   National Technical University of Athens, Greece
Laurent Najman   Université Paris-Est ESIEE, France
Christian Ronse   University of Strasbourg, France
Philippe Salembier   Polytechnic University of Catalonia, Spain

Dan Schonfeld   University of Illinois at Chicago, USA
Pierre Soille   European Commission, Joint Research Centre, Italy
Hugues Talbot   Université Paris-Est ESIEE, France
Michael H.F. Wilkinson   University of Groningen, The Netherlands

Program Committee

Jesús Angulo   MINES ParisTech, France
Akira Asano   Kansai University, Japan
Junior Barrera   Universidade de São Paulo, Brazil
Jon Atli Benediktsson   University of Iceland, Iceland
Isabelle Bloch   Telecom ParisTech, France
Gunilla Borgefors   Uppsala University, Sweden
Michael Buckley   CSIRO, Australia
Bernhard Burgeth   Saarland University, Germany
Bhabatosh Chanda   Indian Statistical Institute, India
Jocelyn Chanussot   Grenoble Institute of Technology, France
David Coeurjolly   CNRS, LIRIS, France
Jean Cousty   Université Paris-Est ESIEE, France
Jose Crespo   Universidad Politécnica de Madrid, Spain
Vladimir Curic   Imint Vidhance, Sweden
Johan Debayle   ENS des Mines de Saint-Etienne, France
Etienne Decenciere   MINES ParisTech, France
Petr Dokladal   MINES ParisTech, France
Abderrahim Elmoataz   University of Caen Basse Normandie, France
Adrian Evans   University of Bath, UK
Thierry Géraud   EPITA Research and Development Lab., France
Marcin Iwanowski   Warsaw University of Technology, Poland
Andrei Jalba   Eindhoven University of Technology, The Netherlands
Dominique Jeulin   MINES ParisTech, France
Ron Kimmel   Technion, Israel
Christer Kiselman   Uppsala University, Sweden
Corinne Lagorre   Université Paris-Est Créteil, France
Anders Landström   Luleå University of Technology, Sweden
Sébastien Lefèvre   Université de Bretagne Sud, France
Cris L. Luengo Hendriks   Flagship Biosciences, USA
Petros Maragos   National Technical University of Athens, Greece
Beatriz Marcotegui   MINES ParisTech, France
Juliette Mattioli   Thales
Petr Matula   Masaryk University, Czech Republic
Fernand Meyer   MINES ParisTech, France
Laurent Najman   Université Paris-Est ESIEE, France
Valery Naranjo   Universidad Politécnica de Valencia, Spain
Georgios Ouzounis   DigitalGlobe, Inc., USA
Nicolas Passat   Université de Reims Champagne-Ardenne, France
Benjamin Perret   Université Paris-Est ESIEE, France

Wilfried Philips	Ghent University, Belgium
Jos Roerdink	University of Groningen, The Netherlands
Christian Ronse	Université de Strasbourg, France
Philippe Salembier	Universitat Politècnica de Catalunya, Spain
Gabriella Sanniti di Baja	Institute of Cybernetics E.Caianiello, CNR, Italy
Katja Schladitz	Fraunhofer ITWM, Germany
Jean Serra	France
B.S. Daya Sagar	Indian Statistical Institute Bangalore Centre, India
Natasa Sladoje	Uppsala University, Sweden
Pierre Soille	European Commission Joint Research Centre, Italy
Robin Strand	Uppsala University, Sweden
Hugues Talbot	Université Paris Est ESIEE, France
Ivan Terol-Villalobos	CIDETEQ, S.C., Mexico
Jasper van de Gronde	University of Groningen, The Netherlands
Marc Van Droogenbroeck	University of Liège, Belgium
Santiago Velasco-Forero	MINES ParisTech, France
Rafael Verdú-Monedero	Technical University of Cartagena, Spain
Martin Welk	UMIT, Austria
Michael Wilkinson	University of Groningen, The Netherlands

Additional Reviewers

Aptoula, Erchan
Asplund, Teo
Cavallaro, Gabriele
Chabardès, Théodore
Dalla Mura, Mauro

Kleefeld, Andreas
Lindblad, Joakim
Marpu, Prashanth Reddy
Roussillon, Tristan

Contents

Trees and Hierarchies

Topological and Graph-Based Clustering, Classification and Filtering

Connected Operators and Attribute Filters

Algebraic Theory, Max-Plus and Max-Min Mathematics

Morphological Perceptrons: Geometry and Training Algorithms

Vasileios Charisopoulos$^{(\boxtimes)}$ and Petros Maragos

School of ECE, National Technical University of Athens, 15773 Athens, Greece
vharisop@gmail.com, maragos@cs.ntua.gr

Abstract. Neural networks have traditionally relied on mostly linear models, such as the multiply-accumulate architecture of a linear perceptron that remains the dominant paradigm of neuronal computation. However, from a biological standpoint, neuron activity may as well involve inherently nonlinear and competitive operations. Mathematical morphology and minimax algebra provide the necessary background in the study of neural networks made up from these kinds of nonlinear units. This paper deals with such a model, called the morphological perceptron. We study some of its geometrical properties and introduce a training algorithm for binary classification. We point out the relationship between morphological classifiers and the recent field of tropical geometry, which enables us to obtain a precise bound on the number of linear regions of the maxout unit, a popular choice for deep neural networks introduced recently. Finally, we present some relevant numerical results.

Keywords: Mathematical morphology · Neural networks · Machine learning · Tropical geometry · Optimization

1 Introduction

In traditional literature on pattern recognition and machine learning, the so-called perceptron, introduced by Rosenblatt [21], has been the dominant model of neuronal computation. A *neuron* is a computational unit whose activation is a "multiply-accumulate" product of the input and a set of associated *synaptic weights*, optionally fed through a non-linearity. This model has been challenged in terms of both biological and mathematical plausibility by the morphological paradigm, widely used in computer vision and related disciplines. This has lately attracted a stronger interest from researchers in computational intelligence motivating further theoretical and practical advances in morphological neural networks, despite the fact that learning methods based on lattice algebra and mathematical morphology can be traced back to at least as far as the 90s (e.g. [6,19]).

In this paper, we re-visit the model of the *morphological perceptron* [24] in Sect. 3 and relate it with the framework of $(\max, +)$ and $(\min, +)$ algebras. In Sect. 3.1, we investigate its potential as a classifier, providing some fundamental geometric insight. We present a training algorithm for binary classification

© Springer International Publishing AG 2017
J. Angulo et al. (Eds.): ISMM 2017, LNCS 10225, pp. 3–15, 2017.
DOI: 10.1007/978-3-319-57240-6_1

that uses the Convex-Concave Procedure and a more robust variant utilizing a simple form of outlier ablation. We also consider more general models such as maxout activations [11], relating the number of linear regions of a maxout unit with the *Newton Polytope* of its activation function, in Sect. 4. Finally, in Sect. 5, we present some experimental results pertinent to the efficiency of our proposed algorithm and provide some insight on the use of morphological layers in multilayer architectures.

We briefly describe the notation that we use. Denoting by $\mathbb{R}$ the line of real numbers, $(-\infty, \infty)$, let $\mathbb{R}_{max} = \mathbb{R} \cup \{-\infty\}$ and $\mathbb{R}_{min} = \mathbb{R} \cup \{\infty\}$. We use lowercase symbols for scalars (like x), lowercase symbols in boldface for vectors (like $\boldsymbol{w}$) and uppercase symbols in boldface for matrices (like $\boldsymbol{A}$). Vectors are assumed to be column vectors, unless explicitly stated otherwise.

We will focus on the (max, +) *semiring*, which is the semiring with underlying set $\mathbb{R}_{max}$, using max as its binary "addition" and + as its binary "multiplication". We may also refer to the (min, +) semiring which has an analogous definition, while the two semirings are actually isomorphic by the trivial mapping $\phi(x) = -x$. Both fall under the category of *idempotent* semirings [10], and are considered examples of so-called *tropical semirings*.[1]

Finally, we will use the symbol $\boxplus$ to refer to matrix and vector "multiplication" in (max, +) algebra and $\boxplus'$ for its dual in (min, +) algebra, following the convention established in [15]. Formally, we can define matrix multiplication as:

$$(\boldsymbol{A} \boxplus \boldsymbol{B})_{ij} = \bigvee_{q=1}^{k} A_{iq} + B_{qj} \qquad (\boldsymbol{A} \boxplus' \boldsymbol{B})_{ij} = \bigwedge_{q=1}^{k} A_{iq} + B_{qj} \qquad (1)$$

for matrices of compatible dimensions.

2   Related Work

In [20], the authors argued about the biological plausibility of nonlinear responses, such as those introduced in Sect. 3. They proposed neurons computing max-sums and min-sums in an effort to mimic the response of a dendrite in a biological system, and showed that networks built from such neurons can approximate any compact region in Euclidean space within any desired degree of accuracy. They also presented a constructive algorithm for binary classification. Sussner and Esmi [24] introduced an algorithm based on competitive learning, combining morphological neurons to enclose training patterns in bounding boxes, achieving low response times and independence from the order by which training patterns are presented to the training procedure.

---

[1] The term "tropical" was playfully introduced by French mathematicians in honor of the Brazilian theoretical computer scientist, Imre Simon. Another example of a tropical semiring is the (max, ×) semiring, also referred to as the subtropical semiring.

Yang and Maragos [29] introduced the class of min-max classifiers, boolean-valued functions appearing as thresholded minima of maximum terms or maxima of minimum terms:

$$f_{\text{max-min}}(x_1, x_2, \ldots x_d) = \bigwedge_j \bigvee_{i \in I_j} l_i, \quad l_i \in \{x_i, 1 - x_i\} \tag{2}$$

and vice-versa for $f_{\text{min-max}}$. In the above, I_j is the set of indices corresponding to term j. These classifiers produce decision regions similar to those formed by a $(\max, +)$ or $(\min, +)$ perceptron.

Barrera et al. [3] tried to tackle the problem of statistically optimal design for set operators on binary images, consisting of morphological operators on sets. They introduced an *interval splitting* procedure for learning boolean concepts and applied it to binary image analysis, such as edge detection or texture recognition.

With the exception of [29], the above introduce constructive training algorithms which may produce complex decision regions, as they fit models precisely to the training set. They may create superfluous decision areas to include outliers that might be disregarded when using gradient based training methods, a fact that motivates the work in Sect. 3.2.

In a recent technical report, Gärtner and Jaggi [8] proposed the concept of a *tropical support vector machine*. Its response and j-th decision region are given by:

$$y(\boldsymbol{x}) = \bigwedge_{i=1}^{n} w_i + x_i , \quad \mathcal{R}^j(\boldsymbol{x}) = \{\boldsymbol{x} : w_j + x_j \leq w_i + x_i, \forall i\} \tag{3}$$

instead of a "classical" decision region (e.g. defined by some discriminant function).

Cuninghame-Green's work on minimax algebra [5] provides much of the matrix-vector framework for the finite-dimensional morphological paradigm. A fundamental result behind Sussner and Valle's article [25] on morphological analogues of classical associative memories such as the Hopfield network, states that the "closest" under-approximation of a target vector $\boldsymbol{b}$ by a max-product in the form $\boldsymbol{A} \boxplus \boldsymbol{x}$ can be found by the so-called *principal solution* of a max-linear equation.

Theorem 1. [5] *If $\boldsymbol{A} \in \mathbb{R}_{\max}^{m \times n}, \boldsymbol{b} \in \mathbb{R}_{\max}^{m}$, then*

$$\overline{\boldsymbol{x}} = \boldsymbol{A}^{\sharp} \boxplus' \boldsymbol{b} \quad (\boldsymbol{A}^{\sharp} \triangleq -\boldsymbol{A}^T) \tag{4}$$

is the greatest solution to $\boldsymbol{A} \boxplus \boldsymbol{x} \leq \boldsymbol{b}$, and furthermore $\boldsymbol{A} \boxplus \boldsymbol{x} = \boldsymbol{b}$ has a solution if and only if $\overline{\boldsymbol{x}}$ is a solution.[2]

---

[2] The matrix $-\boldsymbol{A}^T$, often denoted by $\boldsymbol{A}^{\sharp}$ in the tropical geometry community, is sometimes called the *Cuninghame-Green inverse* of $\boldsymbol{A}$.

3   The Morphological Perceptron

Classical literature defines the perceptron as a computational unit with a linear activation possibly fed into a non-linearity. Its output is the result of the application of an activation function, that is usually nonlinear, to its activation $\phi(\boldsymbol{x})$. Popular examples are the logistic sigmoid function or the rectifier linear unit, which has grown in popularity among deep learning practitioners [17]. For the morphological neuron, in [20], its response to an input $\boldsymbol{x} \in \mathbb{R}^n$ is given by

$$\tau(\boldsymbol{x}) = p \cdot \bigvee_{i=1}^{n} r_i(x_i + w_i) , \qquad \tau'(\boldsymbol{x}) = p \cdot \bigwedge_{i=1}^{n} r_i(x_i + m_i) \qquad (5)$$

for the cases of the $(\max, +)$ and $(\min, +)$ semirings respectively. Parameters r_i and p take values in $\{+1, -1\}$ depending on whether the synapses and the output are excitatory or inhibitory. We adopt a much simpler version:

Definition 1. (Morphological Perceptron). *Given an input vector $\boldsymbol{x} \in \mathbb{R}_{\max}^n$, the morphological perceptron associated with weight vector $\boldsymbol{w} \in \mathbb{R}_{\max}^n$ and activation bias $w_0 \in \mathbb{R}_{\max}$ computes the activation*

$$\tau(\boldsymbol{x}) = w_0 \vee (w_1 + x_1) \vee \cdots \vee (w_n + x_n) = w_0 \vee \left(\bigvee_{i=1}^{n} w_i + x_i \right) \qquad (6)$$

We may define a "dual" model on the $(\min, +)$ semiring, as the perceptron with parameters $\boldsymbol{m} \in \mathbb{R}_{\min}^n, m_0 \in \mathbb{R}_{\min}$ that computes the activation

$$\tau'(\boldsymbol{x}) = m_0 \wedge (m_1 + x_1) \wedge \cdots \wedge (m_n + x_n) = m_0 \wedge \left(\bigwedge_{i=1}^{n} m_i + x_i \right) \qquad (7)$$

The models defined by (6, 7) may also be referred to as $(\max, +)$ and $(\min, +)$ perceptron, respectively. They can be treated as instances of morphological filters [14, 22], as they define a (grayscale) dilation and erosion over a finite window, computed at a certain point in space or time. Note that $\tau(\boldsymbol{x})$ is a nonlinear, convex (as piecewise maximum of affine functions) function of $\boldsymbol{x}, \boldsymbol{w}$ that is continuous everywhere, but not differentiable everywhere (points where multiple terms maximize $\tau(\boldsymbol{x})$ are singular).

3.1   Geometry of a $(\max, +)$ Perceptron for Binary Classification

Let us now put the morphological perceptron into the context of binary classification. We will first try to investigate the perceptron's geometrical properties drawing some background from tropical geometry.

Let $\boldsymbol{X} \in \mathbb{R}_{\max}^{k \times n}$ be a matrix containing the patterns to be classified as its rows, let $\boldsymbol{x}^{(k)}$ denote the k-th pattern (row) and let $\mathcal{C}_1, \mathcal{C}_0$ be the two classes of the relevant decision problem. Without loss of generality, we may choose

$y_k = 1$ if $\boldsymbol{x}^{(k)} \in C_1$ and $y_k = -1$ if $\boldsymbol{x}^{(k)} \in C_0$. Using the notation in (1), the $(\max, +)$ perceptron with parameter vector $\boldsymbol{w}$ computes the output

$$\tau(\boldsymbol{x}) = \boldsymbol{w}^T \boxplus \boldsymbol{x} \tag{8}$$

Note that the variant we study here has no activation bias ($w_0 = -\infty$). If we assign class labels to patterns based on the sign function, we have $\tau(\boldsymbol{x}) > 0 \Rightarrow \boldsymbol{x} \in C_1$, $\tau(\boldsymbol{x}) < 0 \Rightarrow \boldsymbol{x} \in C_0$. Therefore, the decision regions formed by that perceptron have the form

$$\mathcal{R}_1 := \{\boldsymbol{x} \in \mathbb{R}^n_{\max} : \boldsymbol{w}^T \boxplus \boldsymbol{x} \geq 0\}, \ \mathcal{R}_0 := \{\boldsymbol{x} \in \mathbb{R}^n_{\max} : \boldsymbol{w}^T \boxplus \boldsymbol{x} \leq 0\} \tag{9}$$

As it turns out, these inequalities are collections of so called *affine tropical half-spaces* and define *tropical polyhedra* [9,13], which we will now introduce.

Definition 2 (Affine tropical halfspace). *Let $a, b \in \mathbb{R}^{n+1}_{\max}$. An **affine tropical halfspace** is a subset of $\mathbb{R}^n_{\max}$ defined by*

$$\mathcal{T}(\boldsymbol{a}, \boldsymbol{b}) := \left\{ \boldsymbol{x} \in \mathbb{R}^n_{\max} : \left(\bigvee_{i=1}^n a_i + x_i \right) \vee a_{n+1} \geq \left(\bigvee_{i=1}^n b_i + x_i \right) \vee b_{n+1} \right\} \tag{10}$$

We can further assume that $\min(a_i, b_i) = -\infty \ \forall i \in \{1, 2, \ldots, n+1\}$, as per [9, Lemma 1].

A *tropical polyhedron* is the intersection of finitely many tropical halfspaces (and comes in signed and unsigned variants, as in [1]). In our context, we will deal with tropical polyhedra like the following: assume $\boldsymbol{A} \in \mathbb{R}^{m \times n}_{\max}, \boldsymbol{B} \in \mathbb{R}^{k \times n}_{\max}, \boldsymbol{c} \in \mathbb{R}^m_{\max}$ and $\boldsymbol{d} \in \mathbb{R}^k_{\max}$. The inequalities

$$\boldsymbol{A} \boxplus \boldsymbol{x} \geq \boldsymbol{c}, \quad \boldsymbol{B} \boxplus \boldsymbol{x} \leq \boldsymbol{d} \tag{11}$$

define a subset $\mathcal{P} \subseteq \mathbb{R}^n_{\max}$ that is a tropical polyhedron, which can be empty if some of the inequalities cannot be satisfied, leading us to our first remark.

Proposition 1 (Feasible Regions are Tropical Polyhedra). *Let $\boldsymbol{X} \in \mathbb{R}^{k \times n}_{\max}$ be a matrix containing input patterns of dimension n as its rows, partitioned into two distinct matrices $\boldsymbol{X}_{\mathrm{pos}}$ and $\boldsymbol{X}_{\mathrm{neg}}$, which contain all patterns of classes C_1, C_0 respectively. Let $\mathcal{T}$ be the tropical polyhedron defined by*

$$\mathcal{T}(\boldsymbol{X}_{\mathrm{pos}}, \boldsymbol{X}_{\mathrm{neg}}) = \{\boldsymbol{w} \in \mathbb{R}^n_{\max} : \boldsymbol{X}_{\mathrm{pos}} \boxplus \boldsymbol{w} \geq 0, \ \boldsymbol{X}_{\mathrm{neg}} \boxplus \boldsymbol{w} \leq 0\} \tag{12}$$

Patterns $\boldsymbol{X}_{\mathrm{pos}}, \boldsymbol{X}_{\mathrm{neg}}$ can be completely separated by a $(\max, +)$ perceptron if and only if $\mathcal{T}$ is nonempty.

Remark 1. In [9], it has been shown that the question of a tropical polyhedron being nonempty is polynomially equivalent to an associated mean payoff game having a winning initial state.

Using the notion of the *Cuninghame-Green inverse* from Theorem 1, we can restate the separability condition in Proposition 1. As we know that $\overline{\boldsymbol{w}} = \boldsymbol{X}^{\sharp}_{\mathrm{neg}} \boxplus' \boldsymbol{0}$ is the greatest solution to $\boldsymbol{X}_{\mathrm{neg}} \boxplus \boldsymbol{w} \leq \boldsymbol{0}$, that condition is equivalent to

$$\boldsymbol{X}_{\mathrm{pos}} \boxplus (\boldsymbol{X}^{\sharp}_{\mathrm{neg}} \boxplus' \boldsymbol{0}) \geq \boldsymbol{0} \tag{13}$$

3.2   A Training Algorithm Based on the Convex-Concave Procedure

In this section, we present a training algorithm that uses the Convex-Concave Procedure [30] in a manner similar to how traditional Support Vector Machines use convex optimization to determine the optimal weight assignment for a binary classification problem. It is possible to state an optimization problem with a convex cost function and constraints that consist of inequalities of difference-of-convex (DC) functions. Such optimization problems can be solved (at least approximately) by the Convex-Concave Procedure.

$$\text{Minimize } J(\boldsymbol{X}, \boldsymbol{w}) = \sum_{k=1}^{K} \max(\xi_k, 0)$$

$$\text{s. t.} \begin{cases} \bigvee_{i=1}^{n} w_i + x_i^{(k)} \leq \xi_k & \text{if } \boldsymbol{x}^{(k)} \in \mathcal{C}_0 \\ \bigvee_{i=1}^{n} w_i + x_i^{(k)} \geq -\xi_k & \text{if } \boldsymbol{x}^{(k)} \in \mathcal{C}_1 \end{cases} \tag{14}$$

The slack variables ξ_k in the constraints are used to ensure that only misclassified patterns will contribute to J. In our implementation, we use [23, Algorithm 1.1], utilizing the authors' DCCP library that extends CvxPy [7], a modelling language for convex optimization in Python. An application on separable patterns generated from a Gaussian distribution can be seen in Fig. 1.

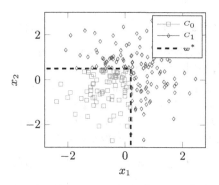

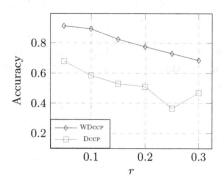

Fig. 1. Decision surface

Fig. 2. Method accuracy

So far, we have not addressed the case where patterns are not separable or contain "abnormal" entries and outliers. Although many ways have been proposed to deal with the presence of outliers [28], the method we used to overcome this was to "penalize" patterns with greater chances of being outliers. We introduce a simple weighting scheme that assigns, to each pattern, a factor that is inversely proportional to its distance (measured by some ℓ_p-norm) from its class's centroid.

$$\boldsymbol{\mu}_i := \frac{1}{|\mathcal{C}_i|} \sum_{\boldsymbol{x}^{(k)} \in \mathcal{C}_i} \boldsymbol{x}^{(k)}, \quad \lambda_k := \frac{1}{||\boldsymbol{x}^{(k)} - \boldsymbol{\mu}_i||_p} \tag{15}$$

$$\nu_k := \frac{\lambda_k}{\max_k \lambda_k} \tag{16}$$

Equation (16) above serves as a normalization step that scales all λ_k in the $(0, 1]$ range. We arrive at a reformulated optimization problem which can be stated as

$$\text{Minimize } J(\boldsymbol{X}, \boldsymbol{w}, \boldsymbol{\nu}) = \sum_{k=1}^{K} \nu_k \cdot \max(\xi_k, 0)$$

$$\text{s. t.} \begin{cases} \bigvee\limits_{i=1}^{n} w_i + x_i^{(k)} \leq \xi_k & \text{if } \boldsymbol{x}^{(k)} \in \mathcal{C}_0 \\ \bigvee\limits_{i=1}^{n} w_i + x_i^{(k)} \geq -\xi_k & \text{if } \boldsymbol{x}^{(k)} \in \mathcal{C}_1 \end{cases} \tag{17}$$

To illustrate the practical benefits of this method (which we will refer to as WDCCP), we use both versions of the optimization problem on a set of randomly generated data which is initially separable but then a percentage r of its class labels is flipped. Comparative results for a series of percentages r are found in Fig. 2. The results for $r = 20\%$ can be seen in Fig. 3, with the dashed line representing the weights found by WDCCP. This weighting method can be extended to complex or heterogeneous data; for example, one could try and fit a set of patterns to a mixture of Gaussians or perform clustering to obtain the coefficients $\boldsymbol{\nu}$.

It is possible to generalize the morphological perceptron to combinations of dilations (max-terms) and erosions (min-terms). In [2], the authors introduce the *Dilation-Erosion Linear Perceptron*, which contains a convex combination of a dilation and an erosion, as:

$$M(\boldsymbol{x}) = \lambda \tau(\boldsymbol{x}) + (1 - \lambda)\tau'(\boldsymbol{x}), \quad \lambda \in [0, 1] \tag{18}$$

plus a linear term, employing gradient descent for training. The formulation in (17) can be used here too, as constraints in difference-of-convex programming can be (assuming f_i convex):

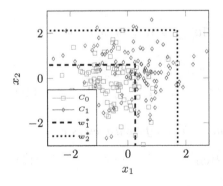

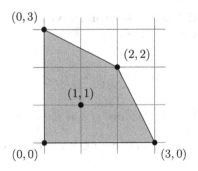

Fig. 3. Optimal weights found   Fig. 4. Newt(p) of Eq. (23)

$$f_i(x) - g_i(x) \leq 0, \ g_i \text{ convex, or } f_i(x) + g'_i(x) \leq 0, \ g'_i \text{ concave} \qquad (19)$$

This observation is exploited in the first experiment of Sect. 5.

4   Geometric Interpretation of Maxout Units

Maxout units were introduced by Goodfellow et al. [11]. A maxout unit is associated with a weight matrix $W \in \mathbb{R}_{max}^{k \times n}$ as well as an activation bias vector $b \in \mathbb{R}_{max}^{k}$. Given an input pattern $x \in \mathbb{R}_{max}^{n}$ and denoting by $W_{j,:}$ the j-th row vector of W, a maxout unit computes the following activation:

$$h(x) = \bigvee_{j=1}^{k} W_{j,:} x + b_j = \bigvee_{j=1}^{k} \left[\left(\sum_{i=1}^{n} W_{ji} x_i \right) + b_j \right] \qquad (20)$$

Essentially, a maxout unit generalizes the morphological perceptron using k terms (referred to as the unit's *rank*) that involve affine expressions. In tropical algebra, such expressions are called *tropical polynomials* [13] or *maxpolynomials* [4] when specifically referring to the $(\max, +)$ semiring. In [16], maxout units are investigated geometrically in an effort to obtain bounds for the number of linear regions of a deep neural network with maxout layers:

Proposition 2 ([16], Proposition 7). *The maximal number of linear regions of a single layer maxout network with n inputs and m outputs of rank k is lower bounded by $k^{\min(n,m)}$ and upper bounded by $\min \left\{ \sum_{j=0}^{n} \binom{k^2 m}{j}, k^m \right\}$.*

This result readily applies to layers consisting of $(\max, +)$ perceptrons, as a $(\max, +)$ perceptron has rank $k = n$.

For a maxout unit of rank k, the authors argued that the number of its linear regions is exactly k if every term is maximal at some point. We provide an exact result using tools from tropical geometry; namely, the *Newton Polytope* of a maxpolynomial. For definitions and fundamental results on polytopes the reader is referred to [31]; we kick off our investigation omitting the presence of the bias term b_j as seen in (20).

Definition 3 (Newton Polytope). *Let $p : \mathbb{R}_{max}^{n} \to \mathbb{R}_{max}$ be a maxpolynomial with k terms, given by*

$$p(x) = \max_{i \in 1,2,\ldots,k} \{ c_{i1} x_1 + c_{i2} x_2 + \cdots + c_{in} x_n \} = \bigvee_{i=1}^{k} c_i^T x \qquad (21)$$

The Newton Polytope of p is the convex hull of the coefficient vectors c_i:

$$\text{Newt}(p) = \text{conv}\{ c_i : i \in 1, \ldots, k \} = \text{conv}\{ (c_{i1}, c_{i2}, \ldots c_{in}) : i \in 1, \ldots, k \} \qquad (22)$$

For an illustrative example, see Fig. 4. The maxpolynomial in question is

$$p(\boldsymbol{x}) = 0 \vee (x + y) \vee 3x \vee (2x + 2y) \vee 3y \qquad (23)$$

and its terms can be matched to the coefficient vectors $(0,0), (1,1), (3,0), (2,2)$ and $(0,3)$ respectively. The Newton Polytope's vertices give us information about the number of linear regions of the associated maxpolynomial:

Proposition 3. *Let $p(\boldsymbol{x})$ be a maxout unit with activation given by (21). The number of p's linear regions is equal to the number of vertices of its Newton Polytope, Newt(p).*

Proof. A proof can be given using the fundamental theorem of Linear Programming [26, Theorem 3.4]. Consider the linear program:

$$\text{Maximize } \boldsymbol{x}^T \boldsymbol{c}$$
$$\text{s.t. } \boldsymbol{c} \in \text{Newt}(p) \qquad (24)$$

Note that, for our purposes, $\boldsymbol{c}$ is the variable to be optimized. Letting $\boldsymbol{c}$ run over assignments of coefficient vectors, we know that for every $\boldsymbol{x}$, Problem (24) is a linear program for which the maximum is attained at one of the vertices of $\text{Newt}(p)$. Therefore, points $\boldsymbol{c}_i \in \text{int}(\text{Newt}(p))$ map to coefficient vectors of non-maximal terms of p.   □

By Proposition 3, we conclude that the term $x + y$ can be omitted from $p(\boldsymbol{x})$ in (23) without altering it as a function of $\boldsymbol{x}$. Proposition 3 can be extended to maxpolynomials with constant terms, such as maxout units with bias terms b_j. Let the extended Newton Polytope be

$$p(\boldsymbol{x}) = \bigvee_{j=1}^{k} b_j + \boldsymbol{c}_j^T \boldsymbol{x} \Rightarrow \text{Newt}(p) = \text{conv}\left\{(b_j, \boldsymbol{c}_j) : j \in 1, \ldots, k\right\} \qquad (25)$$

Let $\boldsymbol{c}' = (b, \boldsymbol{c})$ and $\boldsymbol{x}' = (1, \boldsymbol{x})$. Note that the relevant linear program is now

$$\text{Maximize } (\boldsymbol{x}')^T \boldsymbol{c}'$$
$$\text{s.t. } \boldsymbol{c}' \in \text{Newt}(p) \qquad (26)$$

The optimal solutions of this program lie in the *upper hull* of $\text{Newt}(p)$, $\text{Newt}^{\text{max}}(p)$, with respect to b. For a convex polytope P, its upper hull is

$$P^{\text{max}} := \{(\lambda, \boldsymbol{x}) \in P : (t, \boldsymbol{x}) \in P \Rightarrow t \leq \lambda\} \qquad (27)$$

Therefore, the number of linear regions of a maxout unit given by (20) is equal to the number of vertices on the upper hull of its Newton Polytope. Those results are easily extended for the following models:

Proposition 4. *Let $h_1, \dots h_m$ be a collection of maxpolynomials. Let*

$$g_\vee(x) = \bigvee_{i=1}^{m} h_i(x), \qquad g_+(x) = \sum_{i=1}^{m} h_i(x) \tag{28}$$

The Newton Polytopes of the functions defined above are

$$\text{Newt}(g_\vee) = \text{conv}(\text{Newt}(h_1), \dots \text{Newt}(h_m)) \tag{29}$$

$$\text{Newt}(g_+) = \text{Newt}(h_1) \oplus \text{Newt}(h_2) \cdots \oplus \text{Newt}(h_m) \tag{30}$$

where $\oplus$ denotes the Minkowski sum of the Newton Polytopes.

5   Experiments

In this section, we present results from a few numerical experiments conducted to examine the efficiency of our proposed algorithm and the behavior of morphological units as parts of a multilayer neural network.

5.1   Evaluation of the WDCCP Method

Our first experiment uses a dilation-erosion or *max-min* morphological perceptron, whose response is given by

$$y(x) = \lambda \left(\bigvee_{i=1}^{n} w_i + x_i \right) + (1 - \lambda) \left(\bigwedge_{i=1}^{n} m_i + x_i \right) \tag{31}$$

We set $\lambda = 0.5$ and trained it using both stochastic gradient descent with MSE cost and learning rate η (SGD) as well as the WDCCP method on Ripley's Synthetic Dataset [18] and the Wisconsin Breast Cancer Dataset [27]. Both are 2-class, non-separable datasets. For simplicity, we fixed the number of epochs for the gradient method at 100 and set $\tau_{\max} = 0.01$ and stopping criterion $\epsilon \leq 10^{-3}$ for the WDCCP method. We repeated each experiment 50 times to obtain mean and standard deviation for its classification accuracy, shown in Table 1. On all cases, the WDCCP method required less than 10 iterations to converge and exhibited far better results than gradient descent. The negligible standard deviation of its accuracy hints towards robustness in comparison to other methods.

5.2   Layers Using Morphological Perceptrons

We experimented on the MNIST dataset of handwritten digits [12] to investigate how morphological units behave when incorporated in layers of neural networks. After some unsuccessful attempts using a single-layer network, we settled on the following architecture: a layer of n_1 linear units followed by a $(\max, +)$ output layer of 10 units with softmax activations. The case for $n_1 = 64$ is illuminating,

Table 1. Ripleys/WDBC test set results

η	Ripleys		WDBC	
	SGD	WDCCP	SGD	WDCCP
0.01	0.838 ± 0.011	**0.902** ± 0.001	0.726 ± 0.002	**0.908** ± 0.001
0.02	0.739 ± 0.012		0.763 ± 0.006	
0.03	0.827 ± 0.008		0.726 ± 0.004	
0.04	0.834 ± 0.008		0.751 ± 0.007	
0.05	0.800 ± 0.009		0.783 ± 0.012	
0.06	0.785 ± 0.008		0.768 ± 0.01	
0.07	0.776 ± 0.009		0.729 ± 0.009	
0.08	0.769 ± 0.01		0.732 ± 0.01	
0.09	0.799 ± 0.009		0.730 ± 0.015	
0.1	0.749 ± 0.011		0.729 ± 0.009	

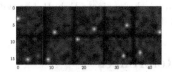

Fig. 5. Dilation layer

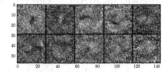

Fig. 6. Active filters

as we decided to plot the morphological filters as grayscale images shown in Fig. 5. Plotting the linear units resulted in noisy images except for those shown in Fig. 6, corresponding to maximal weights in the dilation layer. The dilation layer takes into account just one or two linear activation units per digit (pictured as bright dots), so we re-evaluated the accuracy after "deactivating" the rest of them, obtaining the same accuracy, as shown in Table 2.

Table 2. MNIST results

Layer n_1	Accuracy	Accuracy without "dead" units	# Active filters
24	84.29%	84.28%	17
32	84.84%	84.85%	15
48	84.63%	84.61%	18
64	92.1%	92.07%	10

6   Conclusions and Future Work

In this paper, we examined some properties and the behavior of morphological classifiers and introduced a training algorithm based on a well-studied optimization problem. We aim to further investigate the potential of both ours and other models, such as that proposed in [8]. A natural next step would be to examine their performance as parts of deeper architectures, possibly taking advantage of their tendency towards sparse activations to simplify the resulting networks.

The subtle connections with tropical geometry that we were able to identify make us believe that it could also aid others in the effort to study fundamental properties of deep, nonlinear architectures. We hope that the results of this paper will further motivate researchers active in those areas towards that end.

Acknowledgements. This work was partially supported by the European Union under the projects BabyRobot with grant H2020-687831 and I-SUPPORT with grant H2020-643666.

References

1. Allamigeon, X., Benchimol, P., Gaubert, S., Joswig, M.: Tropicalizing the simplex algorithm. SIAM J. Discret. Math. **29**(2), 751–795 (2015)
2. Araújo, R.D.A., Oliveira, A.L., Meira, S.R: A hybrid neuron with gradient-based learning for binary classification problems. In: Encontro Nacional de Inteligência Artificial-ENIA (2012)
3. Barrera, J., Dougherty, E.R., Tomita, N.S.: Automatic programming of binary morphological machines by design of statistically optimal operators in the context of computational learning theory. J. Electron. Imaging **6**(1), 54–67 (1997)
4. Butkovič, P.: Max-linear Systems: Theory and Algorithms. Springer Science & Business Media, Heidelberg (2010)
5. Cuninghame-Green, R.A.: Minimax Algebra. Lecture Notes in Economics and Mathematical Systems, vol. 166. Springer, Heidelberg (1979)
6. Davidson, J.L., Hummer, F.: Morphology neural networks: an introduction with applications. Circ. Syst. Sig. Process. **12**(2), 177–210 (1993)
7. Diamond, S., Boyd, S.: CVXPY: a Python-embedded modeling language for convex optimization. J. Mach. Learn. Res. **17**(83), 1–5 (2016)
8. Gärtner, B., Jaggi, M.: Tropical support vector machines. Technical report ACS-TR-362502-01 (2008)
9. Gaubert, S., Katz, R.D.: Minimal half-spaces and external representation of tropical polyhedra. J. Algebraic Comb. **33**(3), 325–348 (2011)
10. Gondran, M., Minoux, M.: Graphs, Dioids and Semirings: New Models and Algorithms, vol. 41. Springer Science & Business Media, Heidelberg (2008)
11. Goodfellow, I.J., Warde-Farley, D., Mirza, M., Courville, A.C., Bengio, Y.: Maxout networks. ICML **3**(28), 1319–1327 (2013)
12. LeCun, Y., Cortes, C., Burges, C.J.: The MNIST database of handwritten digits (1998)
13. Maclagan, D., Sturmfels, B.: Introduction to Tropical Geometry, vol. 161. American Mathematical Society, Providence (2015)
14. Maragos, P.: Morphological filtering for image enhancement and feature detection. In: Bovik, A.C. (ed.) The Image and Video Processing Handbook, 2nd edn, pp. 135–156. Elsevier Academic Press, Amsterdam (2005)
15. Maragos, P.: Dynamical systems on weighted lattices: general theory. arXiv preprint arXiv:1606.07347 (2016)
16. Montufar, G.F., Pascanu, R., Cho, K., Bengio, Y.: On the number of linear regions of deep neural networks. In: Advances in Neural Information Processing Systems, pp. 2924–2932 (2014)
17. Nair, V., Hinton, G.E.: Rectified linear units improve restricted Boltzmann machines. In: ICML 2010, pp. 807–814 (2010)
18. Ripley, B.D.: Pattern Recognition and Neural Networks. Cambridge University Press, Cambridge (2007)
19. Ritter, G.X., Sussner, P.: An introduction to morphological neural networks. In: 1996 Proceedings of the 13th International Conference on Pattern Recognition, vol. 4, pp. 709–717. IEEE (1996)

20. Ritter, G.X., Urcid, G.: Lattice algebra approach to single-neuron computation. IEEE Trans. Neural Netw. **14**(2), 282–295 (2003)
21. Rosenblatt, F.: The perceptron: a probabilistic model for information storage and organization in the brain. Psychol. Rev. **65**(6), 386 (1958)
22. Serra, J.: Image Analysis and Mathematical Morphology, vol. 1. Academic Press, Cambridge (1982)
23. Shen, X., Diamond, S., Gu, Y., Boyd, S.: Disciplined convex-concave programming. arXiv preprint arXiv:1604.02639 (2016)
24. Sussner, P., Esmi, E.L.: Morphological perceptrons with competitive learning: lattice-theoretical framework and constructive learning algorithm. Inf. Sci. **181**(10), 1929–1950 (2011)
25. Sussner, P., Valle, M.E.: Gray-scale morphological associative memories. IEEE Trans. Neural Netw. **17**(3), 559–570 (2006)
26. Vanderbei, R.J., et al.: Linear Programming. Springer, Heidelberg (2015)
27. Wolberg, W.H., Mangasarian, O.L.: Multisurface method of pattern separation for medical diagnosis applied to breast cytology. Proc. Nat. Acad. Sci. **87**(23), 9193–9196 (1990)
28. Xu, L., Crammer, K., Schuurmans, D.: Robust support vector machine training via convex outlier ablation. In: AAAI, vol. 6, pp. 536–542 (2006)
29. Yang, P.F., Maragos, P.: Min-max classifiers: learnability, design and application. Pattern Recogn. **28**(6), 879–899 (1995)
30. Yuille, A.L., Rangarajan, A.: The concave-convex procedure. Neural Comput. **15**(4), 915–936 (2003)
31. Ziegler, G.M.: Lectures on Polytopes, vol. 152. Springer Science & Business Media, Heidelberg (1995)

Morphological Links Between Formal Concepts and Hypergraphs

Isabelle Bloch[✉]

LTCI, Télécom ParisTech, Université Paris-Saclay, Paris, France
isabelle.bloch@telecom-paristech.fr

Abstract. Hypergraphs can be built from a formal context, and conversely formal contexts can be derived from a hypergraph. Establishing such links allows exploiting morphological operators developed in one framework to derive new operators in the other one. As an example, the combination of derivation operators on formal concepts leads to closing operators on hypergraphs which are not the composition of dilations and erosions. Several other examples are investigated in this paper, with the aim of processing formal contexts and hypergraphs, and navigating in such structures.

Keywords: Formal concept analysis · Hypergraphs · Mathematical morphology operators

1 Introduction

Mathematical morphology on structured representations of information is an active field of research. Given a structured representation, often represented using a graphical model, the classical way to proceed in the deterministic case is to define a partial ordering inducing a lattice structure on this representation, from which adjunctions and algebraic operators are defined. Operators using structuring elements are defined from relationships or distances on the representation[1]. Since each representation has its own semantics and point of view on the information, different definitions of morphological operators were proposed. In this paper, our aim is to establish relationships between previous works on two types of representations: formal concept analysis on the one hand [1,2,5], and hypergraphs on the other hand [6,7]. The idea is to derive a formal context from a hypergraph and conversely, so as to make each formalism inherit from definitions proposed in the other one. A few examples in each direction will be provided. Note that previous work on simplicial complexes [11,12] could be used, by considering simplicial complexes as particular cases of hypergraphs, but this may not be sufficient for our purpose since in general a concept lattice cannot be fully reconstructed from a simplicial complex, as proved in [14].

---

[1] In this paper we consider only the four basic operators (dilation, erosion, opening, closing).

© Springer International Publishing AG 2017
J. Angulo et al. (Eds.): ISMM 2017, LNCS 10225, pp. 16–27, 2017.
DOI: 10.1007/978-3-319-57240-6_2

In Sect. 2, preliminaries on formal concept analysis and hypergraphs are recalled. A first direction is considered in Sect. 3, with the construction of hypergraphs from a formal context and the derivation of mathematical morphology operators. In Sect. 4, the reverse direction is considered.

2 Preliminaries

In this section, we recall the main definitions and properties of formal concept analysis and hypergraphs, that will be used in this paper.

2.1 Formal Concept Analysis (FCA) [15]

A *formal context* is a triplet $\mathbb{K} = (G, M, I)$, where G is the set of *objects*, M the set of *attributes* or *properties*, and $I \subseteq G \times M$ a *relation* between objects and attributes ($(g, m) \in I$ means that the object g has the attribute m). A *formal concept* of the context $\mathbb{K}$ is a pair (X, Y), with $X \subseteq G$ and $Y \subseteq M$, such that (X, Y) is maximal with the property $X \times Y \subseteq I$. The set X is called the *extent* and the set Y is called the *intent* of the formal concept (X, Y). For any formal concept a, we denote its extent by $e(a)$ and its intent by $i(a)$, i.e. $a = (e(a), i(a))$.

The set of all formal concepts of a given context can be hierarchically ordered by inclusion of their extent (or equivalently by inclusion of their intent):

$$(X_1, Y_1) \preceq (X_2, Y_2) \Leftrightarrow X_1 \subseteq X_2 (\Leftrightarrow Y_2 \subseteq Y_1).$$

This order induces a complete lattice which is called the *concept lattice* of the context (G, M, I), denoted $\mathbb{C}(\mathbb{K})$, or simply $\mathbb{C}$. Infimum and supremum of a family of formal concepts $(X_t, Y_t)_{t \in T}$ are given by:

$$\bigwedge_{t \in T} (X_t, Y_t) = \left(\bigcap_{t \in T} X_t, \alpha(\beta(\bigcup_{t \in T} Y_t)) \right), \tag{1}$$

$$\bigvee_{t \in T} (X_t, Y_t) = \left(\beta(\alpha(\bigcup_{t \in T} X_t)), \bigcap_{t \in T} Y_t \right). \tag{2}$$

For $X \subseteq G$ and $Y \subseteq M$, the *derivation operators* α and β are defined as:

$$\alpha(X) = \{m \in M \mid \forall g \in X, (g, m) \in I\},$$
$$\beta(Y) = \{g \in G \mid \forall m \in Y, (g, m) \in I\}.$$

The pair (α, β) is a Galois connection between the partially ordered power sets $(\mathcal{P}(G), \subseteq)$ and $(\mathcal{P}(M), \subseteq)$ i.e.

$$\forall X \in \mathcal{P}(G), \forall Y \in \mathcal{P}(M), Y \subseteq \alpha(X) \Leftrightarrow X \subseteq \beta(Y).$$

Saying that (X, Y), with $X \subseteq G$ and $Y \subseteq M$, is a formal concept is equivalent to $\alpha(X) = Y$ and $\beta(Y) = X$.

As a running example, we consider in this paper a set of objects which are integers between 1 and 10, and some of their properties, as displayed in Fig. 1. The table defining I and the corresponding lattice are shown. In this example, the pair $(\{1, 9\}, \{o, s\})$ is a formal concept.

$\mathbb{K}$	composite	even	odd	prime	square
1			×		×
2		×		×	
3			×	×	
4	×	×			×
5			×	×	
6	×	×			
7			×	×	
8	×	×			
9	×		×		×
10	×	×			

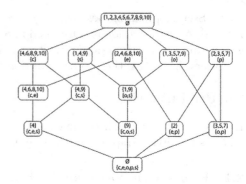

Fig. 1. A simple example of a context and its concept lattice from Wikipedia. Objects are integers from 1 to 10, and attributes are composite (c) (i.e. non prime integer strictly greater than 1), even (e), odd (o), prime (p) and square (s).

2.2   Hypergraphs [3, 9]

A *hypergraph* H, denoted by $H = (V, E)$, is defined by a finite set of *vertices* V and a finite family (which can be a multi-set) E of subsets of V called *hyperedges*. The set of vertices forming a hyperedge e, $e \in E$, is denoted by $v(e)$. It is usual to identify a hyperedge and the corresponding set of vertices. If $\cup_{e \in E} v(e) = V$, the hypergraph is without isolated vertex (a vertex x is isolated if $x \in V \setminus \cup_{e \in E} v(e)$). The set of isolated vertices is denoted by $V_{\setminus E}$. By definition the empty hypergraph is the hypergraph $H_\emptyset$ such that $V = \emptyset$ and $E = \emptyset$.

The incidence graph of a hypergraph $H = (V, E)$ is a bipartite graph $IG(H)$ with a vertex set $S = V \sqcup E$ (where $\sqcup$ stands for the disjoint union), and where $x \in V$ and $e \in E$ are adjacent if and only if $x \in v(e)$. Conversely, to each bipartite graph $\Gamma = (V_1 \sqcup V_2, A)$, we can associate two hypergraphs: a hypergraph $H = (V, E)$, where $V = V_1$ and $E = V_2$ and its dual $H^* = (V^*, E^*)$ by exchanging the roles of vertices and hyperedges, where $V^* = V_2$ and $E^* = V_1$.

3   From Formal Contexts to Hypergraphs

In this section we propose a few ways to build hypergraphs from formal contexts. Morphological operators defined on formal contexts then induce operations on hypergraphs.

3.1   Construction of Hypergraphs from a Formal Context

With any context (G, M, I), we can associate a bipartite graph from the disjoint union of objects and properties, and edges defined by the relation I, i.e. $(G \sqcup M, I)$ [4,16], where $\sqcup$ denotes the disjoint union (an extension to the fuzzy case was proposed in [17]). Two vertices $g \in G$ and $m \in M$ are linked if and only if $(g, m) \in I$. This bipartite graph can be considered as the incidence graph of two dual hypergraphs.

Definition 1. *Let* $\mathbb{K} = (G, M, I)$ *be a formal context. Two hypergraphs are defined from* $\mathbb{K}$ *as:*

1. $H_1 = (V_1, E_1)$ *where the set of vertices* V_1 *is equal to* G *(i.e. the objects), and a hyperedge* $e \in E_1$ *links all objects sharing a given property* $m \in M$, *i.e.* $v(e) = \beta(\{m\})$, *where* $v(e)$ *denotes the set of vertices of* e;
2. $H_2 = (V_2 = M, E_2 = \{\alpha(\{g\}), g \in G\})$, *i.e. the vertices are now properties, and each hyperedge* $e \in E_2$ *corresponds to an object* g *and* $v(e) = \alpha(\{g\})$.

Let us consider the example in Fig. 1, and the two aforementioned hypergraphs associated with this context. We have then $V_1 = \{1, \ldots, 10\}$, and for instance $s \in E_1$ and $v(s) = \{1, 4, 9\}$. Similarly $V_2 = \{c, e, o, p, s\}$. A hyperedge corresponding to object 5 is the subset of vertices $v(5) = \{o, p\}$. Note that in this example we have multiple hyperedges. In particular we also have $v(7) = \{o, p\}$, since objects 5 and 7 have the same set of properties (the two corresponding lines in the table in Fig. 1 are the same). If hypergraphs without repeated hyperedges are considered as preferable, they can be obtained by making the context non redundant, by clarification (removing in particular identical lines and columns in the table for this example).

Instead of considering the bipartite graph defined from the relation I, hyperedges can be built on $G \sqcup M$ from the formal concepts, which provides another interesting hypergraph.

Definition 2. *Let* $\mathbb{C}$ *be the concept lattice associated with the formal context* $\mathbb{K} = (G, M, I)$. *We define a hypergraph associated with* $\mathbb{C}$ *as* $H = (V = G \sqcup M, E = \mathbb{C})$, *i.e. a hyperedge is formed by the subsets* X *and* Y *of* G *and* M *respectively, such that* $(X, Y) \in \mathbb{C}$ *(X and Y are linked if* $\alpha(X) = Y$ *and* $\beta(Y) = X$*). The set of vertices of a hyperedge* e *is then denoted by* $v(e) = \{g \in X\} \sqcup \{m \in Y\}$.

Graphically, the hyperedges of this hypergraph H correspond to the elements of the lattice, as displayed for the number example in Fig. 1. For instance $\{1, 9, o, s\}$ is a hyperedge of H.

3.2 Morphological Operators

As shown in [1, 5] (and previously mentioned in [8]), there are some links between derivation operators and Galois connections on the one hand, and morphological operators and adjunctions on the other hand. This was extended to the fuzzy case in [2]. In particular, the derivation operators α and β are anti-dilations, and the compositions $\alpha\beta$ and $\beta\alpha$ are closings. The Galois connection property between α and β corresponds to the adjunction property between a dilation and an erosion, by reversing the ordering on one of the two spaces. These links are summarized in Table 1[2]. In this section, we further explore how operations on formal concepts, defined from $\mathcal{P}(G)$ into $\mathcal{P}(M)$, from $\mathcal{P}(M)$ into $\mathcal{P}(G)$, or directly on $\mathbb{C}$, lead to operations on hypergraphs.

---

[2] In the table we denote by $Inv(\varphi)$ the set of fixed points of an operator φ (i.e. $x \in Inv(\varphi)$ iff $\varphi(x) = x$).

Table 1. Similarities between some mathematical morphology notions and formal concept analysis [1].

Adjunctions, dilations and erosions	Galois connection, derivation operators
$\delta\colon (\mathcal{L}, \preceq) \to (\mathcal{L}', \preceq'),\ \varepsilon\colon (\mathcal{L}', \preceq') \to (\mathcal{L}, \preceq)$	$\alpha\colon \mathcal{P}(G) \to \mathcal{P}(M),\ \beta\colon \mathcal{P}(M) \to \mathcal{P}(G)$
$\delta(x) \preceq' y \iff x \preceq \varepsilon(y)$	$X \subseteq \beta(Y) \iff Y \subseteq \alpha(X)$
Increasing operators	Decreasing operators
$\varepsilon\delta\varepsilon = \varepsilon,\ \delta\varepsilon\delta = \delta$	$\alpha\beta\alpha = \alpha,\ \beta\alpha\beta = \beta$
$\varepsilon\delta$ = closing (closure operator), $\delta\varepsilon$ = opening (kernel operator)	$\alpha\beta$ and $\beta\alpha$ = both closure operators (closings)
$Inv(\varepsilon\delta) = \varepsilon(\mathcal{L}'),\ Inv(\delta\varepsilon) = \delta(\mathcal{L})$	$Inv(\alpha\beta) = \alpha\big(\mathcal{P}(G)\big),\ Inv(\beta\alpha) = \beta\big(\mathcal{P}(M)\big)$
$\varepsilon(\mathcal{L}')$ is a Moore family, $\delta(\mathcal{L})$ is a dual Moore family	$\alpha\big(\mathcal{P}(G)\big)$ and $\beta\big(\mathcal{P}(M)\big)$ are Moore families (or closure systems)
δ is a dilation: $\delta(\vee x_i) = \vee'\big(\delta(x_i)\big)$	α is an anti-dilation: $\alpha(\cup X_i) = \cap\alpha(X_i)$
ε is an erosion: $\varepsilon(\wedge' y_i) = \wedge\big(\varepsilon(y_i)\big)$	β is an anti-dilation: $\beta(\cup Y_i) = \cap\beta(Y_i)$

Derivation Operators, Dilations and Anti-dilations. Let us first interpret the derivation operators in terms of morphological operators on hypergraphs. Considering H_1, for a singleton $g \in G$, we have $\alpha(\{g\}) = \{m \in M \mid g \in v(m)\}$, in this hypergraph. This means that with each g we associate the set of hyperedges which contain g.

Proposition 1. *The derivation operator α applied on singletons is equivalent to the dilation on hypergraphs introduced in Example 4 of [6], defined from $(\mathcal{P}(V_1), \subseteq)$ into $(\mathcal{P}(E_1), \subseteq)$ as:*

$$\forall g \in V_1, \delta(\{g\}) = \{e \in E_1 \mid g \in v(e)\} \ \ and \ \ \forall X \subseteq V_1, \delta(X) = \bigcup_{g \in X} \delta(\{g\}).$$

We have $\alpha(\{g\}) = \delta(\{g\})$, and for any subset X of G, $\bigcup_{g \in X} \alpha(\{g\}) = \delta(X)$.

As a consequence, since $\alpha(X) = \bigcap_{g \in X} \alpha(\{g\})$, α is the anti-dilation, counterpart of this dilation δ on hypergraphs.

A similar reasoning applies for $Y \in M$, by considering now dilations from $(\mathcal{P}(E_1), \subseteq)$ into $(\mathcal{P}(V_1), \subseteq)$ and corresponding links with β.

By exchanging the roles of G and M, the above interpretations of derivation operators on H_1 are transposed on H_2.

Summarizing, α is interpreted as an anti-dilation from $\mathcal{P}(V_1)$ into $\mathcal{P}(E_1)$ (or equivalently from $\mathcal{P}(E_2)$ into $\mathcal{P}(V_2)$), and β as an anti-dilation from $\mathcal{P}(E_1)$ into $\mathcal{P}(V_1)$ (or equivalently from $\mathcal{P}(V_2)$ into $\mathcal{P}(E_2)$), in the sense of hypergraphs.

The combinations $\beta\alpha$ and $\alpha\beta$ are closings on $\mathcal{P}(G)$ or $\mathcal{P}(M)$, which directly define closings on the derived hypergraphs H_1 and H_2. These are new definitions of closings, enriching the ones previously proposed in [6,7].

To illustrate the effect of $\beta\alpha$, let us consider the example in Fig. 2, where the represented hypergraph is derived from a formal context, as H_1, i.e. the vertices correspond to objects and the hyperedges to properties. In this figure, vertices are represented by dots, and hyperedges by closed lines including the vertices composing them. Let us consider the vertex x_1 (red dot in the figure). We have $\alpha(\{x_1\}) = \delta(\{x_1\}) = \{m \in E_1 \mid (x_1, m) \in I\} = \{e_1, e_5\}$, and $\beta\alpha(\{x_1\}) = \beta(\{e_1, e_5\}) = \{g \in V_1 \mid (g, e_1) \in I \text{ and } (g, e_5) \in I\} = v(e_1) \cap v(e_5) = \{x_1, x_2\}$ (i.e. the red and magenta vertices in the figure). As a comparison, if we consider the adjoint erosion of δ to perform the closing, we have $\varepsilon\delta(\{x_1\}) = \bigcup\{X \in \mathcal{P}(V_1) \mid \forall x \in X, \delta(\{x\}) \subseteq \delta(\{x_1\})\} = v(e_1) \cup v(e_5)$. This shows that the two closings may provide completely different results.

Similarly, let us consider the following example for $\alpha\beta$: $\beta(\{e_1\}) = \{x \in V_1 \mid x \in v(e_1)\} = v(e_1)$, i.e. all colored points (red, magenta and cyan in Fig. 2), and $\alpha\beta(\{e_1\}) = \{m \in E_1 \mid \forall x \in \beta(\{e_1\}), x \in v(m)\} = \{e_1\}$. Considering adjoint erosion and dilation, different results could be obtained.

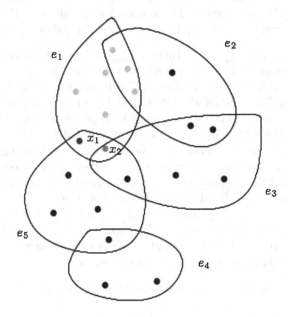

Fig. 2. Example of hypergraph and closing defined from derivation operators: $\alpha(\{x_1\}) = \{e_1, e_5\}$ and $\beta\alpha(\{x_1\}) = \{x_1, x_2\}$; $\alpha\beta(\{e_1\}) = \{e_1\}$. (Color figure online)

Interpretation of Morphological Operations on Formal Concepts in Terms of Hypergraphs. In [1,2], morphological operators were introduced, based on structuring elements defined either from I or from a distance. Let us take as a structuring element centered at $m \in M$, or a neighborhood of m, the set of $g \in G$ such that $(g, m) \in I$ (and conversely the set of $m \in M$ such that $(g, m) \in I$ is a neighborhood of g). Operators δ_I and ε_I^* from $\mathcal{P}(M)$ into $\mathcal{P}(G)$, and δ_I^* and ε_I from $\mathcal{P}(G)$ into $\mathcal{P}(M)$ were defined as:

$\forall X \in \mathcal{P}(G), \forall Y \in \mathcal{P}(M)$

$$\delta_I(Y) = \{g \in G \mid \exists m \in Y, (g, m) \in I\},$$
$$\varepsilon_I(X) = \{m \in M \mid \forall g \in G, (g, m) \in I \Rightarrow g \in X\},$$
$$\delta_I^*(X) = \{m \in M \mid \exists g \in X, (g, m) \in I\},$$
$$\varepsilon_I^*(Y) = \{g \in G \mid \forall m \in M, (g, m) \in I \Rightarrow m \in Y\}.$$

The pairs of operators $(\varepsilon_I, \delta_I)$ and $(\varepsilon_I^*, \delta_I^*)$ are adjunctions (and δ_I and δ_I^* are dilations, ε_I and ε_I^* are erosions). These operators also correspond to possibilistic interpretations of formal concepts, as proposed in [13]. Moreover, the following duality relations hold: $\delta_I(M \setminus Y) = G \setminus \varepsilon_I^*(Y)$ and $\delta_I^*(G \setminus X) = M \setminus \varepsilon_I(X)$.

Interpreting now G and M as sets of vertices and hyperedges of a hypergraph, we come up with morphological operations on hypergraphs, from either the set of vertices to the set of hyperedges or the converse, as also developed in [10] for graphs, and [6,7] for hypergraphs. Let us consider H_1, where vertices correspond to objects and hyperedges to properties. We have the following interpretations:

- $\forall Y \in \mathcal{P}(E_1), \delta_I(Y) = \bigcup_{m \in Y} \beta(\{m\}) = \bigcup_{m \in Y} v(m)$, i.e. all vertices defining the hyperedges in Y. This corresponds to Example 3 in [6].
- $\forall X \in \mathcal{P}(V_1), \varepsilon_I(X) = \{m \in E_1 \mid v(m) \subseteq X\}$, which is the adjoint erosion ε' of the dilation δ' defined as $\forall Y \in \mathcal{P}(E_1), \delta'(Y) = \bigcup_{e \in Y} v(e)$ (see Proposition 11 in [7]). The result is the set of complete hyperedges (i.e. with all vertices contained in the hyperedges) formed by vertices of V_1. The corresponding opening γ' is the set of vertices of these hyperedges. Coming back to the formal concepts, this means that in a given subset X of objects, we remove by this opening the objects which are in incomplete hyperedges, i.e. which have properties shared by objects which are not in X.
- $\forall X \in \mathcal{P}(V_1), \delta_I^*(X) = \bigcup_{g \in X} \alpha(\{g\}) = \bigcup_{g \in X} \{m \in E_1 \mid g \in v(m)\}$. It corresponds to the dilation introduced in Example 3 in [6]. See also Proposition 1.
- $\forall Y \in \mathcal{P}(E_1), \varepsilon_I^*(Y) = \{g \in V_1 \mid \alpha(\{g\}) \subseteq Y\}$, which is the set of vertices such that all hyperedges including these vertices are in Y.

Similar interpretations hold for the second construction $H_2 = (V_2, E_2)$.

Morphological Operations on Hypergraph Representations of Formal Concepts. Morphological operations from distances and neighborhoods, as defined on formal contexts in [1], can be proposed considering the representation as a hypergraph of formal concepts $H = (V = G \sqcup M, E = \mathbb{C})$, as described above.

Let us consider operators defined from a distance on $\mathbb{C}$. Several distances were introduced in [1,2], for instance from valuations ω_G and ω_M defined as the cardinality of the extent ($\omega_G(a) = |e(a)|$) or the intent ($\omega_M(a) = |i(a)|$) of a formal concept a:

$$\forall (a_1, a_2) \in \mathbb{C}^2, d_{\omega_G}(a_1, a_2) = 2\omega_G(a_1 \wedge a_2) - \omega_G(a_1) - \omega_G(a_2),$$
$$\forall (a_1, a_2) \in \mathbb{C}^2, d_{\omega_M}(a_1, a_2) = \omega_M(a_1) + \omega_M(a_2) - 2\omega_M(a_1 \vee a_2),$$

where $\wedge$ and $\vee$ are the infimum and supremum of formal concepts (Eqs. 1 and 2). These two functions are metrics on $\mathbb{C}$ (this also holds in the fuzzy case [2]). Structuring elements defined as balls of these distances can then be used as structuring elements. An example of dilation is illustrated in Fig. 3 (from [2]), using either d_{ω_G} or d_{ω_M}.

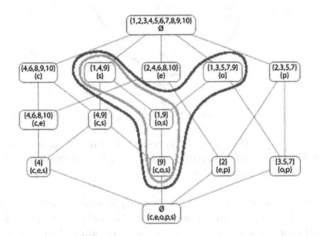

Fig. 3. Dilation of $\{a\} = \{((1,9),\{o,s\})\}$ using a ball of d_{ω_G} (red) and of d_{ω_M} (blue) as structuring element [2]. (Color figure online)

As mentioned above, each element of $\mathbb{C}$ can be interpreted as a hyperedge of a hypergraph $H = (V = G \sqcup M, E = \mathbb{C})$. Hence these dilations induce dilations from $\mathcal{P}(E)$ on $\mathcal{P}(E)$, which are again new operators.

Another definition in [2] relies on the decomposition of each formal concept as the disjunction of join-irreducible elements. Dilating each of these irreducible elements (for instance using balls of d_{ω_G} or d_{ω_M} as structuring elements), defines new operations, which in turn induce new dilations on hypergraphs. An example is reproduced from [2] in Fig. 4, with a direct interpretation in terms of hypergraphs (i.e. dilation from $\mathcal{P}(E)$ into $\mathcal{P}(E)$). In this example, the concept $a_1 = (\{1,4,9\},\{s\})$ is decomposed into irreducible elements as $a_1 = (\{4\},\{c,e,s\}) \vee (\{1,9\},\{o,s\}) \vee (\{9\},\{c,o,s\})$ and each element of the decomposition is dilated using an elementary ball of d_{ω_G} as structuring element.

Another example of dilation from $\mathcal{P}(E)$ into $\mathcal{P}(E)$ can be defined as follows:

$$\forall e \in E, \delta(\{e\}) = \{e' \in E \mid v(e) \sqcap v(e') \neq \emptyset\}, \tag{3}$$

and $\forall A \subseteq E, \delta(A) = \bigcup_{e \in A} \delta(\{e\})$, with the pseudo non empty intersection defined as:

$$v(X,Y) \sqcap v(X',Y') \neq \emptyset \Leftrightarrow X \cap X' \neq \emptyset \text{ and } Y \cap Y' \neq \emptyset.$$

The conjunction of the two constraints allows limiting the neighborhood of a concept and hence the extent of the dilation. For this example, a hypergraph

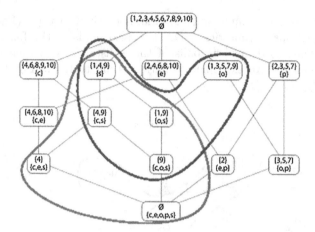

Fig. 4. Dilation of $\{a_1\} = \{(\{1,4,9\}, \{s\})\}$ using a ball of d_{ω_G} as structuring element for each irreducible element of its decomposition [2] (red), and dilation of $(\{1,9\}, \{0,s\})$ using Eq. 3 (blue). (Color figure online)

is first built from a formal context, then morphological operators are defined on the hypergraph, which induce operators on the original concept lattice. This approach can therefore be considered as being in-between the ones in this section and in the next one.

In the example in Fig. 1, the dilation of the formal concept $(\{1,9\}, \{0,s\})$ is:

$$\delta(\{1,9\}, \{0,s\}) = \{(X', Y') \in \mathbb{C} \mid X' \cap \{1,9\} \neq \emptyset \text{ and } Y' \cap \{o,s\} \neq \emptyset\} =$$
$$\{(\{1,4,9\}, \{s\}), (\{1,3,5,7,9\}, \{o\}), (\{4,9\}, \{c,s\}), (\{1,9\}, \{o,s\}), (\{9\}, \{c,o,s\})\}.$$

Note that this dilation, illustrated in Fig. 4 (in blue) is different from the one that can be built on the graph defined by the concept lattice (as depicted in Fig. 1), where the dilation of a concept would include all the concepts linked directly by an edge in the graph (then $(\{4,9\}, \{c,s\})$ would not be included in the dilation of $(\{1,9\}, \{0,s\})$). It is also different from the dilations illustrated in Fig. 3.

We can also limit the extent of dilations by limiting the number of changes to go from one concept to another one.

4   From Hypergraphs to Formal Contexts

4.1   Construction of Formal Contexts from a Hypergraph

Conversely, formal contexts can be defined from a hypergraph $H = (V, E)$.

Definition 3. *Let $H = (V, E)$ be a hypergraph. Two formal contexts are defined from H, by setting*

1. *either $G = V$, $M = E$, and $\forall g \in G, \forall m \in M, (g, m) \in I$ iff $g \in v(m)$,*
2. *or $G = E$, $M = V$, and $\forall g \in G, \forall m \in M, (g, m) \in I$ iff $m \in v(g)$.*

The relation I corresponds to the incidence matrix of H, and the derivation operators α and β can be expressed equivalently using I or using $v(m)$ (or $v(e)$). For instance in the first construction:

$$\forall X \subseteq G = V, \alpha(X) = \{m \in M = E \mid \forall g \in X, g \in v(m)\}; \tag{4}$$

$$\forall Y \subseteq M = E, \beta(Y) = \{g \in G = V \mid \forall m \in Y, g \in v(m)\}. \tag{5}$$

Then, as for any formal context, formal concepts are defined as pairs $(A, B), A \subseteq G = V, B \subseteq M = E$ such that $\alpha(A) = B$ and $\beta(B) = A$. Similar expressions hold for the second construction.

4.2   Morphological Operators

Several morphological operators on hypergraphs have been proposed in [6,7]. Thanks to the two constructions above, they yield directly operators on the derived formal contexts.

Let us consider the operators introduced in the examples of [6], and re-interpret them in terms of formal concepts. In Example 1, the structuring element is defined as the set of hyperedges intersecting the considered one, and we have for each m in M:

$$\delta(\{m\}) = \{m' \in M \mid v(m) \cap v(m') \neq \emptyset\} = \{m' \in M \mid \beta(\{m\}) \cap \beta(\{m'\}) \neq \emptyset\},$$

which represents all properties that have at least one object in common with m. This definition is similar to the one used in Eq. 3. The dilation of any subset of M is defined as the disjunction of the dilations of its elements. The adjoint erosion is given by:

$$\forall Y \in \mathcal{P}(M), \varepsilon(Y) = \bigcup \{Y' \mid \forall m \in Y', \delta(\{m\}) \subseteq Y\}.$$

Note that these operators are defined from $\mathcal{P}(M)$ into $\mathcal{P}(M)$. Operators from $\mathcal{P}(V)$ into $\mathcal{P}(V)$ can be defined in a similar way.

Let us consider the property "composite" in the number example (Fig. 1). We have $\delta(\{c\}) = \{c, e, s, o\}$, $\varepsilon(\{c, e, s, 0\}) = \{c, s\}$ (the dilation of all singletons is provided in Table 2), i.e. the morphological closing of $\{c\}$ is $\varepsilon\delta(\{c\}) = \{c, s\}$.

Table 2. Elementary dilations of the properties in the example of Fig. 1.

m	c	e	o	p	s
$\delta(\{m\})$	$\{c, e, s, o\}$	$\{e, p, c, s\}$	$\{o, s, p, c, s\}$	$\{p, e, o\}$	$\{s, o, c, e\}$

As mentioned in Example 2 in [6], a constraint on the cardinality of the intersection between $v(m)$ and $v(m')$ can be added to limit the extent of the dilation:

$$\delta_k(\{m\}) = \{m' \in M \mid |v(m) \cap v(m')| \geq k\} = \{m' \in M \mid |\beta(\{m\}) \cap \beta(\{m'\})| \geq k\}.$$

This means that we consider in the dilation all properties that have at least k objects in common with m. Such operations can be useful for clustering applications, among others.

We have for instance, for $k = 2$ and $k = 3$, $\delta_2(\{c\}) = \{c, e, s\}, \delta_3(\{c\}) = \{c, e\}$.

As in [6], we can also define similar operations from $\mathcal{P}(M)$ into $\mathcal{P}(V)$ by considering as result of the dilation all vertices forming the hyperedges in the above definitions. The compositions with the adjoint erosion define closings on $\mathcal{P}(M)$ or on $\mathcal{P}(V)$ that are different from $\alpha\beta$ and $\beta\alpha$.

As another example, let us consider the opening γ in Proposition 10 of [7] on the power set of vertices as:

$$\forall X \in \mathcal{P}(V), \gamma(X) = \bigcup\{B_e \mid e \in E, B_e \subseteq X\}, \tag{6}$$

where $\forall e \in E, B_e = \cup\{v(e') \mid e' \in E, v(e') \cap v(e) \neq \emptyset\}$. This opening acts as a filter that removes at least vertices that belong to incomplete hyperedges in X (i.e. for which the set of vertices is not completely included in X). Interpreting the hypergraph as a formal concept leads to a natural interpretation in terms of object filtering: the opening γ applied on a subset of G removes objects that have a property shared by other objects, among which at least one is not in the considered subset. Complete hyperedges can also be removed for this opening (in contrary to γ' above).

5   Outlook

In this paper, we highlighted some straightforward links between formal concepts and hypergraphs. These two ways of representing structured information correspond to different points of view (starting with the graphical representations), that suggest different ways of defining morphological operators. A few examples were given, showing that the links between the two frameworks allow each one to benefit from the other. In particular, operations that may seem very natural in one setting provide new operators in the other one, such enriching the toolbox for manipulating formal concepts and hypergraphs.

This paper presents a preliminary work, that can be developed in several directions, to further explore new morphological operators on the one hand, in particular based on different types of distances, and to derive useful applications on the other hand. Examples of applications include filtering, redundancy elimination, clustering, operations robust to small changes and associated similarities, etc.

Acknowledgments. This work has been partly supported by the French ANR LOGIMA project.

References

1. Atif, J., Bloch, I., Distel, F., Hudelot, C.: Mathematical morphology operators over concept lattices. In: Cellier, P., Distel, F., Ganter, B. (eds.) ICFCA 2013. LNCS (LNAI), vol. 7880, pp. 28–43. Springer, Heidelberg (2013). doi:10.1007/978-3-642-38317-5_2
2. Atif, J., Bloch, I., Hudelot, C.: Some relationships between fuzzy sets, mathematical morphology, rough sets, F-transforms, and formal concept analysis. Int. J. Uncertain. Fuzz. Knowl. Based Syst. 24(2), 1–32 (2016)
3. Berge, C.: Hypergraphs. Elsevier Science Publisher, Amsterdam (1989)
4. Berry, A., Sigayret, A.: Representing a concept lattice by a graph. Discret. Appl. Math. 144(1), 27–42 (2004)
5. Bloch, I.: Mathematical morphology, lattices, and formal concept analysis. In: 8th International Conference on Concept Lattices and Their Applications (CLA 2011) - Invited Conference, p. 1, Nancy, France, October 2011
6. Bloch, I., Bretto, A.: Mathematical morphology on hypergraphs, application to similarity and positive kernel. Comput. Vis. Image Underst. 117(4), 342–354 (2013)
7. Bloch, I., Bretto, A., Leborgne, A.: Robust similarity between hypergraphs based on valuations and mathematical morphology operators. Discret. Appl. Math. 183, 2–19 (2015)
8. Bloch, I., Heijmans, H., Ronse, C.: Mathematical morphology. In: Aiello, M., Pratt-Hartman, I., van Benthem, J. (eds.) Handbook of Spatial Logics, pp. 857–947. Springer, Dordrecht (2007). Chap. 13
9. Bretto, A.: Hypergraph Theory: An Introduction. Mathematical Engineering. Springer, Cham (2013)
10. Cousty, J., Najman, L., Dias, F., Serra, J.: Morphological filtering on graphs. Comput. Vis. Image Underst. 117, 370–385 (2013)
11. Dias, F., Cousty, J., Najman, L.: Some morphological operators on simplicial complex spaces. In: Debled-Rennesson, I., Domenjoud, E., Kerautret, B., Even, P. (eds.) DGCI 2011. LNCS, vol. 6607, pp. 441–452. Springer, Heidelberg (2011). doi:10.1007/978-3-642-19867-0_37
12. Dias, F., Cousty, J., Najman, L.: Dimensional operators for mathematical morphology on simplicial complexes. Pattern Recogn. Lett. 47, 111–119 (2014)
13. Dubois, D., Dupin de Saint-Cyr, F., Prade, H.: A possibility-theoretic view of formal concept analysis. Fundamenta Informaticae 75(1), 195–213 (2007)
14. Freund, A., Andreatta, M., Giavitto, J.L.: Lattice-based and topological representations of binary relations with an application to music. Ann. Math. Artif. Intell. 73(3–4), 311–334 (2015)
15. Ganter, B., Wille, R., Franzke, C.: Formal Concept Analysis: Mathematical Foundations. Springer-Verlag New York, Inc., New York (1997)
16. Gaume, B., Navarro, E., Prade, H.: A parallel between extended formal concept analysis and bipartite graphs analysis. In: Hüllermeier, E., Kruse, R., Hoffmann, F. (eds.) IPMU 2010. LNCS (LNAI), vol. 6178, pp. 270–280. Springer, Heidelberg (2010). doi:10.1007/978-3-642-14049-5_28
17. Ghosh, P., Kundu, K., Sarkar, D.: Fuzzy graph representation of a fuzzy concept lattice. Fuzzy Sets Syst. 161(12), 1669–1675 (2010)

Morphological Semigroups and Scale-Spaces on Ultrametric Spaces

Jesús Angulo$^{(\boxtimes)}$ and Santiago Velasco-Forero

CMM-Centre de Morphologie Mathématique, MINES ParisTech,
PSL-Research University, Fontainebleau, France
jesus.angulo@mines-paristech.fr

Abstract. Ultrametric spaces are the natural mathematical structure to deal with data embedded into a hierarchical representation. This kind of representations is ubiquitous in morphological image processing, from pyramids of nested partitions to more abstract dendrograms from minimum spanning trees. This paper is a formal study of morphological operators for functions defined on ultrametric spaces. First, the notion of ultrametric structuring function is introduced. Then, using as basic ingredient the convolution in (max,min)-algebra, the multi-scale ultrametric dilation and erosion are defined and their semigroup properties are stated. It is proved in particular that they are idempotent operators and consequently they are algebraically ultrametric closing and opening too. Some preliminary examples illustrate the behavior and practical interest of ultrametric dilations/erosions.

Keywords: Ultrametric space · Ultrametric semigroup · Idempotent operator · (max,min)-convolution

1 Introduction

Morphological operators are classically defined for real-valued functions supported on Euclidean or Riemannian spaces [1] and are used for nonlinear image processing. More recently, morphological semigroups for functions on length spaces have been studied [3], whose basic ingredients are the convolution in the $(\max, +)$-algebra (or supremal convolution), the metric distance and a convex shape function. More precisely, given a length space (X, d), a bounded function $f : X \mapsto \mathbb{R}$ and an increasing convex one-dimensional (shape) function $L : \mathbb{R}_+ \to \mathbb{R}_+$ such that $L(0) = 0$, the multiscale dilation $D_{L;\,t}f$ and erosion $E_{L;\,t}f$ operators of f on (X, d) according to L at scale $t > 0$ are defined as

$$D_{L;\,t}f(x) = \sup_{y \in X} \left\{ f(y) - tL\left(\frac{d(x,y)}{t} \right) \right\}, \quad \forall x \in X,$$

$$E_{L;\,t}f(x) = \inf_{y \in X} \left\{ f(y) + tL\left(\frac{d(x,y)}{t} \right) \right\}, \quad \forall x \in X.$$

© Springer International Publishing AG 2017
J. Angulo et al. (Eds.): ISMM 2017, LNCS 10225, pp. 28–39, 2017.
DOI: 10.1007/978-3-319-57240-6_3

A typical example of a shape function is $L(q) = q^P/P$, $P > 1$, such that the canonical shape function corresponds to the case $P = 2$: $L(d(x,y)/t) = d(x,y)^2/2t^2$. The corresponding semigroups are just: $D_{L;\,t}D_{L;\,s}f = D_{L;\,t+s}f$ and $E_{L;\,t}E_{L;\,s}f = E_{L;\,t+s}f$. These semigroups lead to powerful scale-space properties for multiscale filtering, regularization and feature extraction. The goal of this paper is to consider a similar generalization of morphological semigroups to the case of functions on ultrametric spaces.

An ultrametric space is a special kind of metric space in which the triangle inequality is replaced with the stronger condition $d(x, z) \leq \max\{d(x,y), d(y,z)\}$. Several typical properties on the corresponding ultrametric balls are directly derived from this ultrametric triangle inequality, which lead to nested partitions of the space. Related to the latter property, every finite ultrametric space is known to admit a natural hierarchical description called a dendogram, also known as downward tree. Dendrograms represent a tree structure of the data, where the data points are the leaves of the tree and the vertical axis reveals the ordering of the objects into nested clusters of increasing ordering. Datasets endowed with a hierarchical classification structure are nowadays used in many challenging problems; like the case of very high dimensional spaces where the data structure is generally given by cluster-like organization [12]. In the case of morphological image processing, hierarchical representations are ubiquitous [6,10,15].

Processing a function whose domain is such hierarchical representation requires the formulation of filters and operators on ultrametric spaces. The counterpart of Heat kernel and Heat semigroups on ultrametric spaces has been widely studied in recent work [4] (for discrete ultrametric spaces) and [5] (for complete ultrametric spaces). Indeed using the theory of [4,5], diffusion-based signal/image processing techniques can be applied to filter out functions on a hierarchy. A similar counterpart of morphological signal/image processing for such representations is developed in this paper.

Our starting point is the notion convolution of two functions in the (max, min)-algebra. Using this operator, we have recently shown that morphological operators on Euclidean spaces are natural formulated in (max, min)-algebra [2]. We introduce (max, min)-convolution based morphological operators on ultrametric spaces, where the structuring functions are scaled versions of the ultrametric distance (raised to a power $p \geq 1$). We study the corresponding semigroups properties and illustrate their interest in filtering and feature extraction.

In the state-of-the-art on mathematical morphology, there are several research lines related to our work. On the one hand, the theory of adjunctions on the lattice of dendrograms [8]. We remark that in our framework, the operators will be defined on the lattice of functions on the ultrametric space and not in the lattice of dendrograms itself. On the other hand, the various adjunctions on edge or node weighted graphs and their interpretations in terms of flooding [7,9,13] and their application to construct segmentation algorithms from invariants of processed minimum spanning trees with associated morphological operators [10,11].

2   Ultrametric Spaces

Let (X, d) be a metric space. The metric d is called an ultrametric if it satisfies the ultrametric inequality, i.e., $d(x, y) \leq \max\{d(x, z), d(z, y)\}$, that is obviously stronger than the usual triangle inequality. In this case (X, d) is called an ultrametric space. An ultrametric space (X, d) is called discrete if the set X is: (i) countable, (ii) all balls $B_r(x)$ are finite, and (iii) the distance function d takes only integer numbers.

Properties of Ultrametric Balls. Let us consider some well known properties of ultrametric spaces that directly derive from the ultrametric triangle inequality, proofs can be find in any basic reference on the field. The intuition behind such seemingly strange effects is that, due to the strong triangle inequality, distances in ultrametrics do not add up.

(1) The strict ball $B_{<r}(x)$ as well as the non-strict ball $B_{\leq r}(x)$ are both open as well as closed sets for the topology defined by the metric.
(2) Every point inside a ball is its center, i.e., if $d(x, y) < r$ then $B_r(x) = B_r(y)$.
(3) Given three points $x, y, z \in X$,

$$y, z \in B_r(x) \Rightarrow d(y, z) < r,$$
$$y \in B_r(x), \ z \notin B_r(x) \Rightarrow d(y, z) \geq r.$$

(4) For two intersecting balls, one contains the other, i.e., if $B_r(x) \cap B_s(y) \neq \emptyset$ then either $B_r(x) \subseteq B_s(y)$ or $B_s(y) \subseteq B_r(x)$.
(5) Any two ultrametric balls of the same radius r are either disjoint or identical.
(6) The set of all open balls with radius r and center in a closed ball of radius $r > 0$ forms a partition of the latter, and the mutual distance of two distinct open balls is again equal to r.

Consequently, the collection of all distinct balls of the same radius r forms a partition X; for increasing values of r, the balls are also increasing, hence we obtain a family of nested partitions of X which forms a hierarchy.

Examples of Ultrametric Spaces. The p-adic numbers form a complete ultrametric space. The Cantor set, which is a fractal model, is also an ultrametric space. Besides these examples, we are interested for our applications in the duality between discrete ultrametric spaces and downward (or rooted) trees, which are also known as dendrograms.

We can introduce formally a downward tree Γ as follows. Let Γ be a countable connected graph, where the set of vertices of Γ consists of disjoint union of subsets $\{\Gamma_k\}_{k=0}^{\infty}$ with the following properties: (i) from each vertex $v \in \Gamma_k$ there is exactly one edge to Γ_{k+1}; (ii) for each vertex $v \in \Gamma_k$ the number of edges connecting v to Γ_{k-1} is finite and positive, provided $k \geq 1$; (iii) if $|k - l| \neq 1$ then there is no edges between vertices of Γ_k and Γ_l. Let $d_\Gamma(v, w)$ denote the graph distance between the vertices v and w of graph Γ, i.e., the smallest number of edges in a path connecting the two vertices. For two vertices, $x, y \in \Gamma_0$, their nearest common ascestor is the vertex $a \in \Gamma_k$. Note that a is connected to x and y by downward paths of k edges. Then (Γ_0, d) is a discrete ultrametric space.

Dually, any discrete ultrametric space (X, d) admits a representation as the bottom (i.e., set of leaves) of a downward tree Γ. Define the vertices of Γ to be all distinct balls $\{B_k(x)\}$ where $x \in X$ and $k \in \mathbb{Z}_+$. Two balls $B_k(x)$ and $B_l(y)$ are connected by an edge in Γ if $|k - l| \neq 1$ and one of them is subset of the other. That is Γ_0 coincides with the set X, Γ_1 consists of balls of radii 1, etc. Clearly, edges exist only between the vertices of Γ_k and Γ_{k+1}. All balls of a given radius k provide a partition of Γ_0, so that Γ_k consists of the elements of the partition. Each of the balls of radius k is partitioned into finitely many smaller balls of radius $k - 1$ and is contained in exactly one ball of radius $k + 1$. The ultrametric distance can be also defined as $d(x, y) = \min \{k : y \in B_k(x)\}$.

3   Dilation and Erosion Semigroups on Ultrametric Spaces

Functions on Ultrametric Spaces. Given a separable and complete ultrametric space (X, d), let us consider the family of non-negative bounded functions f on (X, d), $f : X \rightarrow [0, M]$. The complement (or negative) function of f, denoted f^c, is obtained by the involution $f^c(x) = M - f(x)$. The set of non-negative bounded functions on ultrametric space is a lattice with respect to the pointwise maximum $\vee$ and minimum $\wedge$.

3.1   Ultrametric Structuring Functions

Definition 1. *A parametric family $\{b_t\}_{t>0}$ of functions $b_t : X \times X \rightarrow (-\infty, M]$ is called by us an ultrametric structuring function in the ultrametric space (X, d) if the following conditions are satisfied for all $x, y \in X$ and for all $t, s > 0$:*

(1) Total mass inequality: $\sup_{y \in X} b_t(x, y) \leq M$
(2) Completeness (or conservative): $b_t(x, x) = M$
(3) Symmetry: $b_t(x, y) = b_t(y, x)$
(4) A structuring function is monotonically decreasing in the ultrametric distance.
(5) The complement of the structuring function, i.e., $b_t^c(x, y) = M - b_t(x, y)$, is an ultrametric distance in (X, b_t^c)
(6) Maxmin semigroup property:

$$b_{\max(t,s)}(x, y) = \sup_{z \in X} \{b_t(x, z) \wedge b_s(z, y)\}. \tag{1}$$

Let us in particular introduce the so-called *natural isotropic structuring function* $b_{P,t}(x, y) = b_{P,t}(d(x, y))$, $P > 0$, in (X, d) as the following strictly monotonically decreasing function whose "shape" depends on power P:

$$b_{P,t}(x, y) = M - \left(\frac{d(x, y)}{t}\right)^P. \tag{2}$$

The case $P = 1$ is considered the canonical ultrametric structuring function.

Proposition 1. *For any $P > 0$, the isotropic P-power function $b_{P,t}(x,y)$ is an ultrametric structuring function.*

Proof. The properties 1 (total mass inequality), 2 (completeness) and 3 (symmetry) are obvious from the definition of ultrametric distance. For property 4, the P-power function is also clearly monotonically increasing in the ultrametric distance and vanishes at 0.

For property 5, we need to prove that if (X, d) is an ultrametric space, then (X, b_t^c) is an ultrametric space too, with $b_{P,t}^c(x,y) = t^{-P}d(x,y)^P$, $t > 0$, $P > 0$. Let $x, y, z \in X$ be given, we have $d(x,y) \leq \max\{d(x,z), d(z,y)\}$. Clearly,

$$t^{-P}d(x,y)^P \leq t^{-P}\max\{d(x,z), d(z,y)\}^P = \max\{t^{-P}d(x,z)^P, t^{-P}d(z,y)^P\}$$

Thus $t^{-P}d(x,y)^P$ is an ultrametric on X.

For property 6 on max-semigroup in the (max, min)-convolution, we will use well-known results from this convolution [2]. First, let write the function f by its strict lower level sets:

$$f(x) = \inf\left\{\lambda : x \in Y_\lambda^-(f)\right\},$$

where $Y_\lambda^-(f) = \{x \in X : f(x) < \lambda\}$. In fact, given $t, s > 0$ we will prove the dual semigroup property

$$b_{P,\max(t,s)}^c(x,y) = \inf_{z \in X}\left\{b_{P,t}^c(x,z) \vee b_{P,s}^c(z,y)\right\}.$$

which is just equivalent to the one we have since:

$$\sup_{z \in X}\left\{b_{P,t}(x,z) \wedge b_{P,s}(z,y)\right\} = M - \inf_{z \in X}\left\{t^{-P}d(x,z)^P \vee s^{-P}d(z,y)^P\right\}.$$

Without the loss of generality, we can fix $P = 1$. By using the classical result from level set representations

$$Y_\lambda^-(\phi_1 \vee \phi_2) = \{x \in X : \phi_1(x) < \lambda \text{ and } \phi_2(x) < \lambda\} = Y_\lambda^-(\phi_1) \cap Y_\lambda^-(\phi_2),$$

the condition, $\forall x, y \in X$,

$$Y_\lambda^-\left(t^{-1}d(x,z) \vee s^{-1}d(z,y)\right) = \left\{\exists z \in X : \left[t^{-1}d(x,z) \vee s^{-1}d(z,y)\right] < \lambda\right\},$$

becomes $\left\{Y_\lambda^-\left(t^{-1}d(x,z)\right) \cap Y_\lambda^-\left(s^{-1}d(z,y)\right) \neq \emptyset\right\}$, or equivalently, $\forall x, y \in X$, $\exists z \in X$ such that $B_{\lambda t}(z) \cap B_{\lambda s}(z) \neq \emptyset$. In addition, $B_{\lambda t}(x) = B_{\lambda t}(z)$, $B_{\lambda s}(y) = B_{\lambda s}(z)$. Using the properties of ultrametric balls, the intersection of two balls centered at z means that there is ball which contains the other of radius $\lambda \max(t, s)$ and that x and y belongs to this ball, i.e.,

$$\{d(x,y) < \lambda \max(t,s)\} \neq \emptyset.$$

In conclusion,

$$Y_\lambda^-\left(t^{-1}d(x,z) \vee s^{-1}d(z,y)\right) = Y_\lambda^-\left(\max(t,s)^{-1}d(x,y)\right),$$

and therefore: $\inf_{z \in X}\left\{b_t^c(x,z) \vee b_s^c(z,y)\right\} = b_{\max(t,s)}^c(x,z)$.

3.2 Ultrametric Dilation and Erosion Multiscale Operators

Definition 2. *Given an ultrametric structuring function* $\{b_t\}_{t>0}$ *in* (X, d), *for any non-negative bounded function* f *the ultrametric dilation* $D_t f$ *and the ultrametric erosion* $E_t f$ *of* f *on* (X, d) *according to* b_t *are defined as*

$$D_t f(x) = \sup_{y \in X} \{f(y) \wedge b_t(x,y)\}, \quad \forall x \in X, \tag{3}$$

$$E_t f(x) = \inf_{y \in X} \{f(y) \vee b_t^c(x,y)\}, \quad \forall x \in X. \tag{4}$$

We can easily identify that the ultrametric dilation is a kind of convolution in (max, min)-algebra of function f by b_t.

Proposition 2. *Ultrametric dilation* $D_t f$ *and erosion* $E_t f$ *have the following properties.*

(1) **Commutation with supremum and infimum.** *Given a set of functions* $\{f_i\}$, $i \in I$ *and* $\forall x \in X$, $\forall t > 0$, *we have*

$$D_t \left(\bigvee_{i \in I} f_i(x) \right) = \bigvee_{i \in I} D_t f_i(x); \quad E_t \left(\bigwedge_{i \in I} f_i(x) \right) = \bigwedge_{i \in I} E_t f_i(x).$$

(2) **Increasingness.** *If* $f(x) \leq g(x)$, $\forall x \in X$, *then*

$$D_t f(x) \leq D_t g(x); \text{ and } E_t f(x) \leq E_t g(x), \forall x \in X, \forall t > 0.$$

(3) **Extensivity and anti-extensivity**

$$D_t f(x) \geq f(x); \text{ and } E_t f(x) \leq f(x), \forall x \in X, \forall t > 0.$$

(4) **Duality by involution.** *For any function* f *and* $\forall x \in X$, *one has*

$$D_t f(x) = [E_t f^c(x)]^c; \text{ and } E_t f(x) = [D_t f^c(x)]^c, \forall t > 0.$$

(5) **Ordering property.** *If* $0 < s < t$ *then* $\forall x \in X$

$$\inf_X f \leq E_t f(x) \leq E_s f(x) \leq f(x) \leq D_s f(x) \leq D_t f(x) \leq \sup_X f.$$

(6) **Semigroup.** *For any function* f *and* $\forall x \in X$, *and for all pair of scales* $s, t > 0$,

$$D_t D_s f = D_{\max(t,s)} f;$$
$$E_t E_s f = E_{\max(t,s)} f.$$

Proof. For property 1, on the distributivity of the operators, we have for all $x \in X$ and for t:

$$D_t \left(\bigvee_{i \in I} f_i(x) \right) = \sup_{y \in X} \left\{ \sup_{i \in I} f_i(y) \wedge b_t(x,y) \right\}$$

$$= \sup_{i \in I} \sup_{y \in X} \{f_i(y) \wedge b_t(x,y)\} = \bigvee_{i \in I} D_t f_i(x).$$

and similarly for the ultrametric erosion.

The properties 2 and 3 of increasingness and extensivity/anti-extensivity are obvious from the properties of supremum/infimum and the property $b_t(x, y) \leq M$, with $b_t(x, x) = M$.

For property 4, on duality by involution, let us prove the first relationship since the other one is obtained by a similar procedure. For all $x \in X$ and for t:

$$[E_t f^c(x)]^c = M - \inf_{y \in X} \{(M - f(y)) \vee b_t^c(x, y)\}$$

$$= M - \inf_{y \in X} \{(M - f(y)) \vee (M - b_t(x, y))\}$$

$$= -\inf_{y \in X} \{-f(y) \vee -b_t(x, y)\} = \sup_{y \in X} \{f(y) \wedge b_t(x, y)\} = D_t f(x).$$

In order to prove the semigroup property 6, let us focus on the ultrametric dilation D_t. For any $x \in X$ and any pair $t, s > 0$, one has:

$$D_t D_s f(x) = \sup_{y \in X} [D_s f(y) \wedge b_t(x, y)]$$

$$= \sup_{y \in X} \left[\sup_{z \in X} \{f(z) \wedge b_s(y, z)\} \wedge b_t(x, y) \right]$$

$$= \sup_{z \in X} \left[f(z) \wedge \sup_{y \in X} \{b_s(y, z) \wedge b_t(x, y)\} \right].$$

Then using the property (6) of semigroup for ultrametric structuring functions, it is obtained that

$$D_t D_s f(x) = \sup_{z \in X} \{f(z) \wedge b_{\max(t,s)}(x, z)\} = D_{\max(t,s)} f(x).$$

The result for the ultrametric erosion is just obtained by duality.

The proof of ordering property 5 for the case $D_t(f)(x) \geq D_s(f)(x)$, $\forall x \in X$ is based on the fact for $t > s > 0$, by the semigroup property on the structuring functions, one has

$$b_t(x, y) = \sup_{z \in X} \{b_t(x, z) \wedge b_s(z, y)\} \Rightarrow b_t(x, y) \geq b_s(x, y),$$

and therefore

$$D_t f(x) = \sup_{y \in X} \{f(y) \wedge b_t(x, y)\} \geq \sup_{y \in X} \{f(y) \wedge b_s(x, y)\} = D_s f(x).$$

Considering the classical algebraic definitions of morphological operators [16] for the case of ultrametric semigroups $\{D_t\}_{t \geq 0}$, resp. $\{E_t\}_{t \geq 0}$, they have the properties of increasingness and commutation with supremum, resp. infimum, which involves that

$$D_t \text{ is a dilation and } E_t \text{ is an erosion.}$$

In addition, they are extensive, resp. anti-extensive, operators and, by the supremal semigroups, both are idempotent operators, i.e., $D_t D_t = D_t$ and $E_t E_t = E_t$, which implies that

$$D_t \text{ is a closing and } E_t \text{ is an opening.}$$

Finally, their semigroups are just the so-called granulometric semigroup [16] and therefore

$\{D_t\}_{t\geq 0}$ is an anti-granulometry and $\{E_t\}_{t\geq 0}$ is a granulometry,

which involve interesting scale-space properties useful for filtering and decomposition.

At first sight, one can be perplexed by this property: ultrametric dilation (resp. ultrametric erosion) is also a closing (resp. opening), since ultrametric dilation commutes with the supremum and the class of invariants of a closing is stable by infimum. However, as these suprema are taken on ultrametric balls of the various partitions, their class is also stable by infimum. The same result was already obtained by Meyer [8] for set operators on partitions.

Note that we do not use the duality by adjunction to link this pair of dilation/erosion since they are already idempotent operators and do not need to compose them to achieve such goal. Reader interested on adjunction in (max, min)-algebra is referred to [2].

3.3 Discrete Ultrametric Dilation and Erosion Semigroups

Let (X, d) be a discrete ultrametric space. Choose a sequence $\{c_k\}_{k=0}^{\infty}$ of positive reals such that $c_0 = 0$ and $c_{k+1} > c_k \geq 0$, $k = 0, 1, \cdots$. Then, given $t > 0$, ones defines the sequence $\{b_{k,t}\}_{k=0}^{\infty}$, such that

$$b_{k,t} = M - t^{-1}c_k. \tag{5}$$

Let us define $\forall k$, $\forall x \in X$, the ultrametric dilation and erosion of radius k on the associated partition as

$$Q_k^{\vee} f(x) = \sup_{y \in B_k(x)} f(y), \tag{6}$$

$$Q_k^{\wedge} f(x) = \inf_{y \in B_k(x)} f(y). \tag{7}$$

Using now (6) and (7), it is straightforward to see that the ultrametric dilation and ultrametric erosion of f by $b_{k,t}$ can be written as

$$D_t f(x) = \sup_{0 \leq k \leq \infty} \{Q_k^{\vee} f(x) \wedge b_{k,t}\}, \tag{8}$$

$$E_t f(x) = \inf_{0 \leq k \leq \infty} \{Q_k^{\wedge} f(x) \vee (M - b_{k,t})\}. \tag{9}$$

It is obvious using this formulation that do not need to compute explicitly the ultrametric distance between all-pairs of points x and y and that $D_t f(x)$ and $E_t f(x)$ are obtained by working on the supremum and infimum mosaics $Q_k^{\vee} f(x)$ and $Q_k^{\wedge} f(x)$ from the set of partitions, which is usually finite, i.e., $k = 0, 1, \cdots, K$.

3.4 Ultrametric ∞-mean and ∞-Laplacian

Ultrametric ∞-mean. Given a set of N points $x_i \in X \subset \mathbb{R}^n$, the L^∞-barycenter, known as 1-center (minimax center), corresponds to the minimizer of max-of-distances function. From a geometric viewpoint, it corresponds to the center of the minimum enclosing ball of the points x_i. In the case of $\mathbb{R}$, which can be called the ∞-mean, the minimax center equals $m_\infty = \frac{1}{2}(\max_{1 \leq i \leq N} x_i + \min_{1 \leq i \leq N} x_i)$. In our framework, the ultrametric dilation and erosion can be used to introduce the notion of ultrametric ∞-mean at scale t just as

$$M_t f(x) = \frac{1}{2}(D_t f(x) + E_t f(x)) \tag{10}$$

This operator is related to the solution of the Tug-of-War stochastic game [14].

Ultrametric ∞-Laplacian. The infinity Laplace (or L^∞-Laplace) operator is a 2nd-order partial differential operator. Viscosity solutions to the equation $\Delta_\infty u = 0$ are known as infinity harmonic functions. More recently, viscosity solutions to the infinity Laplace equation have been identified with the payoff functions from randomized tug-of-war games [14]. In the case of a length spaces, there exists a counterpart, i.e.,

$$\mathcal{L}^\infty u(x) = \max_{y \in \Omega, y \neq x} \left(\frac{u(y) - u(x)}{d(x,y)} \right) + \min_{y \in \Omega, y \neq x} \left(\frac{u(y) - u(x)}{d(x,y)} \right)$$

In the case of ultrametric spaces, we introduce the multi-scale ultrametric ∞-Laplacian, which mimics the idea of the second-order differential operator, as follows:

$$\begin{aligned}
\mathcal{L}_t^\infty f(x) &= (D_t f(x) - f(x)) - (f(x) - E_t f(x)) \\
&= D_t f(x) + E_t f(x) - 2f(x). \tag{11}
\end{aligned}$$

As for the standard laplacian, this operator can be used for enhancement of "edges" of function $f : f \mapsto \tilde{f}_t(x) = f(x) - \mathcal{L}_t^\infty f(x)$.

4 Applications to Image and Data Processing

For the examples that we consider here the ultrametric space (X, d) is built from a minimum spanning tree (MST). First, let G be an edge-weighted undirected neighbor graph with points $x \in X$ as vertices and all edge weights as nonnegative values. An MST of G is a spanning tree that connects all the vertices together with the minimal total weighting for its edges, and let $d(x, y)$ be the largest edge weight in the path of the MST between x and y. Then the vertices of the graph G, with distance measured by d form an ultrametric space. By thresholding the corresponding MST at k, $0 \leq k \leq K$, a set of partitions is obtained which produces all balls $B_k(x)$.

For the case of the discrete images or signals used in the examples, G is 4-connected pixel neighbor graph and the edge weights are the grey-level difference.

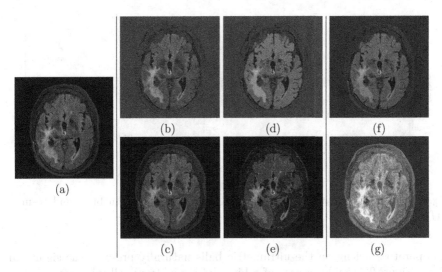

Fig. 1. Ultrametric scale-spaces: (a) original image $f(x)$, (b) and (c) ultrametric dilation $D_t f(x)$ with $t = 0.01$ and $t = 0.1$, (d) and (e) ultrametric erosion $E_t f(x)$ with $t = 0.01$ and $t = 0.1$, (f) ∞-mean with $t = 0.01$, (g) image enhancement by ∞-Laplacian $f(x) - \mathcal{L}_t^\infty f(x)$ with $t = 0.01$.

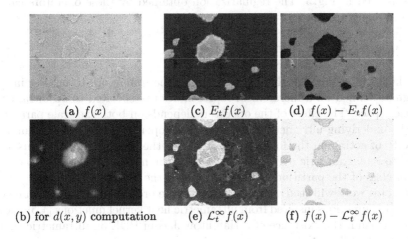

Fig. 2. Ultrametric morphological processing of bimodal image from quantitative phase microscopy: the intensity image in (a) is processed using the ultrametric space derived from the phase image in (b), with scale parameter $t = 0.005$.

In addition, a discrete ultrametric structuring function is always considered, i.e., $b_{k,t} = M - t^{-1}c_k$, with $c_k = k$.

   The first example in Fig. 1 illustrates scale-space of ultrametric dilation $D_t f$ and $E_t f$, for two values of scale t. For $t = 0.01$, the associated ultrametric ∞-mean and enhancement by ∞-Laplacian are also given. One can observe that

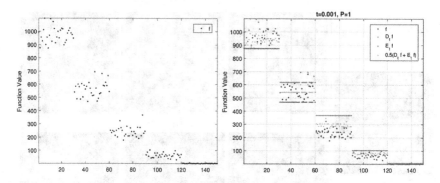

Fig. 3. Ultrametric dilation $D_t f(x)$ (in red), erosion $E_t f(x)$ (in blue) and ∞-mean $M_t f(x)$ (in green) of 1D signal $f(x)$. (Color figure online)

the operators acting on the ultrametric balls naturally preserve the significant edges. Figure 2 provides a case of a bimodal image from cell microscopy, where the quantitative phase image (b) is used to built the ultrametric space, and then, the intensity image (a) is ultrametrically processed using such space. A 1D signal, intrinsically organized into clusters, captured by the ultrametric point space, is used in Fig. 3. The regularization obtained by these operators can be useful in many applications of data processing.

5   Conclusion and Perspectives

The theory introduced in this paper provides the framework to process images or signals defined on a hierarchical representation associated to an ultrametric distance space. The effect of the operators depends on both the scale parameter and the underlying ultrametric distance. These operators have the fundamental property of acting on the function according to the pyramid of partitions associated to its ultrametric domain and therefore the notion of pixel is replaced by that of class of the partition at a given value of the hierarchy.

Ongoing work will study, on the one hand, the properties of other ultrametric structuring functions inspired from ultrametric heat kernel functions [4,5] and on the other hand, the existence of a Hamilton–Jacobi PDE on ultrametric spaces using pseudo-differential operators.

References

1. Angulo, J., Velasco-Forero, S.: Riemannian mathematical morphology. Pattern Recogn. Lett. **47**, 93–101 (2014)
2. Angulo, J.: Convolution in (max, min)-algebra and its role in mathematical morphology. hal-01108121, 59 p. (2014)
3. Angulo, J.: Morphological PDE and dilation/erosion semigroups on length spaces. In: Benediktsson, J.A., Chanussot, J., Najman, L., Talbot, H. (eds.) ISMM 2015. LNCS, vol. 9082, pp. 509–521. Springer, Cham (2015). doi:10.1007/978-3-319-18720-4_43

4. Bendikov, A., Grigor'yan, A., Pittet, C.: On a class of Markov semigroups on discrete ultra-metric spaces. Potential Anal. **37**(2), 125–169 (2012)

5. Bendikov, A., Grigor'yan, A., Pittet, C., Woess, W.: Isotropic Markov semigroups on ultrametric spaces. Uspekhi Mat. Nauk **69**(4), 3–102 (2014)

6. Meyer, F.: Hierarchies of partitions and morphological segmentation. In: Kerckhove, M. (ed.) Scale-Space 2001. LNCS, vol. 2106, pp. 161–182. Springer, Heidelberg (2001). doi:10.1007/3-540-47778-0_14

7. Meyer, F., Stawiaski, J.: Morphology on graphs and minimum spanning trees. In: Wilkinson, M.H.F., Roerdink, J.B.T.M. (eds.) ISMM 2009. LNCS, vol. 5720, pp. 161–170. Springer, Heidelberg (2009). doi:10.1007/978-3-642-03613-2_15

8. Meyer, F.: Adjunctions on the lattice of dendrograms. In: Kropatsch, W.G., Artner, N.M., Haxhimusa, Y., Jiang, X. (eds.) GbRPR 2013. LNCS, vol. 7877, pp. 91–100. Springer, Heidelberg (2013). doi:10.1007/978-3-642-38221-5_10

9. Meyer, F.: Flooding edge or node weighted graphs. In: Hendriks, C.L.L., Borgefors, G., Strand, R. (eds.) ISMM 2013. LNCS, vol. 7883, pp. 341–352. Springer, Heidelberg (2013). doi:10.1007/978-3-642-38294-9_29

10. Meyer, F.: Watersheds on weighted graphs. Pattern Recogn. Lett. **47**, 72–79 (2014)

11. Meyer, F.: The waterfall hierarchy on weighted graphs. In: Benediktsson, J.A., Chanussot, J., Najman, L., Talbot, H. (eds.) ISMM 2015. LNCS, vol. 9082, pp. 325–336. Springer, Cham (2015). doi:10.1007/978-3-319-18720-4_28

12. Murtagh, F., Downs, G., Contreras, P.: Hierarchical clustering of massive, high dimensional data sets by exploiting ultrametric embedding. SIAM J. Sci. Comput. **30**(2), 707–730 (2007)

13. Najman, L., Cousty, J., Perret, B.: Playing with Kruskal: algorithms for morphological trees in edge-weighted graphs. In: Hendriks, C.L.L., Borgefors, G., Strand, R. (eds.) ISMM 2013. LNCS, vol. 7883, pp. 135–146. Springer, Heidelberg (2013). doi:10.1007/978-3-642-38294-9_12

14. Peres, Y., Schramm, O., Sheffield, S., Wilson, D.B.: Tug-of-war and the infinity Laplacian. J. Am. Math. Soc. **22**(1), 167–210 (2009)

15. Salembier, P., Garrido, L.: Binary partition tree as an efficient representation for image processing, segmentation, and information retrieval. IEEE Trans. Image Process. **9**(4), 561–576 (2000)

16. Serra, J.: Image Analysis and Mathematical Morphology. Theoretical Advances, vol. II. Academic Press, London (1988)

Topological Relations Between Bipolar Fuzzy Sets Based on Mathematical Morphology

Isabelle Bloch[✉]

LTCI, Télécom ParisTech, Université Paris-Saclay, Paris, France
isabelle.bloch@telecom-paristech.fr

Abstract. In many domains of information processing, both vagueness, or imprecision, and bipolarity, encompassing positive and negative parts of information, are core features of the information to be modeled and processed. This led to the development of the concept of bipolar fuzzy sets, and of associated models and tools. Here we propose to extend these tools by defining algebraic relations between bipolar fuzzy sets, including intersection, inclusion, adjacency and RCC relations widely used in mereotopology, based on bipolar connectives (in a logical sense) and on mathematical morphology operators.

Keywords: Bipolar fuzzy sets · Topological relations · Adjacency · Mereotopology · Mathematical morphology operators

1 Introduction

In many domains, such as knowledge representation, preference modeling, argumentation, multi-criteria decision analysis, spatial reasoning, both vagueness or imprecision and bipolarity, encompassing positive and negative parts of information, are core features of the information to be modeled and processed. Bipolarity corresponds to a recent trend in contemporary information processing, both from a knowledge representation point of view, and from a processing and reasoning one. It allows distinguishing between (i) positive information, which represents what is guaranteed to be possible, for instance because it has already been observed or experienced, and (ii) negative information, which represents what is impossible or forbidden, or surely false [12]. This domain has recently motivated work in several directions, for instance for applications in knowledge representation, preference modeling, argumentation, multi-criteria decision analysis, cooperative games, among others [12]. Three types of bipolarity are distinguished in [14]: (i) symmetric univariate, where a unique totally ordered scale covers the range from negative (not satisfactory) to positive (satisfactory) information (e.g. modeled by probabilities); (ii) symmetric bivariate, where two separate scales are linked together and concern related information (e.g. modeled by belief functions); (iii) asymmetric or heterogeneous, where two types of information are not necessarily linked together and may come from different sources. This last type is particularly interesting in image interpretation and spatial reasoning.

© Springer International Publishing AG 2017
J. Angulo et al. (Eds.): ISMM 2017, LNCS 10225, pp. 40–51, 2017.
DOI: 10.1007/978-3-319-57240-6_4

In order to include the precise nature of information, fuzzy and possibilistic formalisms for bipolar information have been proposed (see e.g. [14]). This led to the development of the concept of bipolar fuzzy sets, and of associated models and tools, such as fusion and aggregation, similarity and distances, mathematical morphology, etc.

Here we propose to extend this set of tools by defining algebraic relations between bipolar fuzzy sets, including intersection, inclusion, adjacency and relations of region connection calculus (RCC) widely used in mereotopology[1]. Formal definitions are proposed, based on bipolar connectives and on mathematical morphology operators (here we consider only the deterministic part of mathematical morphology and use mostly erosions and dilations). They are shown to have the desired properties and to be consistent with existing definitions on sets and fuzzy sets, while providing an additional bipolar feature. The proposed relations can be used for instance for preference modeling or spatial reasoning, accounting for both bipolarity and imprecision. Any type of bipolar fuzzy set is considered, and the proposed definitions are not restricted to objects with indeterminate or broad boundaries as in the egg-yolk [10] or 9-intersection [9] models.

In Sect. 2, definitions of bipolar fuzzy sets and bipolar connectives are summarized. Extensions of basic set theoretical relations to bipolar fuzzy sets are given in Sect. 3. Mathematical morphology operators are recalled in Sect. 4. Adjacency is addressed in Sect. 5, based on morphological dilation. Finally RCC relations are extended to bipolar fuzzy sets in Sect. 6.[2]

2   Background on Bipolar Fuzzy Sets and Connectives

In this section, we recall some useful definitions on bipolar fuzzy sets and basic connectives (negation, conjunction, disjunction, implication). As mentioned in the introduction, bipolar information has two components, one related to positive information, and one related to negative information. These pieces of information can take different forms, according to the application domain, such as preferences and constraints, observations and rules, possible and forbidden places for an object in space, etc. Let us assume that bipolar information is represented by a pair (μ, ν), where μ represents the positive information and ν the negative information, under a consistency constraint [14], which guarantees that the positive information is compatible with the constraints or rules expressed by the negative information. From a formal point of view, bipolar information can be represented in different settings. Here we consider the representation where μ and ν are membership functions to fuzzy sets, defined over a space $\mathcal{S}$ (for instance the spatial domain, a set of potential options in preference modeling...). As noticed e.g. in [13] and the subsequent discussion, bipolar fuzzy sets are formally equivalent (but with important differences in their semantics) to interval-valued fuzzy

---

[1] Mereology is concerned with part-whole relations, while mereotopology adds topology and studies topological relations where regions (not points) are the primitive objects, useful for qualitative spatial reasoning, see e.g. [1] and the references therein.

[2] All proofs are quite straightforward, and omitted due to lack of space.

sets and to intuitionistic fuzzy sets, and all these are also special cases of L-fuzzy sets introduced in [16]. Despite these formal equivalences, since the semantics are very different, we keep here the terminology of bipolarity.

Definition 1. *A bipolar fuzzy set on S is defined by an ordered pair of functions (μ, ν) from S into $[0, 1]$ such that $\forall x \in S, \mu(x) + \nu(x) \leq 1$ (consistency constraint).*

We consider here that μ and ν are really two different functions, which may represent different types of information or may be issued from different sources (third type bipolarity according to [14]). However, this may also include the symmetric case, reducing the consistency constraint to a duality relation such as $\nu = 1 - \mu$. For each point x, $\mu(x)$ defines the membership degree of x (positive information) and $\nu(x)$ its non-membership degree (negative information). This formalism allows representing both bipolarity and fuzziness. The set of bipolar fuzzy sets defined on S is denoted by $\mathcal{B}$.

Let us denote by $\mathcal{L}$ the set of ordered pairs of numbers (a, b) in $[0, 1]^3$ such that $a + b \leq 1$ (hence $(\mu, \nu) \in \mathcal{B} \Leftrightarrow \forall x \in S, (\mu(x), \nu(x)) \in \mathcal{L}$). In all what follows, for each $(\mu, \nu) \in \mathcal{B}$, we will note $(\mu, \nu)(x) = (\mu(x), \nu(x)) (\in \mathcal{L}), \forall x \in S$. Note that fuzzy sets can be considered as particular cases of bipolar fuzzy sets, either when $\forall x \in S, \nu(x) = 1 - \mu(x)$, or when only one information is available, i.e. $(\mu(x), 0)$ or $(0, 1 - \mu(x))$. Furthermore, if μ (and ν) only takes values 0 and 1, then bipolar fuzzy sets reduce to classical sets.

Let $\preceq$ be a partial ordering on $\mathcal{L}$ such that $(\mathcal{L}, \preceq)$ is a complete lattice. We denote by $\bigvee$ and $\bigwedge$ the supremum and infimum, respectively. The smallest element is denoted by $0_{\mathcal{L}}$ and the largest element by $1_{\mathcal{L}}$. The partial ordering on $\mathcal{L}$ induces a partial ordering on $\mathcal{B}$, also denoted by $\preceq$ for the sake of simplicity:

$$(\mu_1, \nu_1) \preceq (\mu_2, \nu_2) \text{ iff } \forall x \in S, (\mu_1, \nu_1)(x) \preceq (\mu_2, \nu_2)(x). \tag{1}$$

Then $(\mathcal{B}, \preceq)$ is a complete lattice, for which the supremum and infimum are also denoted by $\bigvee$ and $\bigwedge$. The smallest element is the bipolar fuzzy set $0_{\mathcal{B}} = (\mu_0, \nu_0)$ taking value $0_{\mathcal{L}}$ at each point, and the largest element is the bipolar fuzzy set $1_{\mathcal{B}} = (\mu_{\mathbb{I}}, \nu_{\mathbb{I}})$ always equal to $1_{\mathcal{L}}$.

Let us now recall definitions and properties of connectives, that will be useful in the following and that extend to the bipolar case the connectives classically used in fuzzy set theory. In all what follows, increasingness and decreasingness are intended according to the partial ordering $\preceq$. Similar definitions can also be found e.g. in [11] in the case of interval-valued fuzzy sets of intuitionistic fuzzy sets, for a specific partial ordering (Pareto ordering).

Definition 2. *A **negation**, or **complementation**, on $\mathcal{L}$ is a decreasing operator N such that $N(0_{\mathcal{L}}) = 1_{\mathcal{L}}$ and $N(1_{\mathcal{L}}) = 0_{\mathcal{L}}$. In this paper, we restrict ourselves to involutive negations, such that $\forall \mathbf{a} \in \mathcal{L}, N(N(\mathbf{a})) = \mathbf{a}$ (these are the most interesting ones for mathematical morphology).*

---

[3] Note that $[0, 1]$ can be replaced by any poset or complete lattice, in the framework of L-fuzzy sets, and the proposed approach applies in this more general case.

A **conjunction** is an operator C from $\mathcal{L} \times \mathcal{L}$ into $\mathcal{L}$ such that $C(0_{\mathcal{L}}, 0_{\mathcal{L}}) = C(0_{\mathcal{L}}, 1_{\mathcal{L}}) = C(1_{\mathcal{L}}, 0_{\mathcal{L}}) = 0_{\mathcal{L}}$, $C(1_{\mathcal{L}}, 1_{\mathcal{L}}) = 1_{\mathcal{L}}$, and that is increasing in both arguments[4]. A **t-norm** is a commutative and associative bipolar conjunction such that $\forall \mathbf{a} \in \mathcal{L}, C(\mathbf{a}, 1_{\mathcal{L}}) = C(1_{\mathcal{L}}, \mathbf{a}) = \mathbf{a}$ (i.e. the largest element of $\mathcal{L}$ is the unit element of C). If only the property on the unit element holds, then C is called a **semi-norm**.

A **disjunction** is an operator D from $\mathcal{L} \times \mathcal{L}$ into $\mathcal{L}$ such that $D(1_{\mathcal{L}}, 1_{\mathcal{L}}) = D(0_{\mathcal{L}}, 1_{\mathcal{L}}) = D(1_{\mathcal{L}}, 0_{\mathcal{L}}) = 1_{\mathcal{L}}$, $D(0_{\mathcal{L}}, 0_{\mathcal{L}}) = 0_{\mathcal{L}}$, and that is increasing in both arguments. A **t-conorm** is a commutative and associative bipolar disjunction such that $\forall \mathbf{a} \in \mathcal{L}, D(\mathbf{a}, 0_{\mathcal{L}}) = D(0_{\mathcal{L}}, \mathbf{a}) = \mathbf{a}$ (i.e. the smallest element of $\mathcal{L}$ is the unit element of D).

An **implication** is an operator I from $\mathcal{L} \times \mathcal{L}$ into $\mathcal{L}$ such that $I(0_{\mathcal{L}}, 0_{\mathcal{L}}) = I(0_{\mathcal{L}}, 1_{\mathcal{L}}) = I(1_{\mathcal{L}}, 1_{\mathcal{L}}) = 1_{\mathcal{L}}$, $I(1_{\mathcal{L}}, 0_{\mathcal{L}}) = 0_{\mathcal{L}}$ and that is decreasing in the first argument and increasing in the second argument.

In the following, we will call these connectives **bipolar** to make their instantiation on bipolar information explicit. Similarly, elements of $\mathcal{L}$ should be considered as pairs, quantifying the positive and negative parts of information. The properties of these connectives are detailed for instance in [6,11], as well as the links between them. Let us mention of few of them useful in the sequel: given a t-norm C and a negation N, a t-conorm can be defined as $D((a_1, b_1), (a_2, b_2)) = N(C(N((a_1, b_1)), N((a_2, b_2))))$; an implication I induces a negation N defined as $N((a, b)) = I((a, b), 0_{\mathcal{L}})$; an implication can be derived from a negation N and a disjunction D as $I_N((a_1, b_1), (a_2, b_2)) = D(N((a_1, b_1)), (a_2, b_2))$; an implication can also be defined by residuation from a conjunction C such that $\forall (a, b) \in \mathcal{L} \setminus 0_{\mathcal{L}}, C(1_{\mathcal{L}}, (a, b)) \neq 0_{\mathcal{L}}$ as: $I_R((a_1, b_1), (a_2, b_2)) = \bigvee \{(a_3, b_3) \in \mathcal{L} \mid C((a_1, b_1), (a_3, b_3)) \preceq (a_2, b_2)\}$ (and we have $C((a_1, b_1), (a_3, b_3)) \preceq (a_2, b_2) \Leftrightarrow (a_3, b_3) \preceq I((a_1, b_1), (a_2, b_2))$, expressing the adjunction property); if C is a conjunction that admits $1_{\mathcal{L}}$ as unit element, then $C((a, b), (a', b')) \preceq (a, b) \wedge (a', b')$; if I is an implication that admits $1_{\mathcal{L}}$ as unit element on the left, then $(a', b') \preceq I((a, b), (a', b'))$; if I is an implication that admits $0_{\mathcal{L}}$ as unit element on the right, then $(a, b) \preceq I((a, b), (a', b'))$; a residual implication I defined from a bipolar t-norm satisfies $I((a, b), (a', b')) = 1_{\mathcal{L}} \Leftrightarrow (a, b) \preceq (a', b')$. In the following we will mostly consider conjunctions which are bipolar t-norms, and the associated residual implications.

The marginal partial ordering on $\mathcal{L}$, or Pareto ordering (by reversing the scale of negative information) is defined as:

$$(a_1, b_1) \preceq (a_2, b_2) \text{ iff } a_1 \leq a_2 \text{ and } b_1 \geq b_2. \tag{2}$$

This ordering, often used in economics and social choice, has also been used for bipolar information [15], and intuitionistic fuzzy sets (or interval valued fuzzy sets) e.g. in [11]. For this partial ordering, $(\mathcal{L}, \preceq)$ is a complete lattice. The greatest element is $1_{\mathcal{L}} = (1, 0)$ and the smallest element is $0_{\mathcal{L}} = (0, 1)$. The supremum and infimum are respectively defined as:

---

[4] i.e.: $\forall (a_1, a_2, a_1', a_2') \in \mathcal{L}^4, a_1 \preceq a_1'$ and $a_2 \preceq a_2' \Rightarrow C(a_1, a_2) \preceq C(a_1', a_2')$.

$$\bigvee((a_1, b_1), (a_2, b_2)) = (\max(a_1, a_2), \min(b_1, b_2)), \tag{3}$$

$$\bigwedge((a_1, b_1), (a_2, b_2)) = (\min(a_1, a_2), \max(b_1, b_2)). \tag{4}$$

In this paper, we restrict our developments to this partial ordering, as an example. Other partial orderings are discussed in [6], where dilations and erosions based on any ordering are proposed.

Let us now mention a few connectives. In Definition 2, the monotony properties have now to be intended according to the Pareto ordering.

An example of negation, which will be used in the following, is the standard negation, defined by $N((a, b)) = (b, a)$.

Two types of t-norms and t-conorms are considered in [11] (actually in the intuitionistic case) and will be considered here as well in the bipolar case. The first class consists of operators called t-representable bipolar t-norms and t-conorms, which can be expressed using usual t-norms t and t-conorms T as $C((a_1, b_1), (a_2, b_2)) = (t(a_1, a_2), T(b_1, b_2))$, and $D((a_1, b_1), (a_2, b_2)) = (T(a_1, a_2), t(b_1, b_2))$. A typical example is obtained for $t = \min$ and $T = \max$. In the following we will use dual operators t and T. The second class includes bipolar Lukasiewicz operators, which are not t-representable, and are not detailed here.

3   Basic Set Theoretical Relations on Bipolar Fuzzy Sets

Inclusion and intersection can be simply defined as bipolar numbers (i.e. in $\mathcal{L}$) (see e.g. [6] and the references therein):

Definition 3. *A bipolar degree of inclusion of* (μ_1, ν_1) *in* (μ_2, ν_2) *is defined from a bipolar implication I as:*

$$Inc((\mu_1, \nu_1), (\mu_2, \nu_2)) = \bigwedge_{x \in \mathcal{S}} I((\mu_1, \nu_1)(x), (\mu_2, \nu_2)(x)).$$

Definition 4. *A bipolar degree of intersection of* (μ_1, ν_1) *and* (μ_2, ν_2) *is defined from a bipolar conjunction C as:*

$$Int((\mu_1, \nu_1), (\mu_2, \nu_2)) = \bigvee_{x \in \mathcal{S}} C((\mu_1, \nu_1)(x), (\mu_2, \nu_2)(x)).$$

Proposition 1. *The bipolar degrees of inclusion and intersection in Definitions 3 and 4 are consistent with the corresponding definitions in the crisp and fuzzy cases.*

The bipolar degree of inclusion is an element of $\mathcal{L}$ and is decreasing in the first argument and increasing in the second one.

The bipolar degree of intersection is an element of $\mathcal{L}$ and is increasing in both arguments. It is symmetrical in particular if C is a bipolar t-norm.

If the conjunction is t-representable (i.e. $C = (t, T)$ with t a t-norm and T the dual t-conorm), then we have: $Int((\mu_1, \nu_1), (\mu_2, \nu_2)) = (\mu_{int}(\mu_1, \mu_2), 1 - \mu_{int}(1 - \nu_1, 1 - \nu_2))$ where μ_{int} is the degree of intersection between fuzzy sets, defined as $\mu_{int}(\mu, \mu') = \sup_{x \in \mathcal{S}} t(\mu(x), \mu'(x))$.

4   Mathematical Morphology on Bipolar Fuzzy Sets

Mathematical morphology on bipolar fuzzy sets was proposed for the first time in [2], by considering the complete lattice defined from the Pareto ordering. Then it was further developed, with additional properties, geometric aspects and applications to spatial reasoning, in [3,5]. In [6], any partial ordering was considered, and derived operators were also proposed. Similar work has been developed independently, in the setting of intuitionistic fuzzy sets and interval-valued fuzzy sets, also based on Pareto ordering (e.g. [18]). This group proposed an extension to L-fuzzy sets [24], besides its important contribution to connectives (e.g. [11]). Here, while relying on the general algebraic framework of mathematical morphology on the one hand, and on L-fuzzy sets [16] on the other hand, we restrict ourselves to the special case of bipolar fuzzy sets, according to Definition 1, and use the definitions proposed in [2,6], in their particular form involving structuring elements. A structuring element is a binary bipolar relation between elements of $\mathcal{S}$ and its value at "$x - y$" represents the bipolar degree to which this relation is satisfied between x and y. If $\mathcal{S}$ is endowed with a translation (for instance $\mathcal{S}$ is a subset of $\mathbb{R}^n$ or $\mathbb{Z}^n$, representing a spatial domain), then the value of a structuring element at $x - y$ represents the value at point y of the translation of the structuring element at point x.

Definition 5. *Let (μ_B, ν_B) be a bipolar fuzzy structuring element (in $\mathcal{B}$). The erosion of any (μ, ν) in $\mathcal{B}$ by (μ_B, ν_B) is defined from a bipolar implication I as:*

$$\forall x \in \mathcal{S}, \ \varepsilon_{(\mu_B, \nu_B)}((\mu, \nu))(x) = \bigwedge_{y \in \mathcal{S}} I((\mu_B, \nu_B)(y - x), (\mu, \nu)(y)). \tag{5}$$

Definition 6. *Let (μ_B, ν_B) be a bipolar fuzzy structuring element (in $\mathcal{B}$). The dilation of any (μ, ν) in $\mathcal{B}$ by (μ_B, ν_B) is defined from a bipolar conjunction C as:*

$$\forall x \in \mathcal{S}, \ \delta_{(\mu_B, \nu_B)}((\mu, \nu))(x) = \bigvee_{y \in \mathcal{S}} C((\mu_B, \nu_B)(x - y), (\mu, \nu)(y)). \tag{6}$$

These definitions are proved to provide bipolar fuzzy sets, and express erosion (respectively dilation), as a degree of inclusion (respectively intersection) of the translation (if defined on $\mathcal{S}$) of the structuring element and the bipolar fuzzy set to be transformed, according to Definitions 3 and 4.

The properties of these definitions are detailed in [6]. In particular, we will exploit in the following the fact that dilation commutes with the supremum of the lattice, and is increasing. It is extensive if and only if the origin of $\mathcal{S}$ completely belongs to the structuring element (i.e. with bipolar degree $(1, 0)$). We will restrict ourselves to extensive dilations in the following, i.e. such that $(\mu, \nu) \preceq \delta(\mu, \nu)$. The two operations ε and δ form an adjunction if and only if I is the residuated implication of C (i.e. I and C are adjoint). Finally, these definitions are equivalent to the fuzzy definitions if no bipolarity is taken into account (the dilation of a fuzzy set μ by a structuring element μ_B is defined as $\delta_{\mu_B}(\mu)(x) = \sup_{y \in \mathcal{S}} t(\mu_B(x - y), \mu(y))$ where t is a t-norm [4], and a similar expression for erosion).

5 Adjacency Between Bipolar Fuzzy Sets Based on Mathematical Morphology

In this section, we extend our previous work on adjacency between fuzzy sets [8] to the case of bipolar fuzzy sets. The underlying idea is similar, and relies on the fact that two entities (e.g. objects in space) are adjacent if they do not intersect, but as soon as one of them is dilated, they intersect.

Definition 7. *Let δ be a bipolar fuzzy dilation, C a bipolar conjunction, N a bipolar negation, and Int a bipolar degree of intersection (Definition 4). The adjacency between two bipolar fuzzy sets (μ_1, ν_1) and (μ_2, ν_2) is defined as a bipolar number in $\mathcal{L}$ as:*

$$Adj((\mu_1, \nu_1), (\mu_2, \nu_2)) = C(N(Int((\mu_1, \nu_1), (\mu_2, \nu_2))), Int((\mu_1, \nu_1), \delta(\mu_2, \nu_2))).$$

This definition formalizes the conjunction $(C(.))$ between the non intersection of the two entities $(N(Int(.)))$ and the intersection of one entities and the dilation of the other one.

Throughout the dilation can be defined according to the application. For instance for applications in the spatial domain, we may define a structuring element representing the smallest discernable spatial unit, or the imprecision related to objects or points positions. The dilation is then computed using this structuring element. The simplest one in a discrete domain would be composed of one central point and its neighbors according to a pre-defined discrete connectivity. It would then be a classical set.

Let us detail the case where C is a t-representable conjunction [11], i.e. $C((a_1, b_1), (a_2, b_2)) = (t(a_1, a_2), T(b_1, b_2))$ where t is a t-norm and T is a t-conorm. We consider here the Pareto ordering and $N(a, b) = (b, a)$, and we denote by δ^+ and δ^- the positive and negative parts of the dilation. For instance for the dilation of a bipolar fuzzy set (μ_2, ν_2) by a structuring element (μ_B, ν_B) we have:

$$\delta^+(x) = \sup_{y \in \mathcal{S}} t(\mu_B(x - y), \mu_2(y)) = \delta_{\mu_B}(\mu_2)(x),$$

$$\delta^-(x) = \inf_{y \in \mathcal{S}} T(\nu_B(x - y), \nu_2(y)) = \varepsilon_{1-\nu_B}(\nu_2).$$

These equations show that the positive part of the dilation is the fuzzy dilation of μ (positive part of the bipolar fuzzy set) by μ_B (positive part of the structuring element), and its negative part is the fuzzy erosion of ν (negative part of the bipolar fuzzy set) by $1 - \nu_B$ (negation of the negative part of the structuring element).

Proposition 2. *For a t-representable conjunction defined from a t-norm t and its dual t-conorm T, we have (μ_{int} being the classical degree of intersection of fuzzy sets):*

$$Adj((\mu_1, \nu_1), (\mu_2, \nu_2)) =$$

$$(t(1 - \mu_{int}(1 - \nu_1, 1 - \nu_2), \mu_{int}(\mu_1, \delta^+)), T(\mu_{int}(\mu_1, \mu_2), 1 - \mu_{int}(1 - \nu_1, 1 - \delta^-))).$$

This result has an interpretation that corresponds to the intuition: (i) the positive part evaluates to which extent $1 - \nu_1$ and $1 - \nu_2$ do not intersect, and μ_1 and δ^+ intersect; (ii) the negative part evaluates to which extent μ_1 and μ_2 intersect, or $1 - \nu_1$ and $1 - \delta^-$ do not intersect.

The non representable case, for instance for Lukasiewicz bipolar conjunction, can be developed in a similar way. The interpretation is slightly more complicated and may be less directly intuitive.

Other properties are expressed in the following proposition.

Proposition 3. *The adjacency defined in Definition 7 has the following properties: (i) it is symmetrical if C is left continuous (according to a metric on $\mathcal{L}$, e.g. Euclidean distance), (ii) it is invariant under geometric transformations such as translation and rotation (if S is a spatial domain, e.g. a subset of $\mathbb{R}^n$ or $\mathbb{Z}^n$), (iii) it is consistent with the definitions in the binary case and in the fuzzy case.*

Note that this relation is not considered in other works such as [9,10,17,20]. Due to the indetermination in the objects and in the relations, in some cases the adjacency relation may be close to relations such as "nearly overlap" in [9]. However, as mentioned in the introduction, the proposed model is more general since it is not restricted to objects with imprecise or broad boundaries (which are the ones considered in these other works).

A few typical adjacency situations are illustrated in Fig. 1. For the sake of simplicity, degrees in the positive and negative parts are not represented. In case (a) in this figure, the two positive parts are adjacent, and the indeterminate parts are overlapping, thus resulting in a strong indeterminacy in the spatial relation. The proposed definition provides a bipolar number where both positive and negative parts are less than 1. In case (b), the positive part is equal to 0,

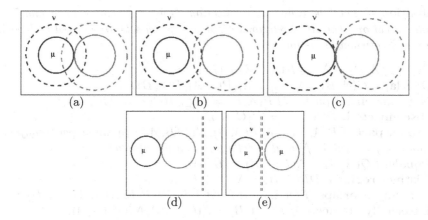

(a)   (b)   (c)

(d)   (e)

Fig. 1. A few typical situations illustrating the adjacency between bipolar fuzzy sets in the spatial domain. The positive part (μ) is inside the plain lines, and the negative part (ν) is outside the dashed lines. For the sake of simplicity, membership degrees are not represented. The blue lines correspond to (μ_1, ν_1) and the red ones to (μ_2, ν_2). (Color figure online)

showing that the relation is less satisfied than in case (a), as expected. The relation is best satisfied in case (c), with no ambiguity, which is reflected by the $(1,0) = 1_{\mathcal{L}}$ bipolar value provided by the proposed definition. In case (d), the two negative parts of the sets are the same. This is an illustration where the bipolarity does not correspond to broad boundaries. Here the resulting bipolar adjacency value is $(0,1) = 0_{\mathcal{L}}$, which corresponds to the fact that the possible region for the two objects is very large (the left half-plane), with potentially no adjacency between them because of overlapping. Finally, in case (d), the negative parts are different half-planes, and this situation is close to the perfect adjacency case. The largest value $(1,0)$ would be obtained if the two positive regions were strictly adjacent. As a concrete example, the spatial entities in the two last situations may correspond to brain structures situated in the same hemisphere (then ν is the contra-lateral hemisphere), or in different hemispheres.

6   RCC Relations Extended to Bipolar Fuzzy Sets

In this section, we propose extensions of the now classical RCC relations [19] to bipolar fuzzy sets. The connection predicate, a reflexive and symmetrical relation[5], is denoted by C, so we will denote conjunctions and bipolar t-norms by $Conj$ in all this section.

Using Connectives. A first approach consists of a direct formal extension of the equations on classical sets, similar to the extension proposed in [22] for the fuzzy case. The proposed extension relies on the connectives and bipolar degrees of intersection and inclusion (see Sects. 2 and 3). Thus we start with a bipolar connection predicate C, which is reflexive and symmetrical. Bipolar fuzzy sets are denoted by capital letters $(A, B... \in \mathcal{B})$.

Definition 8. *Let I be a bipolar implication, $Conj$ a bipolar t-norm and N a bipolar negation. The RCC relations on bipolar fuzzy sets are defined as bipolar degrees of satisfaction as follows:*

- **Part:** $P(A,B) = \bigwedge_{Z \in \mathcal{B}} I(C(A,Z), C(B,Z))$;
- **Overlaps:** $O(A,B) = \bigvee_{Z \in \mathcal{B}} Conj(P(Z,A), P(Z,B))$;
- **Non tangential part:** $NTP(A,B) = \bigwedge_{Z \in \mathcal{B}} I(C(Z,A), O(Z,B))$;
- **Disconnected:** $DC(A,B) = N(C(A,B))$;
- **Proper part:** $PP(A,B) = \bigwedge(P(A,B), N(P(B,A)))$ *(or any bipolar conjunction $Conj$ instead of $\bigwedge$ in this definition and the next ones)*;
- **Equals:** $EQ(A,B) = \bigwedge(P(A,B), P(B,A))$;
- **Distinct regions:** $DR(A,B) = N(O(A,B))$;
- **Partially overlaps:** $PO(A,B) = \bigwedge(O(A,B), N(P(B,A)), N(P(A,B)))$;
- **Externally connected:** $EC(A,B) = \bigwedge(C(A,B), N(O(A,B)))$;
- **Tangential proper part:** $TPP(A,B) = \bigwedge(PP(A,B), N(NTP(A,B)))$;
- **Non tangential proper part:** $NTPP(A,B) = \bigwedge(PP(A,B), NTP(A,B))$.

---

[5] As detailed in [1], approaches for mereotopology differ depending on the interpretation of the connection and the properties of the considered regions (closed, open...).

Using Mathematical Morphology. Another approach relies on the parthood predicate P as starting point. Then we can directly derive the five relations O, PP, EQ, DR, PO as above, thus leading to RCC-5. To get more relations, as in RCC-8 (where the eight relations are DC, EC, TPP, TPP^{-1}, PO, EQ, $NTPP$, $NTPP^{-1}$, where $^{-1}$ denotes the inverse relation), in Definition 8 the predicate C is involved. Here we propose another set of definitions, based on an extensive dilation δ (defined abstractly here, without referring to points of the underlying space). These links between RCC relations and mathematical morphology have been suggested in [7] in the crisp case (on classical sets).

Definition 9. *Given a parthood predicate P (supposed to be reflexive) and an extensive dilation δ, the relations O, PP, EQ, DR, PO are defined as in Definition 8 and the other ones as follows:*

- $DC(A, B) = \bigwedge(DR(A, B), DR(A, \delta(B))) = \bigwedge(DR(A, B), DR(\delta(A), B))$;
- $EC(A, B) = \bigwedge(DR(A, B), O(A, \delta(B))) = \bigwedge(DR(A, B), O(\delta(A), B))$;
- $TPP(A, B) = \bigwedge(PP(A, B), O(\delta(A), N(B)))$;
- $NTPP(A, B) = \bigwedge(P(A, B), P(\delta(A), B)) = P(\delta(A), B)$.

These definitions are simple to implement and provide very concise expressions.

Properties. For crisp and fuzzy sets, the proposed definitions reduce to the existing ones [19, 22], which is a desired consistency. Moreover, the following properties hold.

Proposition 4. *The relations in Definitions 8 and 9 have the following properties:*

- *The relations O, DC, EQ, DR, PO, EC are symmetrical;*
- *P (in Definition 8), O, EQ are reflexive;*
- *$TPP \preceq PP$, $NTPP \preceq PP$, $PP \preceq P$, $EQ \preceq P$, $PO \preceq O$, $P \preceq O$, $EC \preceq DR$, $DC \preceq DR$, and for Definition 8 $O \preceq C$ and $EC \preceq C$;*
- *$D_W(DC, C) = D_W(DR, C) = 1_{\mathcal{L}}$ (for Definition 8), and $D_W(DR, O)$;*
- *if C is increasing in both arguments, then P is decreasing in the first argument and increasing in the second one (for Definition 8), and if it is supposed to have these monotony properties in Definition 9, then O is increasing, DC and DR are decreasing, $NTPP$ and PP are decreasing in the first argument and increasing in the second one.*

Interpretations in Concrete Domains. In Definitions 8 and 9, the relations are abstract and do not make any reference to elements of $\mathcal{S}$ (points). Reasoning can then be performed on any bipolar (spatial) entities. Now, if we move to the concrete domain $\mathcal{S}$, which is often necessary for practical applications, then, we can assign to each abstract bipolar fuzzy set an interpretation in $\mathcal{S}$. Interpretations of the RCC relations then require concrete definitions of C, or P and δ. This can be achieved in different ways: (i) C can be defined as a degree of intersection (Definition 4) and then $O = C$; (ii) can be defined as a closeness relations (as in [21] for the fuzzy case), for instance from a dilation and a degree of intersection; (iii) P can be defined as an inclusion relation (Definition 3) and δ as a dilation with a given structuring element (Definition 6). Note that the morphological definition of EC is then exactly the proposed definition for adjacency in Sect. 5.

7 Conclusion

In this paper, we proposed original definitions of algebraic relations (intersection, inclusion, adjacency and RCC relations) between bipolar fuzzy sets, using mathematical morphology operators, in particular dilation. This problem had never been addressed before, and could be useful in spatial reasoning, but also in other domains, such as preference modeling, where preferences can be considered as bipolar fuzzy sets. Future work aims at exploring such applications, with concrete examples. Other aspects will be investigated as well, such as links with existing works on objects with imprecise, indeterminate or broad boundaries [9,10,17,20] (although our approach is not restricted to such objects). Another interesting direction concerns composition tables and reasoning, extending the work in the crisp and fuzzy cases [19,23]. Finally, other types of spatial relations can be addressed, such as directional relations.

Acknowledgments. This work has been partly supported by the French ANR LOGIMA project.

References

1. Aiello, M., Pratt-Hartmann, I., van Benthem, J. (eds.): Handbook of Spatial Logic. Springer, Netherlands (2007)
2. Bloch, I.: Dilation and erosion of spatial bipolar fuzzy sets. In: Masulli, F., Mitra, S., Pasi, G. (eds.) WILF 2007. LNCS (LNAI), vol. 4578, pp. 385–393. Springer, Heidelberg (2007). doi:10.1007/978-3-540-73400-0_49
3. Bloch, I.: Bipolar fuzzy mathematical morphology for spatial reasoning. In: Wilkinson, M.H.F., Roerdink, J.B.T.M. (eds.) ISMM 2009. LNCS, vol. 5720, pp. 24–34. Springer, Heidelberg (2009). doi:10.1007/978-3-642-03613-2_3
4. Bloch, I.: Duality vs. adjunction for fuzzy mathematical morphology and general form of fuzzy erosions and dilations. Fuzzy Sets Syst. **160**, 1858–1867 (2009)
5. Bloch, I.: Bipolar fuzzy spatial information: geometry, morphology, spatial reasoning. In: Jeansoulin, R., Papini, O., Prade, H., Schockaert, S. (eds.) Methods for Handling Imperfect Spatial Information, vol. 256, pp. 75–102. Springer, Heidelberg (2010)
6. Bloch, I.: Mathematical morphology on bipolar fuzzy sets: general algebraic framework. Int. J. Approx. Reason. **53**, 1031–1061 (2012)
7. Bloch, I., Heijmans, H., Ronse, C.: Mathematical morphology. In: Aiello, M., Pratt-Hartman, I., van Benthem, J. (eds.) Handbook of Spatial Logics, chap. 13, pp. 857–947. Springer, Netherlands (2007)

8. Bloch, I., Maître, H., Anvari, M.: Fuzzy adjacency between image objects. Int. J. Uncertain. Fuzziness Knowl. Based Syst. **5**(6), 615–653 (1997)
9. Clementini, E., Felice, O.D.: Approximate topological relations. Int. J. Approx. Reason. **16**, 173–204 (1997)
10. Cohn, A.G., Gotts, N.M.: The egg-yolk representation of regions with indeterminate boundaries. Geogr. Objects Indeterm. Bound. **2**, 171–187 (1996)
11. Deschrijver, G., Cornelis, C., Kerre, E.: On the representation of intuitionistic fuzzy t-norms and t-conorms. IEEE Trans. Fuzzy Syst. **12**(1), 45–61 (2004)
12. Dubois, D., Prade, H.: Special issue on bipolar representations of information and preference. Int. J. Intell. Syst. **23**(8–10), 999–1152 (2008)
13. Dubois, D., Gottwald, S., Hajek, P., Kacprzyk, J., Prade, H.: Terminology difficulties in fuzzy set theory - the case of "intuitionistic fuzzy sets". Fuzzy Sets Syst. **156**, 485–491 (2005)
14. Dubois, D., Prade, H.: An overview of the asymmetric bipolar representation of positive and negative information in possibility theory. Fuzzy Sets Syst. **160**, 1355–1366 (2009)
15. Fargier, H., Wilson, N.: Algebraic structures for bipolar constraint-based reasoning. In: Mellouli, K. (ed.) ECSQARU 2007. LNCS (LNAI), vol. 4724, pp. 623–634. Springer, Heidelberg (2007). doi:10.1007/978-3-540-75256-1_55
16. Goguen, J.: L-fuzzy sets. J. Math. Anal. Appl. **18**(1), 145–174 (1967)
17. Hazarika, S., Cohn, A.: A taxonomy for spatial vagueness: an alternative egg-yolk interpretation. In: Spatial Vagueness, Uncertainty and Granularity Symposium, Ogunquit, Maine, USA (2001)
18. Mélange, T., Nachtegael, M., Sussner, P., Kerre, E.: Basic properties of the interval-valued fuzzy morphological operators. In: IEEE World Congress on Computational Intelligence, WCCI 2010, Barcelona, Spain, pp. 822–829 (2010)
19. Randell, D., Cui, Z., Cohn, A.: A spatial logic based on regions and connection. In: Principles of Knowledge Representation and Reasoning, KR 1992, Kaufmann, San Mateo, CA, pp. 165–176 (1992)
20. Roy, A.J., Stell, J.G.: Spatial relations between indeterminate regions. Int. J. Approx. Reason. **27**(3), 205–234 (2001)
21. Schockaert, S., De Cock, M., Cornelis, C., Kerre, E.E.: Fuzzy region connection calculus: an interpretation based on closeness. Int. J. Approx. Reason. **48**(1), 332–347 (2008)
22. Schockaert, S., De Cock, M., Cornelis, C., Kerre, E.E.: Fuzzy region connection calculus: representing vague topological information. Int. J. Approx. Reason. **48**(1), 314–331 (2008)
23. Schockaert, S., De Cock, M., Kerre, E.E.: Spatial reasoning in a fuzzy region connection calculus. Artif. Intell. **173**(2), 258–298 (2009)
24. Sussner, P., Nachtegael, M., Mélange, T., Deschrijver, G., Esmi, E., Kerre, E.: Interval-valued and intuitionistic fuzzy mathematical morphologies as special cases of L-fuzzy mathematical morphology. J. Math. Imaging Vis. **43**(1), 50–71 (2012)

Discrete Geometry and Discrete Topology

Introducing the Dahu Pseudo-Distance

Thierry Géraud[1]([✉]), Yongchao Xu[1,2], Edwin Carlinet[1], and Nicolas Boutry[1,3]

[1] EPITA Research and Development Laboratory (LRDE),
Le Kremlin-Bicêtre, France
theo@lrde.epita.fr
[2] Institut Mines Telecom, Telecom ParisTech, Paris, France
[3] Université Paris-Est, LIGM, Équipe A3SI, ESIEE, Champs-sur-Marne, France

Abstract. The minimum barrier (MB) distance is defined as the minimal interval of gray-level values in an image along a path between two points, where the image is considered as a vertex-valued graph. Yet this definition does not fit with the interpretation of an image as an elevation map, i.e. a somehow continuous landscape. In this paper, based on the discrete set-valued continuity setting, we present a new discrete definition for this distance, which is compatible with this interpretation, while being free from digital topology issues. Amazingly, we show that the proposed distance is related to the morphological tree of shapes, which in addition allows for a *fast* and *exact* computation of this distance. That contrasts with the classical definition of the MB distance, where its fast computation is only an approximation.

Keywords: Discrete topology · Minimal path · Minimum barrier distance · Set-valued maps · Tree of shapes

1 Introduction

> *Que la montagne de pixels est belle.* Jean Serrat.

The minimum barrier (MB) distance [20] is defined as the minimal interval of gray-level values in an image along a path between two points.

For instance, the image of Fig. 1(a) can be seen as a graph depicted in Fig. 1(b), and the path between the red points which is depicted in blue is minimal: the sequence of values is $\langle 1, 0, 0, 0, 2 \rangle$ so the interval is $[0, 2]$ and the distance is **2**. Note that, in general, for an image and a couple of points, many minimal paths exist (here the straight path with the sequence of values $\langle 1, 3, 2 \rangle$ gives the interval $[1, 3]$, which also gives the MB distance of 2). If we consider that the discrete function describes a surface such as the one of Fig. 1(c), the blue path depicted in this figure is such that the MB-like distance between the two red dots is now **1**. This paper elaborates on how to define properly this variant of the MB distance which considers that the elevation map is a continuous function.

The MB distance is interesting for several reasons. First, it is an "original" distance which contrasts with classical path-length distances, because it only

---

E. Carlinet is now with DxO, France.

© Springer International Publishing AG 2017
J. Angulo et al. (Eds.): ISMM 2017, LNCS 10225, pp. 55–67, 2017.
DOI: 10.1007/978-3-319-57240-6_5

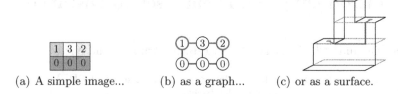

(a) A simple image...   (b) as a graph...   (c) or as a surface.

Fig. 1. Two different minimal barrier distances: (b) gives **2**, while (c) gives **1**. (Color figure online)

relies on the dynamics of the function. Second, it has already successfully been applied to perform salient object segmentation [21,25]. Last, as we will see, its continuous version presented here is related to mathematical morphology. In the example of Fig. 1(c), considering an image as a landscape, we can see that a path can be defined on a steep hillside. We thus have decided to name this particular distance after the legendary (or not [18]) Dahu creature: a mountain goat-like animal with legs of different sides having differing lengths to fit the craggy mountain's side.

To be able to define this continuous version of the MB distance, and to show how to compute it efficiently, we will rely on several tools: set-valued maps, cubical complexes, and the morphological tree of shapes (ToS for short), described in Sect. 2. In Sect. 3, we will start by giving a naive definition of this distance; we will see that it raises some digital topology issue, and we will propose a definition (closely related to the discrete set-valued continuity setting [17]) that overcomes this issue.

Although the definition of the Dahu distance is highly combinatorial, we will see in Sect. 4 that this distance can be efficiently computed thanks to the tree of shapes of an interval-valued map. A remarkable property is that we can efficiently compute the *exact* distance, whereas the only efficient computation of the MB distance remains approximate. We also illustrate the use of the Dahu distance on document image segmentation in Sect. 4.3.

Actually, this paper mainly focus on the new distance definition, on the discrete topology issues related to this distance, and on a method to compute it efficiently. These three subjects are the main contributions detailed hereafter[1].

2   Theoretical Background

This section gives a short tour on the theoretical tools needed in Sects. 3 and 4.

---

[1] As a consequence (and despite how frustrating it might be for the reader), due to limited space, this paper does *not* address the following aspects of our work: practical applications that follow from this new distance definition, a quantitative comparison with the MB distance, considerations about the sampling grid and continuity, implementation details and execution times; last, some proofs have also been omitted.

2.1 Basic Notions

Let us consider a discrete finite set X. The set of all subsets of X is denoted by $\mathcal{P}(X)$, so $E \in \mathcal{P}(X)$ means $E \subseteq X$. X is endowed with a discrete neighborhood relation; the set of neighbors of $x \in X$ is denoted by $\mathcal{N}_X(x) \subset X$, and satisfies: $x' \in \mathcal{N}_X(x)$ iff $x' \neq x$ and $x \in \mathcal{N}_X(x')$.

A discrete path π of X is a sequence $\pi = \langle \pi_1, .., \pi_i, .., \pi_k \rangle$, where k is the length of the sequence, such as $\pi_i \in X$ and $\pi_{i+1} \in \mathcal{N}_X(\pi_i)$. A sequence $\langle x, .., x' \rangle$ is called a path between x and x'; the set of all paths between x and x' is denoted by $\Pi(x, x')$, and we use $\pi(x, x')$ to denote an element of $\Pi(x, x')$.

A connected component Γ of $E \subset X$ is a subset of E which is path-connected ($\forall x \neq x' \in \Gamma$, $\Pi(x, x') \neq \emptyset$) and maximal ($\forall x \in \Gamma$, $\forall x' \in E \backslash \Gamma$, $\Pi(x, x') = \emptyset$). The set of the connected components of E is denoted by $\mathcal{CC}(E)$, and given $x \in X$, the connected component of E containing x is denoted by $\mathcal{C}(E, x) \in \mathcal{CC}(E)$—by convention, if $x \notin E$ then $\mathcal{C}(E, x) = \emptyset$. Given a particular element $x_\infty \in X$, the cavity-fill-in operator (in 2D, hole-filling operator) is denoted by Sat and defined by: $\mathrm{Sat}(E, x_\infty) = X \setminus \mathcal{C}(X \backslash E, x_\infty)$.

Throughout this paper, we will only consider functions having for domain a set X being discrete, finite, and path-connected.

2.2 Scalar vs. Set-Valued Maps

A gray-level image is typically a bounded scalar function such as X is a subset $\mathcal{D} \subset \mathbb{Z}^n$ with $n \geq 2$, and its codomain is $\mathbb{Z}$ or $\mathbb{R}$. It is the case of the image depicted in Fig. 1(a). Now, if we look at the surface of Fig. 1(c), its elevation cannot be defined by a scalar function. Yet it can be a mapping in $\mathbb{R}^2 \to \mathbb{I}_\mathbb{R}$, where $\mathbb{I}_Y$ denotes the space of the intervals of Y. Given two intervals $a = [y_a, y'_a]$ and $b = [y_b, y'_b]$, the interval $[\min(y_a, y_b), \max(y'_a, y'_b)]$ is denoted by span$\{a, b\}$. Now let us recall some definitions related to set-valued maps.

A set-valued map $\mathrm{U} : X \to \mathcal{P}(Y)$ is characterized by its graph, $\mathrm{Gra}(\mathrm{U}) = \{(x, y) \in X \times Y; y \in \mathrm{U}(x)\}$ [1]. One definition of the inverse by a set-valued map U of a subset $M \subset Y$ is $\mathrm{U}^\ominus(M) = \{x \in X; \mathrm{U}(x) \subset M\}$, and it is called the *core* of M by U. Let us assume that X and Y are metric spaces. The "natural" extension of the continuity of single-valued functions to set-valued maps is characterized by the following property: U is continuous iff the core of any open subset of Y is an open set of X.

In [14], the authors have defined the notion of level sets for set-valued maps; at $\lambda \in Y$, the lower and upper level sets of U are respectively:

$$[\mathrm{U} \lhd \lambda] = \{x \in X; \forall y \in \mathrm{U}(x), y < \lambda\} \text{ and } [\mathrm{U} \rhd \lambda] = \{x \in X; \forall y \in \mathrm{U}(x), y > \lambda\}. \quad (1)$$

Such definitions are fundamental because they are some means to build some morphological tools on set-valued maps (we will use them later in this paper in Sects. 2.4 and 4.1).

A scalar function $u : X \to Y$ can be simply translated into a set-valued map $\mathring{u} : X \to \mathbb{I}_Y$ with $\forall x \in X$, $\mathring{u}(x) = \{u(x)\}$. Between set-valued maps of $X \to \mathcal{P}(Y)$ we can define the relation $\preccurlyeq$ by: $\mathrm{U}_1 \preccurlyeq \mathrm{U}_2 \Leftrightarrow \forall x \in X$, $\mathrm{U}_1(x) \subseteq \mathrm{U}_2(x)$.

By abuse of notation, we will write $u \ll \mathtt{U}$ to state that $\overset{\bullet}{u} \ll \mathtt{U}$, meaning that we have $\forall x \in X, u(x) \in \mathtt{U}(x)$. An illustration can be seen later: the map $\overline{u}$ of Fig. 3(d) and the map $\widetilde{u}$ of Fig. 3(c) are such that $\overline{u} \ll \widetilde{u}$. In the following, we will only consider interval-valued maps, that is, maps of $X \to \mathbb{I}_Y$.

2.3   Cubical Complexes

From the sets $H_0^1 = \{\{a\}; a \in \mathbb{Z}\}$ and $H_1^1 = \{\{a, a+1\}; a \in \mathbb{Z}\}$, we can define $\mathbb{H}^1 = H_0^1 \cup H_1^1$ and the set $\mathbb{H}^n$ as the n-ary Cartesian power of $\mathbb{H}^1$. If an element $h \subset \mathbb{Z}^n$ is the Cartesian product of d elements of H_1^1 and $n-d$ elements of H_0^1, we say that h is a d-face of $\mathbb{H}^n$ and that d is the dimension of h; it is denoted $\dim(h)$. The set of all faces, $\mathbb{H}^n$, is called the nD space of cubical complexes. Figure 2(a) depicts a set of faces $\{h, h', h''\} \subset \mathbb{H}^2$ where $h = \{0\} \times \{1\}$, $h' = \{1\} \times \{0, 1\}$, and $h'' = \{0, 1\} \times \{0, 1\}$, the dimension of these faces being respectively 0, 1, and 2; they are depicted as subsets of $\mathbb{Z}^2$ (left) and as geometrical objects (right).

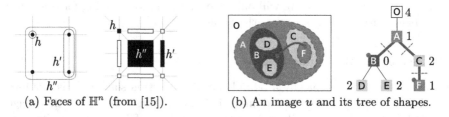

(a) Faces of $\mathbb{H}^n$ (from [15]).   (b) An image u and its tree of shapes.

Fig. 2. A cubical complex for space X (left) and a tree of shapes for u (right). (Color figure online)

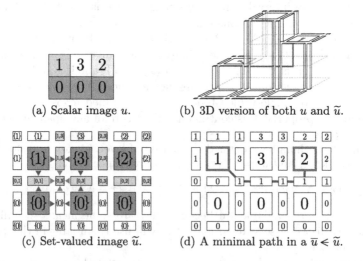

(a) Scalar image u.   (b) 3D version of both u and $\widetilde{u}$.

(c) Set-valued image $\widetilde{u}$.   (d) A minimal path in a $\overline{u} \ll \widetilde{u}$.

Fig. 3. From a gray-scale image (a) to a discrete but continuous representation (c). (Color figure online)

The pair $(\mathbb{H}^n, \subseteq)$ forms a poset, from which can be derived a T0-Alexandroff topology on $\mathbb{H}^n$. The closure operator on subsets of $\mathbb{H}^n$ is denoted by cl. The n-faces are the minimal open sets of $\mathbb{H}^n$, and the set of n-faces, denoted by H_1^n, is the n-Cartesian product of H_1. The faces of $\mathbb{H}^n$ can be arranged onto a grid, called Khalimsky's grid, and the inclusion between faces leads to a neighborhood relationship between them, depicted in orange in Fig. 2(a) (right). This neighborhood relationship will be used in Sect. 3.2 in order to define paths in $\mathbb{H}^n$.

2.4   Tree of Shapes

Given a gray-level image $u : X \to Y$ and any scalar $\lambda \in Y$, the lower level sets are defined as $[u < \lambda] = \{x \in X; u(x) < \lambda\}$, and the upper level sets as $[u \geq \lambda] = \{x \in X; u(x) \geq \lambda\}$. Considering the connected components of these sets, and using the cavity-fill-in operator, the tree of shapes (ToS) of an image u is classically [16] defined by:

$$\mathfrak{S}(u) = \{\, \mathrm{Sat}(\Gamma); \Gamma \in \mathcal{CC}([u < \lambda]) \cup \mathcal{CC}([u \geq \lambda]) \,\}_\lambda. \tag{2}$$

An image and its tree of shapes are depicted in Fig. 2. An element of $\mathfrak{S}(u)$ is called a shape; it is a connected component of X with no cavity, and its boundary is a level line of u. Every shape corresponds to a node of the tree; for instance, in Fig. 2(b) (right), the sub-tree rooted at node "B" corresponds to the shape B ∪ D ∪ E. Keeping the level of every node—such as displayed in Fig. 2(b) (right)—allows to reconstruct the image from its tree. The tree of shapes of an image u is a morphological representation of u which makes it easier to deal with the image contents [10]. Storing [6] and computing [11,14] the tree of shapes can be done very efficiently. Last, let us mention that the tree of shapes is a versatile tool to perform image filtering [23], and a very relevant structure to perform some computer vision tasks [5,9,22].

Let us now recall some results from [17] about defining the tree of shapes on set-valued maps. Given an interval-valued map $U : \mathbb{H}^n \to \mathbb{I}_Y$, using Eq. 1, its set of level sets is defined by $\mathfrak{T}(U) = \bigcup_\lambda \mathcal{CC}([U \lhd \lambda]) \cup \mathcal{CC}([U \rhd \lambda])$. The set of connected components of X defined by:

$$\mathfrak{S}(U) = \{\, \mathrm{Sat}(\Gamma); \Gamma \in \mathfrak{T}(U) \,\} \tag{3}$$

can be arranged into a tree for some very particular maps U. Every interval-valued maps U does not have a tree of shapes. Yet, if all the level sets of U are "well-composed" [3], then the tree exists.

2.5   Minimum Barrier Distance

As defined in [20], the barrier of a path π in a gray-level image u is $\tau_u(\pi) = \max_{\pi_i \in \pi} u(\pi_i) - \min_{\pi_i \in \pi} u(\pi_i)$, and the minimum barrier distance between x and x' in u is:

$$d_u^{\mathrm{MB}}(x, x') = \min_{\pi \in \Pi(x,x')} \tau_u(\pi). \tag{4}$$

It is actually a pseudo-distance since, $\forall u$ and $\forall x, x', x'' \in X$, it verifies: $d_u^{MB}(x) \geq 0$ (non-negativity); $d_u^{MB}(x, x) = 0$ (identity); $d_u^{MB}(x, x') = d_u^{MB}(x', x)$ (symmetry); and $d_u^{MB}(x, x'') \leq d_u^{MB}(x, x') + d_u^{MB}(x', x'')$ (subadditivity). Yet it is not a distance because the positivity property ($x' \neq x \Rightarrow d_u^{MB}(x, x') > 0$) does not hold. Indeed we can have $d_u^{MB}(x, x') = 0$ for some $x \neq x'$, e.g., in flat zones.

This pseudo-distance have been recently used in [25] and [21] to perform salient object segmentation, and it allows to achieve some leading performance (in terms of efficiency and accuracy) as compared to state-of-the-art methods.

3   The Dahu Pseudo-Distance

To tackle the issue explained in the introduction, we need to map a discrete image into a continuous representation such as Fig. 1(c); yet for computations to remain tractable we want to stay in a discrete setting.

3.1   A Convenient Continuous Yet Discrete Representation

This section explain how to represent a gray-value image $u : \mathcal{D} \subset \mathbb{Z}^n \to Y$ by an interval-valued map $\widetilde{u} : \mathcal{D}_H \subset \mathbb{H}^n \to \mathbb{I}_Y$, and by a new tree of shapes $\mathfrak{S}(\widetilde{u})$.

First, to every $x = (x_{(1)}, ..., x_{(n)}) \in \mathcal{D}$, let us associate the n-face $h_x = \{(x_{(1)}, x_{(1)} + 1) \times ... \times (x_{(n)}, x_{(n)} + 1)\} \in H_1^n$. $\mathcal{D}_H \subset \mathbb{H}^n$ is the domain defined by $\mathcal{D}_H = cl(\{h_x ; x \in \mathcal{D}\})$, which will correspond to $\mathcal{D}$ in $\mathbb{H}^n$. From u we define $\widetilde{u}$ by:

$$\forall h \in \mathcal{D}_H, \; \widetilde{u}(h) = \text{span}\{ u(x); x \in \mathcal{D} \text{ and } h \subset h_x \}. \tag{5}$$

A simple example of a transform of a gray-level image u, given in Fig. 3(a), into an interval-valued map $\widetilde{u}$ is given in Fig. 3(c). As a result of Eq. 5, we can see that $\widetilde{u}_{|H_1^n}$ (the pinkish part) looks like the set-valued version $\overset{\bullet}{u}$ of the original image u). The span computation is displayed by the gray and olive triangles, respectively for the dimensions $d = 1$ and $d = 0$. Eventually, only faces with a dimension less than n can be non-degenerated intervals (they are then displayed in orange).

A 3D version of the discrete interval-valued representation $\widetilde{u}$ is depicted in Fig. 3(b); one can easily see that it is a discrete equivalent of the continuous surface of Fig. 1(c). We will define the tree of shapes of such $\widetilde{u}$ maps in Sect. 4.1.

3.2   Definition of the Naive Dahu Pseudo-Distance

Relying on an interval-valued representation $\widetilde{u} : \mathcal{D}_H \subset \mathbb{H}^n \to \mathbb{I}_Y$ of a scalar image $u : \mathcal{D} \subset \mathbb{Z}^n \to Y$, we define the Dahu pseudo-distance by:

$$d_u^{naive}(x, x') = \min_{\overline{u} \lessdot \widetilde{u}} d_{\overline{u}}^{MB}(h_x, h_{x'}). \tag{6}$$

Let us just recall that the notation $\bar{u} \ll \tilde{u}$ means that $\forall x$, $\bar{u}(x) \in \tilde{u}(x)$, so the argument of the minimum operator is a scalar function $\bar{u} : \mathcal{D}_H \subset \mathbb{H}^n \to Y$. It means that we actually search for a minimal path in the cubical complex $\mathcal{D}_H$ (not in $\mathcal{D}$), with the classical definition of the minimum barrier distance d^{MB} (that is, a distance defined on scalar functions), and considering **all** the possible scalar functions $\bar{u}$ that are "included" in the interval-valued map $\tilde{u}$. The definition of the Dahu pseudo-distance is thus combinatorial w.r.t. all the scalar images $\bar{u}$ "included" in $\tilde{u}$ (see the parts highlighted in blue in Eq. 6). The Dahu pseudo-distance can be interpreted as the *best minimum barrier distance that we can have considering that the input function is continuous*.

The Dahu function is a pseudo-distance since it verifies the non-negativity, identity, symmetry, and subadditivity properties. In addition, we have the property: $\forall u$, $\forall x, x'$, $d_u^{\mathrm{naive}}(x, x') \leq d_u^{\mathrm{MB}}(x, x')$.

An example is given in Fig. 3(a) for an input function u, in Fig. 3(c) for its "continuous" representation $\tilde{u}$, and in Fig. 3(d) for a scalar function $\bar{u} \ll \tilde{u}$. A minimal Dahu path between two original points of u is depicted in blue Fig. 3(d); it is actually the same path as the one depicted on the continuous 3D surface in Fig. 1(c).

3.3 Solving a Discrete Topology Issue

In 2D discrete images, we can have saddle points. An example is given in Fig. 4(a), from which we deduce the continuous representation (given in Fig. 4(b)) in order to compute some minimal paths. We have $d_u^{\mathrm{naive}}(x_a, x'_a) = 0$ which is obtained with $\bar{u}_a$ (Fig. 4(c)), and $d_u^{\mathrm{naive}}(x_b, x'_b) = 2$ which is obtained with $\bar{u}_b$ (Fig. 4(d)). An interpretation of this situation "in the continuous world" is the following. On one hand, having $d_u^{\mathrm{naive}}(x_a, x'_a) = 0$ implies not only that the level in $\bar{u}_a$ of the saddle point (the 0-face) is **0**, but also that there is a 0-level line joining x_a and x'_a in $\tilde{u}$. On the other hand, since we have $u(x_b) = 4$ and $u(x_b) = 6$, $\bar{u}_b$ considers that the level **5** exists between x_b and x'_b, which is in contradiction with the previous conclusion. We thus end up with some inconsistencies, if we consider two distinct discrete paths in u, and if we try to interpret these paths in a continuous way in $\tilde{u}$. The main issue, here in 2D, is that the representation $\tilde{u}$ cannot prevent lines to cross at different levels, which is the case at saddle points. From this example, we can draw a negative conclusion: *The representation $\tilde{u}$, as defined in Eq. 5, cannot give us a continuous interpretation of the set of all paths in u without topological flaws.*

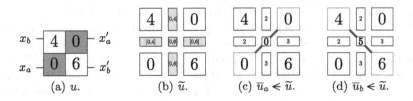

(a) u.   (b) $\tilde{u}$.   (c) $\bar{u}_a \ll \tilde{u}$.   (d) $\bar{u}_b \ll \tilde{u}$.

Fig. 4. The saddle case in 2D as a symptom of a discrete topology issue with $\tilde{u}$.

A solution to this problem is to add an intermediate step which computes an "interpolation" of u that has no pathological parts such as the saddle points as in the 2D case. In the general nD case, these parts are called critical configurations, and an image without any critical configuration is said "digitally well-composed" (DWC) [2].

Let us denote by $u_\square$ an interpolation of the scalar image $u : \mathbb{Z}^n \to Y$, defined on the subdivided space $\left(\frac{\mathbb{Z}}{2}\right)^n$, and taking its values either in Y or in another scalar space Y'. A continuous yet discrete representation of u is now $\widetilde{u_\square}$:

$$(u : \mathbb{Z}^n \to Y) \xrightarrow{\text{step 1}} \left(u_\square : \left(\frac{\mathbb{Z}}{2}\right)^n \to Y'\right) \xrightarrow{\text{step 2}} \left(\widetilde{u_\square} : \left(\frac{\mathbb{H}}{2}\right)^n \to \mathbb{I}_{Y'}\right). \quad (7)$$

Under the assumption of $u_\square$ being digitally well-composed, the interval-valued image $\widetilde{u_\square}$ allows for representing u in a continuous yet discrete way, and without any topological inconsistency. We will say that such maps $\widetilde{u_\square}$ are *path-consistent*.

Now we can provide a new definition of the Dahu pseudo-distance:

$$\forall\, x, x' \in \mathbb{Z}^n, \quad \mathrm{d}_u(x, x') = \min_{\overline{u} \,\preccurlyeq\, \widetilde{u_\square}} d_{\overline{u}}^{\mathrm{MB}}(h_x, h_{x'}). \quad (8)$$

As compared with the initial (naive) definition given by Eq. 6, we have just replaced $\widetilde{u}$ by $\widetilde{u_\square}$.

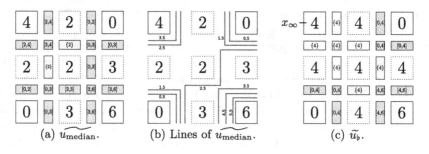

(a) $\widetilde{u_{\text{median}}}$.   (b) Lines of $\widetilde{u_{\text{median}}}$.   (c) $\widetilde{u_b}$.

Fig. 5. Interpolation-based interval-valued maps which are path-consistent.

Several subdivision-based interpolations of nD scalar images are known to produce DWC images. It is the case of the interpolation with the max operator, used in [13] for the quasi-linear computation of the tree of shapes of a scalar image, as defined by Eq. 2. In the 2D case—and not in nD with $n > 2$—the only *self-dual local* interpolation verifying strong invariant properties leading to DWC images is obtained with the median operator [13]. It is depicted in Fig. 5(a) (after the step 2 $u_\square \to \widetilde{u_\square}$ being applied); Fig. 5(b) shows that the level lines do *not* cross each other. For the general nD case, a self-dual non-local interpolation has been defined [3], which is illustrated in Fig. 5(c) (again, after applying the step 2 to get an interval-valued representation).

4   Computing the Dahu Pseudo-Distance

The tree of shapes of an image is also called topographic map. Since the Dahu distance is related to the notion of topography, the tree of shapes may be a good candidate tool to compute any Dahu distance $d_u(x, x')$.

4.1   Tree of Shapes of Interval-Valued Maps

Since the scalar interpolations $u_\square$ are digitally well-composed, the tree of shapes of the interval-valued maps $\widetilde{u_\square}$, namely $\mathfrak{S}(\widetilde{u_\square})$, as defined by Eq. 3, exists. One can easily understand that it is due to the fact that $\widetilde{u_\square}$ is path-consistent: reminding that a level line is the boundary of a shape, since two level lines at different levels cannot cross each other, it implies that shapes are either disjoint or nested. Actually, the set $\mathfrak{S}(\widetilde{u_\square})$ verifies the following properties: (i) Every element of this set is a connected component of $\mathbb{H}^n$, having no cavity, and being a regular open set. (ii) The set $\mathfrak{S}(\widetilde{u_\square})$ is a self-dual decomposition of $\mathbb{H}^n$ with respect to u. (iii) The boundary of every component is an iso-level line of u, so:

$$\forall S \in \mathfrak{S}(\widetilde{u_\square}), \exists \mu \in Y' \text{ such that } \forall h \in cl(S) \setminus S, \mu \in \widetilde{u_\square}(h).$$

(iv) A mapping that associates a level, denoted by $\mu(S)$, to every element $S \in \mathfrak{S}(\widetilde{u_\square})$ makes this set being a representation of $\widetilde{u_\square}$. We will say that $\mu(S)$ is the level of the shape S (we will use this mapping in the next section).

4.2   When the Dahu Pseudo-Distance and the Tree of Shapes Meet

Intuitively it is easy to understand that the minimal path between two points of an image u, w.r.t. the Dahu pseudo-distance, corresponds to a path of nodes on the tree of shapes $\mathfrak{S}(\widetilde{u_\square})$. If we look back at Fig. 2(b) (left), the path between the two points (x, x') indicated by red bullets in u starts from region B, then goes through A and C, and finally ends in region F. When considering the tree of shapes, in Fig. 2(b) (right), this path is exactly the same, and it crosses the level lines, depicted by dashed lines both in the image and on the tree. Such a path is minimal because every path in $\Pi(x, x')$ should at least cross this same set of level lines to go from x to x'. So the Dahu pseudo-distance corresponds to the level dynamics of this set of lines. In both the continuous version in Fig. 1(c) and its discrete representation in Fig. 3(b), we can directly read that the distance between the red points is 1 (the level lines are the dashed green lines).

Given a node t of a tree, let us denote by $par(t)$ the parent node of t in the tree, and by $lca(t, t')$ the lowest common ancestor of the nodes t and t'. Let us denote by t_x the node of $\mathfrak{S}(\widetilde{u_\square})$ corresponding to $x \in X$; this node corresponds to the smallest shape of $\mathfrak{S}(\widetilde{u_\square})$ containing h_x. We can concatenate the two sequences of nodes $\pi_x = \langle t_x, par(t_x), par^2(t_x), ..., lca(t_x, t_{x'}) \rangle$ and $\pi_{x'} = \langle t_{x'}, par(t_{x'}), par^2(t_{x'}), ..., lca(t_x, t_{x'}) \rangle$, to form the new sequence:

$$\pi(t_x, t_{x'}) = \pi_x \frown \pi_{x'}^{-1} = \langle t_x, par(t_x), ..., lca(t_x, t_{x'}), ..., par(t_{x'}), t_{x'} \rangle. \qquad (9)$$

Note that, if $t_{x'}$ is an ancester of t_x (or the reverse way), this writing is slightly different: $\pi(t_x, t_{x'}) = \langle t_x, \mathrm{par}(t_x), ..., t_{x'} \rangle$. In the case of the example of Fig. 2(b), the lca of the nodes t_B and t_F is t_A, and the sequence of nodes is $\langle t_\mathsf{B}, t_\mathsf{A}, t_\mathsf{C}, t_\mathsf{F} \rangle$ (depicted on the right), and it corresponds on the tree to the minimal path in the image space (depicted on the left).

To every shape $S \in \mathfrak{S}(\widetilde{u_\square})$ corresponds a node, say t_S, of the tree of shapes. Let us then write abusively $\mu(t_S) = \mu(S)$, where $\mu(S)$ is the level of the shape as defined in Sect. 4.1. Eventually, the Dahu pseudo-distance can then be re-expressed by:

$$\mathfrak{d}_u(x, x') = \max_{t \in \pi(t_x, t_{x'})} \mu(t) - \min_{t \in \pi(t_x, t_{x'})} \mu(t), \tag{10}$$

where the nodes considered and the mapping μ are related to the tree $\mathfrak{S}(\widetilde{u_\square})$.

Practically, to compute $\mathfrak{d}_u(x, x')$ we just have to update the interval values (min and max of μ) starting respectively from the nodes of x and of x', up to their lca. For instance, on π_x we iteratively compute:

$$\mu_{\min}(\langle t_x, ..., \mathrm{par}^{i+1}(t_x)\rangle) = \min(\mu_{\min}(\langle t_x, ..., \mathrm{par}^i(t_x)\rangle), \mu(\mathrm{par}^{i+1}(t_x))),$$

and likewise for $\mu_{\max}$. Eventually, $\mathfrak{d}_u(x, x')$ is the length of the span of both intervals, that is: $\mathfrak{d}_u(x, x') = \max(\mu_{\max}(\pi_x), \mu_{\max}(\pi_{x'})) - \min(\mu_{\min}(\pi_x), \mu_{\min}(\pi_{x'}))$. On a regular PC, running GNU Linux with a 2.67 GHz CPU, computing one million distances between points randomly taken in the **lena** image takes about 1 s. Some code is available from: http://publications.lrde.epita.fr/geraud.17.ismm.

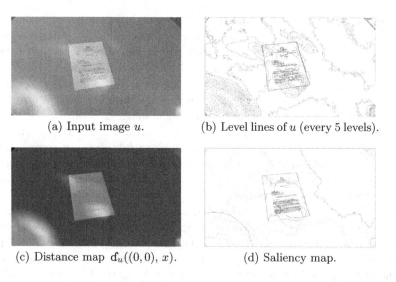

(a) Input image u. (b) Level lines of u (every 5 levels).

(c) Distance map $\mathfrak{d}_u((0, 0), x)$. (d) Saliency map.

Fig. 6. Using the Dahu pseudo-distance on an actual application.

4.3   Illustration on Document Image Segmentation

Figure 6 gives an illustration on a camera-based document image taken from the SmartDoc competition organized at ICDAR 2015 [4]. Let us first recall that the tree of shapes has already been used to such an application [8,22].

From a scalar image u (Fig. 6(a)), we compute its tree of shapes $\mathfrak{S}(\widetilde{u_\square})$ (Fig. 6(b)), which allows for computing very efficiently a distance map from the top left point $(0,0)$ (Fig. 6(c)) using the Dahu distance. One can see that the document is now well contrasted w.r.t. its environment. A saliency map corresponding to the Dahu distance is depicted in Fig. 6(d).

5   Related Works

In addition to the references given throughout the paper, related either to the context of our work or to its theoretical background, several other works shall be mentioned. Actually the notion of path on tree nodes also appear in [7] to assign to each node a label and then perform some scribble-based segmentation, in [8] to count level lines and build a multi-variate tree of shapes, and in [12] to compute some curvilinear variation and separate an object from its background. The notion of barrier used on the tree of shapes is also close to the one of shape saliency used in [24]. Last, about distance transforms and minimal path approaches, the reader can refer to the recent survey [19].

6   Conclusion

We have presented a new pseudo-distance, the Dahu distance, which is a variant of the minimum barrier (MB) distance on a continuous—yet discrete—representation of images. There are two major advantages of this distance as compared to the "classical" MB distance:

- It considers that the input discrete images are actually defined in a continuous world, that is, that their elevation map is a surface.
- Computing the Dahu distance for many couples of points (x, x') is very *efficient*, thanks the pre-computation of a tree of shapes; furthermore, the distances $d_u(x, x')$ obtained using this tree are *exact*. That contrasts with the efficient but approximate computation of the MB distance given in [21].

Some other practical interests of this new distance over its classical definition remain to be proved. Yet we believe that there are numerous perspectives for applications, which are made possible thanks to the fast tree-based computation. Another perspective is to adapt this distance to color images; that can be done using the multi-variate tree of shapes (MToS) defined in [8].

Acknowledgments. The authors want to thank Guillaume Tochon (new evangelist with 🔒) and Laurent W. Najman for their valuable comments on an erstwhile version of this paper, and the reviewers for their helpful remarks. This work has been conducted within the MOBIDEM project, part of the "Systematic Paris-Region" and "Images & Network" Clusters. This project is partially funded by the French Government and its economic development agencies.

References

1. Aubin, J.P., Frankowska, H.: Set-Valued Analysis. Birkhäuser, Basel (2009)
2. Boutry, N., Géraud, T., Najman, L.: On making nD images well-composed by a self-dual local interpolation. In: Barcucci, E., Frosini, A., Rinaldi, S. (eds.) DGCI 2014. LNCS, vol. 8668, pp. 320–331. Springer, Cham (2014). doi:10.1007/978-3-319-09955-2_27
3. Boutry, N., Géraud, T., Najman, L.: How to make nD functions digitally well-composed in a self-dual way. In: Benediktsson, J.A., Chanussot, J., Najman, L., Talbot, H. (eds.) ISMM 2015. LNCS, vol. 9082, pp. 561–572. Springer, Cham (2015). doi:10.1007/978-3-319-18720-4_47
4. Burie, J., Chazalon, J., et al.: ICDAR 2015 competition on smartphone document capture and OCR (SmartDoc). In: Proceedings of ICDAR, pp. 1161–1165 (2015)
5. Cao, F., Lisani, J.L., Morel, J.M., Musé, P., Sur, F.: A Theory of Shape Identification. Lecture Notes in Mathematics, vol. 1948. Springer, Heidelberg (2008)
6. Carlinet, E., Géraud, T.: A comparative review of component tree computation algorithms. IEEE Trans. Image Process. **23**(9), 3885–3895 (2014)
7. Carlinet, E., Géraud, T.: Morphological object picking based on the color tree of shapes. In: Proceedings of IPTA, pp. 125–130 (2015)
8. Carlinet, E., Géraud, T.: MToS: a tree of shapes for multivariate images. IEEE Trans. Image Process. **24**(12), 5330–5342 (2015)
9. Caselles, V., Coll, B., Morel, J.M.: Topographic maps and local contrast changes in natural images. Int. J. Comput. Vis. **33**(1), 5–27 (1999)
10. Caselles, V., Monasse, P.: Geometric Description of Images as Topographic Maps. Lecture Notes in Mathematics, vol. 1984. Springer, Heidelberg (2009)
11. Crozet, S., Géraud, T.: A first parallel algorithm to compute the morphological tree of shapes of nD images. In: Proceedings of ICIP, pp. 2933–2937 (2014)
12. Dubrovina, A., Hershkovitz, R., Kimmel, R.: Image editing using level set trees. In: Proceedings of ICIP, pp. 4442–4446 (2014)
13. Géraud, T., Carlinet, E., Crozet, S.: Self-duality and digital topology: links between the morphological tree of shapes and well-composed gray-level images. In: Benediktsson, J.A., Chanussot, J., Najman, L., Talbot, H. (eds.) ISMM 2015. LNCS, vol. 9082, pp. 573–584. Springer, Cham (2015). doi:10.1007/978-3-319-18720-4_48
14. Géraud, T., Carlinet, E., Crozet, S., Najman, L.: A quasi-linear algorithm to compute the tree of shapes of nD images. In: Hendriks, C.L.L., Borgefors, G., Strand, R. (eds.) ISMM 2013. LNCS, vol. 7883, pp. 98–110. Springer, Heidelberg (2013). doi:10.1007/978-3-642-38294-9_9
15. Mazo, L., Passat, N., Couprie, M., Ronse, C.: Digital imaging: a unified topological framework. J. Math. Imaging Vis. **44**(1), 19–37 (2012)
16. Monasse, P., Guichard, F.: Fast computation of a contrast-invariant image representation. IEEE Trans. Image Process. **9**(5), 860–872 (2000)
17. Najman, L., Géraud, T.: Discrete set-valued continuity and interpolation. In: Hendriks, C.L.L., Borgefors, G., Strand, R. (eds.) ISMM 2013. LNCS, vol. 7883, pp. 37–48. Springer, Heidelberg (2013). doi:10.1007/978-3-642-38294-9_4

18. Najman, L.W.: I met a Dahu (2015). Private communication at ISMM
19. Saha, P.K., Strand, R., Borgefors, G.: Digital topology and geometry in medical imaging: a survey. IEEE Trans. Med. Imaging **34**(9), 1940–1964 (2015)
20. Strand, R., Ciesielski, K.C., Malmberg, F., Saha, P.K.: The minimum barrier distance. Comput. Vis. Image Underst. **117**, 429–437 (2013)
21. Tu, W.C., He, S., Yang, Q., Chien, S.Y.: Real-time salient object detection with a minimum spanning tree. In: Proceedings of IEEE CVPR, pp. 2334–2342 (2016)
22. Xu, Y., Carlinet, E., Géraud, T., Najman, L.: Hierarchical segmentation using tree-based shape spaces. IEEE PAMI (2017). In early access
23. Xu, Y., Géraud, T., Najman, L.: Connected filtering on tree-based shape-spaces. IEEE Trans. Pattern Anal. Mach. Intell. **38**(6), 1126–1140 (2016)
24. Xu, Y., Géraud, T., Najman, L.: Hierarchical image simplification and segmentation based on Mumford-Shah-salient level line selection. Pattern Recogn. Lett. **83**, 278–286 (2016)
25. Zhang, J., Sclaroff, S., Lin, Z., Shen, X., Price, B., Mech, R.: Minimum barrier salient object detection at 80 FPS. In: Proceedings of ICCV, pp. 1404–1412 (2015)

Symmetric Counterparts of Classical 1D Haar Filters for Improved Image Reconstruction via Discrete Back-Projection

Matthew Ceko[✉] and Imants Svalbe

School of Physics and Astronomy, Monash University, Melbourne, Australia
{matthew.ceko,imants.svalbe}@monash.edu

Abstract. A discrete 2D p:q lattice is comprised of known pixel values spaced at regular p:q intervals, where p, q are relatively prime integers. The lattice has zero values elsewhere. Sets of new symmetric convolution masks were constructed recently whose purpose is to interpolate values for all locations around each lattice point. These symmetric masks were found to outperform the traditional asymmetric masks that interpolate in proportion to the area each pixel shares within a p:q neighbourhood. The 1D projection of these new 2D symmetric masks can also be used when reconstructing images via filtered back-projection (FBP). Here the 1D symmetric filters are shown to outperform the traditional Haar filters that are built from the area-based masks. Images reconstructed using FBP with symmetric filters have errors up to 10% smaller than with Haar filters, and prove to be more robust under Poisson noise.

1 Introduction

A p:q lattice is a sparse lattice whereby each regularly spaced pixel value is separated by p rows and q columns. The lattice is constructed by uniquely mapping a pixel at coordinates (x, y) to integer coordinates (x', y') through the affine rotation matrix. An example is shown in Fig. 1.

$$\begin{bmatrix} x' \\ y' \\ 1 \end{bmatrix} = \begin{bmatrix} q & -p & 0 \\ p & q & 0 \\ 0 & 0 & 1 \end{bmatrix} \begin{bmatrix} x \\ y \\ 1 \end{bmatrix} \tag{1}$$

Lines across this lattice are defined by $\theta = \tan^{-1}(p/q)$ where p, q are co-prime integers, satisfying $\gcd(p, q) = 1$. These lattices arise via affine rotations, as well as through 1D projections of 2D images under discrete angle projection schemes [2,11].

Traditionally, area-based convolution masks have been used to in-fill p:q lattices. Area-based masks are constructed on the basis that the value interpolated at each vacant pixel position depends only on the fraction of each pixel area that lies inside the bounding lines of the p:q lattice [10]. The geometry of a p:q lattice leads to these masks generally being asymmetric in shape (see Fig. 1).

© Springer International Publishing AG 2017
J. Angulo et al. (Eds.): ISMM 2017, LNCS 10225, pp. 68–80, 2017.
DOI: 10.1007/978-3-319-57240-6_6

$$
\begin{array}{cccccc}
1 & 3 & 5 & 7 & 9 & 1 \\
9 & 10 & 10 & 10 & 10 & 3 \\
7 & 10 & 10 & 10 & 10 & 5 \\
5 & 10 & 10 & 10 & 10 & 7 \\
3 & 10 & 10 & 10 & 10 & 9 \\
1 & 9 & 7 & 5 & 3 & 1
\end{array}
\qquad
\begin{array}{ccccccc}
1 & 1 & 1 & 1 & 1 & 1 & 1 \\
1 & 2 & 3 & 3 & 3 & 2 & 1 \\
1 & 3 & 4 & 4 & 4 & 3 & 1 \\
1 & 3 & 4 & 4 & 4 & 3 & 1 \\
1 & 3 & 4 & 4 & 4 & 3 & 1 \\
1 & 2 & 3 & 3 & 3 & 2 & 1 \\
1 & 1 & 1 & 1 & 1 & 1 & 1
\end{array}
$$

Fig. 1. $p{:}q = 1{:}5$ lattice (left). The red line is oriented at $\tan^{-1}(1/5)$, and the blue line at $5{:}1$, which can be seen as discrete projected rays. The space around each black pixel can be filled using area-based (centre) and symmetric (right) convolution masks. (Color figure online)

Recent revisions to the construction of such masks has shown that a focus on symmetry provides smoother in-filling. Families of symmetric masks can be constructed which correctly in-fill the $p{:}q$ lattice [1,11]. Since the masks are real and symmetric, so too are their Fourier transforms. This ensures the phase information remains untouched. Due to their ability to in-fill the discrete $p{:}q$ lattice, these symmetric masks are a prime candidate for an alternate pixel model for the Mojette transform[1]. This paper shows that discrete projections interpolated using new symmetric in-fill masks gives improved back-projected reconstructions over the area-based Haar model.

2 The Mojette Transform

The Mojette transform is a discretisation of the Radon transform, where angles are chosen from the discrete set $\theta = \tan^{-1}(p/q)$. Under the Mojette transform, each projection is the sum of the pixel values that intersect the line $b = -qk + pl$. The Dirac-Mojette operator [2] acts on image f by

$$
[M_\delta f](b,p,q) = proj_\delta(b,p,q) = \sum_{k=-\infty}^{+\infty} \sum_{l=-\infty}^{+\infty} f(k,l)\Delta(b + kq - pl) \qquad (2)
$$

where $\Delta(b) = 1$ if $b = 0$, otherwise $\Delta(b) = 0$ and $(b,p,q) \in \mathbb{Z}$. An example of a Dirac-Mojette transform is performed in Fig. 2a. To move from a Dirac projection to a ray of finite width, better approximating practical beams such as x-rays, a pixel model must be chosen to smooth back-projected values. A pixel model maps a discrete $P \times Q$ image $f(k,l)$ onto the object $f_{pm}(x,y)$ in continuous space via

$$
f_{pm}(x,y) = \sum_{k=0}^{P-1} \sum_{l=0}^{Q-1} f(k,l)pm(x - k, y - l) \qquad (3)
$$

where $pm(x,y)$ is the chosen pixel model [6]. Using this, generalised Mojette projections can be defined through arbitrary pixel models as

---

[1] Full library of symmetric masks available at https://github.com/mceko/symmetric _masks_library.

$$proj_{pm}(b,p,q) = \sum_m proj_\delta(b,p,q)pmp(b-m,p,q) = proj_\delta(b,p,q) * pmp(b+qk-pl,p,q) \tag{4}$$

Here, $pmp(b,p,q)$ is the pixel model projection interpolator along the direction (p,q), defined as

$$pmp(b,p,q) = \int_{-\infty}^{+\infty} \int_{-\infty}^{+\infty} pm(x,y)\delta(b+qx-py). \tag{5}$$

Therefore a Mojette projection under any pixel model is given by a discrete convolution of the Dirac-Mojette projection with the (p,q) projection of the chosen pixel model. Typically, the Haar spline basis is used as the chosen pixel model. Here we propose a new pixel model based on projections of symmetric 2D masks.

The Haar pixel model is based on projections of a square pixel [3] of size w, typically set to $w = 1$. The projections are trapezoids except for $\theta = 0$ which has a square projection and $\theta = \frac{\pi}{4}$ where the projection is an isosceles triangle. The intermediate angles can be viewed as an evolution from a square to triangle projection.

Trapezoids can be built via convolutions of square functions. In discrete space, this corresponds to convolutions of the Dirac comb [4]. The Haar pixel model is represented by the spline 0 kernels:

$$pmp_0 = \begin{cases} 1 & \text{if } p=0 \text{ or } q=0 \\ \underbrace{(1,...,1)}_{p} * \underbrace{(1,...,1)}_{q} & \text{if } p+q \text{ is even} \\ \frac{1}{2}(1,1) * \underbrace{(1,...,1)}_{p} * \underbrace{(1,...,1)}_{q} & \text{if } p+q \text{ is odd} \end{cases} \tag{6}$$

Notice the additional convolution with $\frac{1}{2}(1,1)$ when $p+q$ is odd. For two signals of length l_1 and l_2, their convolution has length $l_1 + l_2 - 1$. Hence when $p+q$ is odd, the length of their convolution is even and does not have a well-defined centre. This results in an ambiguous centre of the original bin. The convolution with $\frac{1}{2}(1,1)$ is an averaging kernel which averages the original even sized mask applied to the left and right hand shift with respect to the bin.

When $p+q$ is odd, the averaged pmp_0 kernel is equivalent to the column (or row) sums of the 2D area-based mask [10]. Since symmetric convolution masks are often more favourable in their ability to in-fill a $p{:}q$ lattice, we would like to test a pixel model where the 1D kernels are based on column sums of 2D symmetric masks to see if they are able to improve the accuracy of back-projection.

A second motivation for using symmetric masks comes from an alternate implementation of the Mojette transform. The Dirac-Mojette transform can be equivalently viewed as row and column projections of the affine rotated $p{:}q$ lattice, as shown in Fig. 2b. Hence any interpolation mask that in-fills the $p{:}q$ lattice is a potential candidate for a pixel model kernel.

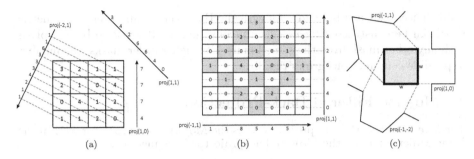

(a)   (b)   (c)

Fig. 2. (a) Dirac-Mojette projections (1,0), (1,1) and (−2,1) for a 4 × 4 array. The origin is taken to be the bottom-left pixel. (b) Dirac-Mojette projections (1,1) and (−1,1) performed through affine rotation. (c) Projections (1,0), (−1,1) and (−1,−2) for a pixel of size w.

3   Construction of 1D Symmetric Masks

Analogous to the construction of the averaged Haar masks (when $p + q$ is odd) via 2D area-based masks, 1D symmetric masks can be constructed by taking the column (or row) sums of 2D symmetric masks. In taking the column sums, approximately 99% of the 2D symmetric masks for $|p|, |q| \leq 11$ with a maximum allowed integer mask coefficient of 16, produce distinct 1D masks. Hence there are many possible symmetric masks to consider as candidates for each $p{:}q$, even though the masks are now used as one-dimensional projections. However, considering only linearly independent masks reduces the set considerably. We refer to this linearly independent set as basis masks. The number of projected basis masks is bounded by the number of basis masks in 2D, although this number is often the same or at least close.

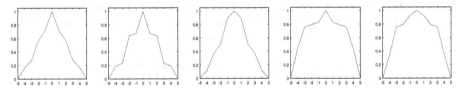

Fig. 3. Basis set of 1D projected masks derived from the 2D basis set for 3:5.

It can be seen in Fig. 3 that the 1D projected basis mask set presents a range of vastly different mask shapes from which we can build masks with a desired shape through linear combinations. The projected views of the symmetric shapes vary from a rough triangular form to more trapezoidal shapes. A β-spline of order n is defined through the recursive relation

$$\beta_{n,w}(b) = \beta_{n-1,w} * \beta_{0,w} = \underbrace{\beta_{0,w} * \beta_{0,w} * \ldots * \beta_{0,w}}_{n+1 \text{ times}} . \tag{7}$$

This process can be performed to create higher order symmetric masks. However, since masks are not unique for a given $p{:}q$ there are a variety of possible valid choices that can be made. As with the Haar masks, an order n symmetric mask can be constructed by $n+1$ auto-convolutions. It can also be convolved with any other mask from the same family. For higher order masks, the number of possible combinations grows rapidly.

4   Mojette Filtered Back-Projection

The inverse of the Mojette projector M is the Mojette backprojector M^*. In the Dirac-Mojette model, the Mojette backprojector is defined as

$$[M^*_\delta proj_{(p,q)}](k,l) = \sum_{i=-\infty}^{+\infty} \sum_{j=-\infty}^{+\infty} \sum_{b \in B} \sum_{(p,q) \in I} \delta(k-i)\delta(l-j) \times proj(b,p,q)\Delta(b+qi-pj) \quad (8)$$

where I is the set of (p,q) directions. The lower-left corner is taken to be the origin $(0,0)$. Using a chosen pixel model, a filtering stage can be applied before back-projection. This involves applying a Ramp filter, which is a high-pass filter that compensates for the uneven tiling by back-projection of 1D views across pixels near the array centre. Mojette filtered back-projection is performed through the set of operators

$$f(k,l) = M^*_{pm}KM_{pm}f(k,l). \quad (9)$$

This algorithm can be applied with any pixel model and an appropriate discretisation of the ramp filter, K. For equally spaced angles, the ramp filter is given in frequency space by $|\nu|$. This is a high-pass filter that suppresses low frequency components that contribute to blurring and keeps high frequency components that provide contrast. Since the Mojette transform does not have equally spaced projections, angle dependent filters must be introduced. This involves sampling the ramp filter using angle dependent sampling. The continuous exact Spline-0 FBP filter is derived in the frequency domain for projection angle θ as [3]

$$K_0(\nu,\theta) = w^2\pi|\nu|\mathrm{sinc}(\nu w\cos\theta)\mathrm{sinc}(\nu w\sin\theta). \quad (10)$$

The inverse Fourier transform of $K_0(\nu,\theta)$ provides an explicit formula in the spatial domain, $k_0(t,\theta)$. θ is restricted to discrete angles $\tan\theta = q/p$ with projection sampling $bw/(p^2+q^2)$ [7].

$$k_0(b,p,q) = \begin{cases} \dfrac{-1}{\pi}\dfrac{2w}{4w^2b^2-1} & (p,q)=(1,0) \\[2ex] \dfrac{p^2+q^2}{2\pi pq}\ln\left|\dfrac{b^2-\left(\frac{p+q}{2}\right)^2}{b^2-\left(\frac{p-q}{2}\right)^2}\right| & \text{otherwise} \end{cases} \quad (11)$$

This filter cannot be implemented in the spatial domain in a straightforward manner due to the discontinuities in the filter where the pixel projection is continuous, but not differentiable. These points occur at $b = \pm\frac{p+q}{2}, \pm\frac{p-q}{2}$, where the ramp filter behaves as a derivative. However these points are able to be computed using a Dirichlet condition, i.e. converge to the mean [3].

5   Trapezoidal Decomposition of Symmetric Masks

Symmetric 1D arrays can be decomposed as linear combinations of trapezoids. Since the 1D symmetric masks are always of odd size, they can be decomposed as linear combinations of Haar masks for which $p + q$ is even. For example, the projected 3:5 mask $(0, 5, 7, 18, 22, 32, 22, 18, 7, 5, 0)$ can be shown as the linear sum of trapezoids in Fig. 4.

To decompose any symmetric mask into trapezoids, we find the vertical and horizontal distance between corresponding symmetric vertices (including endpoints) on each side of the mask. This defines the top and bottom lengths, a and b, and the height of each trapezoid, h. The symmetry of the masks indicates that the trapezoids are always isosceles trapezoids, hence they are fully described by the three parameters a, b and h. The Haar mask for projection angle (p, q) is given by $pmp_0(p, q)$, so we look for $c, p, q \in \mathbb{Z}$ such that $c \cdot pmp_0(p, q)$ produces the trapezoid defined by a, b and h. The scaled mask, $c \cdot pmp_0(p, q)$ constructs an isosceles trapezoid with top length $a = q - p + 1$, bottom length $b = p + q + 1$ and height $h = cp$. This set of equations can be solved for p, q and c.

$$p = \frac{b - a}{2}, \quad q = \frac{a + b}{2} - 1, \quad c = \frac{h}{p} \tag{12}$$

Therefore, any symmetric mask can be written as a linear combination of Haar masks. For example, the symmetric mask in Fig. 4 can be written as $10pmp_0(1, 1) + 4pmp_0(1, 3) + 11pmp_0(1, 5) + 2pmp_0(1, 7) + 5pmp_0(1, 9)$. Using the linearity of the Fourier transform, the derivation of a single k_0 filter can be applied to any arbitrary linear combination of Haar masks. Hence the discretised ramp filter for the symmetric mask in Fig. 4, k_{s0}, is given by $k_{s0}(b) = 10k_0(b, 1, 1) + 4k_0(b, 1, 3) + 11k_0(b, 1, 5) + 2k_0(b, 1, 7) + 5k_0(b, 1, 9)$.

Notice the similarities in k_{s0} when compared to k_0 for the Haar masks. Both filters exhibit discontinuities where the mask is continuous but not differentiable. These discontinuities tend towards $+\infty$ for concave vertices and $-\infty$ for convex points. Again it is preferred to implement the filters in the Fourier domain due to these discontinuities.

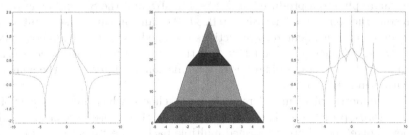

Fig. 4. (3,5) masks: Haar trapezoidal projection mask with k_0 filter (left), symmetric mask trapezoidal decomposition (centre) and symmetric mask with k_{s0} filter (right).

This algorithm can be used to construct equivalent k_{s0} filters for symmetric masks. Mojette FBP can then be performed with the symmetric mask pixel model as an alternative to the Haar basis. Each $p{:}q$ presents a wide range of candidate symmetric masks from which to choose.

6   Comparison of Symmetric and Haar Pixel Model for FBP

We now assess the performance of the Haar and symmetric pixel model for filtered back-projection. Two standard test images were used in this work to assess the quality of back-projections for each pixel model, following the work of Subirats *et al.* [9]. Both images are the same size, 129×129, which has a well-defined centre and can be adequately sampled with relatively small $p{:}q$ angles. The first test image is the square phantom, consisting of a 15×15 pixel square where the pixel intensity is 1, surrounded by a border of value $\frac{1}{2}$. The corners of this border are set to $\frac{1}{4}$ to make it bandlimited, and then zero padded to 129×129 pixels. The second test image is the 129×129 Shepp-Logan phantom, which simulates a brain scan using various sized ellipses at different intensities.

The two 129×129 noise-free phantom images were projected with the set of discrete angles constructed using a Farey sequence. A Farey sequence of order $n \in \mathbb{N}$, denoted F_n, is the set of irreducible rational numbers $\frac{p}{q}$ from 0 $\left(\frac{0}{1}\right)$ to 1 $\left(\frac{1}{1}\right)$ with $0 \le p \le q \le n$ and $\gcd(p,q) = 1$, arranged in increasing order [2]. A Farey sequence of order 10 provides 128 angles, which easily satisfies the Katz Criterion [5]. The Mojette projections are then convolved with a chosen pixel model, after which the corresponding discretisation of the ramp filter is applied. Finally these projections are back-projected using a simple Mojette back-projection scheme to form the reconstructions which can then be compared to the original images for errors.

To compare the symmetric and area-based pixel models, we first take the Haar-Mojette projections of the square and Shepp-Logan phantoms under $I = 128$ projections angles given by the Farey sequence F_{10}. These projections are then back-projected with the k_0 filter. The mean squared error (MSE) of each reconstruction is used as a measure of reconstruction accuracy. Back-projections of the square phantom produced an MSE of 0.0078 and the Shepp-Logan phantom reconstruction showed a MSE of 0.0230. We can compare these results with the accuracy of equivalent reconstructions obtained using the symmetric pixel model. Using this new pixel model, we test the error in the reconstructions using the standard k_0 and trapezoidal linear combination k_{s0} filter after the convolution with symmetric masks.

The symmetric pixel model requires us to choose which mask to convolve for each (p, q) projection. Therefore, to show a "best case" scenario for the symmetric pixel model, we performed multiple back projections to test different masks for each angle. The MSE was recorded for each reconstruction, and the masks which resulted in the lowest error was used for these reconstructions.

In Fig. 5 we observe similarly shaped errors to those produced using the Haar masks, which reflects the distribution of projection angles chosen. However, the symmetric mask MSE is lower than when using the Haar pixel model. A set of symmetric masks can be found that produce a lower MSE than using Haar masks with the standard k_0 filters. This error can be further reduced using corresponding linear combinations of k_0 filters for the trapezoidal decomposition when sampling is sufficient. A 9% improvement in MSE is seen when using the optimal set of symmetric masks with k_{s0} filtering compared to Haar masks with k_0 filters. Linear shift invariance is also satisfied (Fig. 5).

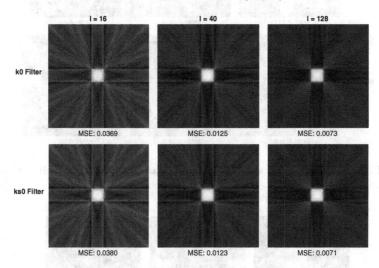

Fig. 5. Filtered back-projection of the square phantom for $I = 16, 40$ and 128 projections with the k_0 and k_{s0} filter for the symmetric pixel model.

Analogous results can be seen again in the reconstructions of the Shepp-Logan phantom in Fig. 6. The reconstruction of this phantom using k_{s0} filters display PSNR values of 6.75 dB, 16.42 dB and 16.77 dB correspondingly under symmetric mask FBP. Again there is a 9% improvement in MSE when using symmetric masks with k_{s0} filters over Haar masks with k_0 filters. Since we observe an improved MSE for both images when using the k_{s0} filter, the discretisation of the ramp filter should be constructed using a trapezoidal decomposition with symmetric masks.

To simulate real, noisy tomographic acquisition, Poisson noise can also be applied to the projections [8]. We require that the same number of photons are present for a different number of projections. By varying the total number of photons along each projection, we are able to simulate different levels of noise. Both pixel models were tested with the application of Poisson noise to each image. The level of noise was varied from $10^6, 10^5, ..., 10^1$ photons per pixel and 1000 trial noise patterns were used for each noise level. The same identical, signal dependent noise pattern was applied to each image for both pixel models. This prevents any bias in the noise generation favouring either model. The MSE was recorded for each noisy projection reconstruction to test if either model handled noise better. The Haar model was implemented with k_0 filtering and the symmetric model with k_{s0} filters. From the 1000 trials, the mean and standard

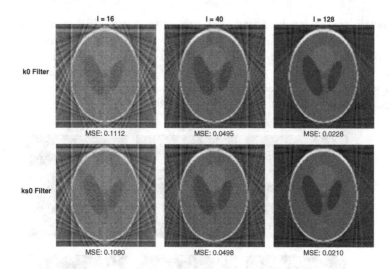

Fig. 6. Filtered back-projection of the Shepp-Logan phantom for $I = 16, 40$ and 128 projections with the k_0 and k_{s0} filter for the symmetric pixel model.

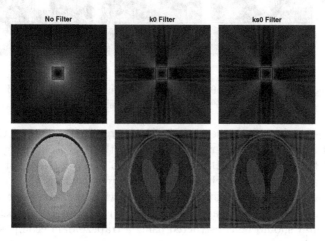

Fig. 7. Errors in the reconstruction of the square and Shepp-Logan phantoms for $I = 128$ projections with no filtering, the k_0 and k_{s0} filter for the symmetric pixel model.

deviation were calculated. For both models, $I = 128$ projections were used, at angles $p{:}q$ generated by the Farey sequence F_{10}.

For both images, and each noise level, the symmetric masks were able to outperform the Haar masks. Hence we list results in terms of the improvement of the symmetric model over the Haar model. This was measured as a percentage difference to act as a normalisation of difference due to the different total intensities of each phantom image. Not only does the symmetric model with k_{s0} filtering handle noise better than the Haar model, but the gain grows with increasing noise relative to the Haar model reconstructions (Table 1).

Table 1. Improvement of the symmetric mask model over the Haar pixel model, measured in percentage difference.

# of photons per pixel	MSE difference of symmetric over Haar (%)	
	Square phantom	Shepp-Logan phantom
10^6	9.15 ± 0.04	4.97 ± 0.10
10^5	9.18 ± 0.14	4.97 ± 0.29
10^4	9.32 ± 0.44	4.96 ± 0.94
10^3	9.67 ± 1.18	4.33 ± 2.77
10^2	11.98 ± 3.41	3.15 ± 4.65
10^1	15.66 ± 6.95	5.52 ± 6.06

These tests, both with and without noise, show that it is possible to find a set of 1D symmetric masks that outperform the traditionally used Haar pixel model. This result advocates for the possibility of Mojette based CT imaging. In practice, a system could be put through "training" with test images that resemble scans the system is expected to encounter to find an averaged optimal set of masks. This set is likely to perform well for any similar gamut of medical images (as shown here in Sect. 8).

7   Reconstruction of a Single Pixel

The Haar masks are constructed by taking (p, q) projections of a unit pixel. Symmetric masks are not built on an explicit pixel, but find application due to their ability to in-fill a $p{:}q$ lattice. We now take the same set of symmetric masks used to obtain the Shepp-Logan phantom reconstructions in Sect. 6, and apply them to back-project a single pixel. This allows us to examine the underlying pixel model that is being used when these symmetric masks are convolved with projection data. We do the same for the Haar masks as a comparison. One would expect the back-projection of Haar masks to produce the same unit square on which they were formulated. However, due to the averaging required for the even Haar masks, this is not the case in practice. Figure 8 demonstrates the reconstruction of a single pixel using Haar masks, and the set of symmetric masks used in the Shepp-Logan phantom reconstructions. The single pixel reconstructions

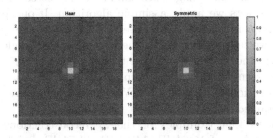

Fig. 8. Single pixel FBP reconstruction using Haar (left) and symmetric (right) masks with k_0 and k_{s0} filters respectively, using 128 projections.

show that the geometry of the symmetric reconstruction more closely matches the symmetry of the square lattice, despite Haar masks being constructed via projections of a square pixel.

8   Mask Sets from Training Images

Although we can find optimal sets of symmetric masks that produce more accurate back-projections than Haar masks, if symmetric masks are not chosen carefully then they may perform poorly. To attain an approximation of the average performance of the symmetric pixel model, random sets of symmetric masks were chosen for 100,000 trial reconstructions under the projection set generated by F_{10}, and for each set FBP was performed on both test images. The square phantom had a mean reconstruction MSE of 0.0117 ± 0.0013 while the Shepp-Logan phantom MSE averaged 0.0290 ± 0.0036. Even within uncertainty, this is noticeably poorer than the Haar pixel model. This does not necessarily imply that our optimal symmetric sets are contrived, and possibly a rare, fortunate exception that happen to perform well. The best performing mask sets for the two test images are similar, and share many of the same masks. This suggests that there may be parameters independent of image data, such as the projection angle set and image size, that dictate the content of this optimum set. To explore this, we construct a set of masks that we find to perform well on a set of training images. We then apply this mask set as our pixel model for a range of images and compare the FBP reconstructions to those produced by the Haar pixel model.

Consider a Mojette-based CT scanning system with detector bins that produce 129×129 reconstructed images from 128 discrete angles constructed using a Farey sequence of order 10. We now calibrate this system by finding optimal mask sets for four test images based on the Shepp-Logan phantom. First we use the same 2D phantom seen throughout Sect. 6, followed by three 2D slices of the 3D Shepp-Logan phantom along the x, y and z-axes of the image. These four phantom images not only represent real CT data, but also have known analytic formulae that describe their construction. We find the optimal set of masks for each test image using the same process used in Sect. 6, by holding the masks constant for every projection angle besides one for which we iterate through all masks to find the individual mask that produces the lowest error. The sets of optimum masks for all four images are similar, as expected. To modulate the differences between sets we apply a simple algorithm. If one particular mask shows up most frequently, then we choose that mask for the given projection angle. Otherwise we take an average of the masks for that angle. Using the set of masks found via the four Shepp-Logan training images, we apply FBP on a sequence of 27 consecutive 129×129 brain MRI slices. This data is taken from MATLAB's 3-dimensional MRI data set. We can then compare the reconstructions to those obtained using Haar-Mojette FBP.

Figure 9 displays four examples of the 27 reconstructions tested, which resulted in PSNR values of 15.64 dB, 14.40 dB, 14.74 dB and 14.93 dB in each

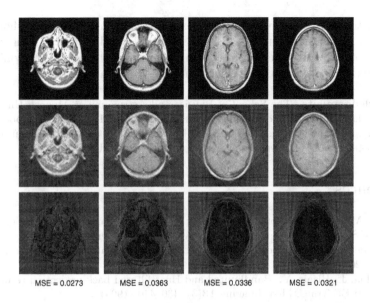

MSE = 0.0273   MSE = 0.0363   MSE = 0.0336   MSE = 0.0321

Fig. 9. Symmetric pixel model FBP reconstructions, for 4 examples from the 27 slices. From top to bottom, rows contain the original images, reconstructions and errors.

respective column. Across the 27 images, we see an average improvement of 5% using the symmetric masks over the Haar model. Here, the symmetric masks were not specifically tailored to suit any of these images. They were merely based on representative test data derived from the Shepp-Logan phantom. There may exist a more refined, analytical technique for finding optimal symmetric mask sets for an imaging system. But here we have shown that it is possible to find a symmetric mask set based on relatively few test images that consistently outperform Haar masks over a range of medical images.

9   Future Work and Conclusion

A large number of symmetric masks exist that, by design, in-fill a $p{:}q$ lattice. The same masks find application in discrete image rotation and back-projection, where they outperform the classic Haar area-based masks. Sets of symmetric masks, with appropriate filtering, can outperform Haar FPB for a range of reconstructions. Although some insight was provided for desired mask properties, there is no single "best" symmetric mask for each $p{:}q$. Hence future work will consist of alleviating this ambiguity by offering a technique of combining basis masks to extract the optimum linear combination for a given application, independent of the image content. It is possible to extend the construction of 2D symmetric interpolation masks to 3D by applying the same algorithm to a 3D affine rotation volume, and so the 2D projections of 3D symmetric masks should be considered for 3D Mojette applications. A focus on symmetry has proven to improve accuracy in discrete imaging applications, therefore the construction of symmetric masks under a similar ruleset for the hexagonal lattice should be examined. A hexagonal lattice is six-fold symmetric, and provides a denser array sampling of pixels. It would be interesting to examine the hexagonal symmetric mask solution space and their performance for similar applications.

Acknowledgements. The School of Physics and Astronomy at Monash University, Australia, has supported and provided funds for this work. M.C. has the support of the Australian government's Research Training Program (RTP) and J.L. William scholarship from the School of Physics and Astronomy at Monash University.

References

1. Ceko, M.: Symmetric convolution masks for in-fill pixel interpolation. B.Sc. (Hons) thesis, School of Physics and Astronomy, Monash University (2016)
2. Guédon, J.: The Mojette Transform: Theory and Applications. ISTE-Wiley, Chichester (2009)
3. Guédon, J., Bizais, Y.: Bandlimited and Haar filtered back-projection reconstructions. IEEE Trans. Med. Imaging **13**(3), 430–440 (1994)
4. Guédon, J., Normand, N.: Spline Mojette transform: applications in tomography and communications. In: 2002 11th European Signal Processing Conference, pp. 1–4. IEEE (2002)
5. Katz, M.B.: Questions of Uniqueness and Resolution in Reconstruction from Projections. Lecture Notes in Biomath, vol. 26. Springer, Heidelberg (1978)
6. Servières, M.: Reconstruction tomographique mojette. Université de Nantes (2005)
7. Servières, M., Guédon, J., Normand, N.: A discrete tomography approach to PET reconstruction. In: Proceedings of 7th International Conference on Fully 3D Reconstruction in Radiology and Nuclear Medicine, Saint-Malo, France (2003)
8. Servières, M., Normand, N., Bizais, Y., Guedon, J.: Noise behavior of spline Mojette FBP reconstruction. In: Medical Imaging, pp. 2100–2109. International Society for Optics and Photonics (2005)
9. Subirats, P., Servières, M.C.J., Normand, N., Guédon, J.: Angular assessment of the Mojette filtered back projection. In: Medical Imaging, pp. 1951–1960. International Society for Optics and Photonics (2004)
10. Svalbe, I.: Exact, scaled image rotations in finite Radon space. Pattern Recogn. Lett. **32**, 1415–1420 (2011)
11. Svalbe, I., Guinard, A.: Symmetric masks for in-fill pixel interpolation on discrete p:q lattices. In: Normand, N., Guédon, J., Autrusseau, F. (eds.) DGCI 2016. LNCS, vol. 9647, pp. 233–243. Springer, Cham (2016). doi:10.1007/978-3-319-32360-2_18

Watershed and Graph-Based
Segmentation

Watershed Segmentation with a Controlled Precision

Fernand Meyer$^{(\boxtimes)}$

CMM-Centre de Morphologie Mathématique, MINES ParisTech,
PSL-Research University, Paris, France
fernand.meyer@mines-paristech.fr

Abstract. Focusing on catchment zones which may overlap rather than on catchment basins which do not overlap allows to devise innovative segmentation strategies.

1 Introduction

Despite some apparent differences, node and edge weighted graphs are deeply similar with respect to their topography [7]. In fact we show here that to any node or edge weighted graph may be associated a weightless digraph conveying most of their topographical features.

This paper takes the complete opposite approach to the watershed. The focus is put not on the catchment basins but on the catchment zones. Catchment basins form a partition, being each the region of influence of the regional minima for a particular distance function (geodesic distance in [4], topographic distance in [6,8], lexicographic distance in [1]). Shortest path algorithms which mimic the flooding of a topographic surface [5,9] are generally used for constructing them.

A catchment zone, on the contrary, is simply the upstream of a regional minimum, without consideration to other regional minima. Catchment zones of neighboring regional minima do generally overlap and do not form a partition. If one assigns to each catchment zone a label and defines a priority between labels, one obtains a dead leaves tessellation where the catchment zones with higher priority labels cover the catchment zones with lower priority labels. Substituting catchment zones to catchment basins opens the way for innovative, local and fast segmentation algorithms.

The outline of the paper is the following. The first section summarizes the results of [7]. The next section is devoted to analyzing the topography of unweighted digraphs. In the last section we show how the catchment zones may be used for segmenting or locally extracting structures of interest in a graph.

2 The Topography of Node or Edge Weighted Graphs

2.1 Definitions and Notations

A *graph* $G = [\mathcal{N}, \mathcal{E}]$ contains a set $\mathcal{N}$ of nodes and a set $\mathcal{E}$ of edges; an edge being an unordered pair of vertices. The nodes are designated with small letters: $p, q, r...$ The edge linking the nodes p and q is designated by e_{pq} or $(p-q)$.

© Springer International Publishing AG 2017
J. Angulo et al. (Eds.): ISMM 2017, LNCS 10225, pp. 83–94, 2017.
DOI: 10.1007/978-3-319-57240-6_7

An oriented *graph, also called digraph* $\vec{G} = \left[\mathcal{N}, \vec{\mathcal{E}} \right]$ contains a set $\mathcal{N}$ of nodes and a set $\vec{\mathcal{E}}$ of arrows; an arrow being an ordered pair of vertices. The arrow from p to q is written $\vec{e_{pq}}$ or $(p \rightarrow q)$ If both arrows $(p \rightarrow q)$ and $(q \rightarrow p)$ exist, we write $(p \longleftrightarrow q)$.

The sub digraph of $\vec{G}$ spanned by a set A of nodes is the graph $\vec{A} = \left[\mathcal{A}, \vec{\mathcal{E}}_A \right]$, where $\vec{\mathcal{E}}_A$ is the subset of arcs of $\vec{\mathcal{E}}$ having both extremities in A.

A *path* $\vec{\pi}$ *of a digraph* is a sequence of vertices and arrows, interweaved in the following way: $\vec{\pi}$ starts with a vertex, say p, followed by an arrow $\vec{e_{ps}}$, incident to p, followed by the other endpoint s of $\vec{e_{ps}}$, and so on.

Edges and/or nodes may be weighted with weights taken in a totally ordered set $\mathcal{T}$:

* $\mathcal{F}_n = \mathrm{Fun}(\mathcal{N}, \mathcal{T})$: the functions defined on the nodes $\mathcal{N}$ with value in $\mathcal{T}$. The function $\nu \in \mathcal{F}_n$ takes the weight ν_p on the node p in G and $\vec{G}$.
* $\mathcal{F}_e = \mathrm{Fun}(\mathcal{E}, \mathcal{T})$: the functions defined on the edges $\mathcal{E}$ with value in $\mathcal{T}$. The function $\eta \in \mathcal{F}_e$ takes the value η_{pq} on the edge e_{pq} of G or the arrow $\vec{e_{pq}}$ of $\vec{G}$.

We indicate whether the nodes, the edges, both of them or none of them are weighted by the following expressions: $G(\nu, nil)$, $G(nil, \eta)$, $G(\nu, \eta)$, $G(nil, nil)$.

The Flowing Adjunction. We define a dilation from the node weights to the edge weights: $\eta_{pq} = (\delta_{en}\nu)_{pq} = \nu_p \vee \nu_q$ and an erosion from the edge weights to the nodes weights: $\nu_p = (\varepsilon_{ne}\eta)_p = \bigwedge_{(s \text{ neighbors of } p)} \eta_{ps}$.

The pair of operators $(\delta_{en}, \varepsilon_{ne})$ form an adjunction: $\forall \eta \in \mathcal{F}_e$, $\forall \nu \in \mathcal{F}_n$: $\delta_{en}\nu < \eta \Leftrightarrow \nu < \varepsilon_{ne}\eta$.

Thus $\gamma_e = \delta_{en}\varepsilon_{ne}$ is an opening from $\mathcal{F}_e$ into $\mathcal{F}_e$ and $\varphi_n = \varepsilon_{ne}\delta_{en}$ is a closing from $\mathcal{F}_n$ into $\mathcal{F}_n$.

2.2   Flowing Edges, Flowing Paths, Flowing Graphs

Consider a drop of water gliding freely from node to node. Assuming that it never glides upwards, we say that e_{pq} is a flowing edge of q, if the drop of water is able to glide from p to q, which is the case:

– in the node weighted graph $G(\nu, nil)$ iff $\nu_p \geq \nu_q$. We say that e_{pq} is an n-flowing edge of p.
– in the edge weighted graph $G(nil, \eta)$ iff e_{pq} is one of the adjacent edges of p with the smallest weight. We say that e_{pq} is an e-flowing edge of p.

A path $p - e_{pq} - q - e_{qs} - s...$ is a **flowing path** if each node except the last one is followed by one of its flowing edges.

Editing Weighted Graphs and Getting Flowing Graphs. There exist differences between node and edge weighted graphs may be suppressed by modifying the graph structure, i.e. adding or suppressing edges without modifying the weights.

Node Weighted Graphs. Each edge of a node weighted graph $G(\nu, nil)$ is a flowing edge of one of its extremities. Take the edge e_{st} for instance. Either $\nu_s \geq \nu_t$ and e_{st} is a flowing edge of s, or $\nu_t \geq \nu_s$ and e_{st} is a flowing edge of t. The edge e_{st} is a flowing edge of both its extremities if $\nu_s = \nu_t$. On the other hand, the nodes without lower neighbors are not the origin of a flowing edge, they are isolated regional minima. If one adds a self-loop edge connecting an isolated regional minimum with itself, one gets the graph $\circlearrowleft G(\nu, nil)$ in which each node is the origin of a flowing edge.

A flowing edge e_{pq} of extremity p verifies $\nu_p \geq \nu_q$. One assigns to e_{pq} the weight η_{pq} of its highest extremity: $\eta_{pq} = (\delta_{en}\nu)_{pq} = \nu_p \vee \nu_q = \nu_p$. All other edges e_{ps} of extremity p get higher or equal weights: $\eta_{ps} = \nu_p \vee \nu_s \geq \nu_p$. It follows that $\nu_p = \bigwedge_{(r\,\text{neighbors of }p)} \eta_{pr} = (\varepsilon_{ne}\eta)_p = (\varepsilon_{ne}\delta_{en}\nu)_p = (\varphi_n\nu)_p$. Thus, whatever the graph G and whatever the distribution of weights ν on its nodes, these nodes verify $\nu = \varphi_n\nu$ in the augmented graph $\circlearrowleft G(\nu, nil)$. Node and edge weights are coupled. Given ν, η is defined as $\eta = \delta_{en}\nu$. And in $\circlearrowleft G(\nu, nil)$, ν verifies $\nu = \varepsilon_{ne}\delta_{en}\nu = \varepsilon_{ne}\eta$.

Edge Weighted Graphs. In a connected edge weighted graph $G(nil, \eta)$ each node has one or several neighboring edges. Those with minimal weight are flowing edges for this node. Thus, each node is the extremity of at least one flowing edge. However, an edge with adjacent edges with a lower weight is not a flowing edge. Such an edge is cut without disturbing the trajectories of a drop of water. In the resulting pruned graph designated by $\downarrow G(nil, \eta)$, each edge is now a flowing edge of one of its extremity (or of both).

If e_{pq} is a flowing edge of the node p, one assigns to p the weight η_{pq} of e_{pq}. As e_{pq} is one of the smallest edges with extremity p, we get: $\nu_p = \eta_{pq} = \bigwedge_{(r\,\text{neighbors of }p)} \eta_{pr} = (\varepsilon_{ne}\eta)_p$. But the edge e_{pq} is not necessarily the smallest edge of q and $\nu_q = \bigwedge_{(r\,\text{neighbors of }q)} \eta_{qr} \leq \nu_p$. Thus $\eta_{pq} = \nu_p = \nu_p \vee \nu_q = (\delta_{en}\nu)_{pq} = (\delta_{en}\varepsilon_{ne}\eta)_{pq} = (\gamma_e\eta)_{pq}$. It follows that, whatever the graph G and whatever the distribution of weights η on its edges, these weights verify $\eta = \gamma_e\eta$ in $\downarrow G(nil, \eta)$. Given η, ν is defined as $\varepsilon_{ne}\eta$. And in $\downarrow G(nil, \eta)$, η verifies $\eta = \delta_{en}\varepsilon_{ne}\eta = \delta_{en}\nu$.

Flowing Graphs. Starting from an arbitrary graph $G(\nu, nil)$ we obtain a node and edge weighted graph $\circlearrowleft G(\nu, \eta)$, where $\eta = \delta_{en}\nu$. Starting from an arbitrary graph $G(nil, \eta)$ we obtain a node and edge weighted graph $\downarrow G(nil, \eta)$, where $\nu = \varepsilon_{ne}\eta$. The resulting node and weighted graphs are called **flowing graphs**: each edge is the flowing edge of one of its extremities, and each node the extremity of at least one flowing edge; node and edge weights are coupled: $\eta = \delta_{en}\nu$ and $\nu = \varepsilon_{ne}\eta$. It follows that $\nu = \varphi_n\nu$ and $\eta = \gamma_e\eta$.

Gravitational Digraphs. By replacing in a flowing graph $G(\nu, \eta)$ each flowing edge by an oriented arc, we obtain a digraph $\overrightarrow{G}(\nu, \eta)$. Each flowing path of G becomes an oriented path in $\overrightarrow{G}$.

Notations: The existence of an oriented path between p and q is expressed by $p \searrow q$. in the digraph. If an arc links two nodes s and u, we write $s \to u$; if there is no arc from u to s, we write $u \nrightarrow s$.

Consider a flowing path from p to q, the node weights are never increasing: $\nu_p \geq \nu_q$. Thus if $\overrightarrow{\pi_{pq}}$ is a flowing path from p to q and $\overrightarrow{\sigma_{pq}}$ a flowing path from q to p, we simultaneously have $\nu_p \geq \nu_q$ and $\nu_q \geq \nu_p$, implying that $\nu_p = \nu_q$ and that all intermediary nodes in both paths have the same weights. It follows that all internal arcs in the paths $\overrightarrow{\pi_{pq}}$ and $\overrightarrow{\sigma_{qp}}$ are bidirectional arcs, i.e. the oriented paths are bidirectional paths $\overleftrightarrow{\pi_{pq}}$ and $\overleftrightarrow{\sigma_{qp}}$.

Flowing digraphs are thus particular gravitational graphs, defined as follows.

Definition 1. *A digraph $\overrightarrow{G}(nil, nil)$ is a gravitational graph if it verifies: if an oriented path links p and q, and another links q and p, then these paths necessarily are bidirectional paths.*

In the next section we study the topography of general gravitational digraphs and justify their denomination.

3   The Topography of Unweighted Gravitational Digraphs

The structure of a digraph $\overrightarrow{G}(nil, nil)$, despite the absence of weights, has a rich topographical structure.

Flat Zones. The relation $p \overleftrightarrow{} q$ i.e. there exists a path, connecting p and q and containing only bidirectional arcs, is an equivalence relation. Its equivalence classes are the flat zones of the digraph $\overrightarrow{G}$.

Black Holes. The gravitational attraction of a black hole in the universe is so high that it catches everything around it and not even the light gets out. We define the black holes of a gravitational digraph by analogy.

Definition 2. *In a gravitational graph $\overrightarrow{G}$, a node without any outgoing arc is an isolated black hole. Non isolated black holes are flat zones without outgoing arcs.*

Property 1. A black hole M of a gravitational graph is characterized by the following equivalence: $p \in M \Leftrightarrow \{\forall q : p \searrow q \Rightarrow q \searrow p\}$

If there exists no node q such that $p \searrow q$, then p is an isolated black hole. If on the contrary, $p \searrow q$ for a given node q, then $q \searrow p$, implying $p \overleftrightarrow{} q$, i.e. p and q belong to the same flat zone: there is no way to escape from this flat zone through an outgoing arc.

An algorithm for detecting black holes may be derived from this property. If p does not belong to a black hole, there exists an oriented path $\overrightarrow{\pi_{pq}}$ connecting p with a node q inside a black hole. There exists necessarily a couple of nodes s, u, along this path, verifying $s \rightarrow u$ and $u \nrightarrow s$ otherwise all edges of $\overrightarrow{\pi_{pq}}$ would be bidirectional and p would belong to the same black hole as the node q. We may suppose that (s, u) is the first couple of nodes if one starts from M and follows this path upstream, verifying $s \rightarrow u$ and $u \nrightarrow s$. Then p belongs to the upstream of the node s.

Lemma 1. *The non black holes are the upstream of all nodes s having a neighbor u verifying $s \rightarrow u$ and $u \nrightarrow s$. The black holes are obtained by taking their complement.*

Catchment Zones

Definition 3. *The catchment zone of a black hole M is the set of nodes linked by an oriented path with M : $CZ(M) = \{p \in \mathcal{N} \mid p \searrow q, q \in M\}$*

If there exists $q \in M$, such that $p \searrow q$, then $p \searrow s$ for each $s \in M$, as $q \leftrightarrow s$.

Each black hole is surrounded by a catchment zone and each node of a gravitational graph belongs to the attraction zone of at least one black hole. These attraction zones may overlap and a node belong to the attraction zones of several black holes.

Authorized Prunings. Reducing the number of oriented paths leading to a black hole reduces its attraction zone. If an arc $\overrightarrow{e_{pq}}$ is suppressed, then all paths passing through this arc are broken. We call authorized pruning of a digraph prunings which do not modify the black holes but reduce their catchment zones. They are characterized by the following property.

Property 2. In an authorized pruning, the arc $p \rightarrow q$ may be suppressed if after pruning there remains a node u such that $p \rightarrow u$ and $u \nrightarrow p$.

3.1 Operators on Digraphs

For a gravitational digraph $\overrightarrow{G}(nil, nil)$ we also define a labeling function λ, adding a third field to the node and edge weights $\overrightarrow{G}(nil, nil; \lambda)$. Typically, the labels will serve to encode the various sub-structures extracted from the graph. The label function λ is defined on the nodes $\mathcal{N}$ with value in $\mathbb{N}$ and belongs to $\mathcal{F}_n = \text{Fun}(\mathcal{N}, \mathbb{N})$.

In addition, one defines a priority between the labels. Like that, if a node belongs to several catchment zones holding distinct labels, it gets the label of the catchment zone with the highest priority. The algorithm for detecting the black holes of $\overrightarrow{G}(nil, nil)$ produces a number of flat zones of the graph which may get distinct labels.

Operators on Unweighted but Labeled Digraphs

Upstream and Downstream Propagation. We define 2 operators for propagating the labels upstream or downstream:

– the upstream operator $\nearrow^{(1)}$ assigns the label λ of a node p to each node q such that $q \to p$. If two nodes s and t hold distinct labels and verify $p \to s$ and $p \to t$, then p takes the most prioritary label. $\nearrow^{(1)} p$

– the downstream operator $\searrow^{(1)}$ assigns the label λ of a node p to each node q such that $p \to q$. If two nodes s and t hold distinct labels and verify $s \to p$ and $t \to p$, then p takes the most prioritary label.

Notations: $\nearrow^{(1)} X$, $\nearrow^{(k)} X$, $\nearrow X$ represent the label distributions after applying the upstream operator to the labels of the nodes in X, respectively 1, k or ∞ times (∞ meaning up to convergence). Similar notations hold for the downstream propagation: $\searrow^{(1)} X$, $\searrow^{(k)} X$, $\searrow X$.

The labels of a family of nodes (p_i) are propagated by $\searrow (p_i)$ along the oriented paths starting from the nodes p_i, and ultimately also label the black holes at the extremity of these paths. A second upstream propagation applied to $\searrow (p_i)$ labels the catchment zones marked by the nodes p_i. In total, one has applied the operator $\nearrow \searrow (p_i)$ to the family of nodes.

Extracting Overlapping Zones. For a family of labeled nodes, the upstream $\nearrow (p_i)$ appears as a dead leaves model. If two node p, q verifying $(p \to q)$ hold distinct labels, the node p belongs to two distinct catchment zones. The label of q could not be propagated to the node p, as the label of q has a lower priority than the label of p. Such nodes are called "overlapping seeds", forming a set S defined by:

$S = \{p \mid p, q \in N : (p \to q) \text{ and } (\lambda_p \neq \lambda_q)\}$. The upstream of the seeds $\nearrow S$ is called "overlapping core" or simply core.

The Topography of Node or Edge Weighted Graphs. Starting from an arbitrary graph $G(\nu, nil)$ we obtain a node and edge weighted graph $\circlearrowleft G(\nu, \eta)$, where $\eta = \delta_{en}\nu$. Starting from an arbitrary graph $G(nil, \eta)$ we obtain a node and edge weighted graph $\downarrow G(nil, \eta)$, where $\nu = \varepsilon_{ne}\eta$. The resulting node and weighted graphs are called **flowing graphs.** Replacing the flowing edges by arcs we obtain flowing digraphs. The topographical structures defined on digraphs may now be interpreted in terms of the initial graph:

– to a flat zone of the digraph corresponds a flat zone of the node or edge weighted graph.

– to a black hole of the digraph corresponds a regional minimum of the node or edge weighted graph.

– to a catchment zone of a black hole in the digraph corresponds the catchment zone of the corresponding regional minimum in the flat zone of the node or edge weighted graph.

Operators on Node Weighted Digraphs. Consider the node weighted flowing digraph $\overrightarrow{G}(\nu, nil)$. A node p of the flowing digraph $\overrightarrow{G}(\nu, nil)$ belongs to two catchment zones if it is linked by an oriented path with two distinct black holes of $\overrightarrow{G}$. After defining the steepness of a path, we introduce a pruning operator able to cut those which are less steep.

The Steepness Order Relation. To each oriented path ending in a regional minimum we associate an infinite list, whose first elements are the weights of the nodes in the path and the last elements an infinite repetition of the weight of the regional minimum. If p is the origin of two oriented paths π and σ with a series of node weights $(\pi_1, \pi_2, ...\pi_k, \pi_{k+1}, ...\pi_n, ...)$ and $(\sigma_1, \sigma_2, ...\sigma_k, \sigma_{k+1}, ...\sigma_n, ...)$, we say that π is steeper than σ and write $\pi \prec \sigma$ if $\pi_1 < \sigma_1$ or if there exists k such that $\{\forall l < k : \pi_i = \sigma_i\}$ or $\pi_k < \sigma_k$.

Among all paths with the same origin, there is at least one which has the highest steepness. If in a graph $\overrightarrow{G}(\nu, nil)$ only the paths with the highest steepness remain, we say that this graph is $\infty - steep$. If the first k nodes of all paths with the same origin have the same weight (which is also the weight of the steepest path), we say that the graph is $k - steep$.

The Pruning Operators. We define two operators on the graph $\overrightarrow{G}(\nu, nil)$:

- the erosion $\overrightarrow{\varepsilon}$ assigns to each node p the minimal weight of its downstream neighbors: $(\overrightarrow{\varepsilon}\nu)_p = \bigwedge\limits_{q|p\to q} \nu_q$. The effect of $\overrightarrow{\varepsilon}$ is to let the node weights move upstream along each steepest flowing path of $\overrightarrow{G}(\nu, nil)$.
- the operator $\downarrow^2 \overrightarrow{G}(\nu, nil)$ keeps all arcs $(p \to q)$ of origin p such that ν_q is minimal and suppresses all others. The graph $\downarrow^2 \overrightarrow{G}$ is $2 - steep$.

The pruning starts by applying $\downarrow^2$ to $\overrightarrow{G}$, producing a $2 - steep$ graph $\downarrow^2 \overrightarrow{G}$. For $k > 2$, a $k - steep$ digraph $\downarrow^k \overrightarrow{G}$ is obtained by applying $(k-2)$ times the operator $\downarrow^2 \overrightarrow{\varepsilon}$ to the graph $\downarrow^2 \overrightarrow{G}$. Thus $\downarrow^k = (\downarrow^2 \overrightarrow{\varepsilon})^{k-2} \downarrow^2$.

Local Pruning. The operators $\downarrow^2 \overrightarrow{\varepsilon}$ consider for each node p the arcs having p as origin and the weights of their extremities. The operator $\downarrow^k$ considers for each node the flowing paths of origin p counting k nodes, and cuts the first arc of all paths which are not at least $k - steep$.

The union of the flowing paths having their origin in a set X of nodes and counting each k nodes is called $comet(X, k)$. The subgraph of $\overrightarrow{G}$ spanned by the nodes of $comet(X, k)$ is called $\overrightarrow{comet(X, k)}$. The preceding analysis shows that applying $\downarrow^k$ to $\overrightarrow{G}$ or to $\overrightarrow{comet(X, k)}$ has the same effect on the arcs having their origin in X, cutting the first arcs of flowing paths which are not at least $k - steep$.

4   Applications

4.1   The n-Steep Downstream of X and Its Regional Minima

Given a digraph $\overrightarrow{G}$ we compare a global and a local method for extracting the downstream of a set X after the pruning $\downarrow^n$ of $\overrightarrow{G}$. We remark that for $n = k+l$:

$$\downarrow^n = (\downarrow^2 \overrightarrow{\varepsilon})^{n-2} \downarrow^2 = (\downarrow^2 \overrightarrow{\varepsilon})^l (\downarrow^2 \overrightarrow{\varepsilon})^{k-2} \downarrow^2 = (\downarrow^2 \overrightarrow{\varepsilon})^l \downarrow^k$$

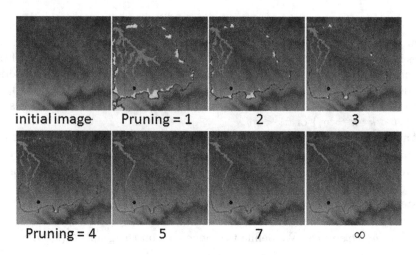

initial image   Pruning = 1   2   3

Pruning = 4   5   7   ∞

Fig. 1. The flowing paths (in violet) and the catchment zone of the dark dot are compared for increasing degrees of pruning $(1, 2, 3, 4, 5, 7, \infty)$. (Color figure online)

1. Global method: apply $\downarrow^n = \left(\downarrow^2 \overrightarrow{\varepsilon}\right)^{n-2} \downarrow^2$ to $\overrightarrow{G}$, and extract $\searrow X$, the downstream of X on the pruned graph $\downarrow^n \overrightarrow{G}$
2. Local method: for $k < n$, apply $\downarrow^k$ to $\overrightarrow{G}$ and extract $\overrightarrow{\text{comet}(X, \infty)} = \searrow X$ on $\downarrow^k \overrightarrow{G}$. Apply $\left(\downarrow^2 \overrightarrow{\varepsilon}\right)^l$ on the subgraph $\overrightarrow{\text{comet}(X, \infty)}$ (which is much smaller than $\downarrow^n \overrightarrow{G}$) and extract $\searrow X$ on the resulting pruned graph $\overrightarrow{H} = \left(\downarrow^2 \overrightarrow{\varepsilon}\right)^l \overrightarrow{\text{comet}(X, \infty)}$. Furthermore, the black holes of the unweighted graph $\overrightarrow{H}$ are the regional minima of the catchment zones containing X in the pruned graph $\downarrow^n \overrightarrow{G}$.

Illustration: Fig. 1 represents a digital elevation model on which a set X is marked with a black dot. The downstream $\searrow X$ of X has been extracted after various degrees of pruning $\downarrow^k \overrightarrow{G}$ $(k = 1, 2, 3, 4, 5, 7, \infty)$. For small values of k the downstream subdivides in various channels; it narrows down for larger values of k, only one branch remains for $k = 5, 7, \infty$. A global pruning $\downarrow^7$ applied to the whole graph would pointlessly prune many arcs in the whole domain, whereas the local approach with $k = 2$ and $l = 5$ is more economical. Like that, only the pruning $\downarrow^2 \overrightarrow{G}$ is done on the complete graph $\overrightarrow{G}$ whereas the subsequent operators are applied to the much smaller graph $\overrightarrow{\text{comet}(X, \infty)}$ constructed on $\downarrow^2 \overrightarrow{G}$.

4.2   Locally Correcting a Dead Leaves Tessellation

Let $M = (m_i)$ be a family of labeled regional minima (not necessarily all of them) of a flowing digraph $\overrightarrow{G}$. The upstream propagation of their labels in a pruned graph $\downarrow^n \overrightarrow{G}$ creates a dead leave tessellation of catchment zones. The stacking order of the CZ reflects the priority relations between labels, the CZ with the highest priority being on the top. If the overlapping zones are large, the resulting partition is highly biased.

The overlapping zones get smaller when the digraph is pruned. A node belongs to distinct CZ, if it is linked with the corresponding regional minima through two flowing paths whose n first nodes hold exactly the same weight. This is more and more unlikely for higher values of n. The value of n for which the dead leave tessellation produces a quasi unbiased partition of CZ depends on the types of images one has to segment like the sharpness and contrast. If the images to analyze or segment are homogeneous, one may fix once and for all a value of n for which the biases, i.e. the overlapping cores, remain within acceptable limits or are minimal.

In Fig. 1 we present the contours of the CZ associated to one regional minimum after increasing degrees of pruning. The overlapping seeds and core are highlighted resp. in blue and green color: they get smaller and smaller with increasing prunings and almost completely vanish for a pruning $k = 7$.

Given a digraph $\overrightarrow{G}$ we want to extract the catchment zones of a family of regional minima (m_i) after the pruning $\downarrow^n$ of $\overrightarrow{G}$. Again, we have the choice between a global suboptimal and a local method. After a few steps of pruning, the catchment zones are correct in large areas of the domain, but there exist limited local overlapping. The global method continues pruning the whole domain:

– apply $\downarrow^n = \left(\downarrow^2 \overrightarrow{\varepsilon}\right)^{n-2} \downarrow^2$ to $\overrightarrow{G}$, and propagate the labels of (m_i) in $\downarrow^n \overrightarrow{G}$.
  The local method focuses on the overlapping zones to correct. For $n = k + l$ we have: $\downarrow^n = \left(\downarrow^2 \overrightarrow{\varepsilon}\right)^{n-2} \downarrow^2 = \left(\downarrow^2 \overrightarrow{\varepsilon}\right)^l \left(\downarrow^2 \overrightarrow{\varepsilon}\right)^{k-2} \downarrow^2 = \left(\downarrow^2 \overrightarrow{\varepsilon}\right)^l \downarrow^k$.
  (a) On $\downarrow^k \overrightarrow{G}$:
– propagate upstream the labels of (m_i), producing a dead leaves tessellation $(CZ)_i$ of catchment zones.
– extract the overlapping seeds (as explained earlier) and their upstream, the overlapping core. This produces the binary set <core>.
– construct comet(<core>, $l + 2$), the nodes belonging to the flowing paths of length $l + 2$ having their origin in <core>.
– construct the subgraph $\overrightarrow{\text{comet}(\text{<core>}, l + 2)}$ spanned by the set comet(<core>, $l + 2$).
  (b) On the graph $\overrightarrow{\text{comet}(\text{<core>}, l + 2)}$:
– apply the operator $\left(\downarrow^2 \overrightarrow{\varepsilon}\right)^l$, trimming the flowing paths having their origin in <core>, and leaving only those which are $(k + l) - steep$
– extract the nodes in the immediate downstream of <core> within comet(<core>, $l + 2$) :
  <rim> = $\{q \in \text{comet}(\text{<core>}, l + 2), q \notin \text{<core>} \mid \exists\, p : p \in \text{<core>}, p \rightarrow q\}$.
– define <halo> = <core> $\cup$ <rim>, the union of the overlapping core and its immediate downstream neighbors.
– construct $\overrightarrow{halo}$, the subgraph of $\overrightarrow{\text{comet}(\text{<core>}, l + 2)}$ spanned by the nodes of <halo> in $\downarrow^k \overrightarrow{G}$.

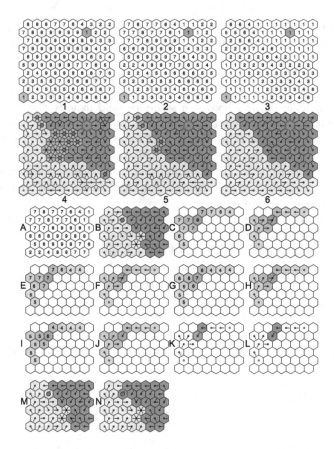

Fig. 2. Prunings of intensity 2,3 and 6 respectively in figures (1,4), (2,5) and (3,6). The overalapping zones are in color. Figures A–N: local correction of an overlapping zone (Color figure online)

(c) On the graph $\overrightarrow{halo}$:

– The nodes of <rim> do not belong to the overlapping core and their labels have been correctly assigned in (a). Therefore they may be propagated upstream in $\overrightarrow{halo}$, assigning corrected labels to the nodes of <core>, which are then substituted to the previous ones in $(CZ)_i$.

Illustration: Fig. 2 presents a digraph $\overrightarrow{G}$ with two regional minima. The node weights and the arrows of the digraph are represented in Fig. 2(1,4), (2,5) and (3,6) respectively for the prunings $k = 2, 3$ and 6. The labels of the minima are propagated upstream, the light gray label covering the dark gray one. The overlapping seeds are highlighted with red dots, their upstream are the overlapping cores, represented in vivid colors. One remarks that the overlappings vanish only for $k = 6$.

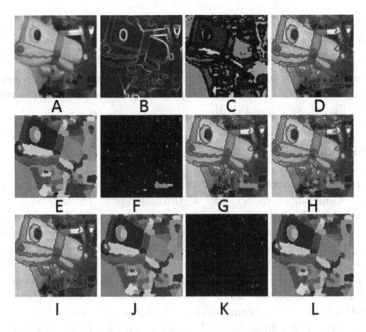

Fig. 3. Local correction of the most prominent errors of a watershed tessellation. (Color figure online)

Figure 2A–N illustrate how the label of the blue node p of Fig. 2(5) may be locally corrected:

- in Fig. 2A–B: the pruned digraph $\downarrow^k \overrightarrow{G}$ and the labeling of its node around the node p.
- in Fig. 2C–D: the subgraph $\overline{\mathrm{comet}(p,5)}$.
- in Fig. 2E–F, G–H, I, J: applying 3 times $(\downarrow^2 \overrightarrow{\varepsilon})$ to $\overline{\mathrm{comet}(p,5)}$
- in Fig. 2K: the subgraph $\overrightarrow{halo}$ spanned by the node p and its downstream node, i.e. the rim
- in Fig. 2L: upstream propagation of the labels of the rim, correcting the label of p
- in Fig. 2M–N: Comparing the labels before and after the correction.

Selecting the Errors to Correct. The method outlined for locally correcting the dead leaves tessellation of catchment zones works as well if it is applied on a subset only of the overlapping core, where the errors are the most significant. Figure 3A, B and C resp. present an image to segment, its gradient considered as a node weighted graph and the regional minima. The labels of the regional minima are up propagated inside the flowing digraph $\downarrow^2 \overrightarrow{G}$ associated to the gradient, producing the contours and the dead leaves tessellation of Fig. 3D, E. Figure 3 shows the discrepancies with the reference partition of Fig. 3L obtained with a classical watershed algorithm [5]. The overlapping core is detected. We chose among the core the largest particles X we wish to correct; they appear in green in Fig. 3G and their downstream comet$(X, 7)$ is represented in Fig. 3H. The operator $\left(\downarrow^2 \overrightarrow{\varepsilon} \right)^5$ is applied to the subgraph $\overrightarrow{\mathrm{comet}(X, 7)}$, and the labels of X corrected within the subgraph $\overrightarrow{halo}$ as explained above. The corrected dead leaves tessellation in Fig. 3I and J presents minimal differences with the reference partition of Fig. 3L as shows Fig. 3K.

5 Conclusion

Stacking catchment zones instead of creating watershed partitions has many advantages. Generality: theory and algorithms apply to any node or edge weighted graph, independently of the dimension of the underlying space. Flexibility: any number of catchment zones may be constructed independently from each other. Controlled precision: the error zones where catchment zones overlap are easy to detect and to correct locally. Adaptability: the errors to correct may be chosen on hand of various criteria, such as its belonging to a type of object, particular boundaries which have to be precise, or simply errors outside a prescribed tolerance. Furthermore, all connected components of errors may be corrected independently from each other and in parallel. Last but not least, the basic operators are elementary neighborhood operators on the graph, easy to implement in devoted hardwares, and in parallel in GPUs.

References

1. Audigier, R., Lotufo, R.: Uniquely-determined thinning of the tie-zone watershed based on label frequency. J. Math. Imaging Vis. **27**(2), 157–173 (2007)
2. Beucher, S., Lantuéjoul, C.: Use of watersheds in contour detection. In: Watersheds of Functions and Picture Segmentation, pp. 1928–1931, Paris, May 1982
3. Cousty, J., Bertrand, G., Najman, L., Couprie, M.: Watershed cuts: minimum spanning forests and the drop of water principle. IEEE Trans. Pattern Anal. Mach. Intell. **31**, 1362–1374 (2009)
4. Lantuéjoul, C., Beucher, S.: On the use of the geodesic metric in image analysis. J. Microsc. (1981)
5. Meyer, F.: Un algorithme optimal de ligne de partage des eaux. In: Proceedings 8$^{\grave{e}me}$ Congrès AFCET, Lyon-Villeurbanne, pp. 847–857 (1991)
6. Meyer, F.: Topographic distance and watershed lines. Sig. Process. **38**(1), 113–125 (1994)
7. Meyer, F.: Watersheds on weighted graphs. Pattern Recogn. Lett. **47**, 72–79 (2014)
8. Najman, L., Schmitt, M.: Watershed of a continuous function. Sig. Process. **38**(1), 99–112 (1994). Mathematical Morphology and its Applications to Signal Processing
9. Soille, P., Vincent, L.: Watersheds in digital spaces: an efficient algorithm based on immersion simulations. IEEE Trans. Pattern Anal. Mach. Intell. **13**(6), 583–598 (1991)

Bandeirantes: A Graph-Based Approach for Curve Tracing and Boundary Tracking

Marcos A.T. Condori, Lucy A.C. Mansilla, and Paulo A.V. Miranda$^{(\boxtimes)}$

Institute of Mathematics and Statistics, University of São Paulo (USP),
São Paulo, SP 05508-090, Brazil
{mtejadac,lucyacm,pmiranda}@vision.ime.usp.br

Abstract. This work presents a novel approach for curve tracing and user-steered boundary tracking in image segmentation, named Bandeirantes. The proposed approach was devised using Image Foresting Transform with unexplored and dynamic connectivity functions, which incorporate the internal energy of the paths, at any curvature scale, resulting in better segmentation of smooth-shaped objects and leading to a better curve tracing algorithm. We analyze its theoretical properties and discuss its relations with other popular methods, such as riverbed, live wire and G-wire. We compare the methods in a curve tracing task of an automatic grading system. The results show that the new method can significantly increase the system accuracy.

Keywords: Live wire · Curve tracing · Boundary tracking · Grading system

1 Introduction

Curve tracing and edge detection/extraction in images are important operations in image processing and computer vision, which are fundamentally different problems [26]. Edge detection [4,7] is usually the process of labeling the locations in the image where the gray level's "rate of change" is high, such as by computing the local image gradient. The emphasis is on the extraction of all pixels that belong to contours and not on the identification or labeling of a particular curve. A Canny edge detector [3] is a simple edge extraction algorithm. On the other hand, curve tracing is an edge integration process that combines local edges, which may be sparse and non-contiguous, into meaningful, long edge curves or closed contours for segmentation. That is, we want to identify an individual curve and label it, by following it across the image, even if it crosses other curves, fades out, and then reappears [26].

In aerial images, curve tracing is used to extract curvilinear features such as roads, railroads and rivers [2], for example, for estimating water levels over time based on airborne SAR/InSAR data [1]. Other applications of curve tracing include object tracking, extraction of features from medical images, such as blood vessels from an X-ray angiogram [6], image segmentation by curve evolution techniques [15,29] and user-steered boundary tracking [9,10,14,22–24].

© Springer International Publishing AG 2017
J. Angulo et al. (Eds.): ISMM 2017, LNCS 10225, pp. 95–106, 2017.
DOI: 10.1007/978-3-319-57240-6_8

In graph-based methods, the edge integration is usually based on a path-cost function that combines a local edge indicator function, in the form of an arc weight. This idea of transforming curve tracing in a problem of finding the minimum-cost path in a weighted and directed graph was initially proposed by Martelli, using the heuristic algorithm A^* [18,19,25]. Optimal results, according to different criteria, were later proposed by the live-wire algorithm [9,10,24], G-wire [14], and riverbed [22,23]. These works have been a foundation for many other papers [5,11,13,16,17,27].

In this work, we present a novel graph-based approach for curve tracing, named *Bandeirantes*. It is based on more complex dynamic connectivity functions, allowing to incorporate the internal energy of the paths, at different curvature scales, and also an intensity adaptive cost. We analyze its theoretical properties and discuss its relations with other popular methods, such as riverbed, live wire and G-wire. The proposed method is evaluated in user-steered boundary tracking for image segmentation, resulting in better segmentation of object's contours with low curvature, and in a curve tracing task of an automatic grading system, leading to a better curve tracing algorithm.

In Sect. 2, we review the related work. Bandeirantes is then proposed in Sect. 3. In Sect. 4, we evaluate the methods and state our conclusions.

2   Related Work

Karthik Raghupathy improved the work by Steger [28] to better suit the goals of curve tracing, emulating the human perception by using both global and local features [26]. The global features are important to handle curve following at junctions and the tracing of faint curves that fade out or disappear and re-emerge again. In [26], the first step involves classifying pixels as being curve points or not depending on whether they lie on some curve in the image (edge detection). Curves are then constructed from a manual starting point by linking the appropriate curve point in the neighborhood to the current curve. As drawbacks, this algorithm requires appropriate selection of eight parameters, the curve point binarization is a critical step subject to failure, and it is not possible to specify a destination point, as in live wire, since the linking terminates when there are no more curve points in the neighborhood of the current pixel.

Active contour model [15,29], also called snakes, is an energy minimizing, deformable spline influenced by constraint and image forces that pull it towards object contours and internal forces (such as curvature) that resist deformation. It deforms an initial curve specified by the user to the target boundary for delineating an object outline from a possibly noisy 2D image. The final shape of the boundary is hard to predict or control. In case of failure, the whole process must be repeated with a new initial curve or using different parameters. As pointed out by Kang [14], in the context of interactive segmentation, live wire provides much tighter control to users since the desired path can be interactively "selected" from multiple candidate paths.

The snake algorithm was later combined with graph-based methods, leading to the geodesically linked active contour model, which basically consists in a set of vertices connected by paths of minimal cost [20,21]. This results in a closed piecewise-smooth curve, over which an edge or region energy functional can be formulated. As a graph-based approach, the method proposed in this work could also be used in the geodesically linked active contour model, but here we are not limiting the scope to closed contours.

The graph-based methods live wire [9,10,24], G-wire [14] and riverbed [22,23] compute paths to all pixels from seed pixels, also known as anchor points. In user-steered image segmentation, the computed path from the previous anchor point to the current mouse position is shown to the user, as he moves the cursor, so that the user can interactively select the desired path and start a new path search from that point. All previous selected paths are kept unchanged (frozen) during the algorithm, so that their nodes cannot be revisited.

Live wire and riverbed are based on the Image Foresting Transform (IFT) framework [8] using different path-cost functions. Riverbed can significantly reduce the number of user interactions (anchor points) as compared to live wire for objects with complex shapes and good contrast, but the opposite is true for objects with low contrast and more regular shape, since live wire can also take into account the spatial distance, which corresponds to the first-order term of the internal energy of snakes. G-wire incorporates the second-order term (curvature term) of the internal energy based on a high-order dynamic programming in a generalized multi-dimensional graph formulation, thus preserving the optimality of the computed paths. However, it is computationally more expensive, requiring a 3-dimensional graph for solving a problem in a 2D image. Moreover, its curvature term is defined on a very local scale, since it analyses only immediately neighboring pixels. As consequence, it can improve the segmentation results in noisy images, but cannot adequately solve long-term curves, as demonstrated in the experiments. Note that, in practice, the multiscale curvature in G-wire is prohibitive, since in the general case with a cost function that spans m arcs the dimensionality of the graph would be $2 + (m - 1)$ [14]. Similar to live wire and riverbed, bandeirantes is also based on the IFT framework [8], conserving its low computational complexity of $O(N \cdot logN)$ for a sparse graph with N nodes.

2.1 Image Foresting Transform (IFT)

An image can be interpreted as a weighted digraph $G = \langle \mathcal{I}, \mathcal{A} \rangle$, whose nodes are the image pixels in its image domain $\mathcal{I} \subset \mathbb{Z}^2$, and whose arcs are the ordered pixel pairs $\langle s, t \rangle \in \mathcal{A}$. In this work, a pixel s is adjacent to a pixel t, if s is in a 8-neighborhood of t.

A path $\pi = \langle t_1, t_2, \ldots, t_n \rangle$ is a sequence of adjacent pixels (i.e., $\langle t_i, t_{i+1} \rangle \in \mathcal{A}$, $i = 1, 2, \ldots, n - 1$) with no repeated vertices ($t_i \neq t_j$ for $i \neq j$). Other greek letters, such as τ, can also be used to denote paths. A path $\pi_t = \langle t_1, t_2, \ldots, t_n = t \rangle$ is a path with terminus at a pixel t. The notation $\pi_{s \rightsquigarrow t} = \langle t_1 = s, t_2, \ldots, t_n = t \rangle$ may also be used, where s stands for the origin and t for the destination node. A path is *trivial* when $\pi_t = \langle t \rangle$. A path $\pi_t = \pi_s \cdot \langle s, t \rangle$ indicates the extension of a path π_s by an arc $\langle s, t \rangle$. $\Pi(G)$ is the set of all possible paths in G.

A *predecessor map* is a function $P : \mathcal{I} \rightarrow \mathcal{I} \cup \{nil\}$ that assigns to each pixel t in $\mathcal{I}$ either some other adjacent pixel in $\mathcal{I}$, or a distinctive marker nil not in $\mathcal{I}$—in which case t is said to be a *root* of the map. A *spanning forest* is a predecessor map which contains no cycles—i.e., one which takes every pixel to nil in a finite number of iterations. For any pixel $t \in \mathcal{I}$, a spanning forest P defines a path π_t^P recursively as $\langle t \rangle$ if $P(t) = nil$, and $\pi_s^P \cdot \langle s, t \rangle$ if $P(t) = s \neq nil$.

A *connectivity function* $f : \Pi(G) \rightarrow \mathcal{V}$ computes a value $f(\pi_t)$ for any path π_t, in some totally ordered set $\mathcal{V}$ of cost values. A path π_t is *optimum* if $f(\pi_t) \leq f(\tau_t)$ for any other path τ_t in G. By taking to each pixel $t \in \mathcal{I}$ one optimum path with terminus at t, we obtain the optimum-path value $V_{opt}(t)$, which is uniquely defined by $V_{opt}(t) = \min_{\pi_t \in \Pi(G)} \{f(\pi_t)\}$.

For a suitable path-cost function f, the *image foresting transform* (IFT) [8] takes an image graph $G = \langle \mathcal{I}, \mathcal{A} \rangle$, and returns an *optimum-path forest* P—i.e., a spanning forest where all paths π_t^P for $t \in \mathcal{I}$ are optimum. The IFT is a generalization of Dijkstra's algorithm to *smooth* path-cost functions [8]. As shown by Frieze [12], the original proof of Dijkstras algorithm is easily generalized to monotonic-incremental (MI) path-cost functions, which conform to $f(\pi_s \cdot \langle s, t \rangle) = f(\pi_s) \odot \langle s, t \rangle$, where $f(\langle t \rangle)$ is given by an arbitrary handicap cost, and $\odot : \mathcal{V} \times \mathcal{A} \rightarrow \mathcal{V}$ is a binary operation that satisfies the conditions: $x\prime \geq x \Rightarrow x\prime \odot \langle s, t \rangle \geq x \odot \langle s, t \rangle$ and $x \odot \langle s, t \rangle \geq x$, for any x, $x\prime \in \mathcal{V}$ and any $\langle s, t \rangle \in \mathcal{A}$.

The live-wire function is given by:

$$f_{LW}(\langle t \rangle) = \begin{cases} 0 & \text{if } t = s^* \\ +\infty & \text{otherwise} \end{cases}$$

$$f_{LW}(\pi_s \cdot \langle s, t \rangle) = f_{LW}(\pi_s) + c(s, t) \qquad (1)$$

where s^* is the seed pixel (starting point). We usually have $c(s, t) = \gamma \cdot c_{ext}(s, t) + \alpha \cdot c_{in}(s, t)$, where c_{ext} and c_{in} denote, respectively, the external and internal energies. The external energy is usually related to the gradient magnitude, such as its complement $c_{ext}(s, t) = K - \frac{\|\nabla I(s)\| + \|\nabla I(t)\|}{2}$, and the internal energy of live wire is a first-order term $c_{in}(s, t) = \|s - t\|$. Note that f_{LW} is a monotonic-incremental path-cost function since $c(s, t) \geq 0$.

The G-wire path-extension rule is given by:

$$f_{GW}(\pi = \langle t_1, t_2, \ldots, t_n \rangle) = f_{GW}(\langle t_1, \ldots, t_{n-1} \rangle) + c(t_{n-2}, t_{n-1}, t_n) \qquad (2)$$

where we usually have $c(t_{n-2}, t_{n-1}, t_n) = \gamma \cdot c_{ext}(t_{n-1}, t_n) + c_{in}(t_{n-2}, t_{n-1}, t_n)$, and its internal energy also includes the second-order term (curvature term) $c_{in}(t_{n-2}, t_{n-1}, t_n) = \alpha \cdot \|t_n - t_{n-1}\|^2 + \beta \cdot \|t_n - 2 \cdot t_{n-1} + t_{n-2}\|^2$. Note that f_{GW} is not a MI function, since its cost increment also depends on t_{n-2}. Actually, it is also not a smooth path-cost function and it requires a high-order dynamic programming in a multi-dimensional graph in order to be optimally solved.

The riverbed path-extension rule is given by: $f_{RB}(\pi_s \cdot \langle s, t \rangle) = c(s, t)$. The function f_{RB} is neither MI nor smooth, and its optimality by IFT is ensured by another criterion of optimality [23].

Algorithm 1 computes a path-cost map V, which converges to V_{opt} for smooth functions [8]. It is also optimized in handling infinite costs, by storing in the priority queue $\mathcal{Q}$ only the nodes with finite-cost path, assuming that $V_{opt}(t) < +\infty$ for all $t \in \mathcal{I}$.

Algorithm 1. IFT Algorithm

INPUT: Image graph $G = \langle \mathcal{I}, \mathcal{A} \rangle$, and function f.
OUTPUT: Optimum-path forest P and the path-cost map V.
AUXILIARY: Priority queue $\mathcal{Q}$, variable tmp, and set $\mathcal{F}$.

1. $\mathcal{F} \leftarrow \emptyset$
2.   **For each** $t \in \mathcal{I}$, **do**
3.   Set $P(t) \leftarrow nil$ and $V(t) \leftarrow f(\langle t \rangle)$.
4.   **If** $V(t) \neq +\infty$, **then** insert t in $\mathcal{Q}$.
5.   **While** $\mathcal{Q} \neq \emptyset$, **do**
6.   Remove s from $\mathcal{Q}$ such that $V(s)$ is minimum.
7.   Add s to $\mathcal{F}$.
8.   **For each** pixel t such that $\langle s, t \rangle \in \mathcal{A}$, **do**
9.   **If** $t \notin \mathcal{F}$, **then**
10.   Compute $tmp \leftarrow f(\pi_s^P \cdot \langle s, t \rangle)$.
11.   **If** $tmp < V(t)$, **then**
12.   Set $P(t) \leftarrow s$, $V(t) \leftarrow tmp$.
13.   **If** $t \notin \mathcal{Q}$, **then** insert t in $\mathcal{Q}$.

3   Bandeirantes Curve Tracing

The main motivation of this work is to overcome G-wire's limitation that its curvature term is defined only in a very local scale, that analyzes only immediately neighboring pixels. The multiscale curvature in G-wire is prohibitive, since in the general case with a cost function that spans m arcs the dimensionality of the graph would be $2 + (m - 1)$ for computing curves in a 2D image [14].

The idea of Bandeirantes is to use path-cost functions satisfying the equation:

$$f_B(\langle t \rangle) = f_{LW}(\langle t \rangle)$$
$$f_B(\pi = \langle t_1, t_2, \ldots, t_n \rangle) = f_B(\langle t_1, \ldots, t_{n-1} \rangle) + c(t_1, \ldots, t_n) \qquad (3)$$

where $c(t_1, \ldots, t_n) \geq 0$ is a function of several variables.

This function f_B is neither MI nor smooth, hence it would violate the principle of optimality in the underlying IFT algorithm. The usage of high-order dynamic programming in a generalized multi-dimensional graph formulation as proposed by G-wire is impracticable for this problem.

Despite f_B not being smooth, the IFT algorithm with f_B is optimum from the following point of view. Each arc $\langle s, t \rangle \in \mathcal{A}$ is processed a single time during Algorithm 1 after s gets removed from $\mathcal{Q}$ on Line 6. Consider a weighted directed graph $G\prime = \langle \mathcal{I}, \mathcal{A} \rangle$, with dynamically created weights during the IFT execution with f_B on G, such that the arc weights $w(s, t)$ for $\langle s, t \rangle \in \mathcal{A}$ are defined inside the loop, just after Line 8, as $w(s, t) \stackrel{\text{def}}{=} c(t_1, \ldots, t_n = t)$, where $\pi_s^P = \langle t_1, \ldots, t_{n-1} = s \rangle$.

The spanning forest P computed by the IFT with f_B on G is an optimum-path forest for f_{LW} on $G\prime$, using $c(s,t) = w(s,t)$ in Eq. 1. So, the IFT with f_B is a live wire optimum result over a dynamically created graph. The name *Bandeirantes*[1] comes from this observation, in allusion to pathfinders exploring an unknown territory that can only be known in real time, similar to the arc weights that are computed on the fly.

In G-wire, the computed paths cannot be stored in a predecessor map $P : \mathcal{I} \to \mathcal{I} \cup \{nil\}$, with a single entry for each image pixel, and its memory consumption becomes impractical for high-order multiscale curvature. In this work, we would like to have the best feasible result for f_B with low memory consumption, using a single predecessor map $P : \mathcal{I} \to \mathcal{I} \cup \{nil\}$ to store the computed paths. Next, we discuss some theoretical properties to give this support to the usage of Bandeirantes. Let's consider the following definitions:

Definition 1 (Hereditarily-optimum path). *For a given path-cost function f, a path $\pi_{t_n} = \langle t_1, t_2, \ldots, t_n \rangle$ is hereditarily optimum if all paths $\pi_{t_i} = \langle t_1, t_2, \ldots, t_i \rangle$, $i = 1, 2, \ldots, n$ are optimum paths.*

Definition 2 (Monotone path-cost function). *A path-cost function f is monotone if for any path $\pi_{t_n} = \langle t_1, t_2, \ldots, t_n \rangle \in \Pi(G)$ we have that $f(\langle t_1, \ldots, t_i \rangle) \leq f(\langle t_1, \ldots, t_j \rangle)$ whenever $i \leq j \leq n$.*

Definition 3 (Replacement property). *A path-cost function f has the replacement property if for any paths π_s and π'_s ending at s such that $f(\pi_s) = f(\pi'_s)$, we have $f(\pi_s \cdot \langle s, t \rangle) = f(\pi'_s \cdot \langle s, t \rangle)$, for any $s \in \mathcal{I}$ and $\langle s, t \rangle \in \mathcal{A}$.*

We have the following theoretical proposition:

Proposition 1. *For a given image graph $G = \langle \mathcal{I}, \mathcal{A} \rangle$, consider a monotone path-cost function f with the replacement property. Let $\mathcal{O}$ be the set of all pixels $t \in \mathcal{I}$, such that there exists a hereditarily-optimum path π_t for f. In any spanning forest P computed in G by Algorithm 1 for f, all the paths τ_t^P with $t \in \mathcal{O}$ are optimum paths.*

Proof. Let $\pi_t = \langle t_1, t_2, \ldots, t_i, \ldots, t_n = t \rangle$ be a given hereditarily-optimum path for f, and $\tau_t^P = \langle s_1, s_2, \ldots, s_m = t \rangle$ be the computed path in the spanning forest P. We have the following proof by mathematical induction:

- **The basis:** Show that the statement holds for $n = 1$. In this case, we have $t = t_1$, and therefore $\pi_t = \langle t_1 \rangle$. Since $f(\langle t_1 \rangle)$ is a finite value, t_1 is inserted into the priority queue $\mathcal{Q}$ (Line 4 of Algorithm 1). Since $f(\langle t_1 \rangle)$ is optimum, we can safely assure that it won't leave $\mathcal{Q}$ (Line 6) with a cost $V(t_1)$ worse than $f(\langle t_1 \rangle)$, therefore, the computed path τ_t^P must be optimum.

---

[1] The Bandeirantes were 17th-century Portuguese settlers in Brazil and fortune hunters from the São Paulo region. They led expeditions called bandeiras which penetrated the interior of Brazil far south and west of the Tordesillas Line of 1494. As they ventured into unmapped regions in search of profit and adventure, they expanded the effective borders of the Brazilian colony.

– **The inductive step:** Assume that the statement is true for the hereditarily-optimum path π_{t_i}, $i \geq 1$. We must prove that it also holds for the hereditarily-optimum path $\pi_{t_{i+1}} = \pi_{t_i} \cdot \langle t_i, t_{i+1} \rangle$. By the hypothesis we have that the path $\tau_{t_i}^P$ in P is optimum. Given that the replacement property is satisfied for f, we may conclude that $f(\tau_{t_i}^P \cdot \langle t_i, t_{i+1} \rangle) = f(\pi_{t_i} \cdot \langle t_i, t_{i+1} \rangle) = f(\pi_{t_{i+1}})$, so we may conclude that $\tau_{t_i}^P \cdot \langle t_i, t_{i+1} \rangle$ is optimum. By algorithm construction, if t_{i+1} was not yet conquered, we have that this extended path $\tau_{t_i}^P \cdot \langle t_i, t_{i+1} \rangle$ will be evaluated by Algorithm 1 (Lines 8–13), and since it is optimum, it won't be posteriorly replaced by any other path (as guaranteed by the strict inequality in Line 11). If t_{i+1} was already conquered by following a different path in P, its cost cannot be worse than $f(\pi_{t_{i+1}})$, because f is a monotone path-cost function. So in the end, $\tau_{t_{i+1}}^P$ will be an optimum path.

As a consequence of Proposition 1, we have that for any monotone path-cost function f with the replacement property, Algorithm 1 only fails to find an optimum-path forest P, if no such forest exists. Note that in any optimum-path forest P, all paths π_t^P are hereditarily-optimum paths by definition. Thus, the existence of an optimum-path forest P implies that $\mathcal{O} = \mathcal{I}$ on Proposition 1.

Function f_B (Eq. 3) is a monotone path-cost function, but it does not have the replacement property. Note, however, that since its cost increment $c(t_1, \ldots, t_n)$ depends on several variables, ties are rarely detected when using f_B in real domain $f_B : \Pi(G) \rightarrow \mathbb{R}$, as demonstrated in the experiments. Note also that it is always possible to add a second lexicographic cost component in order to eliminate the occurrence of ties. Thus, in the absence of ties, the requirement of the replacement property in Definition 3 becomes void and Proposition 1 becomes valid. So, in practice, IFT with f_B will not find an optimum-path forest P only if no such forest exists.

Moreover, the successive application of the previous proposition gives a characterization of Algorithm 1 for a monotone path-cost function f with the replacement property, as the result of a sequence of optimizations, where each optimization step involves a maximal set of elements, in a well-structured way.

Consider the following definitions: Let $\mathcal{E}^{set}(X) = \{\langle s, t \rangle \in \mathcal{A} \mid s \in X \wedge t \in X\}$ denote the set of all arcs interconnecting nodes in the set X, $\mathcal{E}^{path}(\pi)$ denote the set of all arcs in the path π (i.e., $\mathcal{E}^{path}(\pi = \langle t_1, \ldots, t_k \rangle) = \{\langle t_i, t_{i+1} \rangle \mid 1 \leq i \leq k-1\}$), $\mathcal{E}^{cut}(X, Y) = \{\langle s, t \rangle \in \mathcal{A} \mid s \in X \wedge t \in Y\}$, $\mathcal{E}^{pred}(X) = \bigcup_{\forall t \in X} \mathcal{E}^{path}(\pi_t^P)$.

In the first optimization step, optimum paths τ_t^P are computed for all $t \in \mathcal{O}$ by Algorithm 1 (Proposition 1). Let's denote $\mathcal{O}$ as $\mathcal{O}^1$ for this first step of Algorithm 1. Note that when the last pixel p from $\mathcal{O}^1$ leaves the priority queue $\mathcal{Q}$, all pixel with cost values greater than $V(p)$ were not yet processed and $\mathcal{O}^1 \subset \mathcal{F}$ at this point. For all $s \in \mathcal{I} \backslash \mathcal{F}$ we have $V(s) \geq V(p)$. Since the pixels in $\mathcal{O}^1$ are also in $\mathcal{F}$ their paths cannot be further changed by Algorithm 1. So, let's consider a new graph G^2, where the arcs connecting to nodes in $\mathcal{O}^1$ which are not in the forest P are disregarded. That is, $G^2 = \langle \mathcal{I}, \mathcal{E}^{set}(\mathcal{I} \backslash \mathcal{O}^1) \cup \mathcal{E}^{cut}(\mathcal{O}^1, \mathcal{I} \backslash \mathcal{O}^1) \cup \mathcal{E}^{pred}(\mathcal{O}^1) \rangle$. The next optimization step is relative to this subgraph G^2. Since the arcs connecting to nodes in $\mathcal{O}^1$, are reduced to the arcs in the previous forest P (i.e., $\mathcal{E}^{pred}(\mathcal{O}^1)$) in G^2, we have that the optimum paths τ_t^P, computed on the previous step, will remain optimum in the new graph G^2. So the optimum paths τ_t^P with $t \in \mathcal{O}^1$ compete with each other, seeking for their best extensions to the other pixels in $\mathcal{I} \setminus \mathcal{O}^1$.[2] By applying the Proposition 1 on this new optimization problem one more time, we have a new maximal set of pixels $\mathcal{O}^2$ that can be reached by hereditarily-optimum paths from $\mathcal{O}^1$ in G^2. We can then repeat this process over again. The monotonicity property, which is satisfied by the hypothesis, guarantees that at least one new element will be conquered at each step, so that this process will repeat until $\bigcup_{\forall i} \mathcal{O}^i = \mathcal{I}$.

Therefore, Bandeirantes can be seen as a live wire optimum result over a dynamically created graph, or alternatively as a result for f_B using a single predecessor map $P : \mathcal{I} \to \mathcal{I} \cup \{nil\}$ to store the computed paths, which has the following properties: It computes an optimum-path forest P for f_B whenever there is such a forest, otherwise, it computes optimum paths to a maximal set of pixels, that can be reached by hereditarily-optimum paths, and then search for the optimum extension of these prefixes to another maximal set of pixels in a subgraph, and so on and so forth.

4   Experimental Results and Conclusions

Figure 1 shows a Bandeirantes result as a boundary tracking approach compared to other methods. Note that G-wire gives a similar result compared to live wire, but with a slightly smoother contour. For Bandeirantes, we used $c(t_1, \ldots, t_n) = \|t_n - 2 \cdot t_{n-k} + t_{n-2 \cdot k}\|^p + \alpha^p \cdot (W(t_n) + W(t_{n-1}))^p$ in Eq. 3, where $p = 2$, $\alpha = 0.01$, $k = 15$ and $W(t)$ denotes the gradient magnitude complement. Bandeirantes gives the most regular contour, following more closely the dura mater, due to its curvature analysis involving up to the 30th ancestor on the paths. The second example in Fig. 2 shows that G-wire has difficulty tracking noisy contours with oscillations at a local scale, giving preference to a

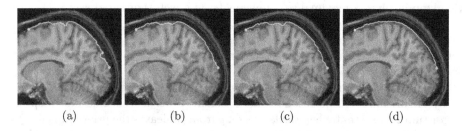

(a)   (b)   (c)   (d)

Fig. 1. The boundary tracking of the brain external surface in a MR-T1 image: (a) riverbed, (b) live wire, (c) G-wire, and (d) Bandeirantes.

---

[2] Consider the moment when the first pixel u not in $\mathcal{O}^1$ leaves the priority queue. Note that τ_u^P is optimum in G^2, because otherwise, since f is monotone, an optimum path $\langle a_0, \ldots, a_m = u \rangle$ to u in G^2 would have a cost $f(\langle a_0, \ldots, a_i \rangle) < f(\tau_u^P)$ for all $i \leq m$. So some ancestor of u in $\langle a_0, \ldots, a_m = u \rangle$ not in $\mathcal{O}^1$ must be removed from $\mathcal{Q}$ prior to u, leading to a contradiction.

smooth curve at the local scale even if it has a high long-term curvature (Fig. 2a). Bandeirantes can track these curves with local disturbances, when they have a low long-term curvature (Fig. 2b).

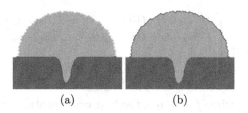

(a)   (b)

Fig. 2. Boundary tracking in a synthetic image by: (a) G-wire, and (b) Bandeirantes.

We conducted quantitative experiments for evaluating the performance of Bandeirantes, compared to live wire, riverbed and G-wire, in a curve tracing task to be used in an automatic grading system. We also tested "Snake with Gradient Vector Flow" [29] of the Active Contour Library in C/C++[3], but it was sensitive to initialization and very slow.

In automatic grading systems, such as *Auto Multiple Choice* (AMC)[4], we usually have multiple choice questionnaires and the students are asked to completely fill the relevant boxes. After the exam, the checked boxes in scanned completed answers sheets can be detected automatically by AMC, using Optical Mark Recognition (OMR). Here, to assess students' computer programming ability, they are asked to link the empty spaces in a given partial code to the correct options from a large list of commands by tracing smooth curves connecting their dots (Fig. 3a).

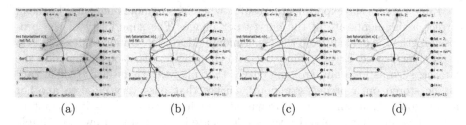

(a)   (b)   (c)   (d)

Fig. 3. (a) Example of a scanned completed answer sheet used in the experiments. The paths computed by Bandeirantes to all options from the: (b) first and (c) second empty spaces. (d) The best matching pairs by the Hungarian algorithm.

For each evaluated method, we first compute its paths from the dots $\{s_1, \ldots, s_M\}$ in the empty spaces to all dots $\{d_1, \ldots, d_N\}$ in the command list

---

[3] https://github.com/HiDiYANG/ActiveContour.
[4] http://home.gna.org/auto-qcm/.

(Fig. 3b and c), then the assignment problem is solved by the Hungarian algorithm[5] on matrix C (Eq. 4), which finds their best matching pair (Fig. 3d). Note that the curve tracing method in [26] could not be used in this evaluation, because it does not allow the specification of the destination pixels.

$$
C_{M,N} = \begin{pmatrix}
f(\pi_{s_1 \rightsquigarrow d_1}) & f(\pi_{s_1 \rightsquigarrow d_2}) & \cdots & f(\pi_{s_1 \rightsquigarrow d_N}) \\
f(\pi_{s_2 \rightsquigarrow d_1}) & f(\pi_{s_2 \rightsquigarrow d_2}) & \cdots & f(\pi_{s_2 \rightsquigarrow d_N}) \\
\vdots & \vdots & \ddots & \vdots \\
f(\pi_{s_M \rightsquigarrow d_1}) & f(\pi_{s_M \rightsquigarrow d_2}) & \cdots & f(\pi_{s_M \rightsquigarrow d_N})
\end{pmatrix}
\tag{4}
$$

In the case of live wire ($f = f_{LW}$), the best configuration found was obtained using $c(s,t) = \alpha^p \cdot (I(s) + I(t))^p + \|s - t\|$ in Eq. 1, where $p = 9$, $\alpha = 0.03$ and $I(t)$ is the image intensity. For riverbed ($f = f_{RB}$), the best setting was to use $c(s,t) = I(s) + I(t)$. For G-wire ($f = f_{GW}$), the best results were obtained using $c(t_{n-2}, t_{n-1}, t_n) = \gamma^p \cdot (I(t_n) + I(t_{n-1}))^p + \alpha \cdot \|t_n - t_{n-1}\| + \|t_n - 2 \cdot t_{n-1} + t_{n-2}\|^p$ in Eq. 2, where $p = 9$, $\gamma = 0.03$ and $\alpha = 0.01$. For Bandeirantes ($f = f_B$), we used $c(t_1, \ldots, t_n) = \|t_n - 2 \cdot t_{n-k} + t_{n-2 \cdot k}\|^p + \alpha^p \cdot (|I(t_n) - \frac{\sum_{i=n-2 \cdot k}^{n-k} I(i)}{k+1}| + I(t_n) + I(t_{n-1}))^p$ in Eq. 3, where $p = 9$, $\alpha = 0.03$ and $k = 16$. The component $|I(t_n) - \frac{\sum_{i=n-2 \cdot k}^{n-k} I(i)}{k+1}|$ in f_B helps Bandeirantes to track weakly drawn curves. Ties were rarely detected (less than 0.01%) in Bandeirantes when using costs in double-precision floating-point format.

Each method is computed using its own path-cost function. The IFT algorithm is used for computing riverbed, live wire and bandeirantes, while G-wire is computed using a high-order dynamic programming in a three-dimensional graph. With regard to filling the entries of matrix C, we consider two versions for each method (riverbed/live wire/G-wire/bandeirantes), one using its own costs ($f_{RB}/f_{LW}/f_{GW}/f_B$), and the other considering in C the costs given by f_B of its computed paths.

Table 1 shows the experimental results, where the numbers indicate the mean correct assignment rate of the curves per image. The superiority of the proposed method is evident from the experiments. Moreover, Bandeirantes was 81% faster than G-wire. Errors mainly occur in challenging situations included in the dataset, such as loop autocrossing curves that intersect themselves (which cannot be stored on a predecessor map), curves that pass very close to another destination option, and curves drawn with high curvature. It is important to state, however, that in practice the students are instructed to draw only simple curves with low curvature, so that the error rate would be lower in practice.

The database composed of 70 scanned images used in the experiments and all the source code in C/C++ are available at the author's website[6].

As future works, we intend to explore Bandeirantes in other applications and test it with other path-cost functions.

---

[5] https://github.com/saebyn/munkres-cpp.
[6] http://www.vision.ime.usp.br/~mtejadac/bandeirantes.html.

Table 1. The mean correct assignment rate of curves per image for different methods.

	Its own cost in C	f_B in C
Riverbed	14.19%	49.88%
Live wire	26.33%	57.60%
G-wire	36.50%	57.03%
Bandeirantes	95.24%	95.24%

Acknowledgements. Thanks to CNPq (308985/2015-0, 486083/2013-6, FINEP 1266/13), FAPESP (2011/50761-2,2014/12236-1), CAPES, and NAP eScience - PRP - USP for funding, and Alexandre Morimitsu and Fabio Albuquerque Dela Antonio for helping to build the dataset.

References

1. Barreto, T.L.M., Almeida, J., Cappabianco, F.A.M.: Estimating accurate water levels for rivers and reservoirs by using SAR products: a multitemporal analysis. Pattern Recogn. Lett. Part 2 **83**, 224–233 (2016)
2. Baumgartner, A., Steger, C., Wiedemann, C., Mayer, H., Eckstein, W., Ebner, H.: Update of roads in GIS from aerial imagery: verification and multi-resolution extraction. Int. Archiv. Photogram. Remote Sens. **31**, 53–58 (1996)
3. Canny, J.: A computational approach to edge detection. IEEE Trans. Pattern Anal. Mach. Intell. (PAMI) **8**(6), 679–698 (1986)
4. Cappabianco, F.A., Lellis, L.S., Miranda, P.A.V., Ide, J.S., Mujica-Parodi, L.: Edge detection robust to intensity inhomogeneity: a 7T MRI case study. In: XXI IberoAmerican Congress on Pattern Recognition (CIARP), Lima, Peru (2016, to appear). Accepted November 2016
5. Chen, X., Udupa, J., Zheng, X., Torigian, D., Alavi, A.: Automatic anatomy recognition via multi object oriented active shape models. In: SPIE on Medical Imaging, Orlando, Florida, USA, vol. 7259, pp. 72594P–72594P–8, February 2009
6. Coppini, C., Demi, M., Poli, R., Valli, G.: An artificial vision system for x-ray images of human coronary arteries. IEEE Trans. Pattern Anal. Mach. Intell. **15**(2), 156–162 (1993)
7. Donoho, D., Huo, X.: Applications of beamlets to detection and extraction of lines, curves and objects in very noisy images. In: Proceedings of NSIP (2001)
8. Falcão, A., Stolfi, J., Lotufo, R.: The image foresting transform: theory, algorithms, and applications. IEEE TPAMI **26**(1), 19–29 (2004)
9. Falcão, A., Udupa, J., Miyazawa, F.: An ultra-fast user-steered image segmentation paradigm: live-wire-on-the-fly. IEEE Trans. Med. Imaging **19**(1), 55–62 (2000)
10. Falcão, A., Udupa, J., Samarasekera, S., Sharma, S., Hirsch, B., Lotufo, R.: User-steered image segmentation paradigms: live-wire and live-lane. Graph. Models Image Process. **60**, 233–260 (1998)
11. Farber, M., Ehrhardt, J., Handels, H.: Live-wire-based segmentation using similarities between corresponding image structures. Comput. Med. Imaging Graph. **31**(7), 549–560 (2007)
12. Frieze, A.: Minimum paths in directed graphs. Oper. Res. Q. **28**(2), 339–346 (1977)

13. He, H., Tian, J., Lin, Y., Lu, K.: A new interactive segmentation scheme based on fuzzy affinity and live-wire. In: Wang, L., Jin, Y. (eds.) FSKD 2005. LNCS (LNAI), vol. 3613, pp. 436–443. Springer, Heidelberg (2005). doi:10.1007/11539506_56

14. Kang, H.: G-wire: a livewire segmentation algorithm based on a generalized graph formulation. Pattern Recogn. Lett. **26**(13), 2042–2051 (2005)

15. Kass, M., Witkin, A., Terzopoulos, D.: Snakes: active contour models. Int. J. Comput. Vis. **1**, 321–331 (1987)

16. Liu, J., Udupa, J.: Oriented active shape models. IEEE Trans. Med. Imaging **28**(4), 571–584 (2009)

17. Malmberg, F., Vidholm, E., Nyström, I.: A 3D live-wire segmentation method for volume images using haptic interaction. In: Kuba, A., Nyúl, L.G., Palágyi, K. (eds.) DGCI 2006. LNCS, vol. 4245, pp. 663–673. Springer, Heidelberg (2006). doi:10.1007/11907350_56

18. Martelli, A.: Edge detection using heuristic search methods. Comput. Graph. Image Process. **1**, 169–182 (1972)

19. Martelli, A.: An application of heuristic search methods to edge and contour detection. Commun. ACM **19**(2), 73–83 (1976)

20. Mille, J., Bougleux, S., Cohen, L.D.: Combination of piecewise-geodesic paths for interactive segmentation. Int. J. Comput. Vis. **112**(1), 1–22 (2015)

21. Mille, J., Cohen, L.D.: Geodesically linked active contours: evolution strategy based on minimal paths. In: Tai, X.-C., Mørken, K., Lysaker, M., Lie, K.-A. (eds.) SSVM 2009. LNCS, vol. 5567, pp. 163–174. Springer, Heidelberg (2009). doi:10.1007/978-3-642-02256-2_14

22. Miranda, P., Falcao, A., Spina, T.: The riverbed approach for user-steered image segmentation. In: 18th IEEE International Conference on Image Processing (ICIP), pp. 3133–3136, September 2011

23. Miranda, P., Falcao, A., Spina, T.: Riverbed: a novel user-steered image segmentation method based on optimum boundary tracking. IEEE Trans. Image Process. **21**(6), 3042–3052 (2012)

24. Mortensen, E., Barrett, W.: Interactive segmentation with intelligent scissors. Graph. Models Image Process. **60**, 349–384 (1998)

25. Nilsson, N.: Principles of Artificial Intelligence. Morgan Kaufmann Publishers Inc., San Francisco (1980)

26. Raghupathy, K.: Curve tracing and curve detection in images. Cornell University. A thesis Presented to the Faculty of the Graduate School of Cornell University, August 2004. https://books.google.com.br/books?id=QlNUAAAAYAAJ

27. Spina, T.V., de Miranda, P.A.V., Falcão, A.X.: Hybrid approaches for interactive image segmentation using the live markers paradigm. IEEE Trans. Image Process. **23**(12), 5756–5769 (2014)

28. Steger, C.: An unbiased detector of curvilinear structures. IEEE Trans. Pattern Anal. Mach. Intell. **20**(2), 113–125 (1998)

29. Xu, C., Prince, J.L.: Snakes, shapes, and gradient vector flow. IEEE Trans. Image Process. **7**(3), 359–369 (1998)

A Supervoxel-Based Solution to Resume Segmentation for Interactive Correction by Differential Image-Foresting Transforms

Anderson Carlos Moreira Tavares[1]([✉]), Paulo André Vechiatto Miranda[1], Thiago Vallin Spina[2], and Alexandre Xavier Falcão[2]

[1] Institute of Mathematics and Statistics (IME), University of Sao Paulo (USP), Rua do Matao, 1010 Sao Paulo/SP, Brazil
{acmt,pmiranda}@ime.usp.br
[2] Institute of Computing (IC), University of Campinas (UNICAMP), Av. Albert Einstein, 1251 - Cidade Universitária, Campinas/SP, Brazil
{tvspina,afalcao}@ic.unicamp.br

Abstract. The foolproof segmentation of 3D anatomical structures in medical images is usually a challenging task, which makes automatic results often far from desirable and interactive repairs necessary. In the past, we introduced a first solution to resume segmentation from third-party software into an initial optimum-path forest for interactive correction by differential image foresting transforms (DIFTs). Here, we present a new method that estimates the initial forest (input segmentation) rooted at more regularly separated seed voxels to facilitate interactive editing. The forest is a supervoxel segmentation from seeds that result from a sequence of image foresting transforms to conform as much as possible the supervoxel boundaries to the boundaries of the object in the input segmentation. We demonstrate the advantages of the new method over the previous one by using a robot user, as an impartial way to correct brain segmentation in MR-T1 images.

Keywords: Interactive editing · Segmentation · Supervoxel · Image-foresting transform

1 Introduction

The task of assigning a distinct label to object and background voxels in 3D medical images, named *segmentation*, is usually challenging due to poorly defined object boundaries, non-standard intensity distribution, field inhomogeneity, noise, partial volume, and the interplay among these factors [19]. As consequence, repairs of automatic segmentation from third-party software (e.g., FreeSurfer [4], SPM2 [8], CLASP[1]) are often necessary. Interactive segmentation from scratch is certainly an undesirable alternative and manual corrections may

---

[1] URL: http://www.bic.mni.mcgill.ca/.

© Springer International Publishing AG 2017
J. Angulo et al. (Eds.): ISMM 2017, LNCS 10225, pp. 107–118, 2017.
DOI: 10.1007/978-3-319-57240-6_9

be time-consuming, tedious, and subject to variations among distinct users. Ideally, one should be able to fix segmentation interactively without destroying its correct parts.

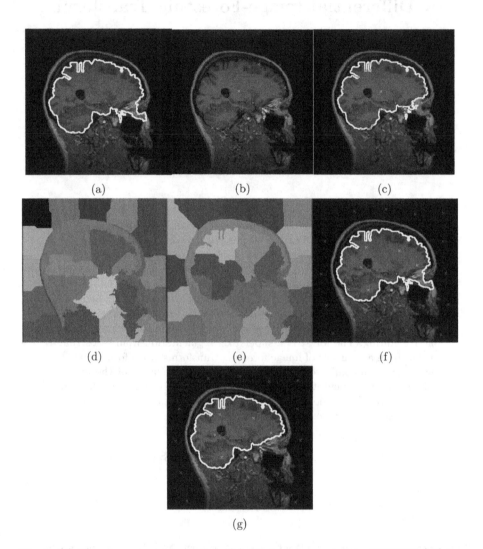

Fig. 1. (a) The given presegmentation. (b) Seed set computed by ISBI2011 [21] has many non-uniformly distributed seeds, and (c) its attempt to fix the segmentation by placing new background markers (red dots) fails. Proposed editing method: (d) Supervoxels by IFT-SLIC to find the seed set. (e) Supervoxels better conforming to the presegmentation are obtained by changing the cost function to f'_D. (f) The union of supervoxels from seeds contained in the presegmentation gives us a starting point to perform corrections. (g) A corrected result is obtained by adding a new background seed (red dot) and running DIFT. (Color figure online)

In the past, we proposed a first solution to resume an input segmentation [21,22] for interactive correction by differential image foresting transforms (DIFTs) [7]. The method interprets a 3D medical image as a graph, whose nodes are the voxels and the arcs are defined by their 6-neighbors, and estimates a minimum set of seed nodes such that the image foresting transform (IFT) algorithm [5] can generate the exact representation of the input segmentation as an optimum-path forest rooted at those seeds. In this forest, the object is defined by the union of trees rooted at its internal seeds. The IFT algorithm executes in time proportional to the nodes (i.e., linear time) and subsequently, the user may add and/or remove seeds (their optimum-path trees) to correct segmentation in sublinear time by using the DIFT algorithm [7]. A drawback is the possible high number of seeds near the object boundaries, making difficult user interaction (Figs. 1a–c).

In this work, we propose an approximate solution in which the seed set consists of more regularly separated nodes, thereby facilitating the interactive correction (Figs. 1d–g). The method starts by estimating uniformly spaced seeds in a region of interest around the input object mask and applies a sequence of DIFTs followed by seed recomputation to conform the boundaries of the supervoxels (i.e., the optimum-path trees) to the object's boundaries, as much as possible. Then, the user can add and/or remove seed nodes at each subsequent iteration of the DIFT algorithm to fix segmentation.

The first step is automatic and it is based on a recent approach for supervoxel segmentation, named *IFT-SLIC* [2], but using here a different choice of parameter (*path-value function*) in the DIFT algorithm to approximate the result to the desired segmentation. IFT-SLIC was inspired in the Simple Linear Iterative Clustering (SLIC) method, that defines supervoxels along a few iterations of the k-means clustering algorithm. In IFT-SLIC, however, the supervoxels are naturally obtained as single connected components and the object is defined by connected paths from its internal seeds. This makes possible the interactive editing of the segmentation.

In the following, we first describe related works on segmentation editing and basic concepts. Then, we detail the proposed method and our experimental results and conclusions.

2   Related Works

In spite of the vast literature on segmentation, only a few works have dealt with the editing issue, usually considering qualitative and highly subjective empirical evaluations [13].

Grady and Funka-Lea [10] apply *Random Walk* (RW) [9] to correct pre-segmented images, optimized with downsampling, which loses important and high frequency information, like small objects, negatively affecting the result. Harrison et al. [12] join discriminative classification and energy minimization with RW for contour-based correction, using GPU training. It inherits disadvantages from contour-based segmentations, like sensitivity to seed placing, lack of texture and region information. It depends on the training set size to propagate the labels to other slices, affecting its accuracy.

Jackowski et al. [14] approximate a digital volume representing the segmented object by a *Rational of Gaussians* (RaG) parametric surface, allowing the user to change the surface by its control points. Advantages are compression for fast transmission, sub-voxel correction and inclusion of graphical effects without a voxelized appearance. But editing non-compact objects by control points is not trivial. Valenzuela et al. [29] use Bézier-based surfaces. The user can modify the curve in one slice, and it propagates to the rest in 3D.

Yang and Choe [30] uses *Graph Cut* (GC) [3], with energy function composed by presegmentation and new user inputs. It considers the presegmentation is almost correct, restricting the user active field. It inherits GC disadvantages. The graph weights are based on the euclidean distance, not effective for non-compact objects, like veins and arteries. To remove parts of the presegmentation, the user must always unnecessarily place background and foreground seeds. Moreover, its conducted evaluation did not include a user effort analysis. Karimov et al. [16] develop a software that suggests correction candidates, based on extraction of region skeletons, which should be similar to ground truth, and histogram similarity analysis. Complex images can affect the candidates number.

Li et al. [17] proposed a user interface to tackle the interactive segmentation problem consisting of two steps. In the first step, the user selects seeds over the foreground and the background for region-based segmentation using Graph Cut on a superpixel-graph, derived from a watershed segmentation of the image. Then, the resulting segmentation boundary is turned into a polygon that can be interactively edited. Changes to the polygon structure serve as soft constraints for local corrections using GC on a pixel-graph. This method was designed for 2D images and the extension to 3D is not straightforward.

Miranda et al. [21,22] proposed an editing solution based on the IFT with experimental analysis in MR-T1 tridimensional images. Contrary to previous methods, it can be applied to multidimensional images and to objects with arbitrary shapes, with low running time and without any special hardware support. It first solves the reverse segmentation problem, with strong theoretical background, reducing the required number of seeds by employing a conservative force [21]. The corrections can then be performed in sublinear time by differential IFT (DIFT) [7]. It is restricted to the max-arc path-cost function over a gradient image, which is usually not the best option to deal with blurred transitions.

Spina et al. [27] proposed a solution with robot users [11], which simulate user interaction by placing brush strokes automatically to iteratively perform the segmentation task resulting in the given presegmentation. It can correct any existing delineation method result [27]. However, it considered a robot user tailored to IFT-based segmentation, since the end goal was to learn the spatial distribution of seeds added to reproduce ground truth training masks, in order to output a statistical seed model of an object of interest to aid in its interactive segmentation. Hence, they were more interested in consistent seed positioning than high accuracy for editing.

Our proposed method is also based on the IFT framework, but it was designed to circumvent the main problems of [21], such as its high number of seeds and non-uniform seed distribution, in order to give more freedom to the user to perform corrections, and using a better path-cost function. The flexibility of the path-cost function of the IFT-SLIC makes it a more general framework than other similar methods that attempt to produce SLIC-like superpixels from watershed segmentation [18, 26].

Lastly, it is worth noting that using boundary-tracking methods in slice by slice fashion to fix 3D segmentations, such as live wire [6], intelligent scissors [25], Riverbed [23], and G-Wire [15], may demand considerable user interaction, proportional to the number of slices with errors, which can be infeasible in many cases. The Live Markers paradigm [28] might mitigate this problem, when coupled with our proposed method, by allowing the propagation of those corrections in 3D, given that the user-selected boundary segments are converted to seeds for competition in a 3D graph.

3   Background

A multidimensional and multispectral digital image is a mapping $I : \mathcal{I} \rightarrow \mathbb{Z}^m$, where $\mathcal{I} \subset \mathbb{Z}^n$ is the image domain and $\mathbb{Z}^m$ is a space of m bands (e.g., color channels). An *adjacency relation* $\mathcal{A}$ is a binary relation on $\mathcal{I}$. We use $t \in \mathcal{A}(s)$ and $(s, t) \in \mathcal{A}$ to indicate that t is adjacent to s. By setting an adjacency relation, I can be represented as a *weighted digraph* $G = (V, E, w)$, where $V = \mathcal{I}$ represents the set of nodes, $E = \mathcal{A}$ is the set of arcs and $w : E \rightarrow \mathbb{R}$ assigns a weight to each arc. In this work, we are interested in the 6-neighborhood relation $\mathcal{A}$ for 3D images.

A *path* $\pi_{s \rightsquigarrow t}$ is a sequence of distinct nodes $\langle v_1 = s, v_2, \ldots, v_n = t \rangle$, with origin s and terminus t, where $(v_i, v_{i+1}) \in \mathcal{A}$ for $i = 1, 2, \ldots, n-1$. π_t represents a path with terminus t from any origin. We use $\pi_t = \pi_s \cdot \langle s, t \rangle$ to denote the concatenation of a path π_s by an arc (s, t). $\pi_t = \langle t \rangle$ is a *trivial* path. $\Pi_t(G)$ is the set of all distinct paths with terminus t, $\Pi_{s \rightsquigarrow t}(G)$ limits $\Pi_t(G)$ for paths with origin s, and $\Pi(G)$ is the set of all distinct paths: $\Pi_{s \rightsquigarrow t}(G) \subseteq \Pi_t(G) \subseteq \Pi(G)$. A *connectivity function* $f : \Pi(G) \rightarrow \mathbb{R}$ assigns a scalar value to any path π in the graph G. A path π_t^* is optimum if $f(\pi_t^*) \leq f(\pi_t), \forall \pi_t \in \Pi_t(G)$.

A *predecessors map* is a function $Pr : V \rightarrow V \cup \{nil\}$ where for $Pr(t) = s$ we have $t \in A(s)$ or $s = nil$. For any pixel $t \in V$, a predecessors map Pr with no cycles defines a path π_t^{Pr} recursively as $\langle t \rangle$ if $Pr(t) = nil$, and $\pi_t^{Pr} \cdot \langle s, t \rangle$ if $Pr(t) = s \neq nil$. Hence, a predecessors map with no cycles defines a *spanning forest*, where all nodes are connected to a set of root nodes $\mathcal{R}(Pr) = \{v \in V : Pr(v) = nil\}$.

3.1   Image Foresting Transform (IFT)

An *Optimal-Path Spanning Forest Problem* (OPSFP) consists on finding a spanning forest Pr, such that π_t^{Pr} are optimal paths, for all $t \in V$, according to a connectivity function f. The IFT is an OPSFP solver by extending Dijkstra shortest path algorithm with multiple sources and different connectivity functions [5]. It uses a dynamic approach by storing the best connectivity values found so far in a map $C : V \rightarrow \mathbb{R}$, which converges to $C(s) = \min_{\pi_s \in \Pi_s(G)} f(\pi_s)$ in the case of *smooth* connectivity functions [5].

In the context of binary interactive segmentation, we usually restrict the optimal paths to paths starting in a set of seed pixels $\mathcal{S} = \mathcal{S}_0 \cup \mathcal{S}_1$, where $\mathcal{S}_0$ and $\mathcal{S}_1$ denote the sets of background and object seeds, respectively. The segmented object is defined by the union of all pixels t that are reached by optimal paths π_t^{Pr} rooted at $\mathcal{S}_1$. Seeds can be added and/or removed to perform corrections to intermediate results by re-executing the algorithm. Falcão et al. [7] proposes *Differential* IFT (DIFT) to compute sequences of IFTs in a differential way, which takes sublinear time complexity for subsequent IFT executions on the same session.

4   IFT-SLIC for Segmentation Editing

A *label map* $L : V \rightarrow \{1, \ldots, c\}$ defines a partition set $\mathcal{P}_L = \{P_1, P_2, \ldots, P_c\}$, where $\bigcup_{i=1}^{c} P_i = V$. A set of *supervoxels* is a partition set composed by regions which share common structural information, like intensity, proximity and texture, and that have uniform size and shape.

The IFT-SLIC [2] combines the benefits of IFT and SLIC [1] to provide a more regular and powerful supervoxel generation. It uses a non-smooth connectivity function f_D (Eq. 1), which is based on the path-cost function $f_{\sum |\Delta I|}$ from [20]. f_D uses the sum of the color distances relative to its root node and the sum of Euclidean distances for encoding the boundary adherence and proximity (compactness), respectively, with a parameter α controlling their trade-off.

Firstly, k equidistant seeds are sampled following a regular grid. Then, two iterative steps are applied: *assignment*, where the nodes are labeled to the closest cluster, according to f_D by computing the IFT using 6-neighborhood, and *update*, where the seed positions and their attribute vectors are moved to their mean values within their respective labeled regions. The assignment and update steps are repeated for a total of 10 iterations. The method outputs a spanning forest in a predecessors map Pr, where each tree defines a supervoxel and its cluster center corresponds to its root r.

$$f_D(\pi_t = \langle t \rangle) = \begin{cases} 0, & \text{if } t \in \mathcal{S} \\ +\infty, & \text{otherwise} \end{cases}$$

$$f_D(\pi_{r \rightsquigarrow s} \cdot \langle s, t \rangle) = f_D(\pi_{r \rightsquigarrow s}) \qquad (1)$$
$$+ \underbrace{(\|\boldsymbol{I}(t) - \boldsymbol{I}_r\| \cdot \alpha)^\beta}_{\text{Boundary Adherence}} + \underbrace{d_{euc}(s, t)}_{\text{Compactness}}$$

where $\boldsymbol{I}_r$ is the mean attribute vector associated to the seed r, and we use in this work $\alpha = 0.04$ and $\beta = 12$, which are values within the range of recommended values in [2].

In order to resume a previous given presegmentation by IFT, we need to devise a seed set that assembles it for the given image. In our proposed method, a first idea is to consider the seeds (cluster centers) obtained by IFT-SLIC for the given image, to obtain a more efficient solution. IFT-SLIC results in a seed set $\mathcal{S} = \{s_1, s_2, \ldots, s_k\}$, where k is the number of supervoxels. If the preseg-mentation object is small relative to the whole graph, we can use its bounding box (with a proper extension margin) and compute the seeds by IFT-SLIC only inside it, in order to reduce the running time. The number of seeds should be proportional to the bounding box size, to keep the seed density constant for different object sizes.

The seeds by IFT-SLIC are then divided in two subsets $\mathcal{S}_0$ and $\mathcal{S}_1$, according to their values in the binary mask B of the presegmentation ($s_i \in \mathcal{S}_0$ if $B(s_i) = 0$, and $s_i \in \mathcal{S}_1$ otherwise). The union of all supervoxels from seeds in $\mathcal{S}_1$, gives us an initial approximation of the presegmentation, denoted as the *initial supervoxel segmentation*, which does not perfectly resemble the presegmentation (Fig. 1d). To further boost the results, we improve the final supervoxel segmentation by changing the connectivity function to f'_D (Fig. 1e) as follows:

$$f'_D(\pi_t = \langle t \rangle) = f_D(\pi_t = \langle t \rangle)$$
$$f'_D(\pi_{r \rightsquigarrow s} \cdot \langle s, t \rangle) = f'_D(\pi_{r \rightsquigarrow s}) + \underbrace{d_{euc}(s,t)}_{\text{Compactness}} \qquad (2)$$
$$+ \underbrace{(\|\mathbf{I}(t) - \mathbf{I}_r\| \cdot \alpha \cdot \gamma^{B(r,t)} + \gamma \cdot B(r,t))^\beta}_{\text{Boundary Adherence}}$$

where $B(r,t) = |B(r) - B(t)|$, that is, $B(r,t)$ captures the transitions in the binary mask B of the presegmentation, and γ plays the same role as the *liberal* and *conservative* forces used in [21]. For higher values of γ, the final supervoxel segmentation better resembles the presegmentation, conserving its fine details. Thus, higher values of γ allow us to reduce the number of supervoxels k, giving more freedom to the user to perform corrections. So we used $k = vol/(200 \cdot \gamma)$, where vol is the number of object voxels in the presegmentation.

The final supervoxel segmentation can then be used as a starting point, so that the user can insert and/or remove seeds from $\mathcal{S}_0$ and $\mathcal{S}_1$ in order to correct the segmentation in a differential way, by using DIFT [7] with function f'_D (Figs. 1f–g). Therefore, the corrections take sublinear time.

5   Experimental Results

In this section, we conducted experiments to measure the user involvement in the editing process of the wrong parts of the presegmentation in real 3 T MRI-T1 images of the brain of size $240 \times 240 \times 180$ voxels with severe inhomogeneity problems. We also quantified the number of estimated seeds, where lower values indicate more flexibility for posterior user corrections. We compared our pro-posed method with the best solution so far by IFT, denoted as ISBI2011 [21]. In all cases, the corrective actions were conducted by a robot user [11], in order to get impartial results, with a spherical brush size of 5 voxels, using an Intel core i3 laptop with 4 GB memory.

Table 1 shows the results of the first experiment (data set D1, composed of ten MRI volumes) to correct the wrong parts of automatic segmentation of the cerebral hemispheres, where the errors are related mainly to the bad positioning of the fuzzy model [24] during the automatic segmentation (Figs. 2a–b). The mean execution time to obtain the initial seeds by the proposed method was 24.0 s and 13.5 s for ISBI2011 [21]. The mean Dice value for the initial supervoxel segmentation using the seeds by IFT-SLIC increased from 89.75% to 99.96% when changing the path cost-function to f'_D for $\gamma = 3$, and from 88.64% to 99.98% for $\gamma = 4$. We noted that lower values of γ ($\gamma < 3$) can lead to a loss of presegmentation details. The proposed method reduced the number of markers required for corrective actions in 68.2% and reduced the total number of initial seeds in 4.3% for $\gamma = 3$. For $\gamma = 4$, we had a reduction of 60.8% for corrective actions and 29.2% for the number of initial seeds.

Table 1. Data set D1: number of markers (nm) required for corrective actions and number of computed initial seeds (ns) per voxels in parts per thousand.

image#	Proposed ($\gamma = 3$)		Proposed ($\gamma = 4$)		ISBI2011	
	nm,	ns (‰),	nm,	ns (‰),	nm,	ns (‰),
01	5,	0.0729	7,	0.0463	46,	0.0657
02	10,	0.0608	13,	0.0463	35,	0.0766
03	12,	0.0729	12,	0.0502	42,	0.0811
04	15,	0.0602	18,	0.0463	33,	0.0949
05	8,	0.0781	11,	0.0648	23,	0.0443
06	6,	0.0677	10,	0.0463	15,	0.0683
07	8,	0.0677	10,	0.0463	26,	0.0750
08	15,	0.0729	16,	0.0501	20,	0.0470
09	6,	0.0501	8,	0.0463	20,	0.0672
10	9,	0.0502	11,	0.0405	36,	0.0631
Mean	9.4,	0.0653	11.6,	0.0483	29.6,	0.0683

On the second experiment (data set D2, composed of ten MRI volumes, in Table 2), we considered a more challenging scenario. We conducted experiments to fix the segmentation of the cortical surface of the brain, where several pronounced errors were intentionally introduced by manual editing along the 3D surface (Figs. 2c-d). The mean Dice value for the initial supervoxel segmentation using the seeds by IFT-SLIC increased from 93.08% to 99.95% when changing the path cost-function to f'_D for $\gamma = 3$, and from 92.48% to 99.95% for $\gamma = 4$. The proposed method reduced the number of markers required for corrective actions in 45% (39.9%) and reduced the total number of initial seeds in 79.4% (84%) for $\gamma = 3$ ($\gamma = 4$).

Table 2. Data set D2: number of markers (nm) required for corrective actions and number of computed initial seeds (ns) per voxels in parts per thousand.

image#	Proposed ($\gamma = 3$)		Proposed ($\gamma = 4$)		ISBI2011	
	nm,	ns (‰),	nm,	ns (‰),	nm,	ns (‰),
01	20,	0.1633	26,	0.1252	33,	0.8230
02	20,	0.1633	22,	0.1379	28,	1.2129
03	23,	0.1516	21,	0.1253	32,	0.9770
04	19,	0.1908	18,	0.1484	37,	0.6807
05	23,	0.1909	22,	0.1485	67,	1.6071
06	19,	0.1379	18,	0.1253	34,	1.0022
07	17,	0.1516	19,	0.1273	31,	0.3774
08	17,	0.1633	23,	0.1253	24,	0.4172
09	21,	0.1633	21,	0.1157	42,	0.4365
10	18,	0.1633	25,	0.0936	30,	0.4303
Mean	19.7,	0.1639	21.5,	0.1272	35.8,	0.7965

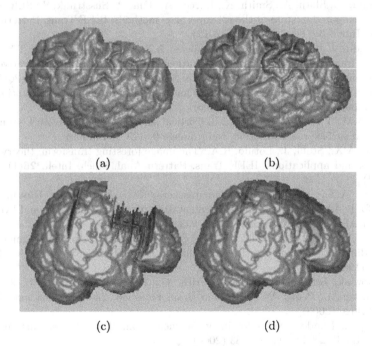

(a)   (b)

(c)   (d)

Fig. 2. 3D renditions of presegmentations with errors (first column) and respective ground truths (second column), with their main differences highlighted in another color. A sample image from each data set composed of ten 3D volumes: (a–b) Data set D1. (c–d) Data set D2 with severe errors. (Color figure online)

6   Conclusions

From the experiments we can conclude that the proposed method can substantially reduce the number of markers required for corrective actions in both scenarios and with a strong reduction of initial seeds in the second case. Our method has better seed distribution over the image than ISBI2011, due to the regular sized supervoxels, avoiding the negative effect of seed concentrations in specific regions, which makes the corrections in these areas to behave like manual segmentation. Moreover, it can be easily extended to multi-class. DIFT runs only within modified trees, thus in sublinear time. As future work, we will investigate other path-cost functions and the applications in other image modalities.

Acknowledgements. Thanks to CNPq (308985/2015-0, 486083/2013-6, FINEP 1266/13), FAPESP (2011/50761-2, 2014/12236-1, 2015/09446-7, 2016/11853-2), CAPES, and NAP eScience - PRP - USP for funding, and Dr. J. K. Udupa (MIPG-UPENN) for the images.

References

1. Achanta, R., Shaji, A., Smith, K., Lucchi, A., Fua, P., Süsstrunk, S.: SLIC superpixels compared to state-of-the-art superpixel methods. IEEE Trans. Pattern Anal. Mach. Intell. **34**(11), 2274–2282 (2012)
2. Alexandre, E.B., Chowdhury, A.S., Falcão, A.X., Miranda, P.A.V.: IFT-SLIC: A general framework for superpixel generation based on simple linear iterative clustering and image foresting transform. In: SIBGRAPI, Salvador, Brazil (2015)
3. Boykov, Y., Funka-Lea, G.: Graph cuts and efficient N-D image segmentation. Int. J. Comput. Vis. **70**(2), 109–131 (2006)
4. Dale, A.M., Fischl, B., Sereno, M.I.: Cortical surface- based analysis: I. segmentation and surface reconstruction. NeuroImage **9**, 179–194 (1999)
5. Falcão, A.X., Stolfi, J., Lotufo, R.A.: The image foresting transform: theory, algorithms, and applications. IEEE Trans. Pattern Anal. Mach. Intell. **26**(1), 19–29 (2004)
6. Falcão, A.X., Udupa, J.K., Samarasekera, S., Sharma, S., Hirsch, B.E., Lotufo, R.A.: User-steered image segmentation paradigms: live-wire and live-lane. Graph. Model. Image Process. **60**(4), 233–260 (1998)
7. Falcão, A., Bergo, F.F., Falcão, A., Bergo, F.F.: Interactive volume segmentation with differential image foresting transforms. IEEE Trans. Med. Imaging **23**(9), 1100–1108 (2004)
8. Frackowiak, R.S.J., Friston, K.J., Frith, C., Dolan, R., Price, C.J., Zeki, S., Ashburner, J., Penny, W.D.: Human Brain Function, 2nd edn. Academic Press, Cambridge (2003)
9. Grady, L.: Random walks for image segmentation. IEEE Trans. Pattern Anal. Mach. Intell. **28**(11), 1768–1783 (2006)
10. Grady, L., Funka-Lea, G.: An energy minimization approach to the data driven editing of presegmented images/volumes. In: Larsen, R., Nielsen, M., Sporring, J. (eds.) MICCAI 2006. LNCS, vol. 4191, pp. 888–895. Springer, Heidelberg (2006). doi:10.1007/11866763_109

11. Gulshan, V., Rother, C., Criminisi, A., Blake, A., Zisserman, A.: Geodesic star convexity for interactive image segmentation. In: CVPR, pp. 3129–3136 (2010)
12. Harrison, A.P., Birkbeck, N., Sofka, M.: IntellEditS: intelligent learning-based editor of segmentations. In: Mori, K., Sakuma, I., Sato, Y., Barillot, C., Navab, N. (eds.) MICCAI 2013. LNCS, vol. 8151, pp. 235–242. Springer, Heidelberg (2013). doi:10.1007/978-3-642-40760-4_30
13. Heckel, F., Moltz, J.H., Tietjen, C., Hahn, H.K.: Sketch-based editing tools for tumour segmentation in 3D medical images. Comput. Graph. Forum 32(8), 144–157 (2013)
14. Jackowski, M., Satter, M., Goshtasby, A.: Approximating digital 3D shapes by rational Gaussian surfaces. IEEE Trans. Vis. Comput. Graphics 9(1), 56–69 (2003)
15. Kang, H.W.: G-wire: a livewire segmentation algorithm based on a generalized graph formulation. Pattern Recogn. Lett. 26(13), 2042–2051 (2005)
16. Karimov, A., Mistelbauer, G., Auzinger, T., Bruckner, S.: Guided volume editing based on histogram dissimilarity. Comput. Graph. Forum 34(3), 91–100 (2015)
17. Li, Y., Sun, J., Tang, C.K., Shum, H.Y.: Lazy snapping. ACM Trans. Graph. 23(3), 303–308 (2004)
18. Machairas, V., Faessel, M., Cárdenas-Peña, D., Chabardes, T., Walter, T., Decencière, E.: Waterpixels. IEEE Trans. Image Process. 24(11), 3707–3716 (2015)
19. Madabhushi, A., Udupa, J.: Interplay between intensity standardization and inhomogeneity correction in MR image processing. IEEE Trans. Med. Imag. 24(5), 561–576 (2005)
20. Mansilla, L.A.C., Miranda, P.A.V., Cappabianco, F.A.: Image segmentation by image foresting transform with non-smooth connectivity functions. In: SIBGRAPI, pp. 147–154, August 2013
21. Miranda, P.A.V., Falcão, A.X., Ruppert, G. C., Cappabianco, F.A.: How to fix any 3D segmentation interactively via image foresting transform and its use in MRI brain segmentation. In: Biomedical Imaging, pp. 2031–2035. IEEE, Chicago, USA, March 2011
22. Miranda, P.A.V., Falcão, A.X., Ruppert, G.C.S.: How to complete any segmentation process interactively via image foresting transform. In: SIBGRAPI, pp. 309–316. IEEE, Gramado, Brazil, August 2010
23. Miranda, P.A.V., Falcão, A.X., Spina, T.V.: Riverbed: a novel user-steered image segmentation method based on optimum boundary tracking. IEEE Trans. Image Process. 21(6), 3042–3052 (2012)
24. Miranda, P.A.V., Falcao, A.X., Udupa, J.K.: Cloud bank: A multiple clouds model and its use in MR brain image segmentation. In: ISBI, pp. 506–509, June 2009
25. Mortensen, E., Barrett, W.: Interactive segmentation with intelligent scissors. Graph. Model. Im. Proc. 60(5), 349–384 (1998)
26. Neubert, P., Protzel, P.: Compact watershed and preemptive slic: On improving trade-offs of superpixel segmentation algorithms. In: ICPR, pp. 996–1001, Stockholm, Sweden, August 2014
27. Spina, T.V., Martins, S.B., Falcão, A.X.: Interactive medical image segmentation by statistical seed models. In: SIBGRAPI, São José dos Campos, Brazil (2016)
28. Spina, T.V., Miranda, P.A.V., Falcão, A.X.: Hybrid approaches for interactive image segmentation using the live markers paradigm. IEEE Trans. Image Process. 23(12), 5756–5769 (2014)

29. Valenzuela, W., Ferguson, S.J., Ignasiak, D., Diserens, G., Vermathen, P., Boesch, C., Reyes, M.: Correction tool for active shape model based lumbar muscle segmentation. In: EMBC, pp. 3033–3036. IEEE, August 2015
30. Yang, H.-F., Choe, Y.: An interactive editing framework for electron microscopy image segmentation. In: Bebis, G., Boyle, R., Parvin, B., Koracin, D., Wang, S., Kyungnam, K., Benes, B., Moreland, K., Borst, C., DiVerdi, S., Yi-Jen, C., Ming, J. (eds.) ISVC 2011. LNCS, vol. 6938, pp. 400–409. Springer, Heidelberg (2011). doi:10.1007/978-3-642-24028-7_37

Seed Robustness of Oriented Image Foresting Transform: Core Computation and the Robustness Coefficient

Anderson Carlos Moreira Tavares[(⊠)], Hans Harley Ccacyahuillca Bejar, and Paulo André Vechiatto Miranda

Institute of Mathematics and Statistics, University of São Paulo (USP), São Paulo, SP 05508-090, Brazil
{acmt,hans,pmiranda}@ime.usp.br

Abstract. In graph-based methods, image segmentation can be seen as a graph partition problem between sets of seed pixels. The core of a seed is the region where it can be moved without altering its segmentation. The larger the core, the greater the robustness of the method in relation to its seed positioning. In this work, we present an algorithm to compute the cores of Oriented Image Foresting Transform (OIFT), an extension of Fuzzy Connectedness and Watersheds to directed weighted graphs, and compare its performance to other methods according to a proposed evaluation measure, the Robustness Coefficient. Our analysis indicates that OIFT has a good balance between accuracy and robustness, being able to overcome several methods in some datasets.

Keywords: Oriented Image Foresting Transform · Fuzzy Connectedness · Graph segmentation

1 Introduction

Image segmentation can be interpreted as a graph partition problem subject to hard constraints, given by seed pixels selected in the image domain [2,4,10, 12,14]. A common framework, sometimes referred to as Generalized Graph Cut (GGC) [6,9], can roughly describe, in a unified manner, several seed-based methods, including Random Walker (RW) [14], shortest path/geodesic [13], voronoi diagram and Power Watershed (PW) [6]. In particular, the *min-cut/max-flow* algorithm, also known simply as *Graph Cut* (GC) [4,5], and some methods by the *Image Foresting Transform* (IFT) framework [13], such as *Watersheds* [12] and *Fuzzy Connectedness* [10], correspond to the ε_1- and ε_∞-minimization problems, respectively, within this framework [9]. The ε_∞-minimization methods have linear-time implementations $O(N)$ with respect to the image size N [8], or $O(N \cdot logN)$ depending on the data structure of the priority queue, while the run time for the ε_1-minimization problem is $O(N^{2.5})$ for sparse graphs [5].

Oriented Image Foresting Transform (OIFT) [17,24] and Oriented Relative Fuzzy Connectedness (ORFC) [3] extend the ε_∞-minimization problem to

© Springer International Publishing AG 2017
J. Angulo et al. (Eds.): ISMM 2017, LNCS 10225, pp. 119–130, 2017.
DOI: 10.1007/978-3-319-57240-6_10

directed weighted graphs. OIFT's energy formulation on digraphs makes it a very versatile method, supporting several high-level priors for object segmentation, including global properties such as connectedness [16,19], shape constraints [18,25] and boundary polarity [17,24], which allow the customization of the segmentation to a given target object [15]. While the introduction of combinatorial graphs with directed edges on other frameworks increases considerably the complexity of the problem [27], OIFT and ORFC still run in linear time.

In this work, we present a formal definition and an algorithm to compute the cores of OIFT seeds, adding another unique feature to this method, since for most segmentation algorithms there are no known efficient ways for computing their cores. The *cores* [1] are the regions where seeds can be moved without altering the segmentation. The cores have several practical applications. In medical research, it is usually desirable to reduce inter- and intra-user variability in image segmentation. In this sense, the cores provide an analytic solution to measure the reproducibility of experiments. In this work, we propose the *Robustness Coefficient* to measure the seed robustness of the methods, allowing a quantitative analysis of this aspect.

The cores can also be used to solve an image segmentation inverse problem, that is, for a given segmented object how to find a set with minimum number of seeds that produces it [1]. This can be used to find a suitable set of seeds that assembles a given automatic segmentation result, in order to allow further changes in the seed set to repair the segmentation interactively [21,22].

The cores can also be employed to build powerful hybrid image segmentation approaches [3,11,28]. For instance, Tavares et al. [28] proposed a hybrid method, denoted as $ORFC_{Core} + GC$, which combines, the strengths of ORFC cores (robustness to the seed choice and low false-positive rate) and Graph Cut (smoother and more regular contours, thus, avoiding "leaking though poorly defined boundary segments").

For the sake of completeness in presentation, Sect. 2 includes an overview of concepts on image graph and a revision of the methods ORFC and OIFT. Section 3 shows the proposed algorithm to compute the cores of OIFT seeds, In Sect. 4, we evaluate OIFT and other methods according to their Robustness Coefficient and state our conclusions.

2   Background

A digital *image* is a mapping $I : \mathcal{I} \to \mathbb{Z}$, assigning an intensity $I(s)$ to a pixel s, where $\mathcal{I} \subset \mathbb{Z}^n$ is the image domain. An image can be interpreted as a weighted digraph $G = \langle V, E, \omega \rangle$, whose nodes V are the image pixels ($V = \mathcal{I}$), the arcs are the ordered pixel pairs $\langle s, t \rangle \in E$, and $\omega : E \to \mathbb{Z}$ assigns a weight to each arc. For example, one can take E to consist of all pairs of ordered pixels $\langle s, t \rangle$ in the Cartesian product $\mathcal{I} \times \mathcal{I}$ such that the Euclidean distance $d_{\mathrm{euc}}(s, t) \leq \rho$ and $s \neq t$, where ρ is a specified constant (e.g., 4-neighborhood, when $\rho = 1$, and 8-neighborhood, when $\rho = \sqrt{2}$, in case of 2D images). The transpose $G^T = \langle V, E^T, \omega^T \rangle$ of G is the unique digraph where $E^T = \{\langle t, s \rangle : \langle s, t \rangle \in E\}$ and

$\omega^T(\langle s,t \rangle) = \omega(\langle t,s \rangle)$. In this work, we consider G as a *symmetric* digraph, where $\langle s,t \rangle \in E \Rightarrow \langle t,s \rangle \in E$.

A *path* $\pi_{s \rightsquigarrow t} = \langle s = t_1, t_2, \ldots, t_n = t \rangle$ is a sequence of adjacent and distinct nodes, where s stands for the *origin* and t for the *terminus*. $\Pi_{s \rightsquigarrow t}$ is the set of all paths in G from s to t, $\Pi_t = \bigcup_{s \in V} \Pi_{s \rightsquigarrow t}$ and $\Pi = \bigcup_{t \in V} \Pi_t$. We use $\Pi(G)$ to explicitly indicate all possible paths in a particular graph G. Let also $\pi_{S \rightsquigarrow t} \in \Pi_{S \rightsquigarrow t} = \{\pi_{s \rightsquigarrow t} : s \in \mathcal{S}\}$, for any $\mathcal{S} \subset V$. A path is *trivial* when $\pi_t = \langle t \rangle$. A path $\pi_t = \pi_s \cdot \langle s,t \rangle$ indicates the extension of a path π_s by an arc $\langle s,t \rangle$. A *predecessors map* is a function $P : V \to V \cup \{nil\}$ where $\forall t \in V, P(t) = s$ where $\langle s,t \rangle \in E$ or $s = nil$. A *spanning forest* is a predecessors map without cycles. The roots of the forest are the nodes $R^P = \{r \in V : P(r) = nil\}$. Let π_t^P be defined recursively as $\langle t \rangle$ if $t \in R^P$, and $\pi_s^P \cdot \langle s,t \rangle$ if $P(t) = s$. Let $DCC_G(s) = \{t \in V : \exists \pi_{s \rightsquigarrow t} \in \Pi(G)\}$ be the *Directed Connected Component* of basepoint $s \in V$, the set of all successors of s, and $SCC_G(s) = \{t \in V : \exists \{\pi_{s \rightsquigarrow t}, \pi_{t \rightsquigarrow s}\} \subseteq \Pi(G)\}$ the *Strongly Connected Component* of s, the set of nodes containing s where any two are connected by paths. We can relate DCC and SCC as: $SCC_G(s) = \{t \in V : s \in DCC_G(t) \text{ and } t \in DCC_G(s)\}$. A *connectivity function* $f : \Pi \to \mathbb{Z}$ assigns a value to any path $\pi \in \Pi$. A path π_t is *optimum* if $f(\pi_t) \geq f(\pi_t')$ for any other path $\pi_t' \in \Pi_t$. We denote an optimum path for t as π_t^*. This generates the *connectivity map* $C_{opt} : V \to \mathbb{Z}$ as $C_{opt}(t) = f(\pi_t^*)$.

Let $\mathcal{X} = \{\mathcal{O} : \mathcal{O} \subseteq V\}$ be the space of all possible binary segmented objects $\mathcal{O}$. A seed-based segmentation uses *seeds* $\mathcal{S} = \mathcal{S}_o \cup \mathcal{S}_b \subseteq V$, where $\mathcal{S}_o$ and $\mathcal{S}_b$ are *object* ($\mathcal{S}_o \subseteq \mathcal{O}$) and *background* ($\mathcal{S}_b \subseteq V \setminus \mathcal{O}$) seed sets, respectively. They restrict $\mathcal{X}$ to $\mathcal{X}(\mathcal{S}_o, \mathcal{S}_b) = \{\mathcal{O} \in \mathcal{X} : \mathcal{S}_o \subseteq \mathcal{O} \subseteq V \setminus \mathcal{S}_b\}$. A *cut* is defined as $\mathcal{C}(\mathcal{O}) = \{\langle s,t \rangle \in E : s \in \mathcal{O} \text{ and } t \notin \mathcal{O}\}$. We can associate an energy value $\varepsilon(\mathcal{O})$ to an object (and its cut), and restrict the set of solutions to those which minimizes it. Let energy $\varepsilon_q(\mathcal{O}) = (\sum_{\langle s,t \rangle \in \mathcal{C}(\mathcal{O})} \omega(\langle s,t \rangle)^q)^{\frac{1}{q}}$. The original Graph Cut algorithm minimizes $\varepsilon_1(\mathcal{O})$, while ORFC and OIFT minimize $\varepsilon_\infty(\mathcal{O}) = \max_{\langle s,t \rangle \in \mathcal{C}(\mathcal{O})} \omega(\langle s,t \rangle)$ [9], as presented next.

2.1 Oriented Image Foresting Transform (OIFT)

The *Image Foresting Transform* (IFT) [13] is an algorithm which takes a graph G, a connectivity function f and computes a connectivity map $C : V \to \mathbb{Z}$ defined by a spanning forest P, as $C(t) = f(\pi_t^P)$, which converges to $C_{opt}(t) = f(\pi_t^*), \forall t \in V$, when f is a *smooth* connectivity function [13].

OIFT is a ε_∞-minimization method [17,24] build upon the IFT framework. It uses the connectivity function $f^{\vec{\sigma}}$ (Eq. 1) in a symmetric digraph.

$$f^{\vec{\sigma}}(\langle t \rangle) = \begin{cases} \infty & \text{if } t \in \mathcal{S}_o \cup \mathcal{S}_b \\ -\infty & \text{otherwise} \end{cases}$$

$$f^{\vec{\sigma}}(\pi_{r \rightsquigarrow s} \cdot \langle s,t \rangle) = \begin{cases} \min\{f^{\vec{\sigma}}(\pi_{r \rightsquigarrow s}), \omega(\langle s,t \rangle) \cdot 2\} & \text{if } r \in \mathcal{S}_o \\ \min\{f^{\vec{\sigma}}(\pi_{r \rightsquigarrow s}), \omega(\langle t,s \rangle) \cdot 2 + 1\} & \text{otherwise} \end{cases} \tag{1}$$

The segmented object $A_{OIFT}(\mathcal{S}_o, \mathcal{S}_b)$ by OIFT is defined from the forest P computed by IFT with $f^{\circlearrowright}$, by taking as object all nodes conquered by paths rooted in $\mathcal{S}_o$, that is, $A_{OIFT}(\mathcal{S}_o, \mathcal{S}_b) = \{t \in V : \pi_t^P \in \Pi_{\mathcal{S}_o \leadsto t}\}$. The optimality of $A_{OIFT}(\mathcal{S}_o, \mathcal{S}_b)$ is given by the ε_∞-minimization problem. Although ORFC (which is described in next section) and OIFT are methods from the same energy class, their outputs are usually different with distinct characteristics (Fig. 1).

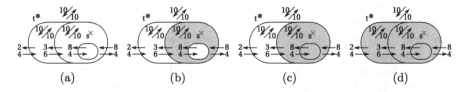

(a)   (b)   (c)   (d)

Fig. 1. (a) Input image graph with $\mathcal{S}_o = \{s\}$ and $\mathcal{S}_b = \{t\}$. (b) ORFC result. (c) A candidate solution. (d) OIFT result. Note that all the three solutions have the same energy $\varepsilon_\infty(\mathcal{O}) = 4$.

In order to explore the boundary orientation/polarity to resolve between very similar nearby boundary segments with opposite transitions (dark to bright/bright to dark), the weight $\omega(\langle s, t \rangle)$ can be defined as:

$$\omega(\langle s, t \rangle) = \begin{cases} \delta(s,t) \times (1 - \alpha) & \text{if } I(s) > I(t) \\ \delta(s,t) \times (1 + \alpha) & \text{if } I(s) < I(t) \\ \delta(s,t) & \text{otherwise} \end{cases} \quad (2)$$

where $\alpha \in [-1, 1]$ is an orientation factor and $\delta(s,t) = \delta(t,s)$ is an undirected similarity measure (e.g., $\delta(s,t) = K - |I(s) - I(t)|$, where K is a maximum intensity variation) [7,23]. Note that we usually have $\omega(\langle s, t \rangle) \neq \omega(\langle t, s \rangle)$ when $\alpha \neq 0$. For $\alpha > 0$, the segmentation by OIFT favors transitions from bright to dark pixels, and $\alpha < 0$ favors the opposite orientation.

2.2   Oriented Relative Fuzzy Connectedness (ORFC)

ORFC [3] is also a ε_∞-minimizer, for arcs from object to background nodes. Let $\varepsilon_\infty^\downarrow(\mathcal{S}_o, \mathcal{S}_b) = \min_{\mathcal{O} \in \mathcal{X}(\mathcal{S}_o, \mathcal{S}_b)} \{\varepsilon_\infty(\mathcal{O})\}$ and $\mathcal{X}_\infty^\downarrow(\mathcal{S}_o, \mathcal{S}_b) = \{\mathcal{O} \in \mathcal{X}(\mathcal{S}_o, \mathcal{S}_b) : \varepsilon_\infty(\mathcal{O}) = \varepsilon_\infty^\downarrow(\mathcal{S}_o, \mathcal{S}_b)\}$. For the seeds $\mathcal{S}_o$ and $\mathcal{S}_b$, A_{ORFC} is defined as follows:

$$A_{ORFC}(\mathcal{S}_o, \mathcal{S}_b) = \left[\bigcup_{s_i \in \mathcal{S}_o} A_{ORFC}(\{s_i\}, \mathcal{S}_b) \right], A_{ORFC}(\{s_i\}, \mathcal{S}_b) = \arg\min_{\mathcal{O} \in \mathcal{X}_\infty^\downarrow(\{s_i\}, \mathcal{S}_b)} |\mathcal{O}| \quad (3)$$

A_{ORFC} uses a connectivity function $f_{\min}^{\leftarrow}$ (Eq. 4), a smooth function which processes reversal (antiparallel) arcs. RFC is a particular case of ORFC when $\alpha = 0$. Algorithm 1 computes the ORFC segmentation in a symmetric digraph, where $C_{opt}(s_i) = \varepsilon_\infty^\downarrow(\{s_i\}, \mathcal{S}_b)$ (see Lemma 1 from Bejar and Miranda [3]).

$$f_{\min}^{\leftarrow}(\langle t \rangle) = \begin{cases} \infty & \text{if } t \in \mathcal{S}_b \\ -\infty & \text{otherwise} \end{cases} \qquad f_{\min}^{\leftarrow}(\pi_{r \leadsto s} \cdot \langle s, t \rangle) = \min\{f_{\min}^{\leftarrow}(\pi_{r \leadsto s}), \omega(\langle t, s \rangle)\}$$

$$(4)$$

Algorithm 1. Computing $A_{ORFC}(\{s_i\}, \mathcal{S}_b)$

---

1 Get connectivity map C_{opt} with $f_{min}^{\leftarrow}$ by IFT;
2 Create $G_> = (V, E', \omega)$ from $G = (V, E, \omega)$ where
  $E' = \{\langle s, t \rangle \in E : \omega(\langle s, t \rangle) > C_{opt}(s_i)\}$;
3 Return $DCC_{G_>}(s_i)$;

---

3 Seed Robustness Analysis of OIFT

Without loss of generality, we will constrain the analysis of robustness only to internal seeds, being the external seeds a completely symmetric problem. In order to define the concept of core, we must introduce the notion of seed equivalence.

Definition 1 *(Equivalent seeds).* *Two internal seeds s_1 and s_2 are said equivalent if they separately produce the same result. That is, for the given external seed set $\mathcal{S}_b$, we have that $A(\{s_1\}, \mathcal{S}_b) = A(\{s_2\}, \mathcal{S}_b)$.*

The notion of equivalent seeds introduced by Definition 1 is a binary relation $\equiv$ on the set of object nodes $A(\mathcal{S}_o, \mathcal{S}_b)$, i.e., $s_1 \equiv s_2$ if and only if s_1 and s_2 are equivalent. This relation is reflexive, symmetric and transitive, hence, it is indeed an *equivalence relation* as defined in mathematics. Therefore, the *core* of a seed s_1 is in fact the *equivalence class* of s_1 under $\equiv$, denoted $[s_1]$, which is defined as $[s_1] = \{t \in A(\{s_1\}, \mathcal{S}_b) : s_1 \equiv t\}$. We use the notation $\mathcal{N}(\{s_1\}, \mathcal{S}_b) = [s_1]$ to indicate the core of s_1 by algorithm A, and we consider $\mathcal{N}(\mathcal{S}_o, \mathcal{S}_b) = \bigcup_{s_i \in \mathcal{S}_o} \mathcal{N}(\{s_i\}, \mathcal{S}_b)$.

Algorithm 2 computes the cores $\mathcal{N}_{ORFC}(\mathcal{S}_o, \mathcal{S}_b)$ of ORFC in linear time, as proposed by Tavares et al. [28], by applying the Tarjan's algorithm in a proper subgraph derived from G, where each core $\mathcal{N}_{ORFC}(\{s_i\}, \mathcal{S}_b)$, $s_i \in \mathcal{S}_o$, corresponds to a SCC.

Algorithm 2. Computing $\mathcal{N}_{ORFC}(\mathcal{S}_o, \mathcal{S}_b)$

---

1 Get connectivity map C_{opt} with $f_{min}^{\leftarrow}$ by IFT;
2 Create $G_> = \langle V, E', \omega \rangle$ from $G = \langle V, E, \omega \rangle$ where
  $E' = \{\langle s, t \rangle \in E : \omega(\langle s, t \rangle) > C_{opt}(s) \wedge C_{opt}(s) = C_{opt}(t)\}$;
3 Apply Tarjan's algorithm in $G_>$;
4 Return only SCCs which contains internal seeds;

From the theoretical relations presented in [28], we know that, for any $s_i \in \mathcal{S}_o$, $\mathcal{N}_{ORFC}(\{s_i\}, \mathcal{S}_b) \subseteq \mathcal{N}_{OIFT}(\{s_i\}, \mathcal{S}_b)$. If a pixel s_1 is equivalent to a pixel s_2 for the OIFT algorithm (i.e., $s_1 \overset{oift}{\equiv} s_2$), and they belong to different ORFC cores (i.e., $\mathcal{N}_{ORFC}(\{s_1\}, \mathcal{S}_b) \neq \mathcal{N}_{ORFC}(\{s_2\}, \mathcal{S}_b)$), then by transitivity we have that $c \overset{oift}{\equiv} d$ for any $c \in \mathcal{N}_{ORFC}(\{s_1\}, \mathcal{S}_b)$ and $d \in \mathcal{N}_{ORFC}(\{s_2\}, \mathcal{S}_b)$. This observation allows us to drastically reduce the complexity of the OIFT core computation, allowing us to work in a Region Adjacency Graph (RAG), composed by the ORFC cores that can be fast computed, rather than working at the pixel level.

Since $\mathcal{N}_{OIFT}(\mathcal{S}_o, \mathcal{S}_b) \subseteq A_{OIFT}(\mathcal{S}_o, \mathcal{S}_b)$, we first compute $A_{OIFT}(\mathcal{S}_o, \mathcal{S}_b)$, then we compute all the ORFC cores inside the OIFT segmentation $A_{OIFT}(\mathcal{S}_o, \mathcal{S}_b)$. Figure 2 illustrates one example, showing all the ORFC cores inside the object for a given image graph (Fig. 2i). Figure 3a shows the resulting RAG, with a node for each ORFC core and one external node x for the background. The arc weights of the RAG are selected as the highest arc values interconnecting their regions.

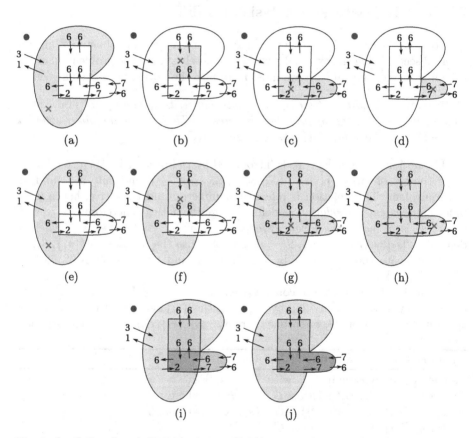

Fig. 2. (a–d) Results of ORFC, where $\omega(\langle s, t \rangle) = 10$ for non-contour edges, with a fixed external seed ● and an internal seed × in different places, (e–h) OIFT results for different internal seeds, (i) ORFC cores and (j) OIFT cores. (Color figure online)

The proposed algorithm to compute the OIFT cores uses a disjoint-set data structure. Initially, each RAG node is its own representative. For each pair $\langle c, d \rangle$, $c \neq x$ and $d \neq x$, of neighboring nodes in the RAG an equivalence test is performed and if the test is satisfied they are joined (union operation). The value $C_c(d) = f(\pi_d^*)$ of an optimum path π_d^* by the connectivity function $f_{\min}^{\leftarrow}$ (Eq. 4) is computed in the induced subgraph $G[V \setminus \{c\}]$ from x to d (Fig. 3b). Similarly we also compute $C_d(c) = f(\pi_c^*)$ as the value of an optimum path π_c^* for $f = f_{\min}^{\leftarrow}$ in the induced subgraph $G[V \setminus \{d\}]$ from x to c (Fig. 3c). If $\omega(\langle c, d \rangle) > C_c(d)$ and $\omega(\langle d, c \rangle) > C_d(c)$ we can conclude that $c \stackrel{\text{oift}}{\equiv} d$ and we perform their union operation (Fig. 3d).

In order to understand the equivalence test performed in RAG, we need to know the following property that distinguishes OIFT from ORFC. From Fig. 1, we can note that in the case of multiple solutions with the same energy, the OIFT result gives preference to boundary pieces with lower energy values. For example, between the border segments with outgoing arcs with values 3 and 2, from Fig. 1c and d, OIFT selects the one with the lowest value in Fig. 1d. This result can be verified theoretically by a proof similar to Theorem 2 (Piecewise optimum property) in [20]. In the equivalence test, $\omega(\langle c, d \rangle)$ and $C_c(d)$ essentially represent the energies of two boundary pieces. Since OIFT gives preference to lower energy values, $\omega(\langle c, d \rangle) > C_c(d)$ implies that a OIFT segmentation from a seed in c would conquer d, and $\omega(\langle d, c \rangle) > C_d(c)$ implies that d would conquer c leading to equivalent seeds. Figure 2j shows the resulting OIFT cores at the pixel level derived from the RAG in Fig. 3d.

Since we have to evaluate the equivalence test, and consequently $C_c(d)$, for all arcs $\langle c, d \rangle$ in the RAG, the final complexity of the algorithm becomes $O(|V|^2 + |E| \cdot |V|)$, where $|V|$ and $|E|$ are the number of nodes and arcs in the RAG. Note that to compute the maps C_c for all $c \in V$ requires $|V|$ IFT's executions and each IFT takes $O(|V| + |E|)$. In practice, the algorithm is fast, because the RAG has a small number of nodes compared to the image graph.

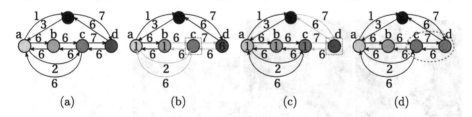

$$\begin{array}{cccc} \text{(a)} & \text{(b)} & \text{(c)} & \text{(d)} \end{array}$$

Fig. 3. (a) Region Adjacency Graph (RAG), composed by the ORFC cores from Fig. 2(b and c). The equivalence test: (b) $\omega(\langle c, d \rangle) = 7 > C_c(d) = 6$, where the values inside the nodes indicate the C_c values, and (c) $\omega(\langle d, c \rangle) = 6 > C_d(c) = 1$, where the values inside the nodes indicate C_d values. (d) The union operation.

4   Experimental Results and Conclusions

Figures 4, 5, 6, 7 and 8 show examples of the incremental computation of the cores by OIFT, from the ORFC cores, for a variety of real images. Next we define a measure to evaluate the robustness of the methods in relation to the seed positioning.

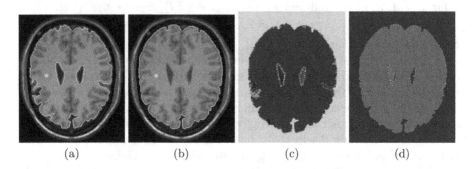

(a)   (b)   (c)   (d)

Fig. 4. A brain image from the BrainWeb - simulated brain database. (a) ORFC segmentation with RC = 99.95%. (b) OIFT segmentation with RC = 96.23%. (c) ORFC cores inside OIFT mask. (d) OIFT cores.

(a)   (b)   (c)   (d)

Fig. 5. Image of a license plate. (a) ORFC segmentation with RC = 97.89%. (b) OIFT segmentation with RC = 89.06%. (c) ORFC cores inside OIFT mask. (d) OIFT cores.

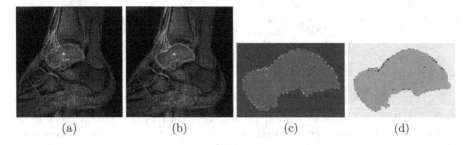

(a)   (b)   (c)   (d)

Fig. 6. MR image of a talus bone with good boundary contrast. (a) ORFC segmentation with RC = 98.60%. (b) OIFT segmentation with RC = 96.01%. (c) ORFC cores inside OIFT mask. (d) OIFT cores.

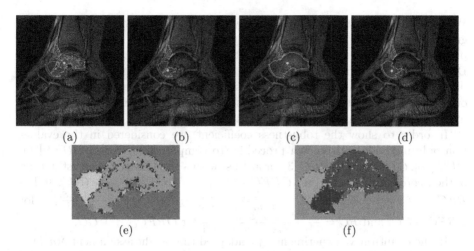

(a)   (b)   (c)   (d)

(e)   (f)

Fig. 7. MR image of a talus bone with poor boundary contrast. (a) ORFC segmentation with RC = 58.87%. (b) Effect of placing the seed outside its core. (c) OIFT segmentation with RC = 52.04%. (d) Effect of placing the seed outside its core. (e) ORFC cores inside the OIFT mask. (f) OIFT cores.

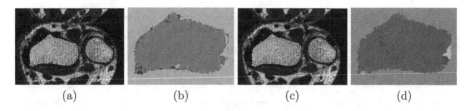

(a)   (b)   (c)   (d)

Fig. 8. An MR image of a wrist with two seed pixels selected inside the bone. (a) ORFC segmentation with RC = 99.17%. (b) OIFT segmentation with RC = 95.40%. (c) ORFC cores inside the OIFT mask. (d) OIFT cores.

For a given segmentation algorithm $A(\mathcal{S}_o, \mathcal{S}_b)$ with cores given by $\mathcal{N}(\mathcal{S}_o, \mathcal{S}_b)$, the *Robustness Coefficient* (RC) is defined as:

$$RC = \frac{|\mathcal{N}(\mathcal{S}_o, \mathcal{S}_b)|}{|A(\mathcal{S}_o, \mathcal{S}_b)|} \tag{5}$$

RC provides an analytic solution to measure the reproducibility of experiments. The higher the RC value, the lower is the sensitivity of the method in relation to inter- and intra-user variability in image segmentation. Note that a high RC value does not imply that the method has a high accuracy, the RC measure only evaluates how easy it is to reproduce the same segmentation, regardless of its accuracy. In this sense, it is a complementary measure to traditional accuracy measures.

In the experiments, we used 40 slice images from real MR images of the foot, to perform the segmentation of the bones talus and calcaneus, and 40 slice images from CT cervical spine studies of 10 subjects to segment the spinal-vertebra.

We used different seed sets automatically obtained by eroding and dilating the ground truth at different radius values. By varying the radius value, we can compute a segmentation for different seed sets and trace accuracy curves, using the Dice coefficient of similarity, and curves of the robustness coefficient. However, in order to generate a more challenging situation, we considered a larger radius of dilation for the external seeds (twice the value of the inner radius), resulting in an asymmetrical arrangement of seeds.

In order to show the robustness coefficient, we considered in the evaluation only methods with known procedure to compute their cores: $IRFC$ [10], RFC [26], $OIFT$ [17], $ORFC$ [3], or at least with a good lower bound estimation of their cores: $RFC + GC$ [11], $ORFC + GC$ [3], and $ORFC_{Core} + GC$ [28]. For $RFC + GC$ we considered $RC = \frac{|\mathcal{N}_{RFC}(\mathcal{S}_o,\mathcal{S}_b)|}{|\mathcal{A}_{RFC+GC}(\mathcal{S}_o,\mathcal{S}_b)|}$, $RC = \frac{|\mathcal{N}_{ORFC}(\mathcal{S}_o,\mathcal{S}_b)|}{|\mathcal{A}_{ORFC+GC}(\mathcal{S}_o,\mathcal{S}_b)|}$ for $ORFC + GC$, and $RC = \frac{|\mathcal{N}_{ORFC}(\mathcal{S}_o,\mathcal{S}_b)|}{|\mathcal{A}_{ORFC_{Core}+GC}(\mathcal{S}_o,\mathcal{S}_b)|}$ for $ORFC_{Core} + GC$.

In the quantitative experiments, we adopted the weight assignment $\delta(a,b) = K - |G(a) + G(b)|$, where $G(a)$ denotes the gradient magnitude of the Sobel operator. For approaches based on directed graphs, we used $\alpha = -0.5$, for the foot bones (transitions from dark to bright pixels) and $\alpha = 0.5$ for the spinal-vertebra; and $\alpha = 0.0$ in the case of undirected approaches.

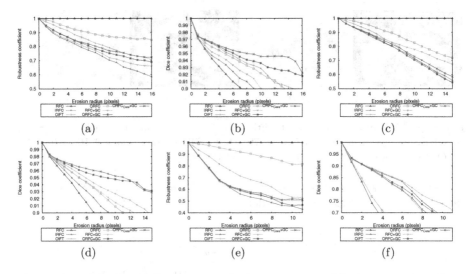

Fig. 9. The mean robustness coefficient curves and the mean accuracy curves (Dice coefficient), using non-equally eroded-dilated seeds, for segmenting: (a–b) talus, (c–d) calcaneus, and (e–f) spinal-vertebra.

Figure 9 shows the experimental results. Note that the robustness coefficient of RFC is always 100%, since $\mathcal{N}_{RFC}(\mathcal{S}_o,\mathcal{S}_b) = \mathcal{A}_{RFC}(\mathcal{S}_o,\mathcal{S}_b)$ [26]. For the bones datasets, with respect to the Dice measure, $OIFT$ is among the first three methods, losing only to the hybrid methods $ORFC + GC$ and $ORFC_{Core} + GC$. However, with respect to the robustness coefficient, $OIFT$ usually gives better results than $ORFC+GC$ and $ORFC_{Core}+GC$, losing only to RFC and $ORFC$. For the spinal-vertebra, the Dice values of all methods decrease rapidly because the object has thin parts and the erosion process rapidly eliminates seeds in several important regions of the object. $OIFT$ has the best Dice values for the spinal-vertebra, and the third best robustness coefficient. So we can conclude that $OIFT$ has a good balance between accuracy and robustness.

As future works, we intend to explore the cores of $OIFT$ to solve the inverse problem of image segmentation, in order to allow the user to interactively perform corrections on any segmentation mask.

Acknowledgements. Thanks to CNPq (308985/2015-0, 486083/2013-6, FINEP 1266/13), FAPESP (2011/50761-2, 2014/12236-1), CAPES, and NAP eScience - PRP - USP for funding, and Dr. J.K. Udupa (MIPG-UPENN) for the images.

References

1. Audigier, R., Lotufo, R.: Seed-relative segmentation robustness of watershed and fuzzy connectedness approaches. In: Proceedings of XX Brazilian Symposium on Computer Graphics and Image Processing (SIBGRAPI), pp. 61–68. IEEE CPS, Belo Horizonte, MG, October 2007
2. Bai, X., Sapiro, G.: Distance cut: interactive segmentation and matting of images and videos. In: Proceedings of IEEE International Conference on Image Processing, vol. 2, pp. II-249–II-252 (2007)
3. Bejar, H.H., Miranda, P.A.: Oriented relative fuzzy connectedness: theory, algorithms, and its applications in hybrid image segmentation methods. EURASIP J. Image Video Process. **2015**, 21 (2015)
4. Boykov, Y., Funka-Lea, G.: Graph cuts and efficient N-D image segmentation. Int. J. Comput. Vis. **70**(2), 109–131 (2006)
5. Boykov, Y., Kolmogorov, V.: An experimental comparison of min-cut/max-flow algorithms for energy minimization in vision. IEEE Trans. Pattern Anal. Mach. Intell. **26**(9), 1124–1137 (2004)
6. Couprie, C., Grady, L., Najman, L., Talbot, H.: Power watersheds: a unifying graph-based optimization framework. IEEE Trans. Pattern Anal. Mach. Intell. **99**(7), 1384–1399 (2010)
7. Ciesielski, K., Udupa, J.: Affinity functions in fuzzy connectedness based image segmentation i: equivalence of affinities. Comput. Vis. Image Underst. **114**(1), 146–154 (2010)
8. Ciesielski, K., Udupa, J., Falcão, A., Miranda, P.: Fuzzy connectedness image segmentation in graph cut formulation: a linear-time algorithm and a comparative analysis. J. Math. Imaging Vis. **44**(3), 375–398 (2012)
9. Ciesielski, K., Udupa, J., Falcão, A., Miranda, P.: A unifying graph-cut image segmentation framework: algorithms it encompasses and equivalences among them. In: Proceedings of SPIE on Medical Imaging: Image Processing, vol. 8314 (2012)
10. Ciesielski, K., Udupa, J., Saha, P., Zhuge, Y.: Iterative relative fuzzy connectedness for multiple objects with multiple seeds. Comput. Vis. Image Underst. **107**(3), 160–182 (2007)
11. Ciesielski, K.C., Miranda, P., Falcão, A., Udupa, J.K.: Joint graph cut and relative fuzzy connectedness image segmentation algorithm. Med. Image Anal. (MEDIA) **17**(8), 1046–1057 (2013)

12. Cousty, J., Bertrand, G., Najman, L., Couprie, M.: Watershed cuts: thinnings, shortest path forests, and topological watersheds. Trans. Pattern Anal. Mach. Intell. **32**, 925–939 (2010)
13. Falcão, A., Stolfi, J., Lotufo, R.: The image foresting transform: theory, algorithms, and applications. IEEE Trans. Pattern Anal. Mach. Intell. **26**(1), 19–29 (2004)
14. Grady, L.: Random walks for image segmentation. IEEE Trans. Pattern Anal. Mach. Intell. **28**(11), 1768–1783 (2006)
15. Lézoray, O., Grady, L.: Image Processing and Analysis with Graphs: Theory and Practice. CRC Press, California (2012)
16. Mansilla, L.A.C., Miranda, P.A.V., Cappabianco, F.A.M.: Oriented image foresting transform segmentation with connectivity constraints. In: IEEE International Conference on Image Processing (ICIP), pp. 2554–2558, September 2016
17. Mansilla, L., Miranda, P.: Image segmentation by oriented image foresting transform: handling ties and colored images. In: 18th International Conference on Digital Signal Processing, Greece, pp. 1–6, July 2013
18. Mansilla, L.A.C., Miranda, P.A.V.: Image segmentation by oriented image foresting transform with geodesic star convexity. In: Wilson, R., Hancock, E., Bors, A., Smith, W. (eds.) CAIP 2013. LNCS, vol. 8047, pp. 572–579. Springer, Heidelberg (2013). doi:10.1007/978-3-642-40261-6_69
19. Mansilla, L.A.C, Miranda, P.A.V: Oriented image foresting transform segmentation: connectivity constraints with adjustable width. In: XXIX Conference on Graphics, Patterns and Images, São José Dos Campos, SP, Brazil, pp. 289–296, October 2016
20. Miranda, P., Falcão, A.: Links between image segmentation based on optimum-path forest and minimum cut in graph. J. Math. Imaging Vis. **35**(2), 128–142 (2009)
21. Miranda, P., Falcão, A., Ruppert, G.: How to complete any segmentation process interactively via image foresting transform. In: 23rd SIBGRAPI: Conference on Graphics, Patterns and Images, pp. 309–316 (2010)
22. Miranda, P., Falcão, A., Ruppert, G., Cappabianco, F.: How to fix any 3D segmentation interactively via image foresting transform and its use in MRI brain segmentation. In: Proceedings of IEEE International Symposium on Biomedical Imaging (ISBI), Chicago, IL, pp. 2031–2035, April 2011
23. Miranda, P., Falcão, A., Udupa, J.: Synergistic arc-weight estimation for interactive image segmentation using graphs. Comput. Vis. Image Underst. **114**(1), 85–99 (2010)
24. Miranda, P., Mansilla, L.: Oriented image foresting transform segmentation by seed competition. IEEE Trans. Image Process. **23**(1), 389–398 (2014)
25. de Moraes Braz, C., Miranda, P.: Image segmentation by image foresting transform with geodesic band constraints. In: 2014 IEEE International Conference on Image Processing (ICIP), pp. 4333–4337, October 2014
26. Saha, P., Udupa, J.: Relative fuzzy connectedness among multiple objects: theory, algorithms, and applications in image segmentation. Comput. Vis. Image Underst. **82**(1), 42–56 (2001)
27. Singaraju, D., Grady, L., Vidal, R.: Interactive image segmentation via minimization of quadratic energies on directed graphs. In: International Conference on Computer Vision and Pattern Recognition, pp. 1–8, June 2008
28. Tavares, A.C.M., Bejar, H.H.C., Miranda, P.A.V.: Seed robustness of oriented relative fuzzy connectedness: core computation and its applications. In: Proceedings of SPIE on Medical Imaging, vol. 10133, Orlando, Florida, US, pp. 1013316–1013316-10, February 2017

Trees and Hierarchies

Evaluation of Combinations of Watershed Hierarchies

Deise Santana Maia[1,2(✉)], Arnaldo de Albuquerque Araujo[2], Jean Cousty[1],
Laurent Najman[1], Benjamin Perret[1], and Hugues Talbot[1]

[1] University Paris-Est, Laboratoire d'Informatique Gaspard-Monge,
A3SI, ESIEE, Noisy-le-Grand, France
deise.santanamaia@esiee.fr
[2] Computer Science Department, Universidade Federal de Minas Gerais,
Belo Horizonte, Brazil

Abstract. The main goal of this paper is to evaluate the potential of
some combinations of watershed hierarchies. We also propose a new combination based on merging level sets of hierarchies. Experiments were
performed on natural image datasets and were based on evaluating the
segmentations extracted from level sets of each hierarchy against the
image ground truths. Our experiments show that most of combinations
studied here are superior to their individual counterparts, which opens
a path for a deeper investigation on combination of hierarchies.

Keywords: Watershed hierarchies · Quasi-flat zones hierarchies ·
Combination of saliency maps

1 Introduction

In this work, we investigate *combinations of watershed hierarchies* through their
saliency maps. Other approaches can be found in the literature, such as [6,14].

Figure 1 provides an example of two different watershed hierarchies, an area-
based one and a dynamics-based one, and their combination, which are built
from successive filterings of an initial watershed segmentation. Hierarchies are
represented thanks to their saliency maps [1,3,4,7,11], *i.e.*, edge-weighted graphs
which comprise information of hierarchical contours. We observe that the sky is
oversegmented at high levels in the first hierarchy and that the beach ground is
oversegmented in the second hierarchy, but this is balanced by their combination.
This observation is general, and no hierarchy is optimal for a whole image. We
expect combinations of hierarchies to perform better, as illustrated for instance
in [4], where the authors have shown that a simple combination of area and
dynamics-based watershed hierarchies produces better visual results than the
individual hierarchies.

This work is partially supported by the Labex Bézout.

J. Angulo et al. (Eds.): ISMM 2017, LNCS 10225, pp. 133–145, 2017.
DOI: 10.1007/978-3-319-57240-6_11

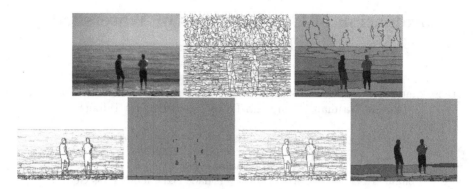

Fig. 1. First row from left to right: original image, saliency map and the 50th highest level set of area based hierarchy. Second row from left to right: saliency map and the 50th highest level sets of dynamics and the combination of area and dynamics based hierarchies.

The main contributions of this paper are:

- Investigation of combinations of hierarchies through their saliency maps using supremum, infimum, linear combination and concatenation functions; and
- Evaluation of supervised and unsupervised combinations of watershed hierarchies.

The plan of the paper is the following. We first review basic notions on hierarchies and saliency maps and present several ways for combining hierarchies (Sect. 2). Then we describe our assessment methodology in Sect. 3. The evaluation results and improvement of combinations compared to the individual watershed hierarchies are described in Sect. 4.

2   Combination of Hierarchies

In this section, we give the notations needed to define combinations of hierarchies through saliency maps and the types of combinations investigated here.

2.1   Hierarchy of Quasi Flat Zones and Saliency Map

This section presents the formal definitions of graphs, hierarchy of partitions, quasi-flat zones hierarchy and saliency map.

A *graph* is a pair $G = (V, E)$, where V is a finite set and E is a set of unordered pairs of distinct elements of V, i.e., $E \subseteq \{\{x, y\} \subseteq V, x \neq y\}$. Each element of V is called a *vertex or a point (of G)*, and each element of E is called an *edge (of G)*.

Let $G = (V, E)$ be a graph. Let X be a subset of V. A sequence $\pi = \langle x_0, \ldots, x_n \rangle$ of elements of X is a *path (in X) from x_0 to x_n* if, for any i in

$\{1, \ldots, n\}$, $\{x_{i-1}, x_i\}$ is an edge of G. The subset X of V is said to be *connected* if for any x and y in X, there exists a path from x to y. A subset X of V is a *connected component* of G if X is connected and, if, for any connected subset Y of V, if $X \subseteq Y$, then we have $X = Y$. The set of connected components of a graph G is denoted by $\mathbf{C}(G)$.

Let V be a set. A *partition* of V is a set $\mathbf{P}$ of non empty disjoint subsets of V whose union is V. If $\mathbf{P}$ is a partition of V, any element of $\mathbf{P}$ is called a *region of* $\mathbf{P}$. Given a graph $G = (V, E)$, the set $\mathbf{C}(G)$ of all connected components of G is a partition of V.

Let V be a set and let $\mathbf{P}_1$ and $\mathbf{P}_2$ be two partitions of V. We say that $\mathbf{P}_1$ is a *refinement* of $\mathbf{P}_2$ if every element of $\mathbf{P}_1$ is included in an element of $\mathbf{P}_2$. A *hierarchy (of partitions)* is a sequence $\mathcal{H} = (\mathbf{P}_0, \ldots, \mathbf{P}_n)$ of partitions of V such that $\mathbf{P}_{i-1}$ is a refinement of $\mathbf{P}_i$, for any i in $\{1, \ldots, n\}$ and such that $\mathbf{P}_n = \{V\}$. A partition of a hierarchy $\mathcal{H}$ is called a *level set* of the hierarchy.

Let G be a graph, if w is a map from the edge set of G to the set $\mathbb{R}^+$ of positive real numbers, then the pair (G, w) is called an *(edge-)weighted graph*. If (G, w) is an edge-weighted graph, for any edge u of G, the value $w(u)$ is called the *weight of u (for w)*.

Important Notation. In the sequel of this article, we consider a weighted graph (G, w). We assume that the vertex set of G is connected. We also denote by $\mathbb{W}$ the range of w, *i.e.*, the set $\{w(u) \mid u \in E\}$ and by $\mathbb{W}^\bullet$ the set $\mathbb{W} \cup \{k+1\}$, where k is the greatest value of $\mathbb{W}$.

Let λ be any element in $\mathbb{R}$. We denote by G_λ the graph (V, E_λ) such that $E_\lambda = \{e \in E \mid w(e) < \lambda\}$. The set $\mathbf{C}(G_\lambda)$ of all connected components of G_λ is called the λ-*level partition* of G. The sequence

$$\mathcal{QFZ}(w) = (\mathbf{C}(G_\lambda) \mid \lambda \in \mathbb{W}^\bullet) \tag{1}$$

is a hierarchy called the *Quasi-Flat Zones hierarchy of w*.

The *saliency map* of a hierarchy $\mathcal{H} = (\mathbf{P}_0, \ldots, \mathbf{P}_n)$ is a map from E to $\{0, \ldots, n\}$, denoted by $\Phi(\mathcal{H})$, such that, for any edge $e = \{x, y\}$ in E, we have $\Phi(\mathcal{H})(e)$ is the greatest value i in $\{0, \ldots, n\}$ such that x and y do not belong to the same region of $\mathbf{P}_i$.

In [4], the authors provide a bijection between saliency maps and hierarchies based on quasi-flat zones hierarchies. Hence, a hierarchy is equivalently represented by its saliency map, a property that is particularly useful in the remaining part of this article.

Note also that, for visualization purposes, when the graph G is associated to a digital image [3,4,7,11], saliency maps can be visualized with images, called *ultrametric contour maps* in [1], in which the contours brightness is proportional to their saliency values.

2.2 Generic Scheme of Combination of Hierarchies

Combining partitions and, *a fortiori*, hierarchies is not straightforward. This problem has been tackled in [2,4,7] thanks to the use of saliency maps and we follow the same approach as used in those papers. More precisely, in order to combine two hierarchies $\mathcal{H}_1$ and $\mathcal{H}_2$, built from the same fine partition, we proceed in three steps [2,4,7]: first the saliency maps of $\mathcal{H}_1$ and $\mathcal{H}_2$ are considered, then the two saliency maps are combined to obtain new weights on the edges of G, and, finally, the combination of hierarchies is the quasi-flat zones hierarchy of the new weight function.

Let $\mathcal{F}$ be the set of all maps from E into $\mathbb{R}^+$. Let n be any positive integer, any map c from $\mathcal{F}^n$ into $\mathcal{F}$ is called a *combining n-weight function*.

Given a sequence of hierarchies $(\mathcal{H}_1, \ldots, \mathcal{H}_n)$ and a combining n-weight function c, the *combinations of* $(\mathcal{H}_1, \ldots, \mathcal{H}_n)$ *by* c is the hierarchy $\mathcal{H}_c(\mathcal{H}_1, \ldots, \mathcal{H}_n)$ defined by:

$$\mathcal{H}_c(\mathcal{H}_1, \ldots, \mathcal{H}_n) = \mathcal{QFZ}(c(\Phi(\mathcal{H}_1), \ldots, \Phi(\mathcal{H}_n))). \tag{2}$$

2.3 Combining n-weight Functions

We consider three classical functions in the instantiation of the combining n-weight function (supremum, infimum and linear combination) and we propose a new type of combination called *concatenation of hierarchies*.

The supremum, infimum and linear combination functions are respectively denoted by $\curlyvee$, $\curlywedge$ and $\boxplus_\Theta$. Given a sequence $(w_1, \ldots, w_n)$ of n saliency maps and a sequence $\Theta = (\alpha_1, \ldots, \alpha_{n-1})$ of $n-1$ values in $\mathbb{R}$ such that $(\alpha_1 + \cdots + \alpha_{n-1}) \leq 1$ and $\alpha_i \geq 0$ for $i \in \{1, \ldots, n-1\}$, the linear combination of $(w_1, \ldots, w_n)$ parametrized by Θ is the sum $\alpha_1 w_1 + \cdots + \alpha_{n-1} w_{n-1} + (1 - \alpha_1 - \cdots - \alpha_{n-1}) w_n$. We denote by A the case where the linear combination is equal to the average. One example of combination of hierarchies by infimum is shown in Fig. 3.

The concatenation of hierarchies is based on merging different level sets of each hierarchy. This type of combination can be useful, for example, when one hierarchy $\mathcal{H}_1$ succeeds at describing the small details of an image at lower level sets, but fails at filtering the small regions to capture the main large objects at higher level sets. Therefore, it can be interesting to concatenate $\mathcal{H}_1$ with another hierarchy $\mathcal{H}_2$ whose high level sets describe well the important regions in the image. This general idea is represented in Fig. 2.

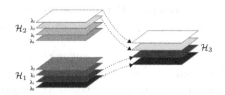

Fig. 2. Concatenation of low levels of $\mathcal{H}_1$ with high levels of $\mathcal{H}_2$.

Given two weight maps w_1 and w_2 and a threshold value λ, the *concatenation of w_1 and w_2* consists in: (i) setting to zero all weights of w_2 lower than λ; (ii) setting to λ all weights of w_1 greater than λ; and (iii) computing the supremum of the two maps obtained at steps (i) and (ii). More generally, given a sequence $(w_1, \ldots, w_n)$ of n weight maps and a series $(\lambda_1, \ldots, \lambda_{n-1})$ of $n - 1$ threshold values in $\mathbb{R}$ such that $\lambda_1 < \lambda_2 < \cdots < \lambda_{n-1}$, we define the *concatenation of* $(w_1, \ldots, w_n)$ *parametrized by* $(\lambda_1, \ldots, \lambda_{n-1})$, thanks to the combining n-weight function $\uplus_\Theta$, by:

$$\forall e \in E, \uplus_\Theta(w_1, \ldots, w_n)(e) = max\{T(w_1(e), 0, \lambda_1), \ldots, T(w_n(e), \lambda_{n-1}, \infty)\} \ (3)$$

where, given a, b and $c \in \mathbb{R}$, we have $T(a, b, c)$ equals 0 if a is lower than b and equals $min(a, c)$ otherwise. Consequently, given a sequence of hierarchies $(\mathcal{H}_1, \ldots, \mathcal{H}_n)$ and threshold values $\Theta = (\lambda_1, \ldots, \lambda_{n-1})$, *the concatenation of* $(\mathcal{H}_1, \ldots, \mathcal{H}_n)$ *with parameter* Θ is $\mathcal{H}_{\uplus_\Theta}(\mathcal{H}_1, \ldots, \mathcal{H}_n)$. One example of concatenation of two hierarchies is shown in Fig. 3.

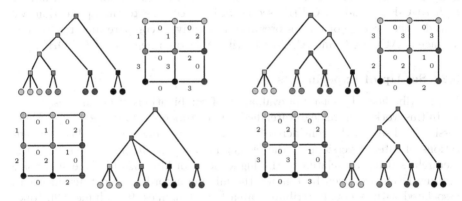

Fig. 3. Illustration of combination by infimum and concatenation of a pair of hierarchies. First row from left to right: $\mathcal{H}_1$, $\Phi(\mathcal{H}_1)$, $\mathcal{H}_2$ and $\Phi(\mathcal{H}_2)$. Second row from left to right: $\curlywedge(\Phi(\mathcal{H}_1), \Phi(\mathcal{H}_2))$, $\mathcal{QFZ}(\curlywedge(\Phi(\mathcal{H}_1), \Phi(\mathcal{H}_2)))$, $\uplus_{(2)}(\Phi(\mathcal{H}_1), \Phi(\mathcal{H}_2))$ and $\mathcal{QFZ}(\uplus_{(2)}(\Phi(\mathcal{H}_1), \Phi(\mathcal{H}_2)))$

3   Assessment Methodology and Set-Up of Experiments

In this section, we present the assessment methodology used to evaluate hierarchies of segmentations.

3.1   Assessment Methodology

In order to account for the performance of the different combinations, we use a supervised assessment strategy developed in [12]. This framework evaluates the possibility of extracting a *good* segmentation from a hierarchy with respect to a given ground-truth, the quality of the extracted segmentation being measured using the *Bidirectional Consistency Error* (BCE) [8]. In order to take account for the hierarchical aspect of the representations, the score of a segmentation is measured against its level of fragmentation, *i.e.*, the ratio between the number of regions in the proposal segmentation compared to the number of regions in the ground-truth segmentation.

Two ways of extracting segmentations from a hierarchy are considered. (1) We compute the cut that maximizes the BCE score for each fragmentation level, leading to the Fragmentation-Optimal Cut score curve (FOC). (2) We compute the BCE score of each level set of the hierarchy, leading to the Fragmentation-Horizontal Cut score curve (FHC). A large difference between the FOC and FHC curves, called here fragmentation curves, suggests that the optimization algorithm has selected regions from various levels of the hierarchy to find the optimal cut: the regions of the ground-truth segmentations are thus spread at different levels in the hierarchy. The normalized area under those curves, denoted respectively by AUC-FOC and AUC-FHC, provides an overall performance summary over a large range of fragmentation levels. Since the importance of having high AUC-FOC and AUC-FHC scores varies according to the application, we consider the average of both scores to quantitatively compare hierarchies. The average of AUC-FOC and AUC-FHC will be denoted here by AUC-FOHC.

3.2 Set-Up of Experiments

We describe here the set-up of evaluation of combinations of hierarchies.

In this work, hierarchical watersheds are considered. More precisely, the successive level sets of the considered hierarchies correspond to watershed segmentations of filtered versions of the weight map w, the higher level sets of the hierarchies being associated to the higher level of filterings [3]. The successive filtering levels are given by ranking the minima according to extinction values associated with regional attributes: area [10], dynamics [9], volume [10], topological height [15], number of minima, number of descendants and diagonal of bounding box [15]. To shorten the notations, we denote those attributes by Area, Dyn, Vol, Height, Min, Desc and DBB.

The evaluations were performed on the 200 test images of the Berkeley Segmentation Dataset and Benchmark 500 (BSDS500) [1].

The hierarchies of segmentation are computed from the image gradients obtained from the Structured Edge (SE) detector [5], which achieved a high contour detection rate on BSDS500.

4 Assessment of Combinations of Hierarchies

In this section we present the experiments with combinations of watershed hierarchies. We then compare the combinations of hierarchies with the individual hierarchies and other two techniques [1,14].

4.1 Baseline

Our baseline is the AUC-FOHC scores of individual watershed hierarchies presented in Table 1. The scores were computed over the test set of BSDS500.

Table 1. AUC-FOC, AUC-FHC and AUC-FOHC scores of individual hierarchies computed over the test set of BSDS500.

	Area	DBB	Dyn	Height	Desc	Min	Vol
AUC-FOC	0.603	0.592	0.541	0.560	0.604	0.609	0.617
AUC-FHC	0.423	0.435	0.480	0.493	0.425	0.453	0.465
AUC-FOHC	0.513	0.514	0.510	0.527	0.514	0.531	0.541

4.2   Evaluation of Parameter-Free Combinations

The evaluation of parameter-free combinations consisted in computing the AUC-FOHC scores of all combinations of pairs of watershed hierarchies over the test set of BSDS500 using Υ, λ and A functions. For each pair of watershed hierarchies, $\mathcal{H}_1$ and $\mathcal{H}_2$, we applied the following combining n-weight functions to their saliency maps: $\Upsilon(\Phi(\mathcal{H}_1), \Phi(\mathcal{H}_2))$, $\lambda(\Phi(\mathcal{H}_1), \Phi(\mathcal{H}_2))$ and $A(\Phi(\mathcal{H}_1), \Phi(\mathcal{H}_2))$.

The highest scores achieved by parameter-free combinations is shown in Table 2. We can observe that, for most pairs of watershed hierarchies, the combination with A presents the highest score. In addition, the highest scores are obtained with combinations using A.

Table 2. Combining n-weight functions and highest AUC-FOHC scores obtained from $c(\Phi(\mathcal{H}_1), \Phi(\mathcal{H}_2))$. For each pair of hierarchies, we have the global combination function which provided the highest AUC-FOHC score and the score obtained from this combination.

$\mathcal{H}_1$	$\mathcal{H}_2$						
	Area	DBB	Dyn	Height	Desc	Min	Vol
Area	-	Υ	A	A	Υ	A	λ
	0.513	0.515	0.566	0.567	0.515	0.529	0.529
DBB		-	A	A	Υ	A	Υ
		0.514	0.566	0.568	0.516	0.526	0.529
Dyn			-	λ	A	A	Υ
			0.510	0.522	0.567	0.563	0.551
Height				-	A	A	A
				0.527	0.568	0.563	0.554
Desc					-	A	λ
					0.514	0.530	0.529
Min						-	Υ
						0.531	0.540
Vol							-
							0.541

4.3   Evaluation of Unsupervised Concatenation of Hierarchies

We present here the evaluation of concatenation of pairs of watershed hierarchies.

To determine the parameter that should be used in the concatenation of each pair of watershed hierarchies, we analyze their fragmentation curves. For each pair of watershed hierarchies, we check which one presents the highest AUC-FOC and AUC-FHC scores for low and high fragmented segmentations. If one of the hierarchies presents the highest scores for both low and high fragmented segmentations, we do not expect to obtain better results from their concatenation. For example, in the curves of Fig. 4, we see that, only for a fragmentation larger than 0 and smaller than approx. 0.65, area outperforms dynamics based watershed hierarchy. Therefore, we conclude that high level sets of area, which are less fragmented, describe an image better than dynamics based hierarchies, and the opposite is true for lower level sets. Hence, the parameters are tuned to concatenate high levels of area to the low levels of dynamics based hierarchy.

In general, the fragmentation curves of concatenations has a smaller difference than the curves of their individual counterparts, which can be seen in Fig. 4. This means that the segmentations extracted from the level sets of concatenations are closer to the optimal cuts for each fragmentation level. Also, half of the concatenations tested here presented higher AUC-FOHC scores than the individual watershed hierarchies, as shown in Table 3.

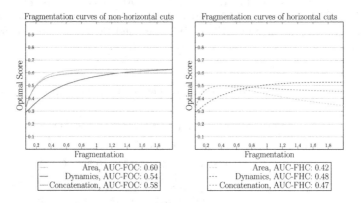

Fig. 4. Fragmentation curves of non-horizontal and horizontal cuts of the concatenation of area and dynamics based watershed hierarchies. The 10th highest levels of area were concatenated to the lower level sets of the dynamics based hierarchy.

4.4   Evaluation of Supervised Linear Combinations

We present here the evaluation of linear combinations of pairs of watershed hierarchies using learned parameters.

For each pair of watershed hierarchies, we determined the linear combination parameter α that optimizes the AUC-FOHC score on the 300 images of the training set of BSDS500.

Table 3. AUC-FOC, AUC-FHC and AUC-FHCO scores of $\uplus_\Theta(\Phi(\mathcal{H}_1), \Phi(\mathcal{H}_2))$, where different values of Θ were used for each concatenation. The AUC-FHCO scores in bold are the ones which are higher than the AUC-FHCO scores of individual $\mathcal{H}_1$ and $\mathcal{H}_2$ hierarchies.

$\mathcal{H}_2$ / $\mathcal{H}_1$	Dynamics					Height				
	Area	DBB	Desc	Min	Vol	Area	DBB	Desc	Min	Vol
AUC-FOC	0.579	0.561	0.586	0.589	0.591	0.579	0.574	0.580	0.582	0.585
AUC-FHC	0.472	0.462	0.462	0.483	0.498	0.472	0.475	0.473	0.485	0.500
AUC-FHCO	**0.525**	0.511	0.526	**0.536**	**0.545**	0.525	0.524	0.527	**0.534**	**0.542**

From the highest scores reached for each combination, we can see that not all combinations produce relevant results, mainly the ones which do not include either dynamics nor topological height based watershed hierarchies. This is shown in Table 4, that contains the best-fitting parameter for each linear combination and their AUC-FOHC scores.

Table 4. Parameters α and AUC-FOHC scores of each linear combination $\boxplus_{(\alpha)}(\Phi(\mathcal{H}_1), \Phi(\mathcal{H}_2))$. The AUC-FOHC scores in bold are the highest scores achieved with linear combination of hierarchies.

$\mathcal{H}_1$	$\mathcal{H}_2$						
	Area	DBB	Dyn	Height	Desc	Min	Vol
Area	-	$\alpha = 0.92$	$\alpha = 0.60$	$\alpha = 0.51$	$\alpha = 0$	$\alpha = 0.11$	$\alpha = 0$
	0.513	0.512	0.568	**0.569**	0.514	0.531	0.541
DBB		-	$\alpha = 0.43$	$\alpha = 0.35$	$\alpha = 0.19$	$\alpha = 0.07$	$\alpha = 0.02$
		0.514	0.566	0.566	0.512	0.531	0.541
Dyn			-	$\alpha = 0.03$	$\alpha = 0.38$	$\alpha = 0.51$	$\alpha = 0.24$
			0.510	0.527	**0.569**	0.564	0.558
Height				-	$\alpha = 0.42$	$\alpha = 0.51$	$\alpha = 0.36$
				0.527	**0.569**	0.560	0.560
Desc					-	$\alpha = 0.25$	$\alpha = 0$
					0.514	0.530	0.541
Min						-	$\alpha = 0.12$
						0.531	0.542
Vol							-
							0.541

One example comparing segmentations extracted from the individual hierarchies based on number of descendants and topological height and their linear combination using the learned parameters is shown in Fig. 5. The linear combination computed for this single image presents a higher AUC-FHC score than

the individual hierarchies (0.604 *versus* 0.465 and 0.551) and a slightly higher
AUC-FOC score (0.802 *versus* 0.801 and 0.708). Based on the AUC-FHC score,
we expect this combination to have better horizontal cuts than the the individual
hierarchies. We can see that the segmentation extracted from the combination
separates better the main regions in this image: sky, mountains and the two sea
regions.

Fig. 5. From left to right: original image, saliency map and the 5th highest level set of
three hierarchies: number of descendants and topological height based hierarchies, and
their linear combination using learned parameters.

4.5   Comparison with Other Techniques

In order to have a more complete evaluation, we have performed a comparison with a state-of-the-art method [14], called Multiscale Combinatorial
Grouping (MCG), and an well-know technique [1], named Ultrametric Contour
Map (UCM), including different assessment measures: Precision-Recall (PR) for
boundaries [1] and the Marked Segmentation [13] (Fig. 6).

The PR for boundaries score is also assessed on BSDS500 and it evaluates the
matching between the boundaries of a given segmentation and the ground-truth
segmentation. The PR curves are built from the F-measure scores of the level
sets of the hierarchies and are summed up in two measures: Optimal Dataset
Scale (ODS) and Optimal Image Scale (OIS).

The Marked Segmentation aims to measure the difficulty of extracting a good
segmentation in a hierarchy given sets of background and foreground markers.
The markers are generated through erosion, skeletonization and frame of the
ground truth, and the score of each segmentation is given by the F-Measure
computed over the Weizmann and Grabcut datasets. Each pair of Background-
Foreground markers are denoted by the method used to compute them. Figure 6
shows the Marked Segmentation results for three pairs of background and foreground markers: Er-Er, Fr-Sk and Sk-Sk, in which Er, Fr and Sk stand for
Erosion, Frame and Skeleton, respectively. The box plots show the quartile distribution of scores on both datasets. The median score of those three pairs of
markers is denoted by ODM. Therefore, the best hierarchies in terms of Marked
Segmentation correspond to the ones with highest ODM scores and most compressed box plots.

MCG explores hierarchies of images at different resolutions based on combined local contours cues. Single segmentations at different resolutions are

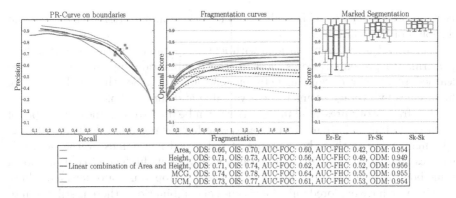

—	Area, ODS: 0.66, OIS: 0.70, AUC-FOC: 0.60, AUC-FHC: 0.42, ODM: 0.954
—	Height, ODS: 0.71, OIS: 0.73, AUC-FOC: 0.56, AUC-FHC: 0.49, ODM: 0.949
— Linear combination of Area and Height	ODS: 0.71, OIS: 0.74, AUC-FOC: 0.62, AUC-FHC: 0.52, ODM: 0.956
—	MCG, ODS: 0.74, OIS: 0.78, AUC-FOC: 0.64, AUC-FHC: 0.55, ODM: 0.955
—	UCM, ODS: 0.73, OIS: 0.77, AUC-FOC: 0.61, AUC-FHC: 0.53, ODM: 0.954

Fig. 6. Comparison of our linear combination of area and topological height with MCG and UCM: Precision-recall (PR) for boundaries, Fragmentation curves of non-horizontal cuts (plain curve) and horizontal cuts (dashed curve), and Marked Segmentation of three pairs of markers.

aligned and combined into a single saliency map. The UCM described in [1] is obtained from the watershed transform of the output of a high quality contour detector.

Our best combination does not achieve the PR and fragmentation scores presented by MCG and UCM, but it outperforms UCM in terms of Fragmentation Curves for non-horizontal cuts and presents competitive Marked Segmentation results compared to MCG and UCM.

5 Discussion and Conclusion

This paper shows the potential of combination of hierarchies through the evaluation of supervised and unsupervised combinations of watershed hierarchies. We evaluated combinations of pairs of watershed hierarchies using classical functions (infimum, supremum and linear combination) and a newly proposed method called concatenation of hierarchies.

All combinations described here were proved to be useful for at least one pair of watershed hierarchies. However, not all pairs of hierarchies presented significant results when compared to the individual hierarchies. For example, the parameter-free combination of area and volume based hierarchies with highest score did not present a score superior to the individual volume based hierarchy. Also, the best-fitting parameter α for the linear combination of area and volume based hierarchies is equal to zero. This means that none of the combinations of area and volume tested here was superior to both individual hierarchies.

We observed that the combinations with highest scores contained either dynamics or topological height based hierarchies, but not both. Both dynamics and topological height are related to *depths*. Therefore, this is not surprising that the combination of hierarchies based on those attributes does not bring new interesting results. This is also valid for other similar attributes as area and diagonal of bounding box.

Among all linear combinations with learned parameters, the ones with best performance presented scores close to the combination by average, with the learned parameter α ranging between 0.39 and 0.51. So, the combination of hierarchies by average seems to be a valuable and simple choice.

The framework presented in this article can be used to combine other types of hierarchies, but we have not investigated whether the results could be interesting. The main point is that the contours of watershed hierarchies overlap and this ensures that none of their combinations will present duplicated contours for a same boundary of the input image, which could happen using other hierarchies.

Since we have explored only a few types of combinations, there is still room to find new functions able produce relevant results. The evaluation of combinations performed here also invites us to go a step further in this topic, for example, by learning how to choose the optimal combination parameters for each image.

References

1. Arbelaez, P., Maire, M., Fowlkes, C., Malik, J.: Contour detection and hierarchical image segmentation. IEEE PAMI **33**(5), 898–916 (2011)
2. Cousty, J., Najman, L., Serra, J.: Raising in watershed lattices. In: 2008 15th IEEE ICIP, pp. 2196–2199 (2008)
3. Cousty, J., Najman, L.: Incremental algorithm for hierarchical minimum spanning forests and saliency of watershed cuts. In: Soille, P., Pesaresi, M., Ouzounis, G.K. (eds.) ISMM 2011. LNCS, vol. 6671, pp. 272–283. Springer, Heidelberg (2011). doi:10.1007/978-3-642-21569-8_24
4. Cousty, J., Najman, L., Kenmochi, Y., Guimarães, S.J.F.: Hierarchical segmentations with graphs: quasi-flat zones, minimum spanning trees, and saliency maps. Technical report (2016). https://hal.archives-ouvertes.fr/hal-01344727/document
5. Dollár, P., Zitnick, C.L.: Structured forests for fast edge detection. In: Proceedings of IEEE ICCV, pp. 1841–1848 (2013)
6. Fehri, A., Velasco-Forero, S., Meyer, F.: Automatic selection of stochastic watershed hierarchies. In: EUSIPCO, pp. 1877–1881. IEEE (2016)
7. Kiran, B.R., Serra, J.: Fusion of ground truths and hierarchies of segmentations. PRL **47**, 63–71 (2014)
8. Martin, D.R.: An empirical approach to grouping and segmentaqtion. Ph.D. thesis, University of California (2003)
9. Meyer, F.: The dynamics of minima and contours. In: Maragos, P., Schafer, R., Butt, M. (eds.) Mathematical Morphology and Its Applications to Image and Signal Processing, pp. 329–336. Kluwer, Boston (1996)
10. Meyer, F., Vachier, C., Oliveras, A., Salembier, P.: Morphological tools for segmentation: connected filters and watersheds. Annales des télécommunications **52**, 367–379 (1997). Springer
11. Najman, L., Schmitt, M.: Geodesic saliency of watershed contours and hierarchical segmentation. IEEE PAMI **18**(12), 1163–1173 (1996)
12. Perret, B., Cousty, J., Guimaraes, S.J.F., Maia, D.S.: Evaluation of hierarchies of watersheds for natural image analysis. Technical report (2016). https://hal.archives-ouvertes.fr/hal-01430865/document

13. Perret, B., Cousty, J., Ura, J.C.R., Guimarães, S.J.F.: Evaluation of morphological hierarchies for supervised segmentation. In: Benediktsson, J.A., Chanussot, J., Najman, L., Talbot, H. (eds.) ISMM 2015. LNCS, vol. 9082, pp. 39–50. Springer, Cham (2015). doi:10.1007/978-3-319-18720-4_4

14. Pont-Tuset, J., Arbelez, P., Barron, J.T., Marques, F., Malik, J.: Multiscale combinatorial grouping for image segmentation and object proposal generation. IEEE PAMI **39**(1), 128–140 (2017)

15. Silva, A.G., de Alencar Lotufo, R.: New extinction values from efficient construction and analysis of extended attribute component tree. In: SIBGRAPI, pp. 204–211. IEEE (2008)

Prior-Based Hierarchical Segmentation Highlighting Structures of Interest

Amin Fehri$^{(\boxtimes)}$, Santiago Velasco-Forero, and Fernand Meyer

Center for Mathematical Morphology, MINES ParisTech,
PSL Research University, Paris, France
{amin.fehri,santiago.velasco,fernand.meyer}@mines-paristech.fr

Abstract. Image segmentation is the process of partitioning an image into a set of meaningful regions according to some criteria. Hierarchical segmentation has emerged as a major trend in this regard as it favors the emergence of important regions at different scales. On the other hand, many methods allow us to have prior information on the position of structures of interest in the images. In this paper, we present a versatile hierarchical segmentation method that takes into account any prior spatial information and outputs a hierarchical segmentation that emphasizes the contours or regions of interest while preserving the important structures in the image. Several applications are presented that illustrate the method versatility and efficiency.

Keywords: Mathematical Morphology · Hierarchies · Segmentation · Prior-based segmentation · Stochastic watershed

1   Introduction

In this paper, we propose a method to take advantage of any prior spatial information previously obtained on an image to get a hierarchical segmentation of this image that emphasizes its regions of interest, allowing us to get more details in the designated regions of interest of an image while still preserving its strong structural information.

Potential applications are numerous. When having a limited storage capacity (for very large images for example), this would allow us to keep details in the regions of interest as a priority. Similarly, in situations of transmission with limited bandwidth, one could first transmit the important information of the image: the details of the face for a video-call, the pitch and the players for a soccer game and so on. One could also use such a tool as a preprocessing one, for example to focus on an individual from one camera view to the next one in video surveillance tasks. Finally, from an artistic point of view, the result is interesting and similar to a combination of focus and cartoon effects. Some of these examples are illustrated in this paper.

Image segmentation has been shown to be inherently a multi-scale problem [10]. That is why hierarchical segmentation has become the major trend in image

© Springer International Publishing AG 2017
J. Angulo et al. (Eds.): ISMM 2017, LNCS 10225, pp. 146–158, 2017.
DOI: 10.1007/978-3-319-57240-6_12

segmentation and most top-performance segmentation techniques [3,17,22,23] fall into this category: hierarchical segmentation does not output a single partition of the image pixels into sets but instead a single multi-scale structure that aims at capturing relevant objects at all scales. Researches on this topic are still vivid as differential area profiles [21], robust segmentation of high-dimensional data [9] as well as theoretical aspects regarding the concept of partition lattice [24,26] and optimal partition in a hierarchy [11,12,31]. Our goal in this paper is to develop a hierarchical segmentation algorithm that focuses on certain predetermined zones of the image. The hierarchical aspect also allows us, for tasks previously described, to very simply tune the level of details wanted depending on the application.

Furthermore, our algorithm is very versatile, as the spatial prior information that it uses can be obtained by any of the numerous learning-based approaches proposed over the last decades to roughly localize objects [13,20,25]. In this regard, our work joins an important research point that consists in designing approaches to incorporate prior knowledge in the segmentation, as shape prior on level sets [4], star-shape prior by graph-cut [29], use of a shape prior hierarchical characterization obtained with deep learning [5], or related work making use of stochastic watershed to perform targeted image segmentation [16].

The remainder of the paper is organized as follows. Section 2 explains how we construct and use graph-based hierarchical segmentation. Then Sect. 3 specifies how we use prior information on the image to obtain hierarchies with regionalized fineness. Several examples of applications of this method are described in Sect. 4. Finally, conclusions and perspectives are presented in Sect. 5.

2   Hierarchies and Partitions

2.1   Graph-Based Hierarchical Segmentation

Obtaining a suitable segmentation directly from an image is very difficult. This is why it is often make use of hierarchies to organize and propose interesting contours by valuating them. In this section, we remind the reader how to construct and use graph-based hierarchical segmentation.

For each image, let us suppose that a fine partition is produced by an initial segmentation (for instance a set of superpixels [1,15] or the basins produced by a classical watershed algorithm [18]) and contains all contours making sense in the image. We define a dissimilarity measure between adjacent tiles of this fine partition. One can then see the image as a graph, the *region adjacency graph* (RAG), in which each node represents a tile of the partition; an edge links two nodes if the corresponding regions are neighbors in the image; the weight of the edge is equal to the dissimilarity between these regions. Working on the RAG is much more efficient than working on the image, as there are far less nodes in the RAG than there are pixels in the image.

Formally, we denote this graph $\mathcal{G} = (\mathcal{V}, \mathcal{E}, \mathbf{W})$, where $\mathcal{V}$ corresponds to the image domain or set of pixels/fine regions, $\mathcal{E} \subset \mathcal{V} \times \mathcal{V}$ is the set of edges linking neighbour regions, $\mathbf{W} : \mathcal{E} \to \mathbb{R}^+$ is the dissimilarity measure usually based on local gradient information (or color or texture), for instance $\mathbf{W}(i,j) \propto |\mathbf{I}(v_i) - \mathbf{I}(v_j)|$ with $\mathbf{I} : \mathcal{V} \to \mathbb{R}$ representing the image intensity.

The edge linking the nodes p and q is designated by e_{pq}. A path is a sequence of nodes and edges: for example the path linking the nodes p and s is the set $\{p, e_{pt}, t, e_{ts}, s\}$. A *connected subgraph* is a subgraph where each pair of nodes is connected by a path. A *cycle* is a path whose extremities coincide. A *tree* is a connected graph without cycle. A *spanning tree* is a tree containing all nodes. A *minimum spanning tree* (MST) $\mathcal{MST}(\mathcal{G})$ of a graph $\mathcal{G}$ is a spanning tree with minimal possible weight, obtained for example using the Boruvka algorithm (the weight of a tree being equal to the sum of the weights of its edges). A *forest* is a collection of trees.

A *partition* π of a set $\mathcal{V}$ is a collection of subsets of $\mathcal{V}$, such that the whole set $\mathcal{V}$ is the disjoint union of the subsets in the partition, i.e., $\pi = \{R_1, R_2, \ldots, R_k\}$, such that $\forall i, R_i \subseteq \mathcal{V}; \forall i \neq j, R_i \cap R_j = \emptyset; \bigcup_i^k R_i = \mathcal{V}$.

Cutting all edges of the $\mathcal{MST}(\mathcal{G})$ having a valuation superior to a threshold λ leads to a minimum spanning forest (MSF) $\mathcal{F}(\mathcal{G})$, i.e. to a partition of the graph. Note that the obtained partition is the same that one would have obtain by cutting edges superior to λ directly on $\mathcal{G}$ [19]. Since working on the $\mathcal{MST}(\mathcal{G})$ is less costly and provides similar results regarding graph-based segmentation, we work only with the $\mathcal{MST}(\mathcal{G})$ in the sequel.

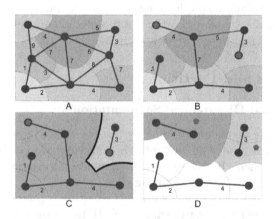

Fig. 1. A: a partition represented by an edge-weighed graph; **B**: a minimum spanning tree of the graph, with 2 markers in blue: the highlighted edge in blue is the highest edge on the path linking the two markers; **C**: the segmentation obtained when cutting this edge; **D**: blue and orange domain are the domains of variation of the two markers generating the same segmentation. (Color figure online)

So cutting edges by decreasing valuations gives an *indexed hierarchy of partitions* $(\mathcal{H}, \boldsymbol{\lambda})$, with $\mathcal{H}$ a *hierarchy of partitions* i.e. a chain of nested partitions $\mathcal{H} = \{\pi_0, \pi_1, \ldots, \pi_n | \forall j, k, \quad 0 \ \le j \le k \le n \Rightarrow \pi_j \sqsubseteq \pi_k\}$, with π_n the single-region partition and π_0 the finest partition on the image, and $\boldsymbol{\lambda} : \mathcal{H} \to \mathbb{R}^+$ being a stratification index verifying $\boldsymbol{\lambda}(\pi) < \boldsymbol{\lambda}(\pi')$ for two nested partitions $\pi \subset \pi'$.

This increasing map allows us to value each contour according to the level of the hierarchy for which it disappears: this is the *saliency* of the contour, and we consider that the higher the saliency, the stronger the contour. For a given hierarchy, the image in which each contour takes as value its saliency is called *Ultrametric Contour Map* (UCM) [3]. Representing a hierarchy by its UCM is an easy way to get an idea of its effect because thresholding an UCM always provides a set of closed curves and so a partition. In this paper, for better visibility, we represent UCM with inverted contrast.

To get a partition for a given hierarchy, there are several possibilities:

- simply thresholding the highest saliency values,
- marking some nodes as important ones and then computing a partition accordingly, which is known as *marker-based segmentation,*
- smartly editing the graph by finding the partition that minimizes an energetic function.

In a complementary approach, we argue that the quality of the obtained partitions highly depends on the hierarchy that we use, and thus that changing the dissimilarity can lead to more suitable partitions. Indeed, if the dissimilarity reflects only a local contrast as in the hierarchy issued by the RAG, the most salient regions in the image are the small contrasted ones. So instead of departing from a simple and rough dissimilarity such as contrast and then use an sophisticated technique to get a good partition out of it, one can also try to obtain a more informative dissimilarity adapted to the content of the image such that the simplest methods are sufficient to compute interesting partitions. This way, the aforementioned techniques lead to segmentations better suited for further exploitation. How can we construct more pertinent and informative dissimilarities?

2.2   Stochastic Watershed Hierarchies

The stochastic watershed (SWS), introduced in [2] on a simulation basis and extended with a graph-based approach in [17], is a versatile tool to construct hierarchies. The seminal idea is to operate multiple times marker-based segmentation with random markers and valuate each edge of the $\mathcal{MST}$ by its frequency of appearance in the resulting segmentations.

Indeed, by spreading markers on the RAG $\mathcal{G}$, one can construct a segmentation as a MSF $\mathcal{F}(\mathcal{G})$ in which each tree takes root in a marked node. Marker-based segmentation directly on the $\mathcal{MST}$ is possible: one must then cut, for each pair of markers, the highest edge on the path linking them. Furthermore, there is a domain of variation in which each marker can move while still leading to the same final segmentation. More details are provided in Fig. 1.

Let us then consider on the $\mathcal{MST}$ an edge e_{st} of weight ω_{st} and compute its probability to be cut. We cut all edges of the $\mathcal{MST}$ with a weight superior or equal to ω_{st}, producing two trees T_s and T_t of roots s and t. If at least one marker falls within the domain R_s of T_s nodes and at least one marker falls within the domain R_t of T_t nodes, then e_{st} will be cut in the final segmentation.

Let denote $\mu(\mathsf{R})$ the number of random markers falling in a region R. We want to attribute to e_{st} the following probability value:

$$\mathbb{P}[(\mu(\mathsf{R}_s) \geq 1) \wedge (\mu(\mathsf{R}_t) \geq 1)] = 1 - \mathbb{P}[(\mu(\mathsf{R}_s) = 0) \vee (\mu(\mathsf{R}_t) = 0)]$$
$$= 1 - \mathbb{P}(\mu(\mathsf{R}_s) = 0) - \mathbb{P}(\mu(\mathsf{R}_t) = 0) + \mathbb{P}(\mu(\mathsf{R}_s \cup \mathsf{R}_t) = 0) \quad (1)$$

If markers are spread following a Poisson distribution, then for a region R:

$$\mathbb{P}(\mu(R) = 0) = \exp^{-\Lambda(R)}, \quad (2)$$

With $\Lambda(R)$ being the expected value (mean value) of the number of markers falling in R. The probability thus becomes:

$$\mathbb{P}(\mu(\mathsf{R}_s) \geq 1 \wedge \mu(\mathsf{R}_t) \geq 1) = 1 - \exp^{-\Lambda(\mathsf{R}_s)} - \exp^{-\Lambda(\mathsf{R}_t)} + \exp^{-\Lambda(\mathsf{R}_s \cup \mathsf{R}_t)} \quad (3)$$

When the Poisson distribution has an homogeneous density λ:

$$\Lambda(R) = \mathtt{area}(R)\lambda, \quad (4)$$

When the Poisson distribution has a non-uniform density λ:

$$\Lambda(R) = \int_{(x,y)\in R} \lambda(x,y)\,\mathrm{d}x\mathrm{d}y \quad (5)$$

The output of the SWS algorithm thus depends on the departure $\mathcal{MST}$ (structure and edges valuations) and of the probabilistic law governing the markers distribution. Furthermore, SWS hierarchies can be chained, leading to a wide exploratory space that can be used in a segmentation workflow [8].

Because of its versatility and good performance, SWS represents a good departure algorithm to modify in order to inject prior information. Indeed, when having a prior information about the image, is it possible to use it in order to have more details in some parts rather than others?

3   Hierarchies Highlighthing Structures of Interest Using Prior Information

3.1   Hierarchy with Regionalized Fineness (HRF)

In the original SWS, a uniform distribution of markers is used (whatever size or form they may have). In order to have stronger contours in a specific region of the image, we adapt the model so that more markers are spread in this region.

Let E be an object or class of interest, for example $E =$ "face of a person", and $\mathbf{I}$ be the studied image. We denote by θ_E the probability density function (PDF) associated with E obtained separately, and defined on the domain D of $\mathbf{I}$, and by $\mathrm{PM}(\mathbf{I}, \theta_E)$ the probabilistic map associated, in which each pixel $p(x,y)$ of $\mathbf{I}$ takes as value $\theta_E(x,y)$ its probability to be part of E. Given such an information on the position of an event in an image, we obtain a hierarchical segmentation focused on this region by modulating the distribution of markers.

If λ is a density defined on D to distribute markers (uniform or not), we set $\theta_E \lambda$ as a new density, thus favoring the emergence of contours within the regions of interest.

Considering a region R of the image, the mean number of markers falling within R is then:

$$\Lambda_E(\mathsf{R}) = \int_{(x,y) \in \mathsf{R}} \theta_E(x, y) \lambda(x, y) \, dx dy \qquad (6)$$

Note that if we want N markers to fall in average within the domain D, we work with a slightly modified density:

$$\hat{\lambda} = \frac{N}{\mu(D)} \lambda \qquad (7)$$

Furthermore, this approach can be easily extended to the case where we want to take advantage of information from multiple sources. Indeed, if θ_{E_1} and θ_{E_2} are the PDF associated with two events E_1 and E_2, we can combine those two sources by using as a new density $(\theta_{E_1} + \theta_{E_2}) \lambda$.

3.2 Methodology

We present here the steps to compute a HRF for an event E given a probabilistic map $\mathrm{PM}(\mathbf{I}, \theta_E)$ providing spatial prior information on an image $\mathbf{I}$:

- compute a fine partition π_0 of the image, define a dissimilarity measure between adjacent regions and compute the RAG $\mathcal{G}$, and then the $\mathcal{MST}(\mathcal{G})$ to easily work with graphs,
- compute a probabilistic map $\pi_\mu = \pi_\mu(\pi_0, \mathrm{PM}(\mathbf{I}, \theta_E))$ with each region of the fine partition π_0 taking as a new value the mean value of $\mathrm{PM}(\mathbf{I}, \theta_E))$ in this region,
- compute new values of edges by a bottom-up approach as described in Sect. 3.1, where for each region R_i of π_0, $\Lambda(\mathsf{R}_i)$ corresponds to the mean value taken by pixels of the region R_i in π_μ. Note that this approach allows a highly efficient implementation using dynamic programming on graphs.

3.3 Modulating the HRF Depending on the Couple of Regions Considered

If we want to favor certain contours to the detriment of others, we can modulate the density of markers in each region by taking into account the strength of the contour separating them but also the relative position of both regions.

We use the same example and notations as in Sect. 3.1, and thus want to modulate the distribution of markers relatively to R_s, R_t and their frontier. For example, to stress the strength of the gradient separating both regions we can locally spread markers following the distribution $\chi(\mathsf{R}_s, \mathsf{R}_t) \lambda$, with $\chi(\mathsf{R}_s, \mathsf{R}_t) = \omega_{st}$. This corresponds to the classical volume-based SWS, which allows to obtain a hierarchy that takes into account both surfaces of regions and contrast between them.

To go further, one can use any prior information in a similar way. Indeed, while using prior information to influence the output of the segmentation workflow, one might also want to choose whether the relevant information to emphasize in resulting segmentations is the foreground, the background or the transitions between them.

For example, having more details in the transition regions between background and foreground allows us to have more precision where the limit between foreground and background is actually unclear. As a matter of fact, the prior information often only provides rough positions of the foreground object with blurry contours, and such a process would allow to get precise contours of this object from the image.

Let us consider this case and define for each couple of regions (R_s, R_t) a suitable $\chi(R_s, R_t)$. We then want $\chi(R_s, R_t)$ to be low if R_s and R_t both are in the background or the foreground, and high if R_s is the background and R_t in the foreground (or the opposite). We use:

$$\begin{cases} \tilde{\lambda} = \chi\lambda \\ \chi(R_s, R_t) = \frac{\max(m(R_s), m(R_t))(1 - \min(m(R_s), m(R_t)))}{0.01 + \sigma(R_s)\sigma(R_t)}, \end{cases} \tag{8}$$

$m(R)$ (resp. $\sigma(R)$) being the normalized mean (resp. normalized standard deviation) of pixels values in the region R of $PM(\mathbf{I})$. Thus the number of markers spread will be higher when the contrast between adjacent regions is high (numerator term) and when these regions are coherent (denominator term).

Then for each edge, its new probability to be cut is:

$$\mathbb{P}[(\mu(R_p) \geq 1) \wedge (\mu(R_q) \geq 1)] = 1 - \exp^{-\chi(R_s, R_t)\Lambda(R_p)} - \exp^{-\chi(R_s, R_t)\Lambda(R_q)} + \exp^{-\chi(R_s, R_t)\Lambda(R_p \cup R_q)} \tag{9}$$

In the spirit of [6], this mechanism provides us with a way to "realign" the hierarchy with respect to the relevant prior information to get more details where the information is blurry. Similar adaptations can be thought of to emphasize details of background or foreground regions.

In the following, we illustrate the methodology exposed with some applications.

4 Application Examples

4.1 Scalable Transmission Favoring Regions of Interest

Let us consider a situation where one emitter wants to transmit an image through a channel with a limited bandwidth, e.g. for a videoconference call. In such a case, the more important informations to transmit are details on the face of the person on the image. Besides, we nowadays have highly efficient face detectors, using for example Haar-wavelets as features in a learning-based vision approach [30]. Considering that for an image in entry, the face can be easily detected, we

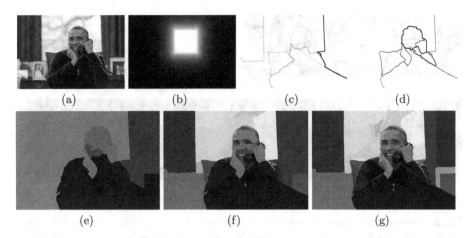

Fig. 2. Hierarchical segmentation of faces. (a) Original image (b) Prior: Probabilistic map obtained thanks to face detection algorithm and (c) Volume-based SWS Hierarchy UCM (d) Volume-based HRF UCM with face position as prior (e)(f)(g) Examples of segmentations obtained with HRF - 10,100,1000 regions.

can use this information to produce a hierarchical segmentation of the image that accentuates the details around the face while giving a good sketch of the image elsewhere. Depending on the bandwidth available, we can then choose the level of the hierarchy to select and obtain the associated partition to transmit, ensuring us to convey the face with as much details as possible. Some results are presented in Fig. 2, with notably a comparison between a classical volume-based SWS UCM and a volume-based HRF UCM.

4.2   Artistic Aspect: Focus and Cartoon Effect

The same method can also be used for artistic purposes. For example, when taking as prior the result of a blur detector [28], we can accentuate the focus effect wanted by the photograph and turn it into a cartoon effect as well - see Fig. 3 for an illustration of the results.

In the same spirit, various methods now exist to automatically roughly localize the principal object in an image. We inspire ourselves from [20] to do so. Using the state-of-the-art convolution neural network (CNN) classifier VGG19 [27] trained on the 1000 classes ImageNet database [7], we first determine what is the main class in the image. Note that this CNN takes as input only images of size 224×224 pixels. Once it is known, we can then, by rescaling the image by a factor $s \in \{0.5, 0.7, 1.0, 1.4, 2.0, 2.8\}$, compute for sub-windows of size 224×224 of the image the probability of appearance of the main class. By simply superimposing the results for all sub-windows, we thus obtain a probabilistic map of the main class for each rescaling factor. By max-pooling, we keep in memory the result of the scale for which the probability is the highest. The *heatmap*

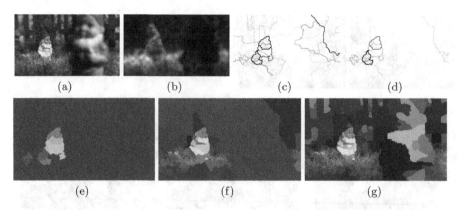

Fig. 3. Hierarchical segmentation of non-blur objects. (a) Original image (b) Prior image obtained with non-blur zones detection algorithm (c) Volume-based SWS Hierarchy UCM (d) Volume-based HRF UCM with non-blur zones as prior (e)(f)(g) Examples of segmentations obtained with HRF - 10,200,2000 regions.

thus produced can then be used to feed our algorithm. This way, we have at our disposal an automatized way to focus on the principal class in the scene. Some results are presented in Fig. 4.

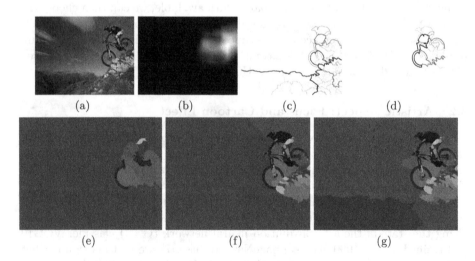

Fig. 4. Hierarchical segmentation of the main class (class "bike") in an image with heatmap issued of a CNN-based method as input. (a) Original image, (b) Heatmap issued by the CNN-based localization method, (c) Volume-based Watershed Hierarchy UCM and (d) Hierarchy with prior UCM. (e)(f)(g) Examples of segmentations obtained with HRF - 10,100,200 regions.

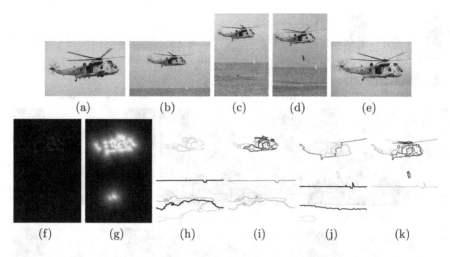

Fig. 5. Hierarchical co-segmentation of matched objects. (a)-(e) Images I_1 to I_5 (f) All matched key-points of I_3 with other images key-points (g) Prior for I_3: probability map generated thanks to morphological distance function (h)(i) Volume-based SWS Hierarchy UCM for I_3 and I_4 (j)(k) Volume-based HRF UCM for I_3 and I_4 with matched key-points as prior.

4.3   Hierarchical Co-segmentation

Another potential application is to co-segment with the same fineness level an object appearing in several different images. For example, when given a list of images of the same object taken from different perspectives/for different conditions, we can follow the state-of-the-art matching procedure [14]: (i) compute all key-points in both images, (ii) compute local descriptors at these key-points, (iii) match those key-points using a spatial coherency algorithm as RANSAC. Once it is done we retain these matched key-points for both images, and generate probability maps of the appearance of the matched objects using a morphological distance function to the matched key-points.

These probability maps can then feed our algorithm, given as result a hierarchical co-segmentation that emphasizes the matched zones of the image. Some results are presented in Fig. 5.

4.4   Example of the Effect of the HRF Highlighting Transitions Between Foreground and Background

We illustrate here the HRF highlighting transitions between foreground and background presented in Sect. 3.3, by presenting its effect in the face detection example presented in Fig. 6.

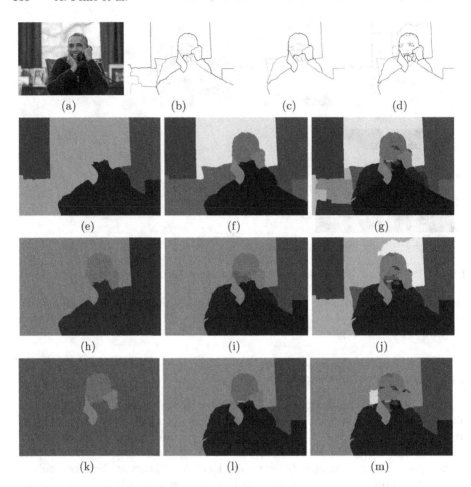

Fig. 6. Hierarchical segmentation of faces. (a) Original image (b)(c)(d) UCM of HRF depending of the couples of regions (Sect. 3.3), of the regions only (Sect. 3.1), and both combined. The rest of the images are segmentations examples with 4,10,25 regions, the hierarchies being presented in the same order.

5   Conclusions and Perspectives

In this paper we have proposed a novel and efficient hierarchical segmentation algorithm that emphasizes the regions of interest in the image by using spatial exogenous information on it. The wide variety of sources for this exogenous information makes our method extremely versatile and its potential applications numerous, as shown by the examples developed in the last section. To go further, we could find a way to efficiently extend this work to videos. One could also imagine a semantic segmentation method that would go back and forth between localization algorithm and HRF to progressively refine the contours of the main objects in the image.

References

1. Achanta, R., Shanji, A., Smith, K., Lucchi, A., Fua, P., Süsstrunk, S.: SLIC super pixels compared to state-of-the-art super pixel methods. IEEE PAMI 34, 2274–2282 (2012)
2. Angulo, J., Jeulin, D.: Stochastic watershed segmentation. ISMM 1, 265–276 (2007)
3. Arbelaez, P., Maire, M., Fowlkes, C., Malik, J.: Contour detection and hierarchical image segmentation. IEEE PAMI 33(5), 898–916 (2011)
4. Chan, T., Zhu, W.: Level set based shape prior segmentation. CVPR 2, 1164–1170 (2005)
5. Chen, F., Yu, H., Hu, R., Zeng, X.: Deep learning shape priors for object segmentation. In: CVPR, pp. 1870–1877 (2013)
6. Chen, Y., Dai, D., Pont-Tuset, J., Van Gool, L.: Scale-aware alignment of hierarchical image segmentation. In: CVPR (2016)
7. Deng, J., Dong, W., Socher, R., Li, L.J., Li, K., Fei-Fei, L.: ImageNet: a large-scale hierarchical image database. In: CVPR (2009)
8. Fehri, A., Velasco, S., Meyer, F.: Automatic selection of stochastic watershed hierarchies. In: EUSIPCO, August 2016
9. Gueguen, L., Velasco-Forero, S., Soille, P.: Local mutual information for dissimilarity-based image segmentation. JMIV 48, 1–20 (2013)
10. Guigues, L., Cocquerez, J.P., Le Men, H.: Scale-sets image analysis. IJCV 68(3), 289–317 (2006)
11. Kiran, B.R., Serra, J.: Ground truth energies for hierarchies of segmentations. In: Hendriks, C.L.L., Borgefors, G., Strand, R. (eds.) ISMM 2013. LNCS, vol. 7883, pp. 123–134. Springer, Heidelberg (2013). doi:10.1007/978-3-642-38294-9_11
12. Kiran, B.R., Serra, J.: Global-local optimizations by hierarchical cuts and climbing energies. Pattern Recogn. 47(1), 12–24 (2014)
13. Lampert, C., Blaschko, M., Hofmann, T.: Beyond sliding windows: object localization by efficient subwindow search. In: CVPR, pp. 1–8 (2008)
14. Lowe, D.G.: Distinctive image features from scale-invariant keypoints. IJCV 60(2), 91–110 (2004)
15. Machairas, V., Faessel, M., Cárdenas-Peña, D., Chabardes, T., Walter, T., Decencière, E.: Waterpixels. IEEE TIP 24(11), 3707–3716 (2015)
16. Malmberg, F., Hendriks, C.L.L., Strand, R.: Exact evaluation of targeted stochastic watershed cuts. Discrete Appl. Math. 216, 449–460 (2017)
17. Meyer, F.: Stochastic watershed hierarchies. In: ICAPR, pp. 1–8 (2015)
18. Meyer, F., Beucher, S.: Morphological segmentation. J. Vis. Commun. Image Represent. 1(1), 21–46 (1990)
19. Najman, L., Cousty, J., Perret, B.: Playing with Kruskal: algorithms for morphological trees in edge-weighted graphs. In: Hendriks, C.L.L., Borgefors, G., Strand, R. (eds.) ISMM 2013. LNCS, vol. 7883, pp. 135–146. Springer, Heidelberg (2013). doi:10.1007/978-3-642-38294-9_12
20. Oquab, M., Bottou, L., Laptev, I., Sivic, J.: Is object localization for free? Weakly-supervised learning with convolutional neural networks. In: CVPR (2015)
21. Ouzounis, G.K., Pesaresi, M., Soille, P.: Differential area profiles: decomposition properties and efficient computation. IEEE PAMI 34(8), 1533–1548 (2012)
22. Pont-Tuset, J., Arbelaez, P., Barron, J., Marques, F., Malik, J.: Multiscale combinatorial grouping for image segmentation and object proposal generation. PAMI (2016)

23. Ren, Z., Shakhnarovich, G.: Image segmentation by cascaded region agglomeration. In: CVPR, pp. 2011–2018 (2013)
24. Ronse, C.: Ordering partial partitions for image segmentation and filtering: Merging, creating and inflating blocks. JMIV, pp. 1–32 (2013)
25. Sermanet, P., Eigen, D., Zhang, X., Mathieu, M., Fergus, R., LeCun, Y.: Overfeat: integrated recognition, localization and detection using convolutional networks. In: ICLR (2013)
26. Serra, J.: Tutorial on connective morphology. IEEE J. Sel. Top. Signal Proces. **6**(7), 739–752 (2012)
27. Simonyan, K., Zisserman, A.: Very deep convolutional networks for large-scale image recognition. arXiv:1409.1556 (2014)
28. Su, B., Lu, S., Tan, C.L.: Blurred image region detection and classification. In: ACM International Conference on Multimedia, pp. 1397–1400, MM 2011 (2011)
29. Veksler, O.: Star Shape prior for graph-cut image segmentation. In: Forsyth, D., Torr, P., Zisserman, A. (eds.) ECCV 2008. LNCS, vol. 5304, pp. 454–467. Springer, Heidelberg (2008). doi:10.1007/978-3-540-88690-7_34
30. Viola, P., Jones, M.: Rapid object detection using a boosted cascade of simple features. In: CVPR, pp. 511–518 (2001)
31. Xu, Y., Géraud, T., Najman, L.: Hierarchical image simplification and segmentation based on mumford-shah-salient level line selection. Pattern Recogn. Lett. **83**, 278–286 (2016)

Morphological Hierarchical Image Decomposition Based on Laplacian 0-Crossings

Lê Duy Huỳnh[1], Yongchao Xu[1,2], and Thierry Géraud[1(✉)]

[1] EPITA Research and Development Laboratory (LRDE),
Le Kremlin-Bicêtre, France
`theo@lrde.epita.fr`
[2] Institut Mines Telecom, Telecom ParisTech, Paris, France

Abstract. A method of text detection in natural images, to be turned into an effective embedded software on a mobile device, shall be both efficient and lightweight. We observed that a simple method based on the morphological Laplace operator is very appropriate: we can construct in quasi-linear time a hierarchical image decomposition/simplification based on its 0-crossings, and search for some text in the resulting tree. Yet, for this decomposition to be sound, we need "0-crossings" to be Jordan curves, and to that aim, we rely on some discrete topology tools. Eventually, the hierarchical representation is the morphological tree of shapes of the Laplacian sign (ToSL). Moreover, we provide an algorithm with linear time complexity to compute this representation. We expect that the proposed hierarchical representation can be useful in some applications other than text detection.

Keywords: Morphological Laplace operator · Well-composed images · Tree of shapes · Hierarchical decomposition · Text detection

1 Introduction

In [8] we have proposed a method to perform robust text segmentation in natural images; see Fig. 1 for an illustration. This method is interesting for several reasons. **1.** To select candidate regions for characters, we rely on the morphological Laplace operator. We can observe that this operator is relatively well robust to poor contrast and uneven illumination; the experimental results show that it outperforms the "more classical" component-based methods—namely the Stroke Width Transform (SWT), the Toggle Mapping Morphological Segmentation (TMMS), and the Maximally Stable Extrema Region (MSER) methods—see [8] for details. **2.** Based on the 0-crossings of the morphological Laplace operator, we compute a hierarchical representation of the image contents, so that regions form an inclusion tree. Such a structure is attractive because we have a strong simplification of the image contents, and because dealing with a simple

Thierry Géraud: This work has been conducted in the context of the MOBIDEM project, part of the "Systematic Paris-Region" and "Images & Network" Clusters (France). This project is partially funded by the French Gov. and its economic development agencies.

© Springer International Publishing AG 2017
J. Angulo et al. (Eds.): ISMM 2017, LNCS 10225, pp. 159–171, 2017.
DOI: 10.1007/978-3-319-57240-6_13

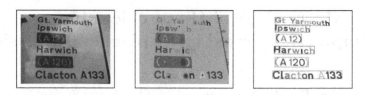

Fig. 1. A hierarchical image decomposition (center) leading to text detection (right).

tree structure allows for some powerful decision taking when grouping regions/ characters into text boxes. Although using the 0-crossings of the morphological Laplace operator as a contour detector is not new [13], getting such a hierarchical representation from it is a novelty. **3.** This text segmentation method is suitable to real-time implementation on mobile devices. Indeed computing the hierarchical representation eventually translates to a simple labeling algorithm that gives both an inclusion tree and a label image; in addition, identifying and grouping text components are achieved by an easy tree-based processing.

In this present paper we focus on two points which are not detailed in [8]: how we can rely on some discrete topology tools to get *a sound definition of a hierarchical decomposition* of an image into regions, and how we take benefit from some properties to get *an efficient algorithm* to compute such a representation.

2   Method Overview

The method that we propose to represent an image by a hierarchical decomposition and to extract text in natural images is very simple: put shortly, text

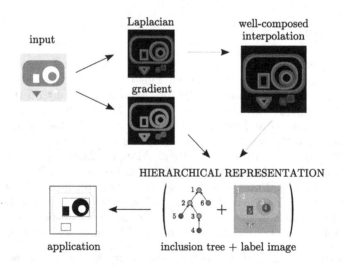

Fig. 2. Overview of the proposed method to get a hierarchical image decomposition. (Color figure online)

components are selected among the 0-crossing lines of the morphological Laplace operator of the gray-level input image. The method is composed of several steps, depicted in Fig. 2, and described and justified just below.

We start with a gray-level input image. If the primary data is a color image, we just take the luminance value of its pixels (we lose color information but it almost never negatively affects text retrieval). We then compute its morphological dilation $\delta_\square$ and erosion $\varepsilon_\square$ to directly deduce two images. First, we obtain the morphological thick gradient $\nabla_\square = \delta_\square - \varepsilon_\square$; it is used later to discard contours that are not enough contrasted. Second, we obtain the morphological Laplace operator $\Delta_\square = \delta_\square + \varepsilon_\square - 2\,\mathrm{id}$ (where id is the identity function). The text character boundaries are expected to belong to the 0-crossing contours of this non-linear operator; actually they are, and their localization is precise.

We *virtually*[1] compute a particular interpolated image, $\Delta_\square^{wc}$, of the Laplacian image $\Delta_\square$, having 4 times more pixels than the original. This resulting image is *well-composed*, meaning that the boundaries of every components of any threshold set are Jordan curves. As a consequence, the (boundaries of the) 0-crossings are simple closed curves: they cannot have the shape of a '8'. In addition, they are disjoint, and this set of curves can be organized in an inclusion tree.

Due to the fact that $\Delta_\square^{wc}$ is well-composed, the regions delimited by the 0-crossings can be labeled very efficiently (by the classical blob labeling algorithm), and their inclusion tree is built. In addition, many 0-crossings are discarded on the fly during the labeling process, because they are not contrasted enough (based on $\nabla_\square$), or because they do no satisfy some geometrical criteria (*e.g.*, when they are too small). The resulting "inclusion tree + label image" is the hierarchical decomposition of the image contents into regions; it is depicted on the bottom-right part of Fig. 2. We end up with two structures: the inclusion tree, encoded by a parenthood relationship between labels, and an image of labels, so that each pixel is assigned to a region. In addition, we have some additional information related to every regions, that are computed during the labeling process: the region area, its bounding box, etc.

From the resulting hierarchical representation, we have derived an application described in [8]: text detection. Indeed, we can group regions, that are separated components, together to form text boxes. For that, we only consider the bottom of this tree (the leaves and sometimes their parent): for each component, we search spatially in the label image what are their left and right components to be grouped into a text box. In this step, we highly take advantage of the tree structure: it allows very easily to discard many regions as non-text, and to determine if a leaf region is a character hole or a plain character.

3   Theoretical Background

3.1   Khalimsky's Grid

From the sets $H_0^1 = \{\{a\}; a \in \mathbb{Z}\}$ and $H_1^1 = \{\{a, a+1\}; a \in \mathbb{Z}\}$, we can define $H^1 = H_0^1 \cup H_1^1$ and the set H^n as the n-ary Cartesian power of H^1.

---

[1] We will see later that we actually do *not* interpolate the Laplacian image, but proceed *as if* there were an interpolation. Practically, it means that we avoid the need of multiplying by 4 the number of pixels in the process.

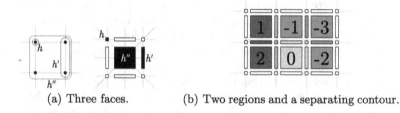

(a) Three faces.   (b) Two regions and a separating contour.

Fig. 3. Representation of images on the Khalimsky's grid. (Color figure online)

If an element $h \subset \mathbb{Z}^n$ is the Cartesian product of d elements of H_1^1 and $n - d$ elements of H_0^1, we say that h is a d-face of H^n and that d is the dimension of h. The set of all faces, H^n, is called the nD space of cubical complexes. Figure 3(a) depicts a set of faces $\{h, h', h''\} \subset H^2$ where $h = \{0\} \times \{1\}$, $h' = \{1\} \times \{0, 1\}$, and $h'' = \{0, 1\} \times \{0, 1\}$; the dimension of those faces are respectively 0, 1, and 2. The set of faces of H^n is arranged onto a grid, called the Khalimsky's grid [9], depicted in orange in Fig. 3(a) (right). Consequently, it means that we just can store values defined on such a grid in a simple matrix structure.

This grid is a way to **1.** get a discrete topology, and **2.** represent what lies in-between pixels, for instance contours. In Fig. 3(b), an image is depicted with integer values. The green and red regions are open sets corresponding to components having respectively positive and negative values. A 1D contour (thus with 0-faces and 1-faces only) is depicted in blue: it is precisely the discrete boundary of the non-positive region on the right.

3.2 Morphological Laplace Operator

In order to detect text in images, many methods first look for candidate regions for characters. A seminal method, based on contour detection, is to consider the 0-crossings of a discrete Laplace operator: given a gray-level image u, the contours of interest are given by $\Delta u = u_{xx} + u_{yy} = 0$. This method is interesting for several reasons: **1.** it is a very simple approach; **2.** it provides closed contours; **3.** labeling the components of the image having the same sign, resp. positive and negative; gives a segmentation; **4.** it is self-dual, *i.e.*, it processes dark objects and bright ones the same way.

The morphological Laplace operator has been defined in [12] by $\Delta_\square = (\delta_\square - \mathrm{id}) - (\mathrm{id} - \varepsilon_\square)$, relying on the elementary dilation (δ) and erosion (ε) morphological operators. Actually, the morphological non-linear version features a much higher fidelity to actual object contours than the linear version; in addition, salient contours hardly depends on the size of the structuring element.

The image depicted in Fig. 4(a) has been created by the authors of [1] to make classical binarization methods fail, due to the presence of uneven illumination. On Fig. 4(c), one can observe that the object boundaries belong to the 0-crossings of the morphological Laplace operator.

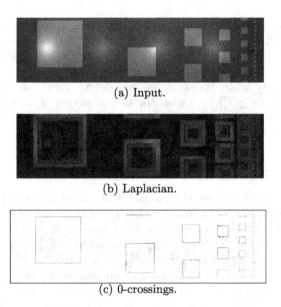

(a) Input.

(b) Laplacian.

(c) 0-crossings.

Fig. 4. Contour characterization by morphological operators; 4(b) is the morphological Laplacian $\Delta_B = \delta_B + \varepsilon_B - 2\mathrm{id}$; 4(c) depicts on $\Delta_B = 0$ the gradient values $\nabla_B = \delta_B - \varepsilon_B$ (inverted) showing that the contours of actual objects are effectively salient.

3.3 Tree of Shapes

The tree of shapes is a morphological self-dual representation of an image; see [7] and [5] for history, implementation, and references. This tree encodes the inclusion of the level sets, *i.e.*, the connected components obtained by thresholding. An illustration is given on a very simple image by Fig. 5(a). The node B lists the pixels of the region B. The contour of this node is a level line, and the component, called *shape*, associated with this node is actually the sub-tree rooted at this node. The node B thus represents the shape given by $B \cup D \cup E$. In [4] the reader can find more details about how to store and process efficiently such a tree structure.

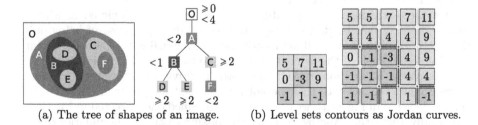

(a) The tree of shapes of an image. (b) Level sets contours as Jordan curves.

Fig. 5. Topological representations and some topological considerations.

With X representing a 2D discrete space, given an image $u : X \to \mathbb{Z}$ and any scalar $\lambda \in \mathbb{Z}$, the lower and upper level sets are respectively defined by $[u < \lambda] = \{x \in X \mid u(x) < \lambda\}$ and $[u \geq \lambda] = \{x \in X \mid u(x) \geq \lambda\}$. Using $\mathcal{CC}$, the operator that takes a set and gives its set of connected components, and Sat, the cavity-fill-in operator, we can define: $\mathfrak{S}(u) = \{ \text{Sat}(\Gamma); \Gamma \in \mathcal{CC}([u < \lambda]) \cup \mathcal{CC}([u \geq \lambda]) \}_\lambda$, called the tree of shapes of u. Note that every λ is considered so we have an indexed family of sets (corresponding to the shapes obtained at level λ).

3.4 Well-Composed Sets and Images

A sub-class of sets defined on the cubical grid, called *well-composed*, has been proposed in [11], where all connectivities are equivalent, thus avoiding many topological problems. A very easy characterization of well-composed sets is based on the notion of "critical configurations"; in the 2D case, a set is well-composed if the configurations ▨ and ▨ do not appear. The notion of well-composedness has been extended in [10] from sets to functions, *i.e.*, gray-level images, and has been generalized in the nD case in [3]. A gray-level image u is well-composed if any set $[u \geq \lambda]$ is well-composed. A straightforward characterization of well-composed 2D gray-level images is that every block $\begin{smallmatrix} a & d \\ c & b \end{smallmatrix}$ should verify: $\text{intvl}(a, b) \cap \text{intvl}(c, d) \neq \emptyset$, where $\text{intvl}(v, w) = [\![\min(v, w), \max(v, w)]\!]$. An image is not *a priori* well-composed. To get a well-composed image from a primary image, one can compute an interpolation of the primary image that is well-composed. Figure 5(b) gives an example of an image which is not well-composed (left), but whose interpolation is well-composed (right). In [2] the authors have proposed a simple method to get a self-dual well-composed interpolation, that makes sense when considering the tree of shapes of an image [6].

3.5 Putting Things Together

As explained in Sect. 2, from a gray-level input image u, we consider the Laplacian image once well-composed; both images are depicted respectively in Figs. 6(a) and (b). The tree of shapes of the Laplacian *sign* image (ToSL), $\mathfrak{S}(\text{sign}(\Delta_\square^{\text{wc}}(u)))$, contains nodes corresponding to positive, negative, and null regions; it is depicted in Fig. 6(d).

Yet the 0-crossings, corresponding to nodes at level 0 (depicted in white), can be "thick", that is, they can contain pixels (2-faces) instead of being only defined by 1D objects (set of 0-faces and 1-faces). In Fig. 6(c) for instance, we can see that some pixels of the 's' contour belong to the 0-crossing region. Since we want to obtain a partition of the image into "positive" and "negative" regions, we merge null regions with their parent regions; that way, instead of the tree of shapes depicted in Fig. 6(d), we consider the simplified one, depicted in Fig. 6(e). Eventually, the actual "0-crossings" we are looking at are the boundaries of the shapes of this final tree, so they are effectively 1D objects. It is illustrated in Fig. 5(b) (and also in Fig. 3(b)): the null region is merged with the negative one,

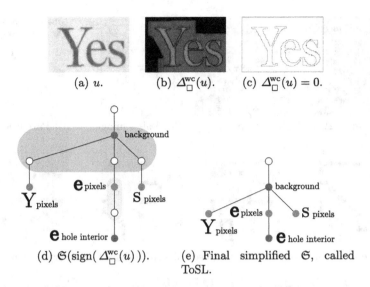

(a) u.   (b) $\Delta_{\square}^{wc}(u)$.   (c) $\Delta_{\square}^{wc}(u) = 0$.

(d) $\mathfrak{S}(\mathrm{sign}(\Delta_{\square}^{wc}(u)))$.   (e) Final simplified $\mathfrak{S}$, called ToSL.

Fig. 6. Tree of shapes of Laplacian sign (ToSL): positive and negative regions are respectively green and red nodes of the ToS, and null regions are white nodes. (Color figure online)

so we consider the blue contour to be the 1D "0-crossing" separating regions having different signs.

In the next section, to compute the hierarchical representation, we do not consider that we have a cubical complex as the space structure: we just ignore that 0-faces and 1-faces exist. Though, from a theoretical point of view, the contours/0-crossings expressed in terms of 0-faces and 1-faces really are Jordan curves.

4   Fast Computation of the Hierarchical Representation

4.1   A Particular Well-Composed Non-local Interpolation

In the following, we do not need to consider the cubical complex, so we only deal with pixels; they are, for instance, the valued 2-faces in Fig. 5(b) (right).

The hierarchical representation is computed from an interpolated image, $\Delta_{\square}^{wc}(u)$, which is a very particular well-composed version of the Laplacian image $\Delta_{\square}(u)$. This particular interpolation takes its origin from the work in [7], and is detailed in [2]. Briefly put, it is a *non-local* interpolation driven by a propagation from the border of the image, which browses the nodes of its tree of shapes from the root to the leaves. The interpolated pixels are assigned with the current gray-level value, which evolves "continuously" during the process. This interpolation has two important features [6]: it is related to the tree of shapes of the input image, so it actually follows the same scheme as a blob labeling algorithm where a blob would be a level set, and its topological behavior is deterministic.

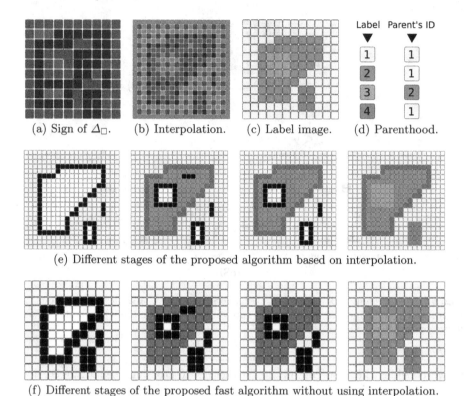

(a) Sign of $\Delta_\square$.   (b) Interpolation.   (c) Label image.   (d) Parenthood.

(e) Different stages of the proposed algorithm based on interpolation.

(f) Different stages of the proposed fast algorithm without using interpolation.

Fig. 7. An example of the proposed labeling algorithm. Black pixels: contours of components inside labeled ones. White pixels: pixels that are not yet labeled and are also not marked as inside contours at the current stage of the labeling process. (Color figure online)

There are two main consequences: this particular interpolation makes sense in the present context, since we want to label regions that are in an inclusion relationship, and we can optimize the computation of the hierarchical representation (the label image and the inclusion tree) by actually *emulating* the interpolation.

An example on the image of Laplacian sign given in Fig. 7(a) is depicted in Fig. 7(b)[2], with the signs -1, 0, and 1 respectively depicted in red, gray, and green. We can observe in Fig. 7(b) that we obtain the desired properties: the boundaries of the regions are Jordan curves, and the regions are in an unambiguous spatial inclusion relationship. The inclusion tree of the Laplacian sign image, called ToSL, is thus a hierarchical image decomposition.

---

[2] Note that the two identical local configurations enclosed by red rectangles in Fig. 7(a) do not lead to the same interpolation; this is due to the non-local interpolation process that depends on the outer region, which is different in the two cases: respectively negative for the top configuration, and positive for the bottom one.

4.2 Labeling Interpolated Sign of $\Delta_\square$ to Construct ToSL

Thanks to the fact that $\Delta_\square^{wc}$ is well-composed, the regions delimited by the 0-crossings can be labeled very efficiently (by the classical blob labeling algorithm), and their inclusion tree is built. This resulting inclusion tree is the tree of shapes of the sign of the Laplacian, a ternary-valued image—with pixels set to −1 (red), 0 (gray), or 1 (green) as depicted in Fig. 7. The whole labeling process is depicted in Fig. 8. It computes a label image $\mathfrak{L}$ (a label is assigned to every region delimited by the 0-crossings of $\Delta_\square$), and a tree structure encoded in an array of parenthood $parent$. Having $parent(l_1) = l_2$ means that the region with label l_1 is included in the region with label l_2, and the particular root label, say l_r, is such that $parent(l_r) = l_r$. Q is a queue of pixels, $border$ is an auxiliary image that marks the inner component contours as active a or inactive $\bar{a}$ state, $nlabels$ is the current number of labels, ℓ is the current label value, and $\mathbb{N}$ represents a neighborhood corresponding to either 4-connectivity ($\mathbb{N}_4$) or 8-connectivity ($\mathbb{N}_8$).

The core of the algorithm is an alternate application of the following two routines (depicted on the right side of Fig. 8): BLOB_LABELING is a classical queue-based "blob labeling" algorithm that labels the underlying connected component with current label value ℓ, and also marks the inner component contours as active state a. Note that actually we do not want regions representing null values in the final tree, so we merge nodes corresponding to 0-crossings with their parent (see line 30 and Fig. 6). FOLLOW_CONTOUR is also a queue-based process which

```
 1 LABELING (Δ□,∇□)                    22 BLOB_LABELING (p,ℓ)
 2 forall p do                         23   𝔏(p) ← ℓ; Q.push(p);
 3 │  𝔏(p) ← 0;                        24 while Q is not empty do
 4 │  border(p) ← undef;               25 │  q ← Q.pop(), N ← N₄;
 5 nlabels ← 1;                        26 │  if border(q) = undef then
 6 forall p do                         27 │  │  N ← N₈; /*optimized*/
 7 │  if 𝔏(p) ≠ 0 then                 28 │  forall n ∈ N(q) do
 8 │  │  continue;                     29 │  │  if 𝔏(n) = 0 and
 9 │  if p = p₀ then                   30 │  │  Δ□(p) × Δ□(n) ≥ 0 then
10 │  │  ℓ ← 1;                        31 │  │  │  𝔏(n) ← ℓ; Q.push(n);
11 │  │  parent[ℓ] = 1;                32 │  else
12 │  else                            33 │  │  border(n) ← a;
13 │  │  𝒫 ←
14 │  │  FOLLOW_CONTOUR(p);            34 FOLLOW_CONTOUR (p)
14 │  │  if evaluate(𝒫) then           35 𝒫.init(); border(p) ← ā; Q.push(p);
15 │  │  │  nlabels ← nlabels + 1;     36 while Q is not empty do
16 │  │  │  ℓ ← nlabels;               37 │  q ← Q.pop(); 𝒫.update(q);
17 │  │  │  parent[ℓ] ← 𝔏(p₋₁);        38 │  forall n ∈ N(q) /*N₈ or N₄*/ do
18 │  │  else                         39 │  │  if border(n) = a then
19 │  │  │  ℓ ← 𝔏(p₋₁);                40 │  │  │  Q.push(n); border(n) ← ā;
20 │  BLOB_LABELING (p,ℓ);             41 return 𝒫;
21 return parent, 𝔏;
```

Fig. 8. Computation of ToSL by labeling interpolated image (black part) and its fast version (adding red part) without using interpolation. (Color figure online)

is similar with previous "blob labeling" algorithm, but applied on the underlying active contour of an unlabeled connected component. Instead of labeling the active contour, this routine browses it and marks it as inactive $\bar{a}$ (see lines 35 and 40), and collects a measure $\mathcal{P}$ characterizing the contour (*e.g.*, its length). Note that both these two routines are very efficient thanks to the queue-based "blob labeling".

More precisely, for the main algorithm (depicted on the left side of Fig. 8), we browse the pixels in raster scan order (main loop, line 6). When we reach an unlabeled pixel p, if this is the first pixel p_0 in the scan order (*i.e.*, the top left pixel), we know that its label value is 1, and the label value of its parent (*i.e.*, itself) is also 1 (lines 9 to 11), because p_0 is in the root node thanks to the added external boundary described in previous section. For all other unlabeled pixel $p \neq p_0$, we follow the contour of the unlabeled region, which is a hole in the label image, thanks to the *border* image. The FOLLOW_CONTOUR routine computes on the fly a contour-based measure $\mathcal{P}$ characterizing the hole, such as the average of gradient's magnitude along the contour and the bounding box of the hole. If this contour-based measure $\mathcal{P}$ does not satisfy some criterion (for instance if it is not contrasted enough), or if it does not satisfy some geometrical criterion (for instance if the hole is too small), we do not create a new label value for this region; this acts as if the region were discarded (in Fig. 7, two regions are discarded between the 2nd and the 4th columns). Let p_{-1} be the pixel just before p in the raster scan order (p_{-1} is guaranteed to be labeled), we assign the label value of p_{-1} to the underlying hole region, which means we merge it with its parent region. If the contour measure $\mathcal{P}$ satisfies the corresponding criterion, we create a new label value to label the hole region, and update the parenthood relationship of this new label value to $\mathfrak{L}(p_{-1})$ (see lines 14 to 19). Then we proceed the routine BLOB_LABELING to label the connected set of pixels having the same Laplacian sign as p or being null and update the auxiliary *border* image. Note that since we use the 4-connectivity neighborhood ($\mathbb{N}_4$) in the routine BLOB_LABELING to label regions and mark neighboring pixels having different sign of $\Delta_\square$, we need to use the 8-connectivity neighborhood ($\mathbb{N}_8$) to follow completely the active contour of an unlabeled region (see also the longest contour in the left image in Fig. 7(e)).

An example of the proposed algorithm on the interpolated image in Fig. 7(b) is depicted in Figs. 7(c–e). Different stages of the algorithm are depicted in Fig. 7(e). Note that the pixels having null Laplacian sign are grouped with the parent region. The two small regions inside cyan and respectively yellow region are also discarded, they are grouped with the parent region. The resulting "label image + tree" are depicted in Figs. 7(c) and (d).

4.3   Optimization of ToSL Construction

The well-composed interpolation described in Sect. 4.1 resolves all the topological issues at critical configurations with the cost of quadrupling the number of pixels. Yet, in practice, we do not need to apply this interpolation, which can be emulated efficiently without subdivising the image domain thanks to the particular non-local way of interpolation described in Sect. 4.1.

The optimized version of the proposed algorithm by emulating the interpolation is depicted in Fig. 8 by the black and red parts. More precisely, the used particular non-local interpolation is based on the inclusion relationship of components. The interpolated values at critical configurations are given by the values of outer region. Besides, the proposed algorithm for such interpolated image labels regions from outside to inside. Consequently, it is equivalent to use 8-connectivity ($\mathbb{N}_8$) when we use blob labeling algorithm to label pixels that are not yet activated thanks to the *border* image (see line 26 and line 27 of BLOB_LABELING in Fig. 8). This makes the other two pixels of critical configurations having a different sign of Laplacian disjoint and marked as active contours. Consequently, when we proceed the FOLLOW_CONTOUR routine in the following, we need to use 4-connectivity ($\mathbb{N}_4$) (see line 38 of FOLLOW_CONTOUR in Fig. 8) to avoid connecting two disjoint unlabeled regions.

An example is given in Fig. 7(f) which shows different stages of the proposed optimized algorithm without using interpolation. Note that the green pixels (resp. red pixels) of the critical configuration in the top (resp. bottom) red rectangle in Fig. 7(a) are considered as connected using 8-connectivity ($\mathbb{N}_8$) when we proceed the routine BLOB_LABELING. The active contours (black pixels) in Fig. 7(f) are processed using 4-connectivity ($\mathbb{N}_4$). The two runs depicted in Figs. 7(e) and (f) result in the same "label image + tree" depicted in Figs. 7(c) and (d).

4.4 Application to Text Segmentation and Detection

As an application, the ToSL has been used for text detection and segmentation; the results have been further detailed in [8]. Briefly put, we compute the ToSL, and then group nodes together to form candidates for text boxes.

During the FOLLOW_CONTOUR process, we discard some regions w.r.t. their contour average gradient magnitude $\overline{\nabla}$ (to keep only contrasted regions), the contour length (to remove noise, *i.e.*, too small regions), and the bounding box

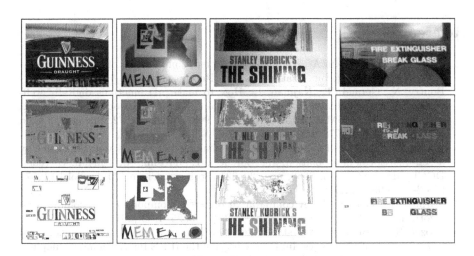

Fig. 9. Qualitative results using "ICDAR 2015 Robust Reading" database: input (top), label image (middle), text boxes (bottom).

size $\mathbb{B}$ (to keep components that reasonably look like characters). The ToSL is such that sibling nodes are objects sharing the same background in the original image. Consequently, the grouping process is performed efficiently because the tree structure helps to limit the sets of candidates that can be grouped together. We found that most of text components are leaves, and sometimes leaf parents (in the case of characters with holes). To group text components together into text boxes, starting from each leaf, we look in the image space, that is in the label image, for a sibling, and we control the grouping thanks to some geometric information. Some results are depicted in Fig. 9; more details and a quantitative evaluation are given in [8].

5   Conclusion

In this paper, we have presented a hierarchical decomposition of the image contents into regions. Although we use the very "classical" idea of relying on the 0-crossings of the Laplace operator image to obtain the objects of interest, our version is innovative for several reasons. First, we rely on the morphological Laplace operator, which performs well in the case of uneven illumination, which is a common defect in natural images. Second, we ensure that the "0-crossings" are really 1D objects. For that, we use a well-composed Laplacian image, we compute the tree of shapes of its sign, and we consider the 1D topological boundary of shapes. As a consequence, the positive and negative regions can be organized into a tree without topological ambiguity. Last we present a linear time complexity algorithm, which is a hardly more sophisticated blob labeling algorithm, to compute the hierarchical structure. We also provide a way, directly during the computation process, to ignore some regions if they are not relevant, and an optimization that mimics well-composedness and then avoids to duplicate pixels.

References

1. Blayvas, I., Bruckstein, A., Kimmel, R.: Efficient computation of adaptive threshold surfaces for image binarization. Pattern Recogn. **39**(1), 89–101 (2006)
2. Boutry, N., Géraud, T., Najman, L.: How to make nD functions digitally well-composed in a self-dual way. In: Benediktsson, J.A., Chanussot, J., Najman, L., Talbot, H. (eds.) ISMM 2015. LNCS, vol. 9082, pp. 561–572. Springer, Cham (2015). doi:10.1007/978-3-319-18720-4_47
3. Boutry, N., Géraud, T., Najman, L.: On making nD images well-composed by a self-dual local interpolation. In: Barcucci, E., Frosini, A., Rinaldi, S. (eds.) DGCI 2014. LNCS, vol. 8668, pp. 320–331. Springer, Cham (2014). doi:10.1007/978-3-319-09955-2_27
4. Carlinet, E., Géraud, T.: A comparative review of component tree computation algorithms. IEEE Trans. Image Process. **23**(9), 3885–3895 (2014)
5. Crozet, S., Géraud, T.: A first parallel algorithm to compute the morphological tree of shapes of nD images. In: Proceedings of the IEEE International Conference on Image Processing (ICIP), pp. 2933–2937 (2014)

6. Géraud, T., Carlinet, E., Crozet, S.: Self-duality and digital topology: links between the morphological tree of shapes and well-composed gray-level images. In: Benediktsson, J.A., Chanussot, J., Najman, L., Talbot, H. (eds.) ISMM 2015. LNCS, vol. 9082, pp. 573–584. Springer, Cham (2015). doi:10.1007/978-3-319-18720-4_48

7. Géraud, T., Carlinet, E., Crozet, S., Najman, L.: A quasi-linear algorithm to compute the tree of shapes of nD images. In: Hendriks, C.L.L., Borgefors, G., Strand, R. (eds.) ISMM 2013. LNCS, vol. 7883, pp. 98–110. Springer, Heidelberg (2013). doi:10.1007/978-3-642-38294-9_9

8. Huỳnh, L.D., Xu, Y., Géraud, T.: Morphology-based hierarchical representation with application to text segmentation in natural images. In: Proceedings of the International Conference on Pattern Recognition (ICPR) (2016, to appear). http://www.lrde.epita.fr/theo/papers/huynh.2016.icpr.pdf

9. Khalimsky, E., Kopperman, R., Meyer, R.: Computer graphics and connected topologies on finite ordered sets. Topol. Appl. **36**, 1–17 (1990)

10. Latecki, L.J.: 3D well-composed pictures. GMIP **59**(3), 164–172 (1997)

11. Latecki, L.J., Eckhardt, U., Rosenfeld, A.: Well-composed sets. Comput. Vis. Image Underst. **61**(1), 70–83 (1995)

12. van Vliet, L., Young, I., Beckers, G.: An edge detection model based on non-linear laplace filtering. In: Proceedings of International Workshop on PRAI, pp. 63–73 (1988)

13. van Vliet, L., Young, I., Beckers, G.: A non-linear laplace operator as edge detector in noisy images. CVGIP **45**(2), 167–195 (1989)

Sparse Stereo Disparity Map Densification Using Hierarchical Image Segmentation

Sébastien Drouyer[1,2(✉)], Serge Beucher[1], Michel Bilodeau[1],
Maxime Moreaud[1,2], and Loïc Sorbier[2]

[1] CMM - Centre de Morphologie Mathématique, Mines ParisTech,
PSL Research University, Fontainebleau, France
{sebastien.drouyer,serge.beucher,michel.bilodeau}@mines-paristech.fr
[2] IFP Energies nouvelles, Rond-point de l'échangeur de Solaize,
BP 3, 69360 Solaize, France
{maxime.moreaud,loic.sorbier}@ifpen.fr

Abstract. We describe a novel method for propagating disparity values
using hierarchical segmentation by waterfall and robust regression models.
High confidence disparity values obtained by state of the art stereo
matching algorithms are interpolated using a coarse to fine approach. We
start from a coarse segmentation of the image and try to fit each region's
disparities using robust regression models. If the fit is not satisfying, the
process is repeated on a finer region's segmentation. Erroneous values in
the initial sparse disparity maps are generally excluded, as we use robust
regressions algorithms and left-right consistency checks. Final disparity
maps are therefore not only denser but can also be more accurate. The
proposed method is general and independent from the sparse disparity
map generation: it can therefore be used as a post-processing step for
any stereo-matching algorithm.

Keywords: Stereo · Hierarchical segmentation · Robust regression
model · Waterfall · Disparity map · Densification

1 Introduction

One of the main research interest in computer vision is the matching of stereoscopic images, as it allows to obtain a 3d reconstruction of the observed scene.
A common practice is to first rectify the pair of images [2], that is to say slightly
distort them so that a point of the scene is projected on the same ordinate
on both images. This rectification allows to simplify the matching process to a
1D search problem. The difference of abscissa of a point's projection between
the rectified pair of images is called disparity. A disparity map estimates the
disparity for each pixel of the rectified pair.

Many different algorithms evaluating these disparity maps already exist in
the literature. They generally use local [16], semi-global [9,12] or global optimization methods [8,15,26] that minimize matching cost of image features between
two images. While some methods, especially the global ones, can produce dense

© Springer International Publishing AG 2017
J. Angulo et al. (Eds.): ISMM 2017, LNCS 10225, pp. 172–184, 2017.
DOI: 10.1007/978-3-319-57240-6_14

disparity maps, other methods produce sparse disparity maps - where there are some pixels with undefined disparity values. This can happen for several reasons. Disparity values can be removed when a confidence measure, such as texture or uniqueness, is under a certain threshold. They can also be removed when a part of the scene is occluded, either by global and semi-global methods, or by using Left-Right Consistency (LRC) checks [11].

Global disparity evaluation methods often feature disparity propagation in order to generate a complete map [1,23], but these propagation techniques are directly linked to the method and are not generalizable.

Apart from diffusion [25] and interpolation [20], a few general propagation methods exist in the literature. From a disparity map where only edges are matched, VMP [19] diffuses disparity values by using predefined masks and a voting process. FGS [17] has been used by the OpenCV library [13] for propagating LRC checked disparity values using an edge-preserving diffusion model. RBS [3] proposes a similar diffusion model, but using a modified version of the bilateral filter.

This work proposes a new densification method that is able to produce a denser and more accurate disparity map from an initial sparse disparity map. We called our method TDSR for Top Down Segmented Regression. From a coarse segmentation of the image, we try to fit each region's disparities using robust regression models. If the fit is not satisfying, the process is repeated on a finer region's segmentation.

This coarse to fine approach allows the algorithm to rely on general purpose parameters. Since each region is processed separately, the algorithm can be easily parallelized to improve efficiency. As we use linear regression models, it is possible to get not only the disparity values but also normals in the disparity space, which can be convenient for multi-view stereo algorithms and object recognition. Finally, the proposed method is general and independent from the sparse disparity map generation: it can therefore be used as a post-processing step for any stereo-matching algorithm.

2 Top Down Segmented Regression Method

2.1 General Concept

We have two rectified stereo images (left and right) depicting the same scene (see example in Fig. 1a). We also have a sparse disparity map between the left and the right images obtained using an existing stereo matching algorithm (see Fig. 1b). Large areas of the disparity map can be undefined (gray pixels), and defined areas can contain errors. Furthermore, we know the camera matrix: if a point is defined in the sparse disparity map we can know its 3D relative position in the scene, and vice versa. From these information, we want to obtain a complete and accurate disparity map (see Fig. 1c).

We will suppose that the scene is a collection of different **objects**. Each object's surface is composed of one or multiple **simple shapes**. We will consider here that those shapes can be represented by a plane: all the 3D points inside

(a) Left image   (b) Sparse disparity map   (c) Objective

Fig. 1. From sparse disparity map to complete one, illustrated by the Vintage pair of the Middlebury dataset [22].

a shape can be modeled using a bivariate linear polynomial, that we will call **model**.

Our method will therefore try to separate all the simple shapes in the reference (left) image. If each region is a simple shape, we can obtain a 3D point cloud estimate of the corresponding region from the sparse disparity map (by converting each disparity value to a 3d point) and modelize it by a bivariate linear polynomial. The model can then be used to correct existing disparity values and fill undefined values in the region, by converting back each 3d point of the model to a disparity value. However, this is not automatically feasible by a segmentation algorithm, so we will instead rely on hierarchical segmentation.

The hierarchical segmentation of an image can be represented by a **partition tree**. The concept is very similar to the binary partition tree [21] presented by Salembier and Garrido, except that a node can have more than two children. The top node represents the whole image, its children represent the most important regions of the image according to the algorithm used, their children represent secondary regions, and so on until a node can't be separated into more sub regions.

An adapted segmentation can't be obtained by simply choosing a specific level of the partition tree: nodes at the lower levels will produce a very coarse segmentation of the image, nodes at the higher levels will produce an over-segmentation. At the middle, some simples shapes are over-segmented and others are merged with neighboring shapes.

We will therefore proceed the following way. We start from the top node, and try to modelize its associated 3D point cloud. If the modelization is satisfying or if the node doesn't have any children, we stop here, associate the model to the node, and delete all the node's children. If it is not satisfying, we retrieve the node's children, and apply the same process individually to each node.

If we combine all the leafs of the resulting tree, we can obtain a **modeled segmentation** of the image, where each region is associated to a model. Some post-processing are done on this modeled segmentation that we will describe in Sect. 2.6. It can then be converted to a complete disparity map.

In order to concretize this general concept, we need to define the following: the sparse disparity map S, the partition tree H, the region modelization and the conditions defining a satisfying modelization.

We will now present those different components. For doing that, we will need to define some parameters. They have been chosen using a grid search to minimize the average error of the disparity maps produced for the Middlebury dataset [22].

2.2   The Sparse Disparity Map S

In order to compute a complete disparity map, we first need an initial sparse disparity map. The quality of our complete disparity map is highly correlated to the quality of the initial disparity map, so choosing an accurate algorithm at this stage is crucial.

We chose to compute this disparity map between the left and right image using the MC-CNN algorithm [27] as it evaluates high quality disparity maps. The algorithm uses a convolutional neural network for computing the matching cost of two 11×11 blocks (one from the left image and one from the right image). The results are then refined using a series of post-processing steps: cross-based cost aggregation, semiglobal matching, a left-right consistency check [11], subpixel enhancement, a median filter, and a bilateral filter. The authors provide two sets of architectures: one optimized for speed, and one for accuracy. We chose the one for accuracy.

To give a sense on how critical the quality of the sparse disparity map is for the performance of our algorithm, we previously used the semi-global MGM [9] algorithm for producing our initial maps. The final disparity maps' average error on the Middlebury dataset [22] was more than 50% higher compared to the current algorithm.

2.3   The Partition Tree H

Obtaining an adapted partition tree for our reference images (left images) is not a straightforward task.

First, the assumption of having simple shapes separated at least at the leafs of the partition tree can be difficult to attain on some images. Segmentation, whether it provides hierarchical information or not, relies on the image gradient. An object occluding another one with similar color can, in the image, have indistinguishable borders with very low or even null gradient.

Secondly, the partition tree must divide the image in a progressive manner. If it does not - for instance a simple shape is merged with another one at a level and divided into twenty parts at the next level of the tree - it can cause our approach to poorly perform in many ways. Since the input disparity map is sparse, it is possible that some regions at the over-segmented level would contain no disparity values making them impossible to modelize. Since regions are much smaller and contain fewer disparity points, the modelization would be more sensitive to errors in the sparse disparity map.

We took these two problems into account when constructing our partition tree. This partition tree will be obtained from the image gradient and defined markers, so we address these issues when computing them.

A first problem that can occur is to have a border with a very progressive change of color between two regions. If we compute a morphological gradient [4] of size 1 in those areas, we will have a very low gradient around those borders, preventing an accurate segmentation. There is a change of color that is occurring, but we need to look at a larger scale. However, we don't want to simply use a morphological gradient with a larger size since we might lose out on small details. That is why we used the multi-scale gradient presented in [7]. We computed this gradient from scale 1 to scale 6.

For markers, we initially used the h-minima transformation of the gradient, with $h = 5$. A second problem is that, when two adjacent regions have similar colors, the gradient can be less than h at some parts of the border between them. This effect is known as gradient leakage [24]. A way to detect these leakage is that markers will get thinner at these locations. Therefore, we want to encourage a split when the markers get thinner. That is why we applied on these markers the adaptive erosion presented in [6] with $\alpha = 0.25$.

We then compute the valued watershed segmentation from the gradient and markers, and apply on it the enhanced waterfalls operator discussed in [5], as it better preserves the important contours than the standard waterfall algorithm [4]. The result is a grayscale image S_h. If $S_h(x, y) = n$ and $n \neq 0$, then the pixel is a border between two or more regions at the level n or higher of the hierarchical segmentation. The higher the level, the coarser the segmentation. See Figs. 2a–c.

Algorithm 1 shows how we transform S_h to a partition tree:

Algorithm 1: From S_h to a partition tree

Function $getPartitionTreeFromSh(S_h)$

    $N \leftarrow \max(S_h)$;   // Set N as the highest level of segmentation

    $T \leftarrow \text{newPartitionTree}()$;

    foreach n in $[N, N-1, ..., 2, 1]$ **do**

        $S_h^n \leftarrow S_h \geq n$;   // set S_h^n as the segmentation at level n

        $L_h^n \leftarrow label(\overline{S_h^n})$;   // Label S_h^n and save it in L_h^n

        $L_h^n \leftarrow unique(L_h^n)$; /* Make each label unique in L_h^n: no similar labels in L_h^n should exist in previously processed levels */

        foreach $label\ l$ in L_h^n **do**

            $R \leftarrow L_h^n == l$; // Set R as the region in L_h^n where label = l

            // Set p as parent id of label l

            if $n == N$ **then**

                $p \leftarrow T.\text{topNode}$;   // The parent is the top node

            else

                $p \leftarrow$ Get label in L_h^{n+1} inside the region R ;   /* Note: no region in L_h^n can belong to two regions in L_h^{n+1} */

            $T.\text{addNode}(id=l, parentId=p, region=R)$;

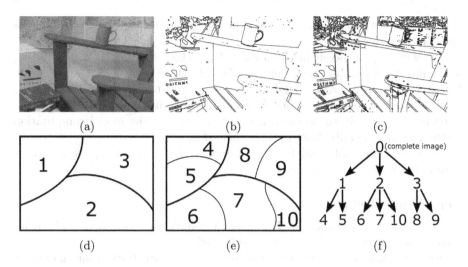

Fig. 2. Waterfall and hierarchy tree. (a) Extract of the left image of the Adirondack pair in the Middlebury dataset. (b) Top level of the enhanced waterfall segmentation of (a). (c) Intermediate level. (d) Example of a top level of waterfall segmentation, where each region has been labeled. (e) Example of a lower level of waterfall segmentation, labeled. (f) Hierarchy tree constructed from (d) and (e).

2.4 Region Modelization

When we process each node of the partition tree, we want to obtain a model that best reflects the real geometry of the region. For doing that, we have an estimated 3D point cloud extracted and converted from the sparse disparity map which contain errors and bias. We need to address two points. First, we have to define whether we take all the 3D points into account or we select a specific subset. Then, we have to specify how the model is computed from this subset.

Points Selection

A common issue is the well-known fattening effect [14]. For computing each point's disparity, stereo algorithms generally match blocks of pixels [9,12,16] instead a single pixels for more robustness. However, around strong edges, the center of the block inherits the disparity of the more contrasted pixels in the block, hence the fattening effect. The worst consequences of this effect take place around occlusions: near the border separating two objects, pixels on the occluded object will inherit the disparity of pixels on the occluding object. Such an issue can seriously impede the modelization process.

Since disparities around borders are not reliable, we decided not to take them into account when computing the model. For doing that, we first separate the region A into two regions: A_b being the border of A and A_i being the inside of A. More specifically $A_i = \varepsilon(A)$, $A_b = A \setminus A_i$, $\varepsilon(A)$ being the morphological erosion of A.

We want here to select all the pixels in A_i whose matching blocks don't intersect with the border. Therefore the retained pixels $\hat{A}_i$ belong to $\hat{A}_i = \varepsilon_{\frac{B}{2}}(A)$, where $\varepsilon_{\frac{B}{2}}$ is the morphological erosion of size $\lceil\frac{B}{2}\rceil$ (on a square grid) and B is the matching block size.

We will keep all pixels in A_b in the subset where the models will be fitted, as $\hat{A}_i$ might contain an insufficient number of disparity values for an accurate fitting. For instance, a region might have a completely textureless inside resulting in an absence of disparity values. Though there might be some outliers in the borders, due notably to occlusions, we are relying on the model computation method we will describe later to filter them out.

As a result, we will modelize only the 3D points from the sparse disparity map in the region $\hat{A} = (\hat{A}_i \cup A_b)$.

Points Modelization

Now that we have selected the 3D points, we can compute our model. There are however some issues that need to be addressed.

First, we have seen previously that some points from A_b can contain outliers which disparity have been contaminated by a nearby object. An other issue is the presence of small areas with disparity values very different from the ground truth [18]. The problem generally appears on areas where there is very little or repetitive texture, as the stereo matching algorithms generally fail on these areas.

We will therefore proceed the following way. We first modelize the points using a linear regression: we try to fit $z = A + Bx + Cy$. If the model is satisfying (see Sect. 2.5), we stop here and the linear equation is affected as the model of the region. If it isn't, we use RANSAC [10] to perform a robust regression on these 3D points: if the proportion of outliers isn't too large, they should be automatically excluded from the model evaluation process thanks to RANSAC. The obtained linear equation is then affected as the model of the region.

If there are no points in the sparse disparity map, we can't compute any model: we will affect to the region an undefined model.

2.5   Conditions Defining a Satisfying Modelization

We have constructed a model from a set of 3D points, and now we want to know if the model is satisfying. By satisfying, we mean that we want a large proportion of points that are close to the model. First, we convert the model to disparity values thanks to the camera matrix. For each point i of the set, we know therefore its value in the sparse disparity map d_i and we also know its disparity value according to the model $\widetilde{d}_i$. A point is an outlier if $|d_i - \widetilde{d}_i| > t_d$, t_d being a custom threshold (we chose $t_d = 2$). We define the fitting score s as the percentage of points in the set that are not outliers. A model is satisfying if $s > t_s$, with $t_s = 70\%$, and if the number of outliers $n_o < 100$.

2.6   Post-processing

We have at this stage a modeled segmentation S_m of our image. To further improve the results obtained, some post-processing are done on S_m. We will note D the disparity map obtained from S_m.

The first problem is that probably some regions have an undefined model because of the absence of points in the sparse disparity map. We have to affect a valid model to these regions, and to do that we will use the modeled segmentation's concept to our advantage. Indeed, these regions are surrounded by regions associated with a model: the main idea is to fill the undefined region with the most probable model in the adjacent regions.

However, since each undefined region might cover multiple simple shapes, we will cut these regions the following way. First, we create a label map ℓ_1 labeling all regions with an undefined model. We create a second label map ℓ_2 labeling all the regions of a similar watershed segmentation than the one computed in Sect. 2.3 but with $h = 12$. We then compute ℓ_f such as $\ell_f(x) = \ell_f(y) \iff ((\ell_1(x) = \ell_1(y)) \wedge (\ell_2(x) = \ell_2(y)))$. Each region labeled in ℓ_f will then be processed independently.

For each label l and associated area A_l, we define its external border as $B_l = (\delta A_l) - A_l$, δA_l being the morphological dilation of A_l of size 1. We list all different models in $S_m(B_l)$. We want to choose the model that creates the largest consensus across adjacent regions separated to the current region by a low gradient. For doing that, we create a list l_p of all positions where $g(B_l) < \min(g(B_l)) + t_g$, g being the gradient of the left image obtained following the process explained in Sect. 2.3. t_g is a threshold to be defined (we chose 10). We affect to $S_m(A_l)$ the model M that maximize the number of points p in l_p such as $(M(p) - D(p)) < t_d$, t_d being the threshold used in Sect. 2.5.

The labels are processed from the one that contain the lowest proportion of undefined models in B_l to the one that contain the highest.

Once S_m and D are filled, we compute S'_m and D' using the same process than for S_m and D but by inverting the left and right images. We remove values in D that are inconsistent according to the LRC check with threshold set to 1. We also remove values at the same positions in S_m. We then fill the values using the same process as described earlier.

3   Results

We benchmarked our TDSR method using the well-known Middlebury dataset [22]. The dataset contains several stereo-images pairs with their associated disparity map ground truth. An executable is also provided allowing to quantitatively compare disparity maps to their ground truth according to various criteria. We chose the following two commonly used criteria in the Middlebury comparative Table: "Bad 2.0", which computes the percentage of disparity values farther than 2.0 from their associated ground truth, and "Avg. error", that computes the average absolute difference between the disparity values and their associated ground truth. These criteria have been separately computed on all values and only on values where no occlusion is occurring. They were computed on full resolution (F).

We compared the results of our TDSR method with three interpolations techniques directly applied on the sparse disparity map: nearest, linear, and cubic.

We also compared our results with the Weighted Least Squares disparity map post filtering method of the OpenCV library [13]. This method uses the Fast Global Smoothing [17] algorithm for diffusing high confidence disparity values using an edge preserving scheme. However, this method is parametric, bringing additional challenges to the comparison. There are three main parameters: "depthDiscontinuityRadius" (or "depthDR"), defining the size of low-confidence regions around depth discontinuities, "lambda", defining the amount of regularization, and "sigma", defining how sensitive the filtering process is to the source image edges. We computed three complete maps: one with the default parameters, one where the parameters have been optimized to minimize "Bad 2.0", and one where the parameters have been optimized to minimize "Avg. error". The parameters have been optimized using a grid search. Tested parameters are shown in Table 1.

Table 1. Tested parameters for FGS.

Parameter	Tested	Default	"Bad 2.0" optimized	"Avg. error" optimized
depthDR	1, 3, 5, 7, 10	5	3	5
lambda ($\times 1000$)	0, 0.1, 0.3, 0.5, 1, 2, 4,6, 8, 10, 12, 14, 16, 20	8	0.1	6
sigma ($/10$)	1, 2.5, 5, 7.5, 10, 12.5, 15, 17.5, 20, 30	10	10	15

Finally, we compared our results with the RBS [3] method, which uses a variant of the bilateral algorithm to propagate disparity values by preserving edges.

Produced disparity maps can be qualitatively compared on the "Vintage" pair of the Middlebury dataset in Fig. 3. They can be quantitatively compared on the whole training set in Table 2.

Compared to other stereo methods in the Middlebury benchmark, on the training set, using the "Avg. error" criterion on all pixels, our solution currently ranks 1st over 59. The results are therefore comparable with state of the art methods.

Interpolation methods are very sensitive to the noise in the sparse disparity map and they don't work well on occlusions. Depending on the chosen parameters, the Fast Global Smoothing method can either produce a less noisy disparity map but with the risk of having the disparity of some objects contaminated by neighboring ones (high "lambda", optimized for "Avg. error"), or more noise but less contamination (low "lambda", optimized for "Bad 2.0"). Default parameters give a good tradeoff between both effects.

Qualitatively, compared to the interpolation, FGS and RBS methods, our approach appears robust to noise and seems to globally manage occlusions. However, some occlusions, notably near the computer screen in the foreground

(a) Left image   (b) Ground truth

(c) Sparse disparity map   (d) Nearest interpolation

(e) FGS with default parameters   (f) TDSR

Fig. 3. Image and disparity map produced for the Vintage pair of the Middlebury dataset.

(see Fig. 3f), are not well evaluated due to lack of data. Some regions can also have their disparity contaminated by nearby objects with similar color.

Quantitatively, compared to the interpolation, FGS and RBS methods, whether we choose to only look at non occluded pixels or to look at all the pixels, our method produces at least similar results according to all chosen criteria. It is important to note that our algorithm's parameters were fixed, though we optimized the parameters of the FGS algorithm for each criterion and compared our algorithm's results to the best cases of the interpolation and FGS methods.

Table 2. Quantitative comparison between interpolation, FGS, RBS, and our proposed TDSR method. *all*: Taking all points into account. *no occ*: Taking only non-occluded points into account.

Method	Bad 2.0 (no occ)	Avg. error (no occ)	Bad 2.0 (all)	Avg. error (all)
Nearest interpolation	10.1	2.83	18.3	7.53
Linear interpolation	10.6	2.71	20.2	7.61
Cubic interpolation	10.6	2.75	20.1	7.70
FGS (Optimized on Bad 2.0)	10.4	2.62	17.4	8.20
FGS (Default)	16.5	2.60	23.1	6.80
FGS (Optimized on Avg. error)	21.3	2.85	27.7	6.61
MC-CNN+RBS [3]	10.8	2.60	19.3	6.66
MC-CNN+TDSR	*10.2*	*2.28*	*15.6*	*4.56*
→ *Improvement to best interp.*	*−1%*	*16%*	*15%*	*40%*
→ *Improvement to best FGS*	*2%*	*12%*	*10%*	*31%*

4   Conclusion

We have presented the Top Down Segmented Regression (TDSR) algorithm that allowed us to densify noisy sparse disparity maps. Our method generated promising results. TDSR is more robust to noise and better preserves the edges than interpolation or diffusion algorithms. The quality of the produced complete disparity map is comparable with state of the art methods.

Our approach can also be used in complement to any stereo matching algorithm as a post-processing step, though its performance will depend on the accuracy of the initial sparse disparity map.

In fact, TDSR could also be adapted to densify other sparse spatial data or to refine dense but noisy data. Dense disparity maps refinement, depth map super-resolution or semantic segmentation post-processing are potential applications that would require very little changes on the proposed approach.

The source code and executable of our implementation are available at: https://hal-mines-paristech.archives-ouvertes.fr/hal-01484143.

References

1. Alvarez, L., Deriche, R., Sánchez, J., Weickert, J.: Dense disparity map estimation respecting image discontinuities: a PDE and scale-space based approach. JVCIR **13**(1–2), 3–21 (2002)
2. Ayache, N., Hansen, C.: Rectification of images for binocular and trinocular stereovision. In: ICPR 1988 (1988)
3. Barron, J.T., Poole, B.: The fast bilateral solver. In: Leibe, B., Matas, J., Sebe, N., Welling, M. (eds.) ECCV 2016. LNCS, vol. 9907, pp. 617–632. Springer, Cham (2016). doi:10.1007/978-3-319-46487-9_38

4. Beucher, S.: Image segmentation and mathematical morphology. Theses, École Nationale Supérieure des Mines de Paris, June 1990
5. Beucher, S.: Towards a unification of waterfalls, standard and P algorithms, working paper or preprint, January 2013
6. Bricola, J.-C., Bilodeau, M., Beucher, S.: A top-down methodology to depth map estimation controlled by morphological segmentation. Technical report (2014)
7. Bricola, J.-C., Bilodeau, M., Beucher, S.: A multi-scale and morphological gradient preserving contrast. In: 14th International Congress for Stereology and Image Analysis, Liège, Belgium, Eric Pirard, July 2015
8. Bricola, J.-C., Bilodeau, M., Beucher, S.: Morphological processing of stereoscopic image superimpositions for disparity map estimation, working paper or preprint, March 2016
9. Facciolo, G., de Franchis, C., Meinhardt, E.: MGM: a significantly more global matching for stereovision. In: BMVC (2015)
10. Fischler, M.A., Bolles, R.C.: Random sample consensus: a paradigm for model fitting with applications to image analysis and automated cartography. Commun. ACM **24**(6), 381–395 (1981)
11. Fua, P.: A parallel stereo algorithm that produces dense depth maps and preserves image features. Mach. Vis. Appl. **6**(1), 35–49 (1993)
12. Hirschmüller, H.: Stereo processing by semiglobal matching and mutual information. IEEE Trans. Pattern Anal. Mach. Intell. **30**(2), 328–341 (2008)
13. Itseez. Open source computer vision library. https://github.com/itseez/opencv (2015)
14. Kanade, T., Okutomi, M.: A stereo matching algorithm with an adaptive window: theory and experiment. In: Proceedings of IEEE ICRA (1991)
15. Kolmogorov, V.: Graph based algorithms for scene reconstruction from two or more views. Ph.D. thesis, Ithaca, NY, USA (2004). AAI3114475
16. Konolige, K.: Small vision systems: hardware and implementation. In: Shirai, Y., Hirose, S. (eds.) Robotics Research, pp. 203–212. Springer Nature, London (1998)
17. Min, D., Choi, S., Lu, J., Ham, B., Sohn, K., Do, M.N.: Fast global image smoothing based on weighted least squares. IEEE Trans. Image Process. **23**(12), 5638–5653 (2014)
18. Moravec, K., Harvey, R., Bangham, J.A.: Improving stereo performance in regions of low texture. In: BMVC (1998)
19. Ralli, J., Díaz, J., Ros, E.: A method for sparse disparity densification using voting mask propagation. J. Vis. Commun. Image Represent. **21**(1), 67–74 (2010)
20. Ralli, J., Pelayo, F., Diaz, J.: Increasing efficiency in disparity calculation. In: Mele, F., Ramella, G., Santillo, S., Ventriglia, F. (eds.) BVAI 2007. LNCS, vol. 4729, pp. 298–307. Springer Nature, Heidelberg (2007). doi:10.1007/978-3-540-75555-5_28
21. Salembier, P., Garrido, L.: Binary partition tree as an efficient representation for filtering, segmentation and information retrieval. In: ICIP 1998 (1998)
22. Scharstein, D., Hirschmüller, H., Kitajima, Y., Krathwohl, G., Nešić, N., Wang, X., Westling, P.: High-resolution stereo datasets with subpixel-accurate ground truth. In: Jiang, X., Hornegger, J., Koch, R. (eds.) GCPR 2014. LNCS, vol. 8753, pp. 31–42. Springer, Cham (2014). doi:10.1007/978-3-319-11752-2_3
23. Scharstein, D., Szeliski, D.: Stereo matching with non-linear diffusion. In: CVPR (1996)

24. Vachier, C., Meyer, F.: The viscous watershed transform. J. Math. Imaging Vis. **22**(2–3), 251–267 (2005)
25. Weickert, J.: Anisotropic diffusion in image processing. Ph.D. thesis (1998)
26. Yang, Q., Wang, L., Yang, R., Stewenius, H., Nister, D.: Stereo matching with color-weighted correlation, hierarchical belief propagation and occlusion handling. In: CVPR (2006)
27. Zbontar, J., LeCun, Y.: Stereo matching by training a convolutional neural network to compare image patches. CoRR, abs/1510.05970 (2015)

Hierarchical Segmentation Based Upon Multi-resolution Approximations and the Watershed Transform

Bruno Figliuzzi[✉], Kaiwen Chang, and Matthieu Faessel

Centre for Mathematical Morphology, Mines ParisTech,
PSL Research University, Paris, France
bruno.figliuzzi@mines-paristech.fr

Abstract. Image segmentation is a classical problem in image processing, which aims at defining an image partition where each identified region corresponds to some object present in the scene. The watershed algorithm is a powerful tool from mathematical morphology to perform this specific task. When applied directly to the gradient of the image to be segmented, it usually yields an over-segmented image. To address this issue, one often uses markers that roughly correspond to the locations of the objects to be segmented. The main challenge associated to marker-controlled segmentation becomes thus the determination of the markers locations. In this article, we present a novel method to select markers for the watershed algorithm based upon multi-resolution approximations. The main principle of the method is to rely on the discrete decimated wavelet transform to obtain successive approximations of the image to be segmented. The minima of the gradient image of each coarse approximation are then propagated back to the original image space and selected as markers for the watershed transform, thus defining a hierarchical structure for the detected contours. The performance of the proposed approach is evaluated by comparing its results to manually segmented images from the Berkeley segmentation database.

Keywords: Watershed transform · Multi-resolution approximations · Hierarchical segmentation · Wavelet transform

1 Introduction

Image segmentation is a classical problem in image processing. Its aim is to provide an image partition where each identified region corresponds to an object present in the image. The watershed transform [1–3] is a popular algorithm based upon mathematical morphology which efficiently performs segmentation tasks. It can easily be understood by making an analogy between an image and a topography relief. In this analogy, the gray value at a given pixel is interpreted as an elevation at some location. The topography relief is then flooded by water coming from the minima of the relief. When water coming from different minima meet at some location, the location is labelled as an edge of the image.

© Springer International Publishing AG 2017
J. Angulo et al. (Eds.): ISMM 2017, LNCS 10225, pp. 185–195, 2017.
DOI: 10.1007/978-3-319-57240-6_15

The watershed algorithm is usually applied to the gradient of the image to be segmented. Each minimum of the gradient therefore gives birth to one region in the resulting segmentation. Due to several factors including noise, quantization error or inherent textures present in the images, gradient operators usually yield a large number of minima. A well-known issue of the watershed algorithm is thus that it usually yields a severely over-segmented image as a result.

To overcome the issue of over-segmentation, a first approach is to apply the gradient operator to images that have previously been filtered. Meyer designed morphological filters, referred to as levelling filters, for this particular task [4,5]. Wavelet based filters have also been used to perform the filtering step. In 2005, Jung and Scharcanski proposed to rely on a redundant wavelet transform to perform image filtering before applying the watershed [6]. The advantage of using the wavelet transform is that its application tends to enhance edges in multiple resolutions, therefore yielding an enhanced version of the gradient. Jung subsequently exploited the multi-scale aspects of the wavelet transform to guide the watershed algorithm toward the detection of edges corresponding to objects of specified sizes [7].

An alternative to overcome the over-segmentation issue is to rely on markers to perform the watershed segmentation. This strategy builds upon the assumption that it is possible to roughly determine the location of the objects of interest to be segmented. The idea is then to perform the flooding from these markers rather than from the minima of the gradient. Another approach that was considered is to select the minima of the gradient according to their importance, by considering for instance h-minima [8,9]. Other approaches including the stochastic watershed rely on stochastic markers that are used to evaluate the frequency at which a contour appear in the segmentation [10]. Finally, following a classical trend in image segmentation [11,12], morphological algorithms have been proposed to perform a bottom-up region merging according to some morphological criteria [13,14].

In this article, our aim is to present a novel method to select markers for the watershed algorithm based upon multi-resolution approximations. The main principle of the method is to rely on the orthogonal wavelet transform to obtain successive approximations of the image to be segmented. The minima of the gradient image of each coarse approximation are then propagated back to the original image space and selected as markers for the watershed transform, thus defining a hierarchical structure for the detected contours. The article is organized as follows. We describe the proposed algorithm and state the main properties of the obtained contours hierarchy in Sect. 2. In Sect. 3, we evaluate the performances of the algorithm on the Berkeley segmentation database. Conclusions are finally drawn in the last section.

2   Multiscale Watershed Segmentation

2.1   Multi-resolution Approximation

A function f from $\mathbb{R}^2$ to $\mathbb{R}$ is said to be square-integrable if and only if the integral

$$\int_{\mathbb{R}^2} |f(x_1, x_2)|^2 dx_1 dx_2 \qquad (1)$$

is finite. We denote by $\mathbf{L}^2(\mathbb{R}^2)$ the set of square-integrable functions. When equipped with the scalar product

$$< f, g >= \int_{\mathbb{R}^2} f(x_1, x_2)g(x_1, x_2)dx_1 dx_2, \qquad (2)$$

it is well-known that $\mathbf{L}^2(\mathbb{R}^2)$ is an Hilbert space of infinite dimension.

Let us first introduce the mathematical notion of multi-resolution approxima-tion [15] which plays a central role in the proposed approach. A multi-resolution of $\mathbf{L}^2(\mathbb{R})$ is a sequence $\{V_j\}_{j\in\mathbb{Z}}$ of closed subspaces of $\mathbf{L}^2(\mathbb{R})$ satisfying the fol-lowing properties

1. $\forall (j, k) \in \mathbb{Z}^2,\ f(.) \in V_j \Leftrightarrow f(. - 2^j k) \in V_j$.
2. $\forall j \in \mathbb{Z},\ V_{j+1} \subset V_j$,
3. $\forall j \in \mathbb{Z},\ f(.) \in V_j \Leftrightarrow f(./2) \in V_{j+1}$,
4. $\lim_{j\to-\infty} V_j = \cap_{j=-\infty}^{j=+\infty} V_j = \{\emptyset\}$,
5. $\lim_{j\to+\infty} V_j = \mathbf{Closure}(\cup_{j=-\infty}^{j=+\infty} V_j) = \mathbf{L}^2(\mathbb{R})$,

In addition, for a sequence $\{V_j\}_{j\in\mathbb{Z}}$ to be a multi-resolution of $\mathbf{L}^2(\mathbb{R})$, there must exist a function θ in $\mathbf{L}^2(\mathbb{R})$ such that the family $\{\theta(t - n)\}_{n\in\mathbb{Z}}$ is a basis of V_0.

Multi-resolution approximations have been extensively used in computer vision since their introduction in the article [16] of Burt and Adelson. From this perspective, a signal of dyadic size 2^J is the orthogonal projection of a func-tion f in $\mathbf{L}^2(\mathbb{R})$ on some space $V_J \subset L^2(\mathbb{R})$. The approximation of the signal at a resolution 2^{-j}, with $j > J$, is defined as its orthogonal projection on a subspace V_j. In higher dimensions D, e.g. for images, multi-resolution approximations of $\mathbf{L}^2(\mathbb{R}^D)$ can be obtained by considering tensorial products between subspaces: $V_j^D = V_j \otimes V_j \otimes \ldots \otimes V_j$.

In our algorithm, we consider a multi-resolution approximation based upon a discrete image wavelet decomposition. It is possible to define a scaling function ϕ from the Riesz basis $\{\theta(t - n)\}_{n\in\mathbb{Z}}$ of V_0. An approximation of the image at scale j is computed by projecting the approximation image at scale $j - 1$ on the family $\{\phi_j(x - n)\}_{n\in\mathbb{Z}}$ of scaling functions, where

$$\phi_j(x) = \frac{1}{\sqrt{2^j}} \phi(\frac{x}{2^j}). \qquad (3)$$

It can be shown that the projection can be computed by relying on iterative convolutions with a low-pass filter, followed by a factor 2 sub-sampling. The dis-crete decimated wavelet transform of the original image is obtained by iteratively filtering the approximation image by tensorial products of a low-pass filter and of a high-pass filter, yielding one approximation image and three details images at each iteration [15].

Fig. 1. Wavelet transform associated to Lena image for two decomposition levels. The size of the image is 512 by 512 pixels. The decomposition was performed using Daubechies wavelet with 12 vanishing moments. The approximation image corresponds to the subimage on the top left quarter. The other three subimages correspond to the projection of each approximation on a family of wavelets. We note that the main objects are relatively well preserved by each approximation image.

2.2   Hierarchical Contour Detection

A multi-resolution approximation of an image is presented in Fig. 1, for two levels of decomposition. The multi-resolution approximation is computed using Daubechies wavelets with 12 vanishing moments. Interestingly, we can note that the main objects and contours of the original image are relatively well preserved in its approximations. However, these images contain considerably less details than the original image, making them of potential interest for segmentation. Several methods have been proposed to handle segmentation using multi-resolution approaches. In 2000, Rezaee *et al.* [17] notably proposed an algorithm combining the pyramid transform and fuzzy clustering, obtaining good segmentation results on magnetic resonance images. In 2003, Kim and Kim [18] proposed a segmentation procedure relying on pyramidal representation and region merging.

It is straightforward to directly apply a watershed algorithm on the approximation images. However, the resolution of these images is significantly lower than the one of the original image. It is therefore difficult to establish a direct

Fig. 2. Original image from the Berkeley segmentation database used to illustrate the algorithm, along with an example of human segmentation.

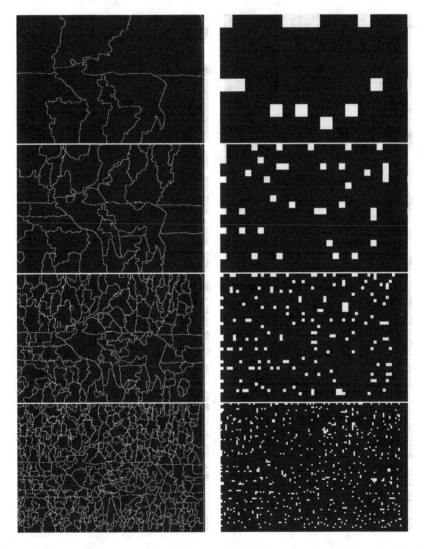

Fig. 3. Contour images corresponding to four decomposition levels (levels 4, 3, 2, 1 respectively) of the image presented in Fig. 2. The markers at each scale are displayed on the right images.

Fig. 4. Original image from the Berkeley segmentation database used to illustrate the algorithm, along with an example of human segmentation.

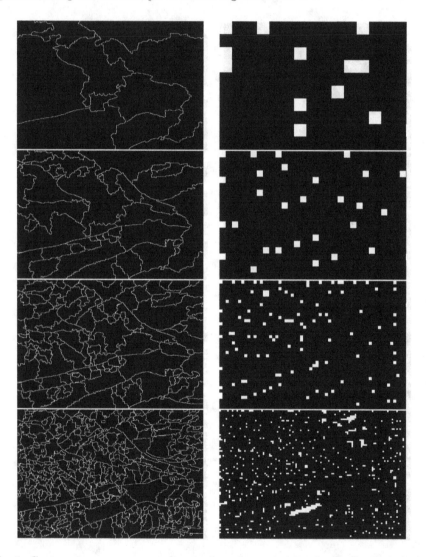

Fig. 5. Contour images corresponding to four decomposition levels (levels 4, 3, 2, 1 respectively) of the image presented in Fig. 4. The markers at each scale are displayed on the right images.

correspondence between the contours of the approximation image and the contours of the original image [18]. In this study, we propose to tackle this issue in a simple manner by defining multi-scale markers for performing the segmentation.

Let us consider an image I of dyadic size 2^J by 2^J pixels. We denote by L the number of decomposition levels. For l between 0 and L, we denote $I^{(l)}$ the approximation of I after l decomposition steps. The size of the image $I^{(l)}$ is therefore 2^{J-l} by 2^{J-l} pixels. By convention, $I^{(0)}$ is the original image I. $I^{(l+1)}$ is obtained by successively applying the low-pass filter associated to the wavelet transform to the rows and the columns of $I^{(l)}$ and down-sampling the result by a factor 2.

To initialize the contour detection algorithm, we first extract the minima of the gradient of the approximation image $I^{(L)}$. We obtain a sequence $\{m_1^{(L)}, ..., m_{K_L}^{(L)}\}$ of K_L markers. The locations of these markers are specified on an image $M^{(L)}$ of size 2^{J-L} by 2^{J-L}. To propagate the markers back to the original image I, we oversample the image $M^{(L)}$ by replacing each pixel $M^{(L)}(p,q)$ by an array of pixels of size 2^L by 2^L whose value is set equal to $M^{(L)}(p,q)$. Then, we apply the watershed algorithm on the image I from the image markers $M^{(L)}$, therefore obtaining a contour image $C^{(L)}$ associated to the approximation image $I^{(L)}$. We then repeat these operations at decomposition level $L-1$ to obtain a contour image $E^{(L-1)}$. We consider then the supremum image $S^{(L-1)}$ defined by

$$S^{(L-1)} = \sup(C^{(L)}, E^{(L-1)}), \tag{4}$$

where the supremum is defined as follows for each pixel $S^{(L-1)}[p,q]$:

$$S^{(L-1)}[p,q] = \max(C^{(L)}[p,q], E^{(L-1)}[p,q]). \tag{5}$$

Finally, we apply a watershed transform to the supremum image $S^{(L-1)}$ to obtain the contour image $C^{(L-1)}$ corresponding to decomposition level $L-1$. This last step is necessary to remove the thick contours that can potentially be created by the supremum between images $C^{(L)}$ and $E^{(L-1)}$. We obtain a multi-scale contour by iteratively applying the procedure described previously for all decomposition levels.

To illustrate the algorithm, the segmentation algorithm is applied to the images displayed in Figs. 2 and 4. The results are displayed in Figs. 3 and 5. We used Daubechies wavelets with 12 vanishing moments to calculate the successive multi-resolution approximations. By construction, the algorithm returns a nested sequence of contours, in the sense that if a pixel of the contour image $C^{(l)}$ is labelled as a contour, then the same pixel in the contour image $C^{(l-1)}$ is also labelled as a contour. The proposed approach therefore results in a hierarchical segmentation. We can see that as expected, the contours of the largest objects tend to be extracted for the approximations with the lowest resolution. By contrast, all contours corresponding to the details of the image to be segmented appear for the approximation images of higher resolution.

3   Experimental Results and Discussion

In Sect. 2, we introduced a hierarchical contour detection algorithm based upon successive multi-resolution approximations of the image to be segmented. It is of interest to assess the validity of the approach by comparing the results of the algorithm to human segmentations. To that end, we rely on the Berkeley Segmentation Database (BSD) [19]. The BSD is a large dataset of natural images that have been manually segmented. At each scale of the contours hierarchy, we compare the results of the detection algorithm to the human annotations.

In our comparison, we consider two criteria. We first estimate the proportion of contours detected by the algorithm and that have also been annotated (precision). To that end, we apply a dilation of size two to the contour image corresponding to the human segmentation, and we consider the intersection between the dilated image and the contour image returned by the algorithm at each decomposition level. The dilation is applied to account for potential inaccuracy of the human contour detection. The number of pixels belonging to the intersection normalized by the number of pixels corresponding to the detected contours provides us with an estimate of the proportion of detected contours that have also been manually segmented. The second criteria that we consider is the proportion of contours that have been detected by humans and that are also detected by the algorithm (recall). At each decomposition level, we apply a dilation of size two to the contour image returned by the algorithm and we consider the intersection between the dilated image and the contour image corresponding to the human segmentation. By counting the number of pixels of the intersection normalized by the number of pixels belonging to the human detected contours, we can determine the proportion of actual contours that are returned by the detection algorithm at each scale.

We estimated both criteria on 200 images from the BSD. The results are presented in Table 1. We can note that on average, the precision increases with the decomposition level in the contours hierarchy. This tends to assess the validity of the proposed multi-scale approach, in the sense that the contours detected with markers obtained after several decomposition levels have a significantly higher probability to correspond to human detected contours than contours directly obtained with the watershed transform. We also note that the recall of the algorithm decreases monotonously on average. This trend was to be expected, since the multi-scale approach inherently removes the contours of the smallest patterns. However, up to three decomposition levels, the recall remains higher than 0.5.

It is finally of interest to compare the results of the wavelet based markers selection to other commonly encountered methods, namely markers selection through h-minima of the image gradient and contour detection after filtering by alternate sequential filters. The difficulty here is to obtain a comparable number of segments between the distinct approaches. To that end, for each image, we select the value of h-minima and the size of structuring element for the alternate sequential filter, respectively, that yield the number of markers the closest from the one used in the wavelet based segmentation. We repeat the process for each decomposition scale. Next, we rely on the aforementioned procedure to estimate the precision and the recall of these methods. The results are summarized in Table 1. We note that, on average, the segmentation based upon the discrete wavelet transform performs significantly better in terms of both precision and recall than the segmentation following an alternate sequential filtering. Our interpretation of this result is that the wavelet transform better preserves the contours of the original image at the highest scales of the transform. By contrast, for structuring elements of large size yielding a number of markers similar to the one obtained with the wavelet decomposition, alternate sequential filters significantly degrade the contours of the image, which explains the poor results that are registered in terms of precision and recall. Marker selection through h-minima values is shown to yield a higher precision than the wavelet based algorithm. However, in terms of recall, the wavelet based algorithm performs significantly better on average. Both approaches remain however significantly distinct and are difficult to compare, since a segmentation based upon markers selection by h-minima is highly sensitive to the local minima of the image gradient.

Table 1. Results of the wavelet based algorithm on the BSD for each decomposition level. Precision corresponds to the average proportion of contours detected by the algorithm that have also been annotated, along with the corresponding standard deviation. Recall corresponds to the average proportion of contours that have been detected by humans and that are also detected by the algorithm, along with the corresponding standard deviation. The proportions are obtained on a database of 200 images. The results of h-minima based segmentation and alternate sequential filtering based segmentation are also presented.

	Wavelet filtering		h-minima		Alternate sequ. filters	
Dec. level	Precision	Recall	Precision	Recall	Precision	Recall
0	0.10 ± 0.05	0.94 ± 0.03	0.12 ± 0.06	0.85 ± 0.16	0.10 ± 0.04	0.91 ± 0.06
1	0.13 ± 0.07	0.82 ± 0.07	0.17 ± 0.08	0.71 ± 0.18	0.12 ± 0.05	0.69 ± 0.09
2	0.17 ± 0.09	0.69 ± 0.11	0.24 ± 0.13	0.51 ± 0.14	0.12 ± 0.05	0.46 ± 0.08
3	0.22 ± 0.12	0.53 ± 0.14	0.30 ± 0.20	0.21 ± 0.08	0.12 ± 0.05	0.22 ± 0.07

4 Conclusion and Perspectives

In this article, we presented a new method to select markers for the watershed algorithm based upon multi-resolution approximations. By relying on the discrete decimated wavelet transform to obtain successive approximations of the image to be segmented, we were able to define a hierarchical structure for the detected contours. We evaluated the performance of the proposed approach by comparing its results to manually segmented images from the Berkeley segmentation database. The comparison provided an empirical evidence that the contours detected for the approximation of lowest resolutions have a higher probability to correspond to human detected contours than contours detected by the classical watershed transform.

An interesting perspective of this study is to use the proposed algorithm in a bottom-up aggregation procedure to obtain a dissimilarity measure between adjacent regions, corresponding to the decomposition scale at which the contour appears in the hierarchical segmentation. The empirical results obtained on the Berkeley Segmentation Database indicate indeed that regions separated by contours appearing early in the hierarchical segmentation procedure are less likely to be merged. We also noted on the comparison performed with the BSD that at the lowest resolutions, the recall of the detection method is around 0.5. A natural extension of this work is therefore to rely on additional statistical features that can eliminate wrong contours while keeping the actual contours in a bottom-up aggregation procedure.

References

1. Beucher, S., Lantuéjoul, C: Use of watersheds in contour detection (1979)
2. Meyer, F., Beucher, S.: Morphological segmentation. J. Vis. Commun. Image Represent. **1**(1), 21–46 (1990)
3. Vincent, L., Soille, P.: Watersheds in digital spaces: an efficient algorithm based on immersion simulations. IEEE Trans. Pattern Anal. Mach. Intell. **13**(6), 583–598 (1991)
4. Meyer, F.: From connected operators to levelings. Comput. Imaging Vis. **12**, 191–198 (1998)
5. Meyer, F.: Levelings, image simplification filters for segmentation. J. Math. Imaging Vis. **20**(1–2), 59–72 (2004)
6. Jung, C.R., Scharcanski, J.: Robust watershed segmentation using wavelets. Image Vis. Comput. **23**(7), 661–669 (2005)
7. Jung, C.R.: Combining wavelets and watersheds for robust multiscale image segmentation. Image Vis. Comput. **25**(1), 24–33 (2007)
8. Soille, P.: Morphological Image Analysis: Principles and Applications. Springer Science & Business Media, Berlin (2013)
9. Cheng, J., Rajapakse, J.C., et al.: Segmentation of clustered nuclei with shape markers and marking function. IEEE Trans. Biomed. Eng. **56**(3), 741–748 (2009)
10. Angulo, J., Jeulin, D.: Stochastic watershed segmentation. In: Proceedings of the 8th International Symposium on Mathematical Morphology, pp. 265–276 (2007)
11. Felzenszwalb, P.F., Huttenlocher, D.P.: Efficient graph-based image segmentation. Int. J. Comput. Vis. **59**(2), 167–181 (2004)
12. Alpert, S., Galun, M., Brandt, A., Basri, R.: Image segmentation by probabilistic bottom-up aggregation and cue integration. IEEE Trans. Pattern Anal. Mach. Intell. **34**(2), 315–327 (2012)
13. Beucher, S.: Watershed, hierarchical segmentation and waterfall algorithm. In: Serra, J., Soille, P. (eds.) Mathematical Morphology and its Applications to Image Processing, pp. 69–76. Springer, Heidelberg (1994)
14. Marcotegui, B., Beucher, S.: Fast implementation of waterfall based on graphs. In: Ronse, C., Najman, L., Decencière, E. (eds.) Mathematical Morphology: 40 Years On, pp. 177–186. Springer, Heidelberg (2005)
15. Mallat, S.: A Wavelet Tour of Signal Processing. Academic Press, Cambridge (1999)
16. Burt, P., Adelson, E.: The Laplacian pyramid as a compact image code. IEEE Trans. Commun. **31**(4), 532–540 (1983)

17. Rezaee, M.R., Van der Zwet, P.M.J., Lelieveldt, B.P.E., Van Der Geest, R.J., Reiber, J.H.C.: A multiresolution image segmentation technique based on pyramidal segmentation and fuzzy clustering. IEEE Trans. Image Process. **9**(7), 1238–1248 (2000)
18. Kim, J.-B., Kim, H.-J.: Multiresolution-based watersheds for efficient image segmentation. Pattern Recogn. Lett. **24**(1), 473–488 (2003)
19. Martin, D., Fowlkes, C., Tal, D., Malik, J.: A database of human segmented natural images and its application to evaluating segmentation algorithms and measuring ecological statistics. In: Proceedings of the Eighth IEEE International Conference on Computer Vision, ICCV, vol. 2, pp. 416–423. IEEE (2001)

Topological and Graph-Based
Clustering, Classification and Filtering

Power Tree Filter: A Theoretical Framework Linking Shortest Path Filters and Minimum Spanning Tree Filters

Sravan Danda[1(✉)], Aditya Challa[1], B.S. Daya Sagar[1], and Laurent Najman[2]

[1] Systems Science and Informatics Unit, Indian Statistical Institute, 8th Mile,
Mysore Road, Bangalore, India
sravan8809@gmail.com
[2] Université Paris-Est, LIGM, Equipe A3SI, ESIEE, Paris, France

Abstract. Edge-preserving image filtering is an important pre-processing step in many filtering applications. In this article, we analyse the basis of edge-preserving filters and also provide theoretical links between the MST filter, which is a recent state-of-art edge-preserving filter, and filters based on geodesics. We define *shortest path filters*, which are closely related to adaptive kernel based filters, and show that MST filter is an approximation to the $\Gamma-$limit of the shortest path filters. We also propose a different approximation for the $\Gamma-$limit that is based on union of all MSTs and show that it yields better results than that of MST approximation by reducing the leaks across object boundaries. We demonstrate the effectiveness of the proposed filter in edge-preserving smoothing by comparing it with the tree filter.

1 Introduction

Image filtering is a prominent problem in computer vision which has been studied for several years. Along with the relevant information, images often contain noise as well as irrelevant information such as texture. Image filtering is usually the first step of several applications such as image abstraction, texture removal, texture editing, stereo matching, optical flow etc. One needs to make sure that relevant information does not get 'smoothed' out during the filtering step, and thus edge preserving filters were developed. There exists several filters for edge-preserving smoothing such as bilateral filter (BF) [18], guided filter (GF) [11], weighted least squares filter (WLS) [7] and L_0 smoothing [19]. Among the other noted techniques are propagated image filter [2] and relative total variation (RTV) filter [20].

Minimum Spanning Tree (MST) filters are one of the edge preserving filters which have been proved to be effective in applications such as scene simplification, texture editing and stereo matching. To the best of our knowledge, Stawiaski et al. [17] first used a MST for filtering. Later, tree filter (TF) [1] was developed as a non-local weighted-average filter on similar idea. The TF can be implemented in linear time [21] (linear in number of pixels) and yields promising

© Springer International Publishing AG 2017
J. Angulo et al. (Eds.): ISMM 2017, LNCS 10225, pp. 199–210, 2017.
DOI: 10.1007/978-3-319-57240-6_16

results. Although the TF yields good results, it admits a drawback (termed as leak problem) due to the fact that any MST has some higher weight edges leading to high collaboration across the object boundaries. As a result, the edges across these boundaries are not well preserved in the filtered image and cause a leakage. Also, these filters are based on heuristics and exploring theoretical links with other existing filters sheds more light on why these work well.

In this article, we aim to provide a theoretical justification for the MST filtering as described in [1] using the concept of Γ-limit (defined in Sect. 3). Loosely speaking, we show that the MST filtering is an approximate solution to the Γ-limit of the shortest path filters which are morphological amoeba-like filters. We also describe another approximation, which reduces the 'leak' problem.

2 Background

2.1 Tree Filter

In this section we describe the MST filtering as in [1]. Let I be a given image and let S denote the filtered image. Let I_i denote the intensity or color of pixel i in the image. Note that given an image I, one can construct a 4-adjacency graph, with weights between adjacent pixels reflecting the absolute difference of intensities or color. More formally, for adjacent pixels i and j, with intensities I_i and I_j respectively, we can define

$$w_{ij} = 1 + \|I_i - I_j\| \tag{1}$$

Note that adding a positive constant ensures that $\Delta(i, j)$ in Sect. 3 is well defined and the MSTs of the graph are invariant. With slight abuse of notation, we refer to this graph by I as well. One can then construct an MST on this graph, say I_{MST}. Note that on any given MST, there exists a unique path between any two pixels. Let $D(i, j)$ denote the number of edges on the path between the pixels i and j on I_{MST}. Define $t_i(j)$ by

$$t_i(j) = \frac{exp(-\frac{D(i,j)}{\sigma})}{\sum_q exp(-\frac{D(i,q)}{\sigma})} \tag{2}$$

Then tree filter is defined as, for each pixel i

$$S_i = \sum_j t_i(j)I_j \tag{3}$$

where $t_i(j)$ denotes the collaborative weights between pixels i and pixel j, σ denotes the falling rate and the summations in Eqs. (2) and (3) are over all the pixels in the image.

The main advantage of the tree filter is the edge preserving property. Intuitively, any MST includes only a few high weighted edges and thus reduces the collaborations across most of the object boundaries. However, as any MST is connected, the existence of at least one edge on the boundary of every object is guaranteed. These edges lead to high collaborative weights and the 'leaks' arise. Apart from the 'leak' problem, the above method does not have a theoretical justification for its working. Another drawback of this method is: since MST of a graph is not unique, the filtering results depends on the choice of MST which is not desired.

2.2   Gamma Convergence

The main idea of Γ−convergence [13] is to calculate a limit of minimizers of a family of minimum problems. This limit is referred to as a Γ−limit. The main advantages of such a calculation are - (1) Γ−limits share a few properties with the minimizers and are sometimes easier to calculate (2) Γ−limits also exhibit interesting 'new' properties and usually result in novel methods. For instance, Γ−limit of methods described in [4,5,9,14] was calculated in [3] and is shown to have interesting properties. In Sect. 3, we show that the MST filter described above can be seen as an approximation to the Γ−limit of shortest path filters. This then would provide theoretical links to justify the edge-preserving capability of the MST filter.

3   Power Tree Filter

Let $\mathcal{G} = (V, E, W)$ be an edge-weighted graph, where V denotes the set of pixels, E denotes the set of edges given by the 4-adjacency relation and w_{ij} indicates the weight on the edge between pixels i and j. Let $W^{(p)}$ denote the weights in Eq. (1) raised to power p, $\{w_{ij}^p\}$. The graph $\mathcal{G}^{(p)} = (V, E, W^{(p)})$ is called an *exponentiated graph*. We also assume that there are k distinct weights $0 < w_1 < w_2 < \cdots < w_k$.

Note that to every path between pixels i and j, $P(i, j)$, one can assign a tuple $(p_1, p_2, \cdots, p_k)$, where p_i denotes the number of edges of weights defined in Eq. (1) raised to power p. We denote this tuple as the *edge-weight distribution* of the path. Let $(p_1, p_2, \cdots, p_k)$ and $(q_1, q_2, \cdots, q_k)$ be edge-weight distributions of two paths P and Q respectively. Let $A = \{1 \le i \le k : p_i \ne q_i\}$, We say that

$$P \ge Q \Leftrightarrow \text{if } A = \emptyset \text{ or } p_{sup(A)} > q_{sup(A)} \tag{4}$$

This is referred to as the *dictionary ordering*. Note that this induces a complete ordering on the set of paths. We call $P(i, j) = (p_1, p_2, \cdots, p_k)$ a *shortest path* if $\sum w_l p_l$ is smallest for P compared to any other path $Q(i, j)$. Let $\Pi(P(i, j))$ denote the number of edges on the path $P(i, j)$.

The overview of the rest of the section is - We first define *shortest path filters*, and give a heuristic explanation as to their advantage. We then explore their relation to morphological amoeba filters and show that these filters can be seen as minimizers to a cost function. We then calculate the Γ−limit of these minimizers, and show that it is the shortest path filter on the union of minimum spanning trees (UMST).

3.1 Shortest Path Filters

Let i and j be any two pixels. Define

$$\Theta(i,j) = \inf\{\Pi(P(i,j)) \text{ where } P(i,j) \text{ is a shortest path.}\} \qquad (5)$$

Now, consider the cost function

$$Q_i(x) = \sum_j exp\left(-\frac{\Theta(i,j)}{\sigma}\right)(x - I_j)^2 \qquad (6)$$

Then the filtered value at pixel i is given by the minimizer of $Q_i(x)$. From Eq. (6), it can easily be seen that

$$S_i = \arg\min Q_i(x) = \sum_j \frac{exp\left(-\frac{\Theta(i,j)}{\sigma}\right)}{\sum_k exp\left(-\frac{\Theta(i,k)}{\sigma}\right)} I_j \qquad (7)$$

where S denotes the filtered image. Shortest path filters (SPF) can be seen as an edge aware extension to the gaussian convolution. Note that gaussian filter considers exponential decay over spatial distances without considering the edge weights. To adapt the gaussian filter to be edge aware, one might consider taking number of edges on a shortest path instead of spatial distances. In the special case of all pixel values being equal in the image, shortest path filter is exactly gaussian filter with l_1 metric. Adapting the Floyd-Warshall algorithm, one can calculate the SPF which takes $\mathcal{O}(|V|^3)$ time and $\mathcal{O}(|V|^2)$ space. Edge-preserving filters using geodesics are also discussed in [10]. Shortest path filters are also related to the adaptive kernel filter such as morphological amoebas [12], as discussed below.

Relation to Morphological Amoebas. In [12] the authors introduced morphological amoebas (adaptive kernels). The kernel is dictated by

$$\kappa(i,j) = \min_{P(i,j)} L_\lambda(P(i,j)) \qquad (8)$$

where $P(i,j)$ is a path between pixels i and j, $<i = x_0, x_1, \cdots, x_n = j>$ and

$$L_\lambda(P(i,j)) = \sum_{i=0}^{n-1}\left(1 + \lambda\|I_{x_{i+1}} - I_{x_i}\|\right) \qquad (9)$$

Then the kernel at pixel i is given by $\{j, \kappa(i,j) < r\}$, where r denotes the radius of the kernel chosen according to the level of smoothing required. An alternative to the the $\kappa(.,.)$ above is the smallest number of edges on a shortest path, denoted by $\Theta(.,.)$ as in Eq. (5). It is easy to see that

$$\Theta(i,j) = \min L_0(P(i,j)) \text{ subject to } P(i,j) \in \arg\min L_1(P(i,j)) \qquad (10)$$

The results obtained by the shortest path filters are shown in Fig. 1. Observe that, even though edges are better preserved in Fig. 1(c) and (d), compared to gaussian filter Fig. 1(b), there is still a halo effect on the boundaries. Taking the gamma limit would reduce these effects, as shown later. See Fig. 2(b) and (c).

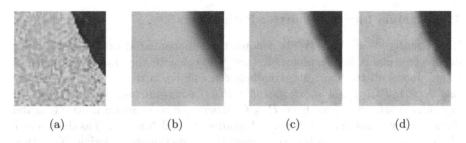

Fig. 1. (a) Original image, (b) Gaussian convolution on (a), (c) Mean filter with shortest path kernel on (a), and (d) Shortest path filter on (a)

Dictionary Ordering, Pass Values and Watershed Cuts. Recall that we are interested in calculating collaborative weights for every pair of pixels, and have concluded that spatial distances alone are not sufficient. Thus for a given pair of pixels, one has to consider the number of edges on a path between them. However, there exists several paths and the choice dictates the filtering method. The minimum of the l^∞ norm over all the paths between a given pair of pixels is called *pass value* [5] or the minimax distance [6] of the pair. Thus, one of the choices is to consider a path with its l^∞ norm equal to the pass value. To the best of our knowledge, this measure was first suggested for adaptive filtering in [17]. The number of edges on one such path might differ from the other creating an ambiguity on which among them to choose. On the other hand, a path with smallest dictionary ordering between a given pair of pixels would also have its l^∞ norm equal to pass value. Moreover, any such path has the same number of edges. The dictionary ordering thus reduces the ambiguity compared to the filters which choose a path with respect to any other norm.

Pass values and shortest paths have also been used in the context of seeded segmentation [5,6,15]. The authors in [4,5] have proved that given a set of seeds, the segments generated by any watershed cut are also the segments generated by minimum spanning forest cuts relative to the seeds and vice versa. Further, they have proved that segments corresponding to watershed cuts have the optimality property that they are connected components in some shortest path forest (SPF) relative to the seeds. The converse does not hold true and the authors thus have provided one way of choosing 'good' SPF cuts.

Now, if a non-seed pixel admits multiple shortest paths to different seeds, to which seed would the pixel be associated? An obvious answer would be the path which has lowest number of l^∞ norm weighted edges. In case of a tie on the number of such edges, the next choice would be to consider the second highest weight, and so on. This exactly reflects the definition of the dictionary ordering in Sect. 3. Thus, in the segmentation aspect, the dictionary ordering provides another way of selecting a SPF cut among all the SPF cuts.

3.2   Gamma Limit of Shortest Path Filters

Recall that $\mathcal{G}^{(p)} = (V, E, W^{(p)})$ denotes the exponentiated graph. Let i and j denote two pixels in the image. Let $\Theta^{(p)}(i,j)$ denote the smallest number of edges among all the shortest paths between i and j in the graph $\mathcal{G}^{(p)}$. We say that a path $P(i,j)$ is a *shortest path with respect to dictionary ordering* if for any other path $Q(i,j)$ we have $P \leq Q$, where the comparison is based on the dictionary ordering defined in Eq. 4. Accordingly we define $\Delta(i,j)$ as the number of edges on a shortest path with respect to the dictionary ordering. Note that $\Delta(i,j)$ is well defined since if for any two paths, P and Q if $P \leq Q$ and $Q \leq P$, then $p_i = q_i$ for all i and hence both the paths have same number of edges. The following is the main theorem of this paper.

Theorem 1. *As $p \to \infty$, we have $\Theta^{(p)}(i,j) \to \Delta(i,j)$ for all pairs i and j. Moreover, any shortest path with respect to the dictionary ordering lies on a MST.*

The above theorem states that the shortest paths converge to the shortest paths with respect to dictionary ordering in the limit. An important consequence of the above theorem is that one can now calculate the Γ-limit of the shortest path filters as described in the following proposition. The proof of the proposition follows from Theorem 1.

Proposition 1. *As $p \to \infty$ the shortest path filter converges to*

$$S_i = \sum_j \frac{exp\left(-\frac{\Delta(i,j)}{\sigma}\right)}{\sum_k exp\left(-\frac{\Delta(i,k)}{\sigma}\right)} I_j \tag{11}$$

We refer to the above limit as *Power Tree Filter* (PTF). As we shall soon see, calculation of the above filter is computationally expensive and an approximation is desired.

Recall that from Eqs. (2) and (3) we have $D(i,j)$ that denotes the path length on a minimum spanning tree, which can be seen as an approximation for $\Delta(i,j)$. This provides a justification for the MST filter - *MST filter is nothing but an approximation to the gamma limit of the shortest path filters.*

3.3   Calculation of the Gamma Limit

In this part, we provide an algorithm to calculate the Γ-limit and give another approximation. Union minimum spanning tree (UMST) is defined as the sub-graph generated by the edges of all the minimum spanning trees. Theorem 1 states that any shortest path with respect to dictionary ordering lies on an MST, and hence on UMST. Thus we need to only concern ourselves with UMST instead of the whole graph. Thanks to the characterization of UMST in [13] which is stated below, we develop an efficient algorithm based on the result.

Theorem 2. *Let $\mathcal{G} = (V, E, W)$ be an edge-weighted graph. Let $\mathcal{G}_{<w}$ denote the graph with the vertex set V and all the edges e_{ij} whose weight $w_{ij} < w$. Let $\mathcal{G}_{UMST}$ denote the UMST of $\mathcal{G}$. Then an edge e with weight $w(e)$ belongs to the $\mathcal{G}_{UMST}$ if and only if the edge e joins two connected components in $\mathcal{G}_{<w(e)}$.*

Theorem 2 is equivalent to the *cut property* of the MST [16]. With a suitable choice of the data structure, the above theorem gives an $\mathcal{O}(|E|)$ (which in our case is $\mathcal{O}(|V|)$) algorithm to calculate the UMST.

Note that the calculation of $\Delta(i, j)$ for each pair (i, j) takes $\mathcal{O}(|V|)$ and hence the crudest algorithm takes $\mathcal{O}(|V|^4)$ time. Using a slightly different version of Floyd-Warshall [8] allows us to calculate the $\Delta(i, j)$ for all pairs in $\mathcal{O}(|V|^3)$, which is still expensive. Thus, one needs an approximation to calculate the Power Tree Filter.

Instead of considering UMST, if one considers a single MST as in MST filter described above, the calculation can be made efficient using top-down and bottom-up calculations in $\mathcal{O}(|V|)$ time [21]. The following steps give an overview of the bottom-up algorithm as described in [21]. (1) Pick a root node i. (2) Starting from the root nodes, update the values recursively using

$$S_p = I_p + \sum_{q \in \text{children of } p} exp(\frac{-1}{\sigma})S_q$$

At the end of these calculations we are ensured to get the filtered value at the root node i. The top-down procedure is similar. For further details see [21]. This justifies taking an MST as an approximation to UMST. A different method is described below, based on a few observations.

Proposition 2. *Given a pixel i, there exists an MST, M, such that the shortest path with respect to dictionary ordering between pixels i and any other pixel j is in M.*

The MST in Proposition 2 is referred to as *Power MST*. If one can calculate the MST as in Proposition 2 for each pixel i, the bottom-up procedure as in [21] can be followed for each pixel separately and we get the PTF filtered value at that pixel. This however is still $\mathcal{O}(|V|^2)$. The algorithm to calculate Power MST is given in Algorithm 1. $dist(e, i, M)$ denotes the distance of the edge e from the vertex i on the MST M, and is defined as the number of edges on the path between i and the farthest node in e (from i) on the MST M.

Note that we can rewrite Eq. (11) as

$$S_i = \frac{1}{NC} \sum_{l} exp(-\frac{l}{\sigma}) \sum_{j, \Delta(i,j)=l} I_j \tag{12}$$

where NC is the normalizing constant. In the above equation observe that the exponential term quickly converges to 0 and hence one can approximate the above expression by

---

Algorithm 1. Approximate Algorithm to calculate MST at pixel
___
Input: An UMST of the graph, depth d and pixel i
Output: Power MST
  1: Set $X = \{i\}$ and an MST $M = (i, \emptyset)$
  2: **while** True **do**
  3:     break = **true**
  4:     **for** e in shortest edges from X to X^c **do**
  5:        **if** $dist(e, i, M) < d$ **then**
  6:          $M \cup e$
  7:          break = **false**
  8:        **end if**
  9:     **end for**
10:     **if** break = **true then**
11:        return M.
12:     **end if**
13: **end while**

Algorithm 2. Approximate Algorithm to calculate PTF
___
Input: A 4-adjancecy graph of the image I, $\mathcal{G}$
Output: Filtered image S.
  1: Construct the UMST of $\mathcal{G}$
  2: **for all** pixels i **do**
  3:     Construct Power MST M_i, using algorithm 1.
  4:     Use *bottom-up* to get the filtered value S_i.
  5: **end for**

$$S_i \approx \frac{1}{NC} \sum_{l=1}^{D} exp(-\frac{l}{\sigma}) \sum_{j, \Delta(i,j)=l} I_j, \tag{13}$$

where D is a parameter indicating a fixed depth. This approximation implies that, theoretically the calculation of the PTF filtered value at each pixel reduces to constant time and hence we get an approximate algorithm which is $\mathcal{O}(|V|)$. The pseudocode is given in Algorithm 2.

4 Analysis and Experiments

4.1 Comparison with MST Filter

Before proceeding to the analysis of the Power Tree filter, we provide the results of the Power Tree filter and compare with the MST filter. Note that in [1] the MST filter results were post-processed with the bilateral filter as well. We provide a visual comparison of the Power Tree filter and MST filter in Fig. 2 with and without the usage of bilateral filter as a post-processing step. Note that the 'leak' issue is reduced in the Power Tree filter compared to the MST filter, as in Fig. 2(b) and (d). Also in Fig. 2(q), (r), (s) and (t) the black border is better preserved in the Power Tree filter compared to the MST filter, irrespective of the usage of bilateral filtering. We further discuss the heuristic reason for this behaviour in the next section.

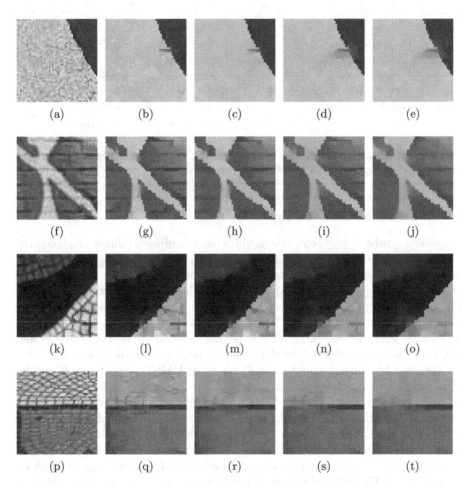

Fig. 2. (a), (f), (k), (p) are the original noisy images. (b), (g), (l), (q) are obtained by Power Tree filtering. (c), (h), (m), (r) are obtained by Power Tree filtering followed by bilateral filtering. (d), (i), (n), (s) are obtained by MST filtering. (e), (j), (o), (t) are obtained by MST filtering followed by bilateral filtering.

In general there would be a trade-off between the amount of smoothing and the level of edge-preserving in an image. This also can be seen from various examples in Fig. 2. The parameters used to generate Fig. 2 are $\sigma = 10$ for both Power Tree filter and MST filter, and $depth = 15$ for the Power Tree filter.

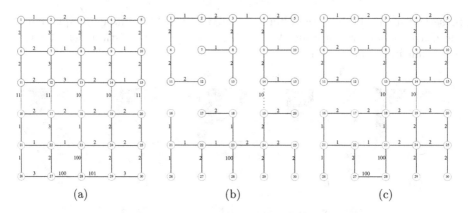

Fig. 3. (a) 4-adjacency graph of a synthetic image, (b) An MST obtained from (a), and (c) UMST obtained from (a) (Color figure online)

4.2   Analysis of the Leak Problem

Consider a synthetic grey scale image which has two objects (whose pixels colored in red and green for easy identification, see Fig. 3(a)). Assume that the pixel values within each of the objects are similar and different across objects. The pixels are numbered 1 to 30 and the edge weights are displayed on the edges. The edges corresponding to object boundaries are represented as dotted edges. In order to illustrate that UMST filter yields better results, it should perform at least as good as tree filter for - removing noise at pixel numbered 28 and reducing the leak at object boundaries say at pixel numbered 13 and 14.

Figures 3(b) and (c) denote a MST and the UMST respectively. Consider pixel numbered 28. One can see that both the edges of weights 100 incident on this pixel are present in the UMST, the noise removal is enhanced due to higher collaboration with the neighbouring pixels when compared to that of tree filter where MST had only one of the edges with weight 100. Now consider the pixel numbered 13. We see that although an extra boundary edge (edge 13−18) appears in the UMST, the presence of an additional interior edge incident on 13 in the UMST nullifies the effect of the boundary edge collaboration. At pixel numbered 14, the UMST filter performs better than tree filter due the presence of the additional interior edge 13−14.

4.3   Error Analysis of the Approximation

In Eq. 13, we have considered an approximation of the PTF. In this section, we analyze the appropriateness of the approximation. In particular we empirically show that as the depth, D, increases the filtered value of the pixel remains constant. In Fig. 4, several pixels are randomly chosen from an image and the filtered values with varying depths are calculated. Then for each pixel, the first differences of the filtered values with respect to the depth are plotted in Fig. 4,

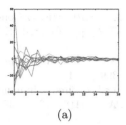 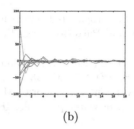 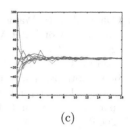

(a)   (b)   (c)

Fig. 4. First difference of the Power Tree filtered value as a function of depth for the 3 bands of RGB image. Each curve corresponds to a randomly chosen pixel in the image. Note that the differences stabilize at a depth of 15 indicating that Eq. 13 yields good approximation to PTF. (Color figure online)

for all the 3 bands. Observe that the first differences tend to 0, and hence the filtered values are stabilized. From the figure, at a depth of 15, the filtered values are more or less stable and hence this is taken for all the experiments in Fig. 2.

5   Conclusion and Future Work

We have shown that MST filter can be seen as an approximation to the Γ−limit of shortest path filters, referred to as Power Tree filter, which provides theoretical links between MST filter and geodesic based filters. Also, we have provided an alternate approximation to Power Tree filter and have shown that it yields better results. We further validated this approximation empirically. The Power Tree filter being closely related to MST filter is expected to yield good results in applications such as stereo-matching, optical-flow, image-abstraction, texture removal and editing, depth up sampling etc.

Note that although Algorithm 2 is theoretically linear time, the constant is quite high and thus one might be able to find faster/better approximations to the PTF. For example, Algorithm 2 can be parallelized since each pixel is processed independently of the other. This is a topic of further research.

Acknowledgments. Sravan Danda and Aditya Challa are thankful for the financial support provided by the Indian Statistical Institute. B.S. Daya Sagar would like to acknowledge the partial support received from EMR/2015/000853 SERB and ISRO/SSPO/Ch-1/2016-17 ISRO research grants. Laurent Najman would like acknowledge the partial support received from ANR-15-CE40-0006 CoMeDiC and ANR-14-CE27-0001 GRAPHSIP research grants.

References

1. Bao, L., Song, Y., Yang, Q., Yuan, H., Wang, G.: Tree filtering: efficient structure-preserving smoothing with a minimum spanning tree. IEEE TIP **23**(2), 555–569 (2014)

2. Chang, J.H.R., Wang, Y.C.F.: Propagated image filtering. In: 2015 IEEE Conference on Computer Vision and Pattern Recognition (CVPR), pp. 10–18. IEEE (2015)

3. Couprie, C., Grady, L., Najman, L., Talbot, H.: Power watershed: a unifying graph-based optimization framework. IEEE PAMI **33**(7), 1384–1399 (2011)

4. Cousty, J., Bertrand, G., Najman, L., Couprie, M.: Watershed cuts: minimum spanning forests and the drop of water principle. IEEE PAMI **31**(8), 1362–1374 (2009)

5. Cousty, J., Bertrand, G., Najman, L., Couprie, M.: Watershed cuts: thinnings, shortest path forests, and topological watersheds. IEEE PAMI **32**(5), 925–939 (2010)

6. Falcao, A.X., Stolfi, J., de Alencar Lotufo, R.: The image foresting transform: theory, algorithms, and applications. IEEE PAMI **26**(1), 19 (2004)

7. Farbman, Z., Fattal, R., Lischinski, D., Szeliski, R.: Edge-preserving decompositions for multi-scale tone and detail manipulation. ACM Trans. Graph. (TOG) **27**, 67 (2008). ACM

8. Floyd, R.W.: Algorithm 97: shortest path. Commun. ACM **5**(6), 345 (1962)

9. Grady, L.: Random walks for image segmentation. IEEE PAMI **28**(11), 1768–1783 (2006)

10. Grazzini, J., Soille, P.: Edge-preserving smoothing using a similarity measure in adaptive geodesic neighbourhoods. Pattern Recogn. **42**(10), 2306–2316 (2009)

11. He, K., Sun, J., Tang, X.: Guided image filtering. In: Daniilidis, K., Maragos, P., Paragios, N. (eds.) ECCV 2010. LNCS, vol. 6311, pp. 1–14. Springer, Heidelberg (2010). doi:10.1007/978-3-642-15549-9_1

12. Lerallut, R., Decencière, É., Meyer, F.: Image filtering using morphological amoebas. Image Vis. Comput. **25**(4), 395–404 (2007)

13. Najman, L.: Extending the PowerWatershed framework thanks to Γ-convergence. Technical report, Université Paris-Est, LIGM, ESIEE Paris (2017). https://hal-upec-upem.archives-ouvertes.fr/hal-01428875

14. Sinop, A.K., Grady, L.: A seeded image segmentation framework unifying graph cuts and random walker which yields a new algorithm. In: 2007 IEEE 11th International Conference on Computer Vision, ICCV 2007, pp. 1–8. IEEE (2007)

15. Soille, P.: Constrained connectivity for hierarchical image partitioning and simplification. IEEE PAMI **30**(7), 1132–1145 (2008)

16. StackOverFlow: Cut property. http://stackoverflow.com/questions/3327708/minimum-spanning-tree-what-exactly-is-the-cut-property. Accessed 05 Jan 2017

17. Stawiaski, J., Meyer, F.: Minimum spanning tree adaptive image filtering. In: 2009 16th IEEE ICIP, pp. 2245–2248. IEEE (2009)

18. Tomasi, C., Manduchi, R.: Bilateral filtering for gray and color images. In: 1998 Sixth International Conference on Computer Vision, ICCV 1998, pp. 839–846. IEEE (1998)

19. Xu, L., Lu, C., Xu, Y., Jia, J.: Image smoothing via L 0 gradient minimization. ACM Trans. Graph. (TOG) **30**, 174 (2011). ACM

20. Xu, L., Yan, Q., Xia, Y., Jia, J.: Structure extraction from texture via relative total variation. ACM Trans. Graph. (TOG) **31**(6), 139 (2012)

21. Yang, Q.: Stereo matching using tree filtering. IEEE PAMI **37**(4), 834–846 (2015)

Watersheds on Hypergraphs for Data Clustering

Fabio Dias[1]([⊠]), Moussa R. Mansour[2,3], Paola Valdivia[2,4], Jean Cousty[5],
and Laurent Najman[5]

[1] Tandon School of Engineering, New York University, New York, USA
`fabio.dias@nyu.edu`
[2] Instituto de Ciências Matemáticas e de Computação,
Universidade de São Paulo, São Carlos, Brazil
[3] Jack's Ventures, Perth, Australia
[4] Inria, Saclay, France
[5] Université Paris-Est, LIGM, Equipe A3SI, Esiee, Champs-sur-Marne, France

Abstract. We present a novel extension of watershed cuts to hypergraphs, allowing the clustering of data represented as an hypergraph, in the context of data sciences. Contrarily to the methods in the literature, instances of data are not represented as nodes, but as edges of the hypergraph. The properties associated with each instance are used to define nodes and feature vectors associated to the edges. This rich representation is unexplored and leads to a data clustering algorithm that considers the induced topology and data similarity concomitantly. We illustrate the capabilities of our method considering a dataset of movies, demonstrating that knowledge from mathematical morphology can be used beyond image processing, for the visual analytics of network data. More results, the data, and the source code used in this work are available at https://github.com/015988/hypershed.

Keywords: Data clustering · Hypergraphs · Watershed algorithm

1 Introduction

Data clustering is one of the most fundamental operations for the exploration of large amounts of information, allowing the identification of similarities and the highlight of differences, reducing the amount of cognitive effort required to gain gist information using visual analytics. Therefore, several methods for clustering data exist in the literature, including methods using graph clustering.

Our interest lies on network data, defined as data that includes relationships between its portions, usually modelled as digital structures, enabling the representation of more detailed nuances. Indeed, when the relationships are not derived from similarities in the data, but represent a different facet of the information, a clustering method needs to consider both to obtain meaningful results.

Interestingly, this is the exact context in which image segmentation methods are developed, considering both the information on the pixels and the neighboring relationship between them. In particular, the watershed algorithm is

© Springer International Publishing AG 2017
J. Angulo et al. (Eds.): ISMM 2017, LNCS 10225, pp. 211–221, 2017.
DOI: 10.1007/978-3-319-57240-6_17

fast, easy to implement, and was recently extended to several digital structures, including graphs. However, to improve the flexibility for data representation, we adopted hypergraphs as base digital structure, and we introduce a trivial extension of the watershed algorithm for hypergraphs. However, our method aims to cluster the edges of the hypergraph, not its nodes. This slight, but crucial, difference is more suitable to represent data relationships where one data point, represented as an edge, is related to several entities, represented as nodes.

In summary, the main contributions of this work are:

- An extension of the watershed algorithm to hypergraphs.
- A novel framework to represent and cluster data represented as an hypergraph.
- An application of the watershed algorithm outside of image processing.

2   Related Work

Since clustering is a crucial step for data sciences, several methods have been proposed [12]. Traditionally, the data itself is composed of points, and the objective is to identify clusters of similar points, considering some metric. Some methods build a similarity graph, where the data points are represented as nodes and the similarity as the weights on the edges [1]. This approach then leverages graph clustering, where the objective is to separate the graph structure into strongly connected clusters [10,21]. However, seldom additional data is considered [25], and the clusters reflect only the topology induced by the similarity function. When available, this application dependent data may lead to a finer, more accurate, clustering result.

Of course, hypergraphs can also be considered, when the relationships in the data cannot be accurately expressed using only pairwise links [13,14]. Several different clustering methods have been proposed, including random walks [9], spectral clustering [24], game theory [4], among others.

However, data clustering using a topology that is not derived from data similarity is not nearly as explored [22], at least not with this interpretation. Indeed, image segmentation is an equivalent problem, where both the data and its relationships need to be concomitantly considered; the objective is to identify portions of pixels (data) that are similar and connected (linked).

While a myriad of methods for image segmentation have been proposed in the literature, including the use hypergraphs for image representation [3,9], the watershed algorithm [17,18,23], and the family of derived methods [2,6,19,20], is one of the most used, because of its simplicity and robust results. Moreover, the algorithm was extended to digital structures as well [5,7,8,16].

Despite the adequacy of the watershed algorithm for the clustering of network data, its use is not properly explored outside of image processing. This is the exact context of this work, where we explore the watershed algorithm for the visual analytics of network data. Visual analytics methods are generally used to systematically explore unknown datasets, combining the perception of an operator with the numerical capabilities of a computer.

Moreover, our method can be implemented without constructing a matricial representation of the digital structure, increasing its scalability and allowing the processing of massive datasets, while maintaining the low computational cost of watershed cuts.

3   Watershed on Hypergraphs

Our objective is to partition the edges of an hypergraph into groups such that the edges in each group have similar data associated with them and a connected through one or more nodes. Between the several possible definitions for the watershed operator, we follow the framework of watershed-cuts [8], adapted where necessary to hypergraphs, including the semantic difference of clustering the edges instead of the nodes.

3.1   Hypergraphs

We define an hypergraph as $H = (V, E, \mathcal{D})$, where V is a finite set of vertices, E is the set of edges such that $\forall e \in E, e \subseteq V$, and $\mathcal{D}$ is a function that associates data to the edges of the hypergraph, in the form of a representative *feature vector*: $\mathcal{D} : E \rightarrow \mathbb{R}^m$, with $m \in \mathbb{N}^+$. While we assume that the feature vector is an array of numbers, any information can be considered, as long as a distance metric between two instances can be defined. Two of the possible visual representations of this structure are illustrated in Fig. 1, considering a small hypergraph with six edges and five nodes. Each edge is associated to a two dimensional real vector that characterizes the data point. This definition is different of the common practice, where data points are represented as nodes and the edges have associated weights.

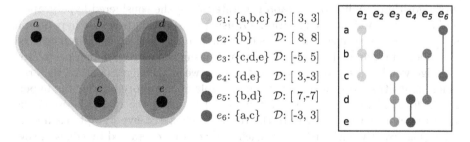

e_1: {a,b,c} $\mathcal{D}$: [3, 3]
e_2: {b} $\mathcal{D}$: [8, 8]
e_3: {c,d,e} $\mathcal{D}$: [-5, 5]
e_4: {d,e} $\mathcal{D}$: [3,-3]
e_5: {b,d} $\mathcal{D}$: [7,-7]
e_6: {a,c} $\mathcal{D}$: [-3, 3]

Fig. 1. Graphical representations of an hypergraph with data associated to its edges.

Let H be an hypergraph. Two distinct edges are *neighbors* if they share a vertex, that is, $N(e_i, e_j) \leftrightarrow ((e_i \cap e_j) \neq \emptyset)$, for any $e_i, e_j \in E$. The *set of neighbors* of an edge e is defined as $N(e) = \{u \in E \mid (e \cap u \neq \emptyset) \wedge (e \neq u)\}$. For instance, in the example depicted in Fig. 1, edges e_1 and e_3 are neighbors, since both include node c, but e_1 and e_4 are not.

We define a *path* π as an ordered sequence of distinct edges of H, $\pi = \langle e_0, e_1, \ldots, e_\ell \rangle$ such that any two consecutive edges are neighbors. For instance, considering the hypergraph in Fig. 1, $\langle e_1, e_2, e_5 \rangle$ is a valid path, where $\langle e_1, e_4, e_6 \rangle$ is not. If $\ell = 0$, π is a *trivial path* $\pi = \langle e_0 \rangle$. If there is a path between any two edges of H, the hypergraph is *connected*.

We define the distance between edges e_i and e_j as $d(e_i, e_j)$, where d is a distance metric between the feature vectors, which can be any distance metric suitable for the considered problem. This distance is analogous to the gradient information in the traditional watershed [18], representing the height to be surmounted by the water in the relief map of the data.

A *descending path* is a path π in which the distance between consecutive edges do not increase, $d(e_{i-1}, e_i) \geq d(e_i, e_{i+1})$, for any $i \in [1, \ell - 1]$. Intuitively, a *steepest descent path* is a descending path where the distance values decrease the most at each step. Therefore, each consecutive edge corresponds to smallest possible distance from the previous edge, $d(e_i, e_{i+1}) = \min\{d(e_i, u), u \in N(e_i)\}$, for any $i \in [0, \ell - 1]$. To simplify the notation, we define a function to represent this minimal distance as $d^{\ominus}(e) = \min\{d(e, u), u \in N(e)\}$, therefore, in a path of steepest descent, $d(e_i, e_{i+1}) = d^{\ominus}(e_i)$, for any $i \in [0, \ell - 1]$.

Definition 1. Watershed clustering. *Let $H = (V, E, \mathcal{D})$ be an hypergraph and $\Pi = \{\pi_0, \pi_1, \ldots, \pi_{|E|}\}$ be a collection of steepest descent paths, such that every edge of H is the first edge of one path, that is, $\pi_i = \langle e_i, \ldots \rangle$, for any $i \in [0, |E|]$. Then a watershed clustering of the edges of H is a function $\Psi : E \rightarrow \mathbb{N}$ that attributes labels to the edges according to the last edge of the path, that is $\Psi(e_i) = \Psi(e_j) \leftrightarrow \exists \pi_i, \pi_j \in \Pi, e_z \in E \mid (\pi_i = \langle e_i, \ldots, e_z \rangle) \wedge (\pi_j = \langle e_j, \ldots, e_z \rangle)$.*

3.2 Relationship to Watershed Cuts

While Definition 1 characterizes watershed clustering on hypergraphs, it does not provide a way of obtaining one. To this end, we directly leverage the watershed cuts algorithm, applied on a graph generated from the considered hypergraph.

We create a weighted-edge graph G, using information from the hypergraph H. Since we aim to cluster the edges of the hypergraph, each edge is represented by a node of G, and edges are placed representing the corresponding neighbors, the edge weights are given by the distance between the edges (feature vectors), using an arbitrary distance metric. An example of this procedure for the hypergraph depicted in Fig. 1 is illustrated in Fig. 2, using the cosine between the feature vector as the distance. In the resulting clustered hypergraph, each edge belongs to exactly one cluster, but each node can be contained by edges of several clusters, since it can be contained by edges on different clusters. Indeed, in the result illustrated on the last panel of Fig. 2 all nodes belong to edges in two distinct clusters. This information is not as easily obtained on the graph constructed from the edges, since the nodes are not explicitly represented.

A node with edges in two or more clusters can be seen as the equivalent of the watershed lines in the classic image processing framework, acting as a barrier between them. For data analysis this aspect can be interesting, because it elicits patterns of behavior.

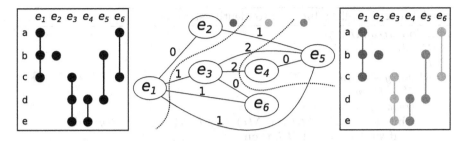

Fig. 2. Hypergraph from Fig. 1, the corresponding weighted edge graph with the result of watershed cuts, and the clustered hypergraph. The distance used for the edge weights is the cosine between the vectors.

Property 1. A watershed cut of the nodes of the graph G is a watershed clustering of the edges of the hypergraph H.

This property is self evident, because there is a bijection between the edges of the hypergraph H and the nodes of the graph G, the neighborhood relationships are preserved, as well as the distances.

3.3   Algorithm for Watershed on Hypergraphs

While we used the close relationship between watershed on hypergraphs and watershed cuts in graphs to provide an easy way to compute the clustering, to explicitly construct another structure is inefficient, particularly when considering large amounts of data. To avoid this reconstruction, we adapt the original watershed-cuts algorithm [8] to hypergraphs.

Algorithm 1. Watershed on hypergraphs
---

   function $watershed(H)$
      for all $e \in E$ **do** $\Psi(e) \leftarrow NO\_LABEL$
      $nb\_labs \leftarrow 0$
      for all $e \in E$ such that $\Psi(e) = NO\_LABEL$ **do**
         $[L, lab] \leftarrow stream(H, \Psi, e)$
         if lab=-1 **then**
            $nb\_labs \leftarrow nb\_labs + 1$
            for all $y \in L$ **do** $\Psi(y) \leftarrow nb\_labs$
         else
            for all $y \in L$ **do** $\Psi(y) \leftarrow lab$
      return Ψ

The two functions introduced in Algorithms 1 and 2 are identical to the original watershed cuts algorithm [8], edges are considered instead of nodes and the neighborhood relation is changed; we refer to the original work for an in depth analysis of the algorithm and its clustering performance.

Algorithm 2. Auxiliary function to identify streams

```
function stream(H, Ψ, e)
    L ← {e}
    L' ← {e}
    while ∃y ∈ L' do
        L' ← L' \ y
        breadth_first ← True
        while breadth_first ∧ (∃z ∈ N(y) | (z ∉ L) ∧ (d(y, z) = d⁻(y))) do
            if Ψ(z) ≠ NO_LABEL then
                return [L, Ψ(z)]
            else
                if d⁻(z) < d⁻(y) then
                    L ← L ∪ {z}
                    L' ← {z}
                    breadth_first ← False
                else
                    L ← L ∪ {z}
                    L' ← L' ∪ {z}
    return [L, −1]
```

Implementation Considerations. While the algorithm is identical to watershed cuts, there are implementation details that are more relevant when considering hypergraphs. For instance, the weights of each edge of the weighted graph are usually precomputed; the algorithm can simply access these values, without any increase in the computational time. Since our algorithm does not explicitly construct the graph, the distances are calculated on-demand, as the algorithm explores the hypergraph. Moreover, the adopted metric can be computationally expensive, so repeated computations of the distance between two edges should be avoided. This can be easily accomplished using memoization, which can use less memory than the explicit computation of the distance between all edges.

If deterministic results are desired, the method needs to have a stable sorting method, where ties in the distance values are broken in the same way. In any case, all possible results are valid clusters, satisfying Definition 1.

4   Illustration of the Method

To illustrate the behavior of our method on real data, we considered information from the The Movie Database (www.themoviedb.org), using a breadth-first search, alternating movies and actors, starting on *Lord of the Rings: The fellowship of the ring*. Each movie is represented as an edge and the involved actors as nodes. By construction, this hypergraph is connected. The feature vectors are derived from they keywords and genres associated to each movie; the distance metric used was the cosine between the two feature vectors. Actors are represented as nodes and movies as edges, including the whole cast.

Therefore, both the characteristics of the movies and the relationships between movies and actors are expressed in the structure, allowing the differentiation between identical movies with different actors. For instance, it might not be desirable to aggregate Peter Jackson's Lord of the Rings trilogy with Ralph Bakshi's 1978 movie, despite the fact that the feature vectors of both should be remarkably similar. This distinction is a direct result of including the topology into the clustering process, and this effect can be avoided by using a clustering method that ignores the topology.

Further, the objective of this section is only to illustrate the behavior of the method, demonstrating that it can be used for network data exploration. Since our method is effectively a translation of watershed cuts into hypergraphs, we refer the reader to the works by Cousty et al. [7,8] for a performance comparison against other segmentation methods.

In this example, we considered a dataset with 100 movies and 1,487 actors. Our implementation of the method was done in Python and is freely available at https://github.com/015988/hypershed. The processing time was approximately 0.1 s on a regular i7 computer. For reference, the processing time for a dataset with 5,000 movies and 39,029 actors took 50 s. Our clustering result is illustrated in Fig. 3. However, due to size constraints, we depict only the 87 actors related to two or more clusters.

Between the 17 resulting clusters, the blue cluster on the leftmost part of the figure groups movies from the Tolkien universe, including, however, two movies classified as documentaries: *The watchmaker's apprentice* and *Slacker uprising*, which can be considered as significantly different when compared to the other movies in the cluster. While similar between themselves, these two movies did not form a separate cluster because they are not neighbors, there is no overlap in the casting. However, both are neighbors of the movies in the Lord of the Rings trilogy, in the former, John Rhys-Davies is the narrator, while in the latter Viggo Mortensen is part of the cast. Similarly, *Dracula*, starred by Christopher Lee, is also included in this cluster.

Similarly, the movies from the *Pirates of the Caribbean* franchise were grouped together in the brown cluster, as well as two movies from the *Indiana Jones* and *James Bond* franchises, in the two gray clusters. Most of the movies from the *X-Men* universe were grouped in the light green cluster, with two movies, *The Wolverine* and *X-Men: Apocalypse* separated into another cluster, most likely because the feature vectors of these two movies are very similar, creating a new minimum. Moreover, the feature vectors of some of the movies are very small, some movies in this dataset contain only one genre and no keywords, compromising the representability of the distances between the movies.

The plot on the bottom left of Fig. 3 illustrates a T-SNE [15] projection of the feature vectors associated with the movies. This method consider each feature vector as a point in a high dimensional space and aims to find a two dimensional embedding of the vectors that preserves the distances between the points in the original high dimensional space. The color of the circles correspond to the clusters of the top part of the figure. This plot ignores the topological

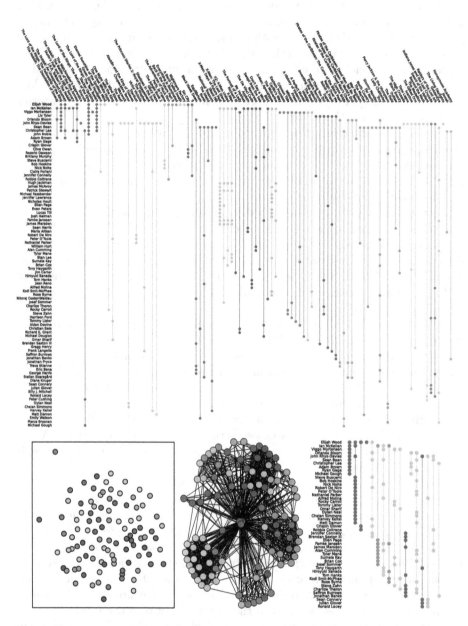

Fig. 3. Clustering result for TMDB dataset with 100 movies/edges. Top: Initial clustering. Bottom left: T-SNE projection of the feature vectors of the movies from its natural high dimensional space into $\mathbb{R}^2$, with the colors corresponding to the clusters. Bottom middle: Force layout on the equivalent graph. Bottom right: Clustering of the initial clusters. (Color figure online)

connections induced by the actors, illustrating that, while there is some correspondence between the watershed clustering and the projection of the points, there is no clear separation of the clusters when only the feature vectors are considered. Therefore, the topology of the hypergraph is tremendously significant to the watershed results, as expected.

Moreover, the plot on the bottom middle of Fig. 3 illustrates an unweighted graph which topology is equivalent to the neighborhood relationships between the edges of the hypergraph, as defined in Sect. 3.2. The graph is depicted using a force layout [11], with the colors also corresponding to the clusters in the top part of Fig. 3. This layout aims to group heavily connected nodes, and clearly depicts several different groupings, interconnected by the blue nodes in the middle. These blue nodes correspond to the *The Lord of the Rings* movies, including the seed movie used to generate the dataset, which explains this topology. As expected, the clustering result is more similar, due to the influence of the induced topology, but not quite equivalent to the strongly connected nodes of the graph, because the feature vectors are also considered.

By representing each cluster as an edge, the clustering can be recursively applied, as illustrated on the bottom right of Fig. 3, where each edge represents a cluster of the top visualization, in the same order. The feature vector of these new edges is defined as the average of the feature vectors of its composing edges. The colors represent the three new clusters. The orange cluster in this clustering corresponds to the groups containing movies of the *X-Men* franchise, the blue group corresponds to, in general, fantasy movies and the light blue cluster to action movies and dramas. Interestingly, Sir Ian McKellen is the only actor in this small dataset to have movies on all three clusters, illustrating his well known versatility. Similarly, the nodes that are not depicted in these figures, nodes whose edges belong to a single cluster, could be used to identify "niche" actors, considering a more comprehensive dataset.

5   Discussion and Conclusions

In this work, we presented a novel way to represent and cluster network data, using hypergraphs to represent relationships between portions of the data.

While data clustering with and without the topology may seem similar on an abstract level, these are two very different problems, and methods that consider the topology cannot be directly compared to the classic methods that consider only the data points. Similarly, neither can be directly compared to network clustering methods that do not consider data associated with the elements. The same argument applies to most clustering score methods as well. The included results aim only to illustrate the use of watershed cuts on hypergraphs as a tool for visual analytics on network data.

We adopted hypergraphs because they are a natural extension of graphs, increasing the applicability of the method. Moreover, edges and clusters are more conceptually similar, both can be interpreted as sets of nodes, leading to naturally hierarchical structure, when the method is recurrently applied. We did not explore this option beyond what is illustrated in Fig. 3 because the visual analytics interface needed to properly represent these results is challenging and beyond the scope of this work; a potentially interesting future work.

Our work leverages the advantages of the watershed algorithm, but it also includes its disadvantages as well, including over segmentation, as illustrated by the two separate clusters containing *X-Men* movies in Fig. 3. Further, not all edges on the same cluster are necessarily similar, the watershed can group edges in a chain of similarity, where two sequential edges are similar, but edges far apart in the chain are not, which can be counterintuitive for data sciences. However, we believe that this effect would be less pronounced on massive, more connected, datasets, with plenty of data points to properly compose each cluster. In this context, the number of clusters would be massive as well, and an our hierarchical approach could present a viable alternative for the visual exploration of such data.

While our proposed method can be considered, quite correctly, as a simple reinterpretation of watershed cuts into a new digital structure, these subtle semantic differences are novel and unexplored in the literature, leading to crucially different results. Further, they allow the application of this method for data sciences, illustrating that the knowledge from image processing can be transported to this context, which was the inspiration for this work.

Acknowledgments. Grants 2016/04391-2, 2014/12815-1, 2015/14426-5, 2013/21779-6, 2013/14089-3, 2011/22749-8 São Paulo Research Foundation (FAPESP). The views expressed are those of the authors and do not reflect the official policy or position of the São Paulo Research Foundation.

References

1. Aggarwal, C.C., Reddy, C.K.: Data Clustering: Algorithms and Applications. Chapman and Hall/CRC, Boca Raton (2013). ISBN 9781466558212
2. Bertrand, G.: On topological watersheds. J. Math. Imaging Vis. **22**(2–3), 217–230 (2005)
3. Bretto, A., Gillibert, L.: Hypergraph-Based Image Representation, pp. 1–11. Springer, Heidelberg (2005)
4. Bulò, S.R., Pelillo, M.: A game-theoretic approach to hypergraph clustering. In: Bengio, Y., Schuurmans, D., Lafferty, J.D., Williams, C.K.I., Culotta, A. (eds.) Advances in Neural Information Processing Systems 22, pp. 1571–1579. Curran Associates Inc., Red Hook (2009)
5. Couprie, C., Grady, L.J., Najman, L., Talbot, H.: Power watershed: a unifying graph-based optimization framework. IEEE Trans. Pattern Anal. Mach. Intell. **33**, 1384–1399 (2011)
6. Couprie, M., Najman, L., Bertrand, G.: Quasi-linear algorithms for the topological watershed. J. Math. Imaging Vis. **22**, 231–249 (2005)
7. Cousty, J., Bertrand, G., Couprie, M., Najman, L.: Collapses and watersheds in pseudomanifolds of arbitrary dimension. J. Math. Imaging Vis. **50**(3), 261–285 (2014)

8. Cousty, J., Bertrand, G., Najman, L., Couprie, M.: Watershed cuts: minimum spanning forests and the drop of water principle. IEEE Trans. Pattern Anal. Mach. Intell. **31**(8), 1362–1374 (2009)

9. Ducournau, A., Bretto, A.: Random walks in directed hypergraphs and application to semi-supervised image segmentation. Comput. Vis. Image Underst. **120**, 91–102 (2014)

10. Fortunato, S.: Community detection in graphs. Phys. Rep. **486**(3), 75–174 (2010)

11. Fruchterman, T.M.J., Reingold, E.M.: Graph drawing by force-directed placement. Softw. Pract. Exp. **21**(11), 1129–1164 (1991)

12. Jain, A.K.: Data clustering: 50 years beyond k-means. Pattern Recogn. Lett. **31**(8), 651–666 (2010)

13. Leordeanu, M., Sminchisescu, C.: Efficient hypergraph clustering. In: AISTATS (2012)

14. Lotfifar, F., Johnson, M.: A serial multilevel hypergraph partitioning algorithm. arXiv preprint arXiv:1601.01336 (2016)

15. Maaten, L.V.D., Hinton, G.: Visualizing data using t-SNE. J. Mach. Learn. Res. **9**, 2579–2605 (2008)

16. Meyer, F.: Watersheds on weighted graphs. Pattern Recogn. Lett. **47**, 72–79 (2014)

17. Meyer, F., Beucher, S.: Morphological segmentation. J. Vis. Commun. Image Represent. **1**(1), 21–46 (1990)

18. Najman, L., Talbot, H. (eds.): Mathematical Morphology. Wiley-Blackwell, Hoboken (2013)

19. Passat, N., Ronse, C., Baruthio, J., Armspach, J.-P., Foucher, J.: Watershed and multimodal data for brain vessel segmentation: application to the superior sagittal sinus. Image Vis. Comput. **25**(4), 512–521 (2007)

20. Roerdink, J.B.T.M., Meijster, A.: The watershed transform: definitions, algorithms and parallelization strategies. Fundam. Inform. **41**, 187–228 (2000)

21. Schaeffer, S.E.: Graph clustering. Comput. Sci. Rev. **1**(1), 27–64 (2007)

22. Villarreal, S.E.G., Schaeffer, S.E.: Local bilateral clustering for identifying research topics and groups from bibliographical data. Knowl. Inf. Syst. **48**(1), 179–199 (2016)

23. Vincent, L., Soille, P.: Watersheds in digital spaces: an efficient algorithm based on immersion simulations. IEEE Trans. Pattern Anal. Mach. Intell. **13**, 583–598 (1991)

24. Zhou, D., Huang, J., Schölkopf, B.: Learning with hypergraphs: clustering, classification, and embedding. In: NIPS (2006)

25. Zhou, Y., Cheng, H., Yu, J.X.: Graph clustering based on structural/attribute similarities. PVLDB **2**, 718–729 (2009)

Cost-Complexity Pruning of Random Forests

B. Ravi Kiran[1]([✉]) and Jean Serra[2]

[1] CRIStAL Lab, UMR 9189, Université Charles de Gaulle, Lille 3,
Villeneuve-d'Ascq, France
kiran.ravi@univ-lille3.fr
[2] Université Paris-Est, A3SI-ESIEE LIGM, Champs-sur-Marne, France
jean.serra@esiee.fr

Abstract. Random forests perform boostrap-aggregation by sampling the training samples with replacement. This enables the evaluation of out-of-bag error which serves as a internal cross-validation mechanism. Our motivation lies in the using of the unsampled training samples to improve the ensemble of decision trees. In this paper we study the effect of using the out-of-bag samples to improve the generalization error first of the decision trees and second the random forest by post-pruning. A preliminary empirical study on four UCI repository datasets show consistent decrease in the size of the forests without considerable loss in accuracy.

Keywords: Random forests · Cost-complexity pruning · Out-of-bag

1 Introduction

Random Forests [5] is an ensemble method which predicts by averaging over multiple instances of classifiers/regressors created by randomized feature selection and bootstrap aggregation (Bagging). The model is one of the most consistently performing predictor in many real world applications [6]. Random forests use CART decision tree classifiers [2] as weak learners. Random forests combine two methods: Bootstrap aggregation [3] (subsampling input samples with replacement) and Random subspace [11] (subsampling the variables without replacement). There has been continued work during the last decade on new randomized ensemble of trees. Extremely randomized trees [9] where instead of choosing the best split among a subset of variables under search for maximum information gain, a random split is chosen. This improves the prediction accuracy. In furthering the understanding of random forests [7] split the training set points into structure points: which decide split points but are not involved in prediction, estimation points: which are used for estimation. The partition into two sets are done randomly to keep consistency of the classifier.

Over-fitting occurs when the statistical model fits noise or misleading points in the input distribution, leading to poor generalization error and performance. In individual decision tree classifiers grown deep, until each input sample can be fit into a leaf, the predictions generalizes poorly on unseen data-points. To

© Springer International Publishing AG 2017
J. Angulo et al. (Eds.): ISMM 2017, LNCS 10225, pp. 222–232, 2017.
DOI: 10.1007/978-3-319-57240-6_18

handle this decision trees are pruned. There has been a decade of study on the different pruning methods, error functions and measures [14,16]. The common procedure follow is: 1. Generate a set in "interesting trees", 2. Estimate the true performance of each of these trees, 3. Choose the best tree. This is called post-pruning since we grow complete decision trees and then generate a set of interesting trees. CART uses cost-complexity pruning by associating with each cost-complexity parameter a nested subtree [10].

Though there has been extensive study on the different error functions to perform post-pruning [13,17], there have been very few studies performed on pruning random forests and tree ensembles. In practice Random forests are quite stable with respect to parameter of number of tree estimators. They are shown to converge asymptotically to the true mean value of the distribution. [10] (page 596) perform an elementary study to show the effect tree size on prediction performance by fixing minimum node size (smaller it is the deeper the tree). This choice of the minimum node size are difficult to justify in practice for a given application. Furthermore [10] discuss that rarity of over-fitting in random forests is a claim, and state that this asymptotic limit can over-fit the dataset; the average of fully grown trees can result in too rich a model, and incur unnecessary variance. [15] demonstrates small gains in performance by controlling the depths of the individual trees grown in random forest.

Finally random forests and tree ensembles are generated by multiple randomization methods. There is no optimization of an explicit loss functions. The core principle in these methods might be interpolation, as shown in this excellent study [18]. Though another important principle is the use of non-parametric density estimation in these recursive procedures [1].

In this paper we are primarily motivated by the internal cross-validation mechanism of random forests. The out-of-the-bag (OOB) samples are the set of data points that were not sampled during the creation of the bootstrap samples to build the individual trees. Our main contribution is the evaluation of the predictive performance cost-complexity pruning on random forest and other tree ensembles under two scenarios: 1. Setting the cost-complexity parameter by minimizing the individual tree prediction error on OOB samples for each tree. 2. Setting the cost-complexity parameter by minimizing average OOB prediction error by the forest on all training samples.

In this paper we do not study ensemble pruning, where the idea is to prune complete instances of decision trees away if they do not improve the accuracy on unseen data.

1.1   Notation and Formulation

Let $Z = \{\mathbf{x}^i, y^i\}_N$ be set of N (input, output) pairs to be used in the creation of a predictor. Supervised learning consists of two types of prediction tasks: regression and classification problem, where in the former we predict continuous target variables, while in the latter we predictor categorical variables. The inputs are assumed to belong to space $X := \mathbb{R}^d$ while $Y := \mathbb{R}$ for regression and $Y := \{C_i\}_K$

with K different abstract classes. A supervised learning problem aims to infer the function $f : X \rightarrow Y$ using the empirical samples Z that "generalizes" well.

Decision trees fundamentally perform data adaptive non-parametric density estimation to achieve classification and regression tasks. Decision trees evaluate the density function of the joint distribution $P(X, Y)$ by recursively splitting the feature space X greedily, such that after each split or subspace, the Ys in the children become "concentrated" or in some sense well partitioned. The best split is chosen by evaluating the information gain (change in entropy) produced before and after a split. Finally at the leaves of the decision trees one is able to evaluate the class/value by observing the subspace (like the bin for histograms) and predicting the majority class respectively [8].

Given a classification target variable with $C_k = \{1, 2, 3, ...K\}$ classes, we denote the proportion of class k in node t as:

$$\hat{p}_{tk} = \frac{1}{n_t} \sum_{x_i \in R_t} I(y_i = k) \tag{1}$$

which represents proportion of classifications in node t in decision region R_t with n_t observations. The prediction in case of classification is performed by taking the majority vote in a leaf, i.e. $\hat{y}_i = \text{argmax}_k \hat{p}_{tk}$. The misclassification error is given by:

$$l(y, \hat{y}) = \frac{1}{N_t} \sum_{i \in R_t} I(y_i \neq \hat{y}_i) = 1 - \hat{p}_{tk}(m) \tag{2}$$

The general idea in classification trees is the recursive partition of $\mathbb{R}^d$ by axis parallel splits which maximizing the gini coefficient:

$$\sum_{k \neq k'} \hat{p}_{tk}\hat{p}_{tk'} = \sum_{k=1}^{K} \hat{p}_{tk}(1 - \hat{p}_{tk}) \tag{3}$$

In a decision split the parameters are the split dimension denoted by j and the split threshold c. Given an input decision region S we are looking for the dimension (here in three dimensions) that minimizes entropy. Since we are splitting along d unordered variables, there are $2^{d-1} - 1$ possible partitions of the d values into two groups (splits) and is computationally prohibitive. We greedily decide the best split on a subset of variables. We apply this procedure iteratively till the termination condition.

As shown in Fig. 1 the set of splits over which the splitting measure is minimized is determined by the coordinates of the training set points. The number of variables or dimension d can be very large (100s-1000 in bio-informatics). Most frequently in CART one considers the sorted coordinates and from them the split points where the class y change and finally one picks the split that minimizes the purity measure best (Fig. 2).

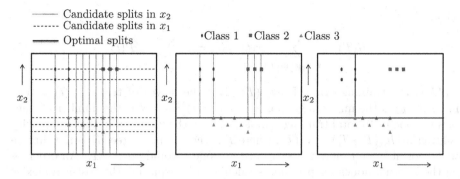

Fig. 1. Choosing the axis and splits. Given $\mathbb{R}^2$ feature space, we need to chose the best split given we chose a single axis. This consists in choosing all the coordinates along the axis, at which there are points. The optimal split is one that separates the classes the best, according to the impurity measure, entropy or other splitting measures. Here we show the sequence of two splits. Since there are finite number of points and feature pairs, there are finite number of splits possible.

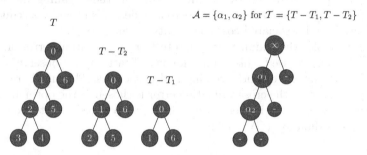

Fig. 2. Figure shows a sequence of nested subtrees $\mathcal{T}$ and the values of cost-complexity parameters associated with these subtrees $\mathcal{A}$ calculated by Eq. (6) and algorithm (2). Here $\alpha_2 < \alpha_1 \implies T - T_1 \subset T - T_2$

1.2   Cost-Complexity Pruning

The decision splits near the leaves often provide pure nodes with very narrow decision regions that are over-fitting to a small set of points. This over-fitting problem is resolved in decision trees by performing pruning [2]. There are several ways to perform pruning: we study the cost-complexity pruning here. Pruning is usually not performed in decision tree ensembles, for example in random forest since bagging takes care of the variance produced by unstable decision trees. Random subspace produces decorrelated decision tree predictions, which explore different sets of predictor/feature interactions.

The basic idea of cost-complexity pruning is to calculate a cost function for each internal node. An internal node is all nodes that are not the leaves nor the root node in a tree. The cost function is given by [10]:

$$R_\alpha(T) = R(T) + \alpha \cdot |\text{Leaves}(T)| \tag{4}$$

where

$$R(T) = \sum_{t \in \text{Leaves}(T)} r(t) \cdot p(t) = \sum_{t \in \text{Leaves}(T)} R(t) \tag{5}$$

$R(T)$ is the training error, Leaves(T) gives the leaves of tree T, $r(t) = 1 - \max_k p(C_k)$ is the misclassification rate and $p(t) = n_t/N$ is the number of samples in node n_t to total training samples N. Now the variation in cost complexity is given by $R_\alpha(T - T_t) - R_\alpha(T)$, where T is the complete tree, T_t is the subtree with root at node t, and a tree pruned at node t would be $T - T_t$. An ordering on the internal nodes for pruning is calculated by equating the cost-complexity function R_α of pruned subtree $T - T_t$ to that of the branch at node t:

$$g(t) = \frac{R(t) - R(T_t)}{|\text{Leaves}(T_t)| - 1} \tag{6}$$

The final step is to choose the weakest link to prune by calculating $\arg\min g(t)$. This calculation of $g(t)$ in Eq. (6) and then pruning the weakest link is repeated until we are left with the root node. This provides a sequence of nested trees $\mathcal{T}$ and associated cost-complexity parameters $\mathcal{A}$.

In Fig. 3 we plot the training error and test(cross-validation) error on 5 folds (usually 20 folds are used, this is only for visualization). We observe a deterioration in performance of both training and test errors. The small tree with 1 SE(standard error) of the cross-validation error is chosen as the optimal subtree. In our studies we use the simpler option which simply the smallest tree with the smallest cross validation (CV) error.

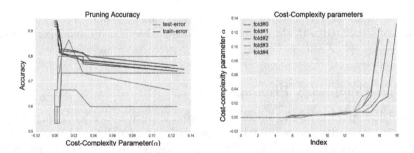

Fig. 3. Training error and averaged cross-validation error on 5 folds as cost-complexity parameter varies with its index. The index here refers to the number of subtrees.

2   Out-of-Bag (OOB) Cost Complexity Pruning

In Random forests, for each tree grown, $\frac{1}{e}N$ samples are not selected in bootstrap, and are called out of bag (OOB) samples. The value $\frac{1}{e}$ refers to the probability of choosing an out-of-bag sample when $N \to \infty$. The OOB samples are used to provide an improved estimate of node probabilities and node error rate in decision trees. They are also a good proxy for generalization error in bagging predictors [4]. OOB data is usually used to get an unbiased estimate of the classification error as trees are added to the forest.

The out-of-bag (OOB) error is the average error on the training set Z predicted such that, samples from the OOB-set $Z \setminus Z_j$ that do not belong to the set of trees $\{T_j\}$ are predicted with as an ensemble, using majority voting (using the sum of their class probabilities).

In our study (see Fig. 4) we use the OOB samples corresponding to a given tree T_j in the random forest ensemble, to calculate the optimal subtree T_j^* by cross-validation. There are two ways we propose to evaluate the optimal cost-complexity parameter, and thus the optimal subtree:

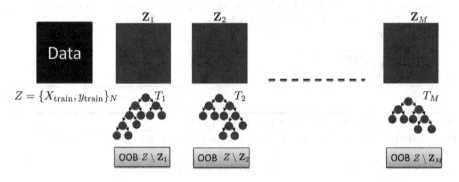

Fig. 4. Figure shows the bagging procedure, OOB samples and its use for cost-complexity pruning on each decision tree in the ensemble. There are two ways to choose the optimal subtree: one uses the OOB samples as a cross-validation (CV) set to evaluate prediction error. The optimal subtree is the smallest subtree that minimizes the prediction error on the CV-set.

– Independent tree pruning: calculate the optimal subtree by evaluating

$$T_j^* = \operatorname*{argmin}_{\alpha \in \mathcal{A}_j} \mathbb{E}\left[\|Y_{\text{OOB}} - T_j^{(\alpha)}(X_{\text{OOB}}^j)\|^2\right] \tag{7}$$

where $X_{\text{OOB}}^j = X_{\text{train}} \setminus X_j$, and X_j being the samples used in the creation of tree j.

– Global threshold pruning: calculate the optimal subtree by evaluating

$$\{T_j^*\}_{j=1}^M = \operatorname*{argmin}_{\alpha \in \cup_j \mathcal{A}_j} \mathbb{E}\left[\|Y_{\text{train}} - \frac{1}{M}\sum_{j=1}^M T_j^{(\alpha)}(X_{\text{OOB}}^j)\|^2\right] \tag{8}$$

where the cross-validation uses the out-of-bag prediction error as to evaluate the optimal $\{\alpha_j\}$ values. This basically considers a single threshold of cost-complexity parameters, which chooses a forest of subtrees for each threshold. The optimal threshold is calculated by cross-validating over the training set.

The independent tree pruning and global threshold pruning are demonstrated in algorithmic form in Fig. 4 as functions, BestTree_byCrossValidation_Tree and BestTree_byCrossValidation_Forest. The main difference between them lies in the cross-validation samples and predictor (tree vs forest) used.

Algorithm 1. Creating random forest

Precondition: $X_{\text{train}} \in \mathbb{R}^{N \times d}, Y_{\text{train}} \in \{C_k\}_1^K$, M-trees

1: **function** CREATEFOREST($\{T_i\}_{i=1}^M, X_{\text{train}}, y_{\text{train}}$)
2:     **for** $j \in [1, \text{M}]$ **do**
3:         $\mathbf{Z}_j \leftarrow$ BootStrap($X_{\text{train}}, y_{\text{train}}, \text{N}$)
4:         $T_j \leftarrow$ GrowDecisionTree($\mathbf{Z}_j$)   ▷ Recursively repeat : 1. select $m = \sqrt{(d)}$ variables 2. pick best split point among m 3. Split node into daughter nodes.
5:         $Y_{\text{pred}} \leftarrow \text{argmax}_{C_k} \frac{1}{M} \sum_{j=1}^M T_j(X_{\text{test}})$
6:     **end for**
7:     Error $\leftarrow \|Y_{\text{pred}} - Y_{\text{test}}\|^2$
8:     **return** $\{T_j\}_{j=1}^M$
9: **end function**

Algorithm 2. Cost complexity pruning on DTs

Precondition: T_j a DT, r, re-substitution error, p node occupancy proportion

1: **function** COSTCOMPLEXITYPRUNE_TREE(T_j, p, r)
2:     $\mathcal{T}_j \leftarrow \emptyset$
3:     $\mathcal{A}_j \leftarrow \emptyset$ ▷ $\mathcal{T}_j$ for set of pruned trees & cost-complexity parameter set $\mathcal{A}_j$.
4:     **while** $\mathcal{T}_j \neq \{0\}$ **do**   ▷ While tree is not pruned to root node.
5:         **for** $t \in T_i \setminus \text{Leaves}(T_i) \setminus \{0\}$ **do** ▷ For each internal node i.e. not a leaf/root
6:             $R(t) \leftarrow r(t) * p(t)$
7:             $R(T_t) \leftarrow \sum_{t \in \text{Leaves}(n)} r(t) * p(t)$
8:             $g(t) \leftarrow \frac{R(t) - R(T_t)}{|\text{Leaves}(n)| - 1}$
9:             $t_\alpha, \alpha \leftarrow \text{argmin } g(t)$ ▷ α^* is decided by cross-validation
10:        **end for**
11:        $\mathcal{T}_j \leftarrow \mathcal{T}_j \cup \text{Prune}(T_j, t_\alpha)$
12:        $\mathcal{A}_j \leftarrow \mathcal{A}_j \cup \alpha$
13:    **end while**
14:    **return** $\mathcal{T}_j, \mathcal{A}_j$
15: **end function**

The decision function of the decision tree (also denoted by the same symbol) T_j would ideally map an input vector $\mathbf{x} \in \mathbb{R}^d$ to any of the C_k classes to be predicted. To perform the prediction for a given sample, we find nodes hit by the sample until it reaches each leaf in the DT, and predict its majority vote. The class-probability weighted majority vote across trees is frequently used since it provides a confidence score on the majority vote in each node across the different trees.

Algorithm 3. Pruning DT ensembles

Precondition: $\{T_j\}_{j=1}^M$ are M DTs from **RF**, **BT** or **ET**

1: **function** COSTCOMPLEXITYPRUNE_FOREST($\{T_j\}_{i=1}^M, X_{\text{train}}, y_{\text{train}}$)
2:     **for** $i \in [1, \text{n\_iter}]$ **do**
3:         **for** $t \in \{T_j\}_{j=1}^M$ **do**
4:             $\mathcal{T}_j, \mathcal{A}_j \leftarrow$ CostComplexityPrune_Tree(T_j, X_{OOB}^j)
5:             $T_j^* \leftarrow$ SmallestTree_byCrossValidation($\mathcal{T}_j, \mathcal{A}_j, Z \setminus Z_j$) $\triangleright Z \setminus Z_j$ is
    replaced by Z (the complete training set) to have a larger CV set(2nd algorithm).
6:         **end for**
7:         $T_j^{**} \leftarrow$ SmallestTree_byCrossValidation_Forest($\mathcal{T}, \mathcal{A}, \{Z \setminus Z_j\}$)
8:         $Y_{\text{pred}} \leftarrow \text{argmax}_{C_k} \frac{1}{M} \sum_{j=1}^M T_j^*(X_{\text{OOB}}^j)$
9:         Error(i) $\leftarrow \|Y_{\text{pred}} - Y_j^{\text{CV}}\|^2$
10:     **end for**
11:     **return** T_j^*
12: **end function**

In algorithm (3) we evaluate the cost complexity pruning across the M different trees $\{T_j\}$ in the ensemble, and obtain the optimal subtrees $\{T_j^*\}$ which minimize the prediction error on the OOB sample set $Z \setminus Z_j$.

One of the dangers of using the OOB set to evelute optimal subtrees individually, is that in small datasets the OOB samples might no more be representative of the original training samples distribution, and might produce large cross-validation errors. Though it remains to be studied whether using the OOB samples as a cross-validation set would effectively reduce the generalization error for the forest, even if we observe reasonable performance.

3   Experiments and Evaluation

Here we evaluate the Random Forest (RF), Extremely randomized tree (Extra-Trees, ET) and Bagged Trees (Bagger, BTs) models from scikit-learn on datasets from the UCI machine learning repository [12]. The data sets of different sizes are chosen. Datasets chosen were: Fisher's Iris (150 samples, 4 features, 3 classes), red wine (1599 samples, 11 features, 6 classes), white wine (4898 samples, 11 features, 6 classes), digits dataset (1797 samples, 64 features, 10 classes).

In Fig. 5 we demonstrate the effect of pruning RFs, BTs and ETs on the different datasets. We observe that random forests an extra trees are often compressed by factors of 0.6 the original size, while maintaining test accuracies, while this is not the case with BTs. To understand the effect of pruning we plot in Fig. 6 the values $\mathcal{A}_j$ for the different trees in each of the ensembles. We observe that more randomization in RFs and ETs provide a larger set of potential subtrees to cross-validate over.

Algorithm 4. Best subtree minimizing CV error on OOB-set

Precondition: $\mathcal{T}_j$ set of nested subtrees of DT T_j, indexed by their cost-complexity parameter $\alpha \in \mathcal{A}_j$

1: **function BestTree_byCrossValidation_Tree**($\mathcal{T}_j, X_{\text{OOB}}^j$)
2:    **for** $\alpha \in \mathcal{A}_j$ **do**
3:       $Y_{\text{pred}}(\alpha) \leftarrow \text{argmax}_{C_k} T_j^\alpha(X_{\text{OOB}}^j)$
4:       $\text{CV-Error}(\alpha) \leftarrow \|Y_{\text{pred}}(\alpha) - Y_{\text{OOB}}^j\|^2$
5:    **end for**
6:    $\alpha_j^* \leftarrow \text{argmin}_{\alpha \in \mathcal{A}_j} \text{CV-Error}$
7:    **return** $\mathcal{T}_j(\alpha_j^*)$ ▷ Returns the best subtree of T_j
8: **end function**

Precondition: $\{\mathcal{T}_j\}_{j=1}^M$ are M DTs T_j and their nested subtrees indexed by their cost-complexity parameter $\alpha \in \mathcal{A}_j$

1: **function BestTree_byCrossValidation_Forest**($\{\mathcal{T}_j\}^M, \{\mathcal{A}_j\}^M, \{X_{\text{OOB}}^j\}^M$)
2:    $\text{Unique\_alpha} \leftarrow \text{Unique}(\bigcup_{j=1}^M \mathcal{A}_j)$
3:    **for** $a \in \text{Unique\_alpha}$ **do**
4:       $\{\alpha_j\} \leftarrow \{\alpha \in \mathcal{A}_j | \alpha \leq a\}$
5:       $Y_{\text{pred}}(\{\alpha_j\}) \leftarrow \text{argmax}_{C_k} \frac{1}{M}\sum_{j=1}^M T_j^\alpha(X_{\text{OOB}}^j)$
6:       $\text{CV-Error}(\{\alpha_j\}) \leftarrow \|Y_{\text{pred}}(\{\alpha_j\}) - Y_{\text{OOB}}^j\|^2$
7:    **end for**
8:    $\{T_j^*\} \leftarrow \text{argmin}_{\{\alpha_j\}} \text{CV-Error}(\{\alpha_j\})$
9:    **return** $\{T_j^*\}$ ▷ Returns the M-best subtrees of $\{T_j\}_1^M$
10: **end function**

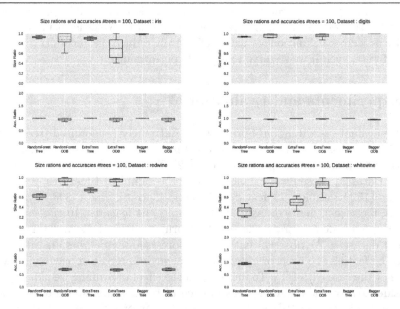

Fig. 5. Performance of pruning on the IRIS dataset and digits dataset for the three models: RandomForest, Bagged trees and Extra Trees. For each model we show the performance measures for minimizing the tree prediction error and the forest OOB prediction error.

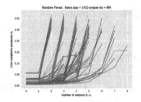 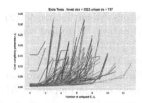 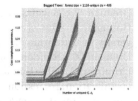

Fig. 6. Plot of the cost-complexity parameters for 100-tree ensemble for RFs, ETs and BTs. The distribution of $\mathcal{A}_j$ across trees j shows BTs constitute of similar trees and thus contain similar cost complexity parameter values, while RFs and furthermore ETs have a larger range of parameter values, reflecting the known fact that these ensembles are further randomized. The consequence for pruning is that RFs and more so ETs on average produce subtrees of different depths, and achieve better prediction accuracy and size ratios as compared to BTs.

4   Conclusions

In this preliminary study of pruning of forests, we studied cost-complexity pruning of decision trees in bagged trees, random forest and extremely randomized trees. In our experiments we observe a reduction in the size of the forest which is dependent on the distribution of points in the dataset. ETs and RFs were shown to perform better than BTs, and were observed to provide a larger set of subtrees to cross-validate. This is the main observation and contribution of the paper.

Our study shows that the out-of-bag samples can be a possible candidate to set the cost-complexity parameter and thus an determine the best subtree for all DTs within ensemble. This combines the two ideas originally introduced by Breiman OOB estimates [4] and bagging predictors [5], while using the internal cross-validation OOB score of random forests to set the optimal cost-complexity parameters for each tree.

The speed of calculation of the forest of subtrees is an issue. In the calculation of the forest of subtrees $\{\mathcal{T}\}_{j=1}^{M}$ we evaluate the predictions at Unique $(\cup_j\{\mathcal{A}_j\})$ different values of the cost-complexity parameter, which represents the number of subtrees in the forest. In future work we propose to calculate the cost complexity parameter for the forest instead of individual trees.

Though these performance results are marginal, the future scope and goal of this study is to identify the sources of over-fitting in random forests and reduce this by post-pruning. This idea might not be incompatible with smooth-spiked averaged decision function provided by random forests [18].

References

1. Biau, G., Devroye, L., Lugosi, G.: Consistency of random forests and other averaging classifiers. J. Mach. Learn. Res. **9**, 2015–2033 (2008)
2. Breiman, L., Friedman, J., Olshen, R.A., Stone, C.J.: Classification and Regression Trees. Chapman and Hall, New York (1984)

3. Breiman, L.: Bagging predictors. Mach. Learn. **24**(2), 123–140 (1996)
4. Breiman, L.: Out-of-bag estimation. Technical report Statistics Department, University of California Berkeley (1996)
5. Breiman, L.: Random forests. Mach. Learn. **45**(1), 5–32 (2001)
6. Criminisi, A., Shotton, J., Konukoglu, E.: Decision forests: a unified framework for classification, regression, density estimation, manifold learning and semi-supervised learning. Found. Trends. Comput. Graph. Vis. **7**, 81–227 (2012)
7. Denil, M., Matheson, D., De Freitas, N.: Narrowing the gap: random forests in theory and in practice. In: ICML, pp. 665–673 (2014)
8. Devroye, L., Gyrfi, L., Lugosi, G.: A Probabilistic Theory of Pattern Recognition. Applications of Mathematics. Springer, New York (1996)
9. Geurts, P., Ernst, D., Wehenkel, L.: Extremely randomized trees. Mach. Learn. **63**(1), 3–42 (2006)
10. Hastie, T., Tibshirani, R., Friedman, J.: The Elements of Statistical Learning: Data Mining, Inference and Prediction, 2nd edn. Springer, Berlin (2009)
11. Ho, T.K.: The random subspace method for constructing decision forests. IEEE Trans. Pattern Anal. Mach. Intell. **20**(8), 832–844 (1998)
12. Lichman, M.: UCI machine learning repository (2013). http://archive.ics.uci.edu/ml
13. Mingers, J.: An empirical comparison of pruning methods for decision tree induction. Mach. Learn. **4**(2), 227–243 (1989)
14. Rakotomalala, R.: Graphes d'induction. Ph.D. thesis, L'Universit Claude Bernard - Lyon I (1997)
15. Segal, M.R.: Machine learning benchmarks and random forest regression. Center for Bioinformatics and Molecular Biostatistics (2004). http://escholarship.org/uc/item/35x3v9t4
16. Torgo, L.: Inductive learning of tree-based regression models. Ph.D. thesis, Universidade do Porto (1999)
17. Weiss, S.M., Indurkhya, N.: Decision tree pruning: biased or optimal? In: AAAI, pp. 626–632 (1994)
18. Wyner, A.J., Olson, M., Bleich, J., Mease, D.: Explaining the success of adaboost and random forests as interpolating classifiers. arXiv preprint arXiv:1504.07676 (2015)

Connected Operators and Attribute Filters

Implicit Component-Graph: A Discussion

Nicolas Passat[1]([✉]), Benoît Naegel[2], and Camille Kurtz[3]

[1] CReSTIC, Université de Reims Champagne-Ardenne, Reims, France
nicolas.passat@univ-reims.fr
[2] CNRS, ICube, Université de Strasbourg, Strasbourg, France
[3] LIPADE, Université Paris-Descartes, Paris, France

Abstract. Component-graphs are defined as the generalization of component-trees to images taking their values in partially ordered sets. Similarly to component-trees, component-graphs are a lossless image model, and can allow for the development of various image processing approaches (antiextensive filtering, segmentation by node selection). However, component-graphs are not trees, but directed acyclic graphs. This induces a structural complexity associated to a higher combinatorial cost. These properties make the handling of component-graphs a non-trivial task. We propose a preliminary discussion on a new way of building and manipulating component-graphs, with the purpose of reaching reasonable space and time costs. Tackling these complexity issues is indeed required for actually involving component-graphs in efficient image processing approaches.

Keywords: Component-graph · Algorithmics · Data-structure

1 Introduction

In mathematical morphology, connected operators [1] are often defined from hierarchical image models, namely trees, that represent an image by considering simultaneously its spatial and spectral properties. Among the most popular tree structures modelling images, one can cite the component-tree (CT) [2], the tree of shapes (ToS) [3], and the binary partition tree (BPT) [4]. The CT and ToS —by contrast with the BPT, that derives from both an image and extrinsic criteria— are intrinsic models, that depend only on the image pixel values, and are built in a deterministic way.

Initially, the CT and the ToS (which can be seen as an autodual version of the CT) were defined for grey-level images, i.e. images whose values are totally ordered. Different ways to extend the notions of CT and ToS to multivalued images, i.e. images whose values are partially ordered, were investigated during the last years. The purpose was, in particular, to allow for the use of such image models in a wider range of image processing applications. The first attempt

This work was partly funded by the French program *Investissement d'Avenir* run by the *Agence Nationale pour la Recherche* (Grant reference ANR-11-INBS-0006).

© Springer International Publishing AG 2017
J. Angulo et al. (Eds.): ISMM 2017, LNCS 10225, pp. 235–248, 2017.
DOI: 10.1007/978-3-319-57240-6_19

of extending the notion of CT to partially-ordered values was proposed in [5], leading to the pioneering notion of component-graph (CG). The main structural properties of CGs were further established [6]. From an algorithmic point of view, first strategies for building CGs were investigated. Except in a specific case —where the partially ordered set of values is structured itself as a tree [7]— these first attempts emphasised the high computational cost of the construction process, and the high spatial cost of the directed acyclic graph (DAG) explicitly modelling a CG [8,9]. By relaxing certain constraints, leading to an improved complexity, the notion of CG was successfully involved in the development of an efficient extension of ToS to multivalued images [10]. From an applicative point of view, the notion of CG, coupled with the recent notion of shaping [11], also led to preliminary, yet promising, results for multimodal image processing [12].

We propose a discussion on a new way of building and manipulating CGs. A CG is complex due to its size (number of nodes and edges) and its structural complexity (as a DAG), which induce high space and time computational costs. Our strategy is to take advantage of the structural properties of a CG in order to build dedicated data-structures gathering information useful for handling it, without explicitly computing its whole DAG. We then hope to obtain a reasonable trade-off between the cost of modelling the CG, and the cost of navigating within it for further image processing purposes.

The following discussion is mainly theoretical, providing preliminary ideas; in particular, we will present neither experimental studies nor application cases. The remainder of the paper is organized as follows. Sections 2 and 3 provide notations and recall basic notions on CGs. In Sect. 4, we enumerate the different functionalities that should be offered by implicit CG data-structures. Sections 5–11 constitute the core of the paper, where we develop our discussion. Section 12 provides concluding remarks.

2   Notations

The used notations are the same as in [6–9]. We only recall here the non-standard ones.

For any symbol further used to denote an order relation ($\subseteq$, $\leqslant$, $\trianglelefteq$, etc.), the inverse symbol ($\supseteq$, $\geqslant$, $\trianglerighteq$, etc.) denotes the associated dual order, while the symbol without lower bar ($\subset$, $<$, $\triangleleft$, etc.) denotes the associated strict order.

If $(X, \leqslant)$ is an ordered set and $x \in X$, we note $x^{\uparrow} = \{y \in X \mid y \geqslant x\}$ and $x^{\downarrow} = \{y \in X \mid y \leqslant x\}$, namely the sets of the elements greater and lower than x, respectively. If $Y \subseteq X$, the sets of all the maximal and minimal elements of Y are noted $\bigtriangledown^{\leqslant} Y$ and $\bigtriangleup^{\leqslant} Y$, respectively.

3   Basic Notions on Component-Graphs

Let Ω be a nonempty finite set. Let V be a nonempty finite set equipped with an order (i.e. reflexive, transitive, antisymmetric) relation $\leqslant$. We assume that $(V, \leqslant)$ admits a minimum, noted $\perp$.

We call an *image* any function

$$\left| \begin{array}{l} I : \Omega \longrightarrow V \\ \quad x \longmapsto v \end{array} \right. \tag{1}$$

Without loss of generality, we assume that $I^{-1}(\{\bot\}) = \{x \in \Omega \mid I(x) = \bot\} \neq \emptyset$. For any $v \in V$, the thresholding function at value v is defined by

$$\left| \begin{array}{l} \lambda_v : V^\Omega \longrightarrow 2^\Omega \\ \quad I \longmapsto \{x \in \Omega \mid v \leqslant I(x)\} \end{array} \right. \tag{2}$$

Let $\frown$ be an adjacency (i.e. irreflexive, symmetric) relation on Ω (the restriction of $\frown$ to any subset of Ω is also noted $\frown$). For any $X \subseteq \Omega$, the set of the connected components of the graph $(X, \frown)$, i.e. the equivalence classes of X with respect to the connectedness relation (i.e. the reflexive–transitive closure) induced by $\frown$, is noted $\mathcal{C}[X]$. Without loss of generality, we assume that $(\Omega, \frown)$ is connected, i.e. $\mathcal{C}[\Omega] = \{\Omega\}$.

Definition 1 (Valued connected component). *Let* $v \in V$ *and* $X \in \mathcal{C}[\lambda_v(I)]$. *The couple* $K = (X, v)$ *is called a* valued connected component; X *is the* support *of K while v is its* value. *We define the set* Θ *of all the valued connected components of I as*

$$\Theta = \bigcup_{v \in V} \mathcal{C}[\lambda_v(I)] \times \{v\} \tag{3}$$

From the order relation $\leqslant$ on V, and the inclusion relation $\subseteq$ on 2^Ω (the power-set of Ω), we define the order $\unlhd$ on Θ as

$$(X_1, v_1) \unlhd (X_2, v_2) \Longleftrightarrow (X_1 \subset X_2) \vee (X_1 = X_2 \wedge v_2 \leqslant v_1) \tag{4}$$

In other words, we enrich the standard inclusion relation, by considering the order on V whenever two valued connected components have the same support. We note $\blacktriangleleft$ the cover relation associated to $\unlhd$, i.e. for all $K_1 \neq K_2 \in \Theta$, we have $K_1 \blacktriangleleft K_2$ iff $K_1 \unlhd K_2$ and there is no $K_3 \in \Theta$ distinct from K_1, K_2 such that $K_1 \unlhd K_3 \unlhd K_2$.

We then have the following definition for the component-graphs.

Definition 2 (Component-graph [5,6]). *The Θ-component-graph (or simply, the* component-graph*) of I is the Hasse diagram* $\mathfrak{G} = (\Theta, \blacktriangleleft)$ *of the ordered set* $(\Theta, \unlhd)$. *The elements of Θ are called* nodes; *the elements of $\blacktriangleleft$ are called* edges; $(\Omega, \bot)$ *is called the* root; *the elements of* $\triangle^{\unlhd} \Theta$ *are called the* leaves *of the component-graph.*

The component-graph $\mathfrak{G}$ of I is then the Hasse diagram of the ordered set $(\Theta, \unlhd)$ (see Fig. 1, for an example). Two other (simpler) variants[1] of CGs were also proposed in [6]. They will not be considered in this study for the sake of concision.

---

[1] These two variants, called $\dot\Theta$- and $\ddot\Theta$-component-graphs, rely on sets of nodes defined as subsets of Θ. A $\ddot\Theta$-component graph only contains nodes that are necessary to model I in a lossless way, while the $\dot\Theta$-component-graph contains nodes with maximal values for a given support.

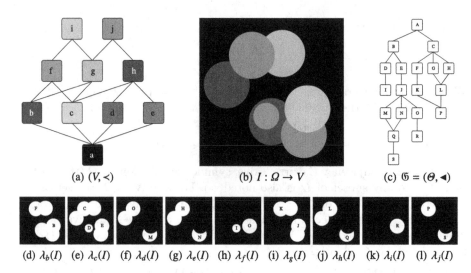

(a) $(V, \prec)$ (b) $I : \Omega \to V$ (c) $\mathfrak{G} = (\Theta, \blacktriangleleft)$

(d) $\lambda_b(I)$ (e) $\lambda_c(I)$ (f) $\lambda_d(I)$ (g) $\lambda_e(I)$ (h) $\lambda_f(I)$ (i) $\lambda_g(I)$ (j) $\lambda_h(I)$ (k) $\lambda_i(I)$ (l) $\lambda_j(I)$

Fig. 1. (a) The Hasse diagram $(V, \prec)$ of an ordered set $(V, \leqslant)$, with $V = \{a, b, \dots, j\}$ (for the sake of readability, each value of V is associated to an arbitrary colour). (b) An image $I : \Omega \to V$. (c) The component-graph $\mathfrak{G}$ of I. (d–l) Thresholded images $\lambda_v(I)$ for $v \in V$ (the threshold image $\lambda_a(I)$ is not depicted: it is composed of an unique connected component A equal to Ω). The letters (A–S) in nodes (c) correspond to the associated connected components in (d–l) [6].

4   Problem Statement

Let $I : \Omega \to V$ be an image defined on a set Ω endowed with an adjacency $\frown$, and taking its values in a set V equipped with an order $\leqslant$. Our purpose is to build dedicated data-structures that model the component-graph $\mathfrak{G}$ of I without explicitly building its whole DAG, and to be able to use them in an efficient way for further image processing purposes. In particular, the main questions that should be easily answered thanks to such data-structures are the following:

(i) Which are the nodes of $\mathfrak{G}$ (i.e. how to identify them)?
(ii) What is a node of $\mathfrak{G}$ (i.e. what are its support and its value)?
(iii) Is a node of $\mathfrak{G}$ lower, greater, or non-comparable to another, with respect to $\trianglelefteq$?

The following sections aim at discussing about the construction of ad hoc data-structures that could allow for answering these questions.

5   Flat Zones and Leaves

5.1   Flat Zone Image

Let $x, y \in \Omega$ be two points of the image I. If x and y are adjacent, i.e. $x \frown y$, and share the same value, i.e. $I(x) = I(y)$, then it is plain that they belong the same valued connected components of $\mathfrak{G}$, i.e. for any $K = (X, v) \in \Theta$, we have $x \in X$ iff $y \in X$. As an immediate corollary, the component-graph obtained from I is isomorphic to the component-graph of the flat zone image[2] associated to I.

A linear time cost $\mathcal{O}(|\frown|)$ flat zone computation can then allow us to simplify the image I into its flat zone analogue, thus reducing the space complexity of the image. From now on, and without loss of generality, we will work on such flat zone images, still noted $I : \Omega \to V$ for the sake of readability, and since they are indeed equivalent for CG building. In particular, under this hypothesis, we have the following property.

Property 3. *Let $x, y \in \Omega$. If x and y are adjacent, then their values are distinct, i.e. $x \frown y \Rightarrow I(x) \neq I(y)$.*

5.2   Detection of the Leaves

Since Ω is finite and $\leqslant$ is antisymmetric, there exist $n \geq 1$ points $x \in \Omega$ such that for all $y \frown x$, we have $I(x) \not< I(y)$, i.e. $I(x) \in V$ is a locally maximal value of the image I. Then, it is plain that $K_x = (\{x\}, I(x))$ is a node of $\mathfrak{G}$, i.e. $K_x \in \Theta$. More precisely, K_x is a minimal element of $\mathfrak{G}$, i.e. $K_x \in \bigwedge^{\trianglelefteq}\Theta$, and is then a leaf of $\mathfrak{G}$ (see Definition 2). As an example, the leaves of the component-graph $\mathfrak{G}$ depicted on Fig. 1(c) correspond to the nodes I, S, R, and P.

The characterization of the leaves relying on a local criterion, we can then detect all of them by an exhaustive scanning of Ω, with a linear time cost[3] $\mathcal{O}(|\frown|)$. In the sequel, we will denote by $\Lambda \subseteq \Omega$ the set of all the points of Ω that correspond to supports of leaves. In other words, we have $\bigwedge^{\trianglelefteq}\Theta = \{K_x = (\{x\}, I(x)) \mid x \in \Lambda\}$.

6   Node Encoding

Let $K = (X, v) \in \Theta$ be a node of $\mathfrak{G}$. If K is not a leaf, i.e. $K \notin \bigwedge^{\trianglelefteq}\Theta$, then there exists a leaf $K_x = (\{x\}, I(x)) \in \bigwedge^{\trianglelefteq}\Theta$ such that $K_x \trianglelefteq K$. In particular, Eq. (4) implies that $x \in X$ and $I(x) > v$. In [6, Property 1], we observed that for a given leaf $K_x = (\{x\}, I(x)) \in \bigwedge^{\trianglelefteq}\Theta$, and for any value $v \leqslant I(x)$, there exists exactly one node $K = (X, v) \in \Theta$ such that $x \in X$.

---

[2] The image where each flat zone (i.e. maximal connected region of constant value) is replaced by a single point, and where the adjacency relation between these flat zones is inherited from $\frown$ (i.e. two flat zones are adjacent iff one point of the first is adjacent to one point of the second).

[3] In digital imaging, we have $|\frown| = \mathcal{O}(|\Omega|)$ (e.g. with 4-, 8-, 6- and 26-adjacencies on $\mathbb{Z}^2$ and $\mathbb{Z}^3$, we have $|\frown| \simeq k.|\Omega|$, with $k = 2$, 4, 3 and 13, respectively), and this generally still holds for the induced flat zone images. Under such hypotheses, the detection of leaves can be performed in linear time $\mathcal{O}(|\Omega|)$ with respect to the size of the image. For the sake of concision, we will assume, from now on, that we have indeed $|\frown| = \mathcal{O}(|\Omega|)$.

This allows us to define the function[4]

$$
\begin{vmatrix}
\kappa : \Lambda \times V \longrightarrow \Theta \cup \{K_\top\} \\
(x, v) \longmapsto K = (X, v) \in \Theta \text{ s.t. } x \in X \quad \text{if } v \leqslant I(x) \\
(x, v) \longmapsto K_\top \qquad\qquad\qquad\qquad\qquad \text{otherwise}
\end{vmatrix}
\tag{5}
$$

This function encodes each node $(X, v) \in \Theta$ of $\mathfrak{G}$ based on two kinds of information: (1) the value v of the node, and (2) a point $x \in \Lambda \cap X$ of locally maximal value, lying within the support of the node.

The function κ is obviously surjective[5]: each node $K = (X, v) \in \Theta$ corresponds to a couple $(x, v) \in \Lambda \times V$ such that $\kappa((x, v)) = K$. This justifies the following property.

Property 4. *The set Θ of the nodes of $\mathfrak{G}$ is defined as $\Theta = \kappa(\Lambda \times V) \setminus \{K_\top\}$.*

However, κ is generally not injective. It is then necessary to handle the synonymy of nodes with respect to κ.

7   Node Synonymy Handling

Let $K_1 = (X_1, v_1), K_2 = (X_2, v_2) \in \Theta$ be nodes of $\mathfrak{G}$. From Property 4, there exist $x_1, x_2 \in \Lambda$ such that $\kappa((x_1, v_1)) = K_1$ and $\kappa((x_2, v_2)) = K_2$. If $v_1 \neq v_2$, it is plain that $K_1 \neq K_2$. However, when $v_1 = v_2$, we may have $x_1 \neq x_2$ while $K_1 = K_2$, i.e. $X_1 = X_2$.

In other words, a node $K = (X, v) \in \Theta$ —and more precisely its support X— can be represented by *any* point $x \in \Lambda \cap X$, that also corresponds to the support of a leaf $K_x = (\{x\}, I(x)) \in \Lambda^{\trianglelefteq} \Theta$. In order to provide an actual modelling of the nodes of $\mathfrak{G}$ from κ, we then have to gather the elements of $\Lambda \times V$ that encode a same node.

To this end, we define the equivalence relation $\sim$ on $\Lambda \times V$ as $(x_1, v_1) \sim (x_2, v_2) \Leftrightarrow \kappa((x_1, v_1)) = \kappa((x_2, v_2))$. From this relation and the function κ, we derive a new function

$$
\begin{vmatrix}
\kappa_\sim : (\Lambda \times V)/\!\sim \longrightarrow \Theta \cup \{K_\top\} \\
[(x, v)]_\sim \longmapsto \kappa((x, v))
\end{vmatrix}
\tag{6}
$$

that inherits the surjectivity of κ while guaranteeing, by construction, injectivity. In order to obtain a modelling of the set of nodes Θ of $\mathfrak{G}$ based on a $\Lambda \times V$ encoding, it is then sufficient to compute the equivalence classes induced by the relation $\sim$.

---

[4] By convention, we define a supplementary node $K_\top$ for handling the cases when the considered value v is greater than (or non-comparable with) the value $I(x)$ associated to x, in order to define κ on the whole Cartesian set $\Lambda \times V$. This does not induce any algorithmic nor structural issue.

[5] Actually, it is surjective iff $\kappa^{-1}(\{K_\top\}) \neq \emptyset$. If $\kappa^{-1}(\{K_\top\}) = \emptyset$, we can simply define $\kappa : \Lambda \times V \to \Theta$, and then the surjectivity property still holds.

It is plain that for any $(x, v) \in \Lambda \times V$, the equivalence class $[(x, v)]_\sim$ is composed of elements (x', v) with different $x' \in \Lambda$, but a same value v. Then, it is possible to partition the set of equivalence classes of $\sim$ with respect to the different values of V. Indeed, for any $v \in V$, we can define the equivalence relation $\sim_v$ on Λ as $x_1 \sim_v x_2 \Leftrightarrow \kappa((x_1, v)) = \kappa((x_2, v))$. In particular, we have $[(x, v)]_\sim = [x]_{\sim_v} \times \{v\}$.

Another important property is the increasingness of these equivalence classes with respect to the decreasingness of $\leqslant$, that is the fact that for any $x \in \Lambda$, we have $v_1 \leqslant v_2 \Rightarrow [x]_{\sim_{v_2}} \subseteq [x]_{\sim_{v_1}}$. This property can also be written as

$$v_1 \leqslant v_2 \Longrightarrow ((x \sim_{v_2} y) \Rightarrow (x \sim_{v_1} y)) \tag{7}$$

then justifying the following property.

Property 5. *The characterization of $\sim$ only requires to define, for each $v \in V$, the relations $x \sim_v y$ such that for $\forall v' > v, x \not\sim_{v'} y$.*

Practically, this property states that the issue of synonymy between nodes could take advantage of the monotony of $\sim_v$ ($v \in V$) with respect to $\leqslant$, to avoid the storage of information already carried by the structure of $(V, \leqslant)$ (see Sect. 12).

The next step is now to define a way to actually build these equivalence classes.

8   Reachable Zones

In order to build the equivalence relation $\sim$, let us come back to the image I and its adjacency graph $(\Omega, \frown)$. Each point $x \in \Omega$ has a given value $I(x) \in V$. It is also adjacent to other points $y \in \Omega$ (i.e. $x \frown y$) with values $I(y)$. From Property 3, we have either $I(x) < I(y)$ or $I(x) > I(y)$ or $I(x), I(y)$ non-comparable.

From a point $x \in \Lambda$, we can reach certain points $y \in \Omega$ by a descent paradigm. More precisely, for such points y, there exists a path $x = x_0 \frown \ldots \frown x_i \frown \ldots \frown x_t = y$ ($t \geq 0$) in Ω such that for any $i \in [\![0, t-1]\!]$, we have $I(x_i) > I(x_{i+1})$. In such case, we note $x \searrow y$. This leads to the following notion of a reachable zone.

Definition 6 (Reachable zone). *Let $x \in \Lambda$. The* reachable zone *of x (in I) is the set $\rho(x) = \{y \in \Omega \mid x \searrow y\}$.*

When a point $y \in \Omega$ belongs to the reachable zone $\rho(x)$ of a point $x \in \Lambda$, the adjacency path from x to y and the $>$ relation between its successive points imply that the node $(Y, I(y)) \in \Theta$ with $y \in Y$ satisfies $x \in Y$. This justifies the following property.

Property 7. *Let $x \in \Lambda$. Let $v \in V$ such that $v \leqslant I(x)$. Let $K = \kappa((x, v)) = (X, v) \in \Theta$. We have*

$$\{y \in \rho(x) \mid v \leqslant I(y)\} \subseteq X \tag{8}$$

This property states that the supports of the nodes $K = (X, v) \in \Theta$ of $\mathfrak{G}$ can be *partially computed* from the reachable zones of the points of Λ. These computed supports may be yet incomplete, since X can lie within the reachable zones of several points of Λ within a same equivalence class $[x]_{\sim_v}$.

However, the following property, that derives from Definition 6, guarantees that no point of Ω will be omitted when computing the nodes of Θ from unions of reachable zones.

Property 8. *The set $\{\rho(x) \mid x \in \Lambda\}$ is a cover of Ω.*

The important fact of this property is that $\bigcup_{x \in \Lambda} \rho(x) = \Omega$. However, the set of reachable zones is *not* a partition of Ω, in general, due to possible overlaps. For instance, let $x_1, x_2 \in \Lambda$, $y \in \Omega$, and let us suppose that $x_1 \frown y \frown x_2$ and $x_1 > y < x_2$; then we have $y \in \rho(x_1) \cap \rho(x_2) \neq \emptyset$.

Remark 9. *The computation of the reachable zones can be carried out by a seeded region-growing, by considering Λ as set of seeds. The time cost is output-dependent, since it is linear with respect to the ratio of overlap between the different reachable zones. More precisely, it is $\mathcal{O}(\sum_{x \in \Lambda} |\rho(x)|) = \mathcal{O}((1 + \gamma).|\Omega|)$, with $\gamma \in [0, |\Omega|/4] \subset \mathbb{R}$ the overlap ratio[6], varying between 0 (no overlap between reachable zones, i.e. $\{\rho(x) \mid x \in \Lambda\}$ is a partition of Ω) and $|\Omega|/4$ (all reachable zones are maximally overlapped).*

The fact that two reachable zones may be adjacent and / or overlap will allow us to complete the construction of the nodes $K \in \Theta$ of $\mathfrak{G}$.

9   Reachable Zone Graph

9.1   Reachable Zone Adjacency

When two reachable zones are adjacent —a fortiori when they overlap— they contribute to the definition of common supports for nodes of the component-graph.

Let $x_1, x_2 \in \Lambda$ ($x_1 \neq x_2$), and let us consider their reachable zones $\rho(x_1)$ and $\rho(x_2)$. Let $y_1 \in \rho(x_1)$ and $y_2 \in \rho(x_2)$ be two points such that $y_1 \frown y_2$. Let us consider a value $v \in V$ such that $v \leqslant I(y_1)$ and $v \leqslant I(y_2)$. Since we have $y_1 \frown y_2$, and from the very definition of reachable zones (Definition 6), there exists a path $x_1 = z_0 \frown \ldots \frown z_i \frown \ldots \frown z_t = x_2$ ($t \geq 1$) within $\rho(x_1) \cup \rho(x_2)$ such that $v \leqslant I(z_i)$ for any $i \in [\![0, t]\!]$. This justifies the following property.

Property 10. *Let $x_1, x_2 \in \Lambda$ with $x_1 \neq x_2$. Let us suppose that there exists $y_1 \frown y_2$ with $y_1 \in \rho(x_1)$ and $y_2 \in \rho(x_2)$. Then, for any $v \in V$ such that $v \leqslant I(y_1)$ and $v \leqslant I(y_2)$, we have $\kappa((x_1, v)) = \kappa((x_2, v))$, i.e. $(x_1, v) \sim (x_2, v)$, i.e. $x_1 \sim_v x_2$.*

---

[6] The exact upper bound of γ is actually $(|\Omega| - 1)^2/4|\Omega|$, and is reached when $|\Lambda| = (|\Omega| + 1)/2$.

This property provides a way to build the equivalence classes of $\sim_v$ ($v \in V$) and then $\sim$. In particular, we can derive a notion of adjacency between the points of Λ or, equivalently, between their reachable zones, leading to a notion of reachable zone graph.

Definition 11 (Reachable zone graph). *Let $\frown_\Lambda$ be the adjacency relation defined on $\Lambda \subseteq \Omega$ as*

$$x_1 \frown_\Lambda x_2 \iff \exists (y_1, y_2) \in \rho(x_1) \times \rho(x_2), y_1 \frown y_2 \tag{9}$$

The graph $\mathfrak{R} = (\Lambda, \frown_\Lambda)$ is called reachable zone graph *(of I).*

9.2 Reachable Zone Graph Valuation

The structural description provided by the reachable zone graph $\mathfrak{R}$ of I is necessary, but not sufficient for building the equivalence classes of $\sim$, and then actually building the nodes of the component-graph $\mathfrak{G}$ via the function κ. Indeed, as stated in Property 10, we also need information about the values of V associated to the adjacency links. This leads us to define a notion of valuation on $\frown_L$ or, more generally[7], on $\Lambda \times \Lambda$.

Definition 12 (Valued reachable zone graph). *Let $\mathfrak{R} = (\Lambda, \frown_\Lambda)$ be the reachable zone graph of an image $I : \Omega \to V$. We define the valuation function σ on the edges of $\mathfrak{R}$ as*

$$\left| \begin{array}{ll} \sigma : \Lambda \times \Lambda \longrightarrow 2^V & \\ (x_1, x_2) \longmapsto \{v \mid \exists (y_1, y_2) \in \rho(x_1) \times \rho(x_2), y_1 \frown y_2 \wedge I(y_1) \geqslant v \leqslant I(y_2)\} & \text{if } x_1 \frown_\Lambda x_2 \\ (x_1, x_2) \longmapsto \emptyset & \text{if } x_1 \not\frown_\Lambda x_2 \end{array} \right. \tag{10}$$

The couple $(\mathfrak{R}, \sigma)$ is called valued reachable zone graph *(of I).*

Remark 13. *For any $x_1, x_2 \in \Lambda$, we have $\sigma((x_1, x_1)) = \emptyset$ and $\sigma((x_1, x_2)) = \sigma((x_2, x_1))$.*

Actually, the definition of σ only requires to consider the couples of points *located at the borders* of reachable zones, as stated by the following reasoning. Let $x_1, x_2 \in \Lambda$ ($x_1 \neq x_2$). Let $v \in \sigma((x_1, x_2))$. Then, there exist $y_1 \in \rho(x_1)$ and $y_2 \in \rho(x_2)$ with $y_1 \frown y_2$, such that $I(y_1) \geqslant v \leqslant I(y_2)$. Now, let us assume that we also have $y_1 \in \rho(x_2)$ and $y_2 \in \rho(x_1)$, i.e. $y_1, y_2 \in \rho(x_1) \cap \rho(x_2)$. There exists a path $x_1 = z_0 \frown \dots \frown z_i \frown \dots \frown z_t = y_1$ ($t \geq 1$) within $\rho(x_1)$, with $I(z_i) > I(z_{i+1})$ for any $i \in [\![0, t-1]\!]$. Let $j = \max\{i \in [\![0, t]\!] \mid z_i \notin \rho(x_2)\}$ (we necessarily have $j < t$). Let $v' = I(z_{j+1})$. By construction, we have $(z_j, z_{j+1}) \in \rho(x_1) \times \rho(x_2)$, $z_j \frown z_{j+1}$ and $I(z_j) \geqslant v' \geqslant v \leqslant v' \leqslant I(z_{j+1})$. Then v could be characterized by considering a couple of points (z_j, z_{j+1}) with $z_j \in \rho(x_1) \backslash \rho(x_2)$ and $z_{j+1} \in \rho(x_2)$, i.e. with z_{j+1} at the border of $\rho(x_2)$. This justifies the following property.

---

[7] The adjacency relation $\frown_L$ is indeed a subset of the Cartesian product $\Lambda \times \Lambda$; so it would be sufficient to define the function $\sigma : \frown_L \to 2^V$. However, such notation is unusual and probably confusing for many readers; then, we prefer to define, without loss of correctness, $\sigma : \Lambda \times \Lambda \to 2^V$, with useless empty values for the couples $(x_1, x_2) \in \Lambda \times \Lambda$ such that $x_1 \not\frown_\Lambda x_2$.

Property 14. *Let $x_1, x_2 \in \Lambda$. Let $v \in V$. We have $v \in \sigma((x_1, x_2))$ iff there exists $y_1 \in \rho(x_1) \setminus \rho(x_2)$ and $y_2 \in \rho(x_2)$ such that $y_1 \frown y_2$ and $I(y_1) \geqslant v \leqslant I(y_2)$.*

Based on this property, the construction of the sets of values $\sigma((x_1, x_2)) \in 2^V$ for each couple $(x_1, x_2) \in \Lambda \times \Lambda$ can be performed by considering only a reduced subset of edges of $\frown$, namely the couples $y_1 \frown y_2$ such that $y_1 \in \rho(x_1) \setminus \rho(x_2)$ and $y_2 \in \rho(x_2)$, i.e. the border of the reachable zone of x_2 adjacent to / within the reachable zone of x_1. In particular, these specific edges can be identified during the computation of the reachable zones, without increasing the time complexity of this step.

Now, let us focus on the determination of the values $v \in V$ to be stored in $\sigma((x_1, x_2))$, when identifying a couple $y_1 \frown y_2$ such that $y_1 \in \rho(x_1) \setminus \rho(x_2)$ and $y_2 \in \rho(x_2)$. Practically, two cases, can occur. **Case 1:** $I(y_1) > I(y_2)$, i.e. $y_2 \in \rho(x_1) \cap \rho(x_2)$. Then all the values $v \leqslant I(y_2)$ are such that $I(y_1) \geqslant v \leqslant I(y_2)$; in other words, we have $I(y_2)^\downarrow \subseteq \sigma((x_1, x_2))$. **Case 2:** $I(y_2)$ and $I(y_1)$ are non-comparable, i.e. $y_2 \in \rho(x_2) \setminus \rho(x_1)$. Then, the values that belong to $\sigma((x_1, x_2))$ with respect to this specific edge are those simultaneously below $I(y_1)$ and $I(y_2)$; in other words, we have $(I(y_1)^\downarrow \cap I(y_2)^\downarrow) \subseteq \sigma((x_1, x_2))$.

Remark 15. *Instead of exhaustively computing the whole set of values $\sigma((x_1, x_2)) \in 2^V$ associated to the edge $x_1 \frown_\Lambda x_2$, it is sufficient to determine the maximal values of such a subset of V. We can then consider the function $\sigma_\nabla : \Lambda \times \Lambda \to 2^V$, associated to σ, and defined by $\sigma_\nabla((x_1, x_2)) = \nabla^\leqslant \sigma((x_1, x_2))$. In particular for any $x_1 \frown_\Lambda x_2$, we have $v \in \sigma((x_1, x_2))$ iff there exists $v' \in \sigma_\nabla((x_1, x_2))$ such that $v \leqslant v'$.*

10   Algorithmic Sketch

Based on the above discussion, we now recall the main steps of the algorithmic process for the construction of data-structures allowing us to handle a component-graph.

10.1   Input / Output

The input of the algorithm is an image I, i.e. a function $I : \Omega \to V$, defined on a graph $(\Omega, \frown)$ and taking its values in an ordered set $(V, \leqslant)$. Note that I is first preprocessed in order to obtain a flat zone image (see Sect. 5.1); this step presents no algorithmic difficulty and allows for reducing the space cost of the image without altering the structure of the component-graph to be further built.

The output of the algorithm, i.e. the set of data-structures required to manipulate the component-graph implicitly modelled, is basically composed of:

- the ordered set $(V, \leqslant)$ (and / or its Hasse diagram $(V, \prec)$);
- the initial graph $(\Omega, \frown)$, equipped with the function $I : \Omega \to V$;
- the set of leaves $\Lambda \subseteq \Omega$;

- the function $\rho : \Lambda \to 2^{\Omega}$ that maps each leaf to its reachable zone (and / or its "inverse" function $\rho^{-1} : \Omega \to 2^{\Lambda}$ that indicates, for each point of Ω, the reachable zone(s) where it lies);
- the reachable zone graph $\mathfrak{R} = (\Lambda, \frown_\Lambda)$;
- the valuation function $\sigma : \Lambda \times \Lambda \to 2^V$ (practically, $\sigma : \frown_\Lambda \to 2^V$) or its "compact" version $\sigma_{\triangledown}$.

10.2   Algorithmics and Complexity

The initial graph $(\Omega, \frown)$ and the ordered set $(V, \leqslant)$ are provided as input. The space cost of $(\Omega, \frown)$ is $\mathcal{O}(|\Omega|)$ (we assume that $|\frown| = \mathcal{O}(|\Omega|)$). The space cost of $(V, \leqslant)$ is $\mathcal{O}(|V|)$ (by modelling $\leqslant$ via its Hasse diagram, i.e. its transitive reduction, we can assume that $|\leqslant| = \mathcal{O}(|V|)$). The space cost of I is $\mathcal{O}(|\Omega|)$.

As stated in Sect. 5.2, the set of leaves $\Lambda \subseteq \Omega$ is computed by a simple scanning of Ω, with respect to the image I. This process has a time cost $\mathcal{O}(|\Omega|)$. The space cost of Λ is $\mathcal{O}(|\Omega|)$ (with, in general, $|\Lambda| \ll |\Omega|$).

As stated in Sect. 8, the function ρ (and equivalently, ρ^{-1}) is computed by a seeded region-growing. This process has a time cost which is output-dependent, and in particular equal to the space cost of the output; this cost is $\mathcal{O}(|\Omega|^\alpha)$ with $\alpha \in [1, 2]$. One may expect that, for standard images, we have $\alpha \simeq 1$.

The adjacency $\frown_\Lambda$ can be computed on the fly, during the construction of the reachable zones. By assuming that, at a given time, the reachable zones of Λ^+ are already constructed while those of Λ^- are not yet ($\Lambda^+ \cup \Lambda^- = \Lambda$), when building the reachable zone of $x \in \Lambda^-$, if a point $y \in \Omega$ is added to $\rho(x)$, we add the couple (x, x') to $\frown_\Lambda$ if there exists $z \in \rho(x')$ in the neighbourhood of y, i.e. $z \frown y$. This process induces no extra time cost to that of the above seeded region-growing. The space cost of $\frown_\Lambda$ is $\mathcal{O}(|\Lambda|^\beta)$ with $\beta \in [1, 2]$. One may expect that, for standard images, we have $\beta \simeq 1$.

The valuation function $\sigma : \Lambda \times \Lambda \to 2^V$ is built by scanning the adjacency links of $\frown$ located on the borders of the reachable zones[8]. This subset of adjacency links has a space cost $\mathcal{O}(|\Omega|^\delta)$, with $\delta \in (0, 1]$. One may expect that $\delta \ll 1$, due to the fact that we consider borders of regions[9]. For each adjacency link of this subset of $\frown$, one or several values (their number can be expected to be, in average, a low constant value k) are stored in σ. This whole storage then has a time cost $\mathcal{O}(k.|\Omega|^\delta)$. Then, some of these values are removed, since they do not belong to $\sigma_{\triangledown}$. If we assume that we are able to compare the values of $(V, \leqslant)$ in constant time[10] $\mathcal{O}(1)$, this removal process, that finally leads to $\sigma_{\triangledown}$, has in average, a time cost of $\mathcal{O}((k.|\Omega|^\delta)^2/|\Lambda|^\beta)$. The space cost of $\sigma_{\triangledown}$ is $\mathcal{O}(k.|\Omega|^\delta)$.

---

[8] Actually, half of these borders, due to the symmetry of the configurations, see Property 14.

[9] For instance, for a digital images in $\mathbb{Z}^3$ (resp. $\mathbb{Z}^2$) of size $|\Omega| = N^3$ (resp. N^2), we may expect a space cost of $\mathcal{O}(N^2)$ (resp. $\mathcal{O}(N^1)$), i.e. $\delta = 2/3$ (resp. $\delta = 1/2$).

[10] A sufficient condition is to model $(V, \leqslant)$ as an intermediate data-structure of space cost $\mathcal{O}(|V|^2)$.

11   Data-Structure Manipulation

Once all these data-structures are built, it is possible to use them in order to manipulate an implicit model of component-graph. In particular, let us come back to the three questions stated in Sect. 4, that should be easily answered for an efficient use of component-graphs.

Which are the nodes of $\mathfrak{G}$ (i.e. how to identify them)? A node $K = (X, v) \in \Theta$ can be identified by a leave $K_x \in \Lambda \trianglelefteq \Theta$ (i.e. a flat zone $x \in \Lambda$) and its value $v \in V$, by considering the κ encoding, that is "$\kappa((x, v))$, the node of value v whose support contains the flat zone x". More generally, since the ρ^{-1} function provides the set of reachable zones where each point of Ω lies, it is also possible to identify the node $K = (X, v) \in \Theta$ by any point of X and its value v, that is "the node of value v whose support contains x".

What is a node of $\mathfrak{G}$ (i.e. what are its support and its value)? A node $K = (X, v) \in \Theta$ is identified by at least one of the flat zones $x \in \Lambda$ within its support X, and its value v. The access to v is then immediate. By contrast, the set X is constructed on the fly, by first computing the set of all the leaves forming the connected component $\Lambda_v^x \ni x$ of the thresholded graph $(\Lambda, \lambda_v(\frown_\Lambda))$, where $\lambda_v(\frown_\Lambda) = \{x_1 \frown_\Lambda x_2 \mid v \in \sigma((x_1, x_2))\}$. This process indeed corresponds to a seeded region-growing on a multivalued map. Once the subset of leaves $\Lambda_v^x \subseteq \Lambda$ is obtained, the support X is computed as the union of the threshold sets of the corresponding reachable zones, i.e. $X = \bigcup_{y \in \Lambda_v^x} \lambda_v(I_{|\rho(y)})$.

Is a node of $\mathfrak{G}$ lower, greater, or non-comparable to another, with respect to $\trianglelefteq$? Let us consider two nodes defined as $K_1 = \kappa(x_1, v_1)$ and $K_2 = \kappa(x_2, v_2)$. Let us first suppose that $v_1 \leqslant v_2$. Then, two cases can occur: (1) $\kappa(x_1, v_1) = \kappa(x_2, v_1)$ or (2) $\kappa(x_1, v_1) \neq \kappa(x_2, v_1)$. In case 1, we have $K_2 \trianglelefteq K_1$; in case 2, they are non-comparable. To decide between these two cases, it is indeed sufficient to compute —as above— the set of all the leaves forming the connected component $\Lambda_{v_1}^{x_1}$ of the thresholded graph $(\Lambda, \lambda_{v_1}(\frown_\Lambda))$, and to check if $x_2 \in \Lambda_{v_1}^{x_1}$ to conclude. Let us now suppose that v_1 and v_2 are non-comparable. A necessary condition for having $K_2 \trianglelefteq K_1$ is that $\Lambda_{v_2}^{x_2} \subseteq \Lambda_{v_1}^{x_1}$. We first compute these two sets; if the inclusion is not satisfied, then K_1 and K_2 are non-comparable, or $K_1 \trianglelefteq K_2$. If the inclusion is satisfied, we have to check, for each leaf $x \in \Lambda_{v_2}^{x_2}$, whether $\lambda_{v_2}(I) \cap \rho(x_2) \subseteq \lambda_{v_1}(I) \cap \rho(x_1)$; this iterative process can be interrupted as soon as a negative answer is obtained. If all these inclusions are finally satisfied, then we have $K_2 \trianglelefteq K_1$.

Remark 16. *By contrast with a standard component-graph —that explicitly models all the nodes Θ and edges $\blacktriangleleft$ of the Hasse diagram— we provide here an implicit representation of $\mathfrak{G}$ that requires to compute on the fly certain information. This computation is however reduced as much as possible, by subdividing the support of the nodes as reachable regions, and manipulating them in a symbolic way, i.e. via their associated leaves and the induced reachable region graph,* *whenever a —more costly— handling of the "real" regions of Ω is not required. Indeed, the actual use of these regions of Ω is considered only when spatial information is mandatory, or when the comparison between nodes no longer depends on information related to the value part of $\trianglelefteq$, i.e. $\leqslant$, but to its spatial part, i.e. $\subseteq$ (see Eq. (4)).*

12   Concluding Remarks

This paper has presented a preliminary discussion about a way to handle component-graphs without explicitly building them. In previous works, we had observed that the computation of a whole component-graph led not only to a high time cost, but also to a data-structure whose space cost forbade an efficient use for image processing. Based on these considerations, our purpose was here to build, with a reasonable time cost, some data-structures of reasonable space cost, allowing us to manipulate an implicit model of component-graph, in particular by computing on the fly, the required information.

This is a work in progress, and we do not yet pretend that the proposed solutions are relevant. We think that the above properties are correct, and that the proposed algorithmic processes lead to the expected results (implementation in progress). At this stage, our main uncertainty is related to the real space and time cost of these data-structures, their construction and their handling. It is plain that these costs are input/output-dependent, and in particular correlated with the nature of the order $\leqslant$, and the size of the value set V. Further experiments will aim at clarifying these points.

Our perspective works are related to (i) the opportunities offered by our paradigm to take advantage of distributed algorithmics (in particular via the notion of reachable zones, that subdivide Ω); (ii) the development of a cache data-structure (e.g., based on the Hasse diagram $(V, \prec)$), in order to reuse some information computed on the fly, taking advantage in particular of Property 5; and (iii) investigation of the links between this approach and the concepts developed on directed graphs [13], of interest with respect to the notion of descending path used to build reachable zones.

References

1. Salembier, P., Wilkinson, M.H.F.: Connected operators: a review of region-based morphological image processing techniques. IEEE Signal Process. Mag. **26**, 136–157 (2009)
2. Salembier, P., Oliveras, A., Garrido, L.: Anti-extensive connected operators for image and sequence processing. IEEE Trans. Image Process. **7**, 555–570 (1998)
3. Monasse, P., Guichard, F.: Scale-space from a level lines tree. J. Vis. Commun. Image Represent. **11**, 224–236 (2000)
4. Salembier, P., Garrido, L.: Binary partition tree as an efficient representation for image processing, segmentation and information retrieval. IEEE Trans. Image Process. **9**, 561–576 (2000)

5. Passat, N., Naegel, B.: An extension of component-trees to partial orders. In: ICIP, 3981–3984. (2009)

6. Passat, N., Naegel, B.: Component-trees and multivalued images: structural properties. J. Math. Imaging Vis. **49**, 37–50 (2014)

7. Kurtz, C., Naegel, B., Passat, N.: Connected filtering based on multivalued component-trees. IEEE Trans. Image Process. **23**, 5152–5164 (2014)

8. Naegel, B., Passat, N.: Towards connected filtering based on component-graphs. In: Hendriks, C.L.L., Borgefors, G., Strand, R. (eds.) ISMM 2013. LNCS, vol. 7883, pp. 353–364. Springer, Heidelberg (2013). doi:10.1007/978-3-642-38294-9_30

9. Naegel, B., Passat, N.: Colour image filtering with component-graphs. In: ICPR, pp. 1621–1626 (2014)

10. Carlinet, E., Géraud, T.: MToS: a tree of shapes for multivariate images. IEEE Trans. Image Process. **24**, 5330–5342 (2015)

11. Xu, Y., Géraud, T., Najman, L.: Connected filtering on tree-based shape-spaces. IEEE Trans. Pattern Anal. Mach. Intell. **38**, 1126–1140 (2016)

12. Grossiord, É., Naegel, B., Talbot, H., Passat, N., Najman, L.: Shape-based analysis on component-graphs for multivalued image processing. In: Benediktsson, J.A., Chanussot, J., Najman, L., Talbot, H. (eds.) ISMM 2015. LNCS, vol. 9082, pp. 446–457. Springer, Cham (2015). doi:10.1007/978-3-319-18720-4_38

13. Perret, B., Cousty, J., Tankyevych, O., Talbot, H., Passat, N.: Directed connected operators: asymmetric hierarchies for image filtering and segmentation. IEEE Trans. Pattern Anal. Mach. Intell. **37**, 1162–1176 (2015)

Attribute Operators for Color Images: Image Segmentation Improved by the Use of Unsupervised Segmentation Evaluation Methods

Sérgio Sousa Filho$^{(\boxtimes)}$ and Franklin César Flores

Department of Informatics, State University of Maringá, Maringá, Paraná, Brazil
pg48344@uem.br, fcflores@din.uem.br

Abstract. Attribute openings and thinnings are morphological connected operators that remove structures from images according to a given criterion. These operators were successfully extended from binary to grayscale images, but such extension to color images is not straightforward. This paper proposes color attribute operators by a combination of color gradients and thresholding decomposition. In this approach, not only structural criteria may be applied, but also criteria based on color features and statistics. This work proposes, in a segmentation framework, two criteria based on unsupervised segmentation evaluation for improvement of color segmentation. Segmentation using our operators performed better than two state-of-the-art methods in 80% of the experiments done using 300 images.

Keywords: Attribute filter · Image segmentation · Thresholding decomposition · Color processing

1 Introduction

Mathematical Morphology (MM) is very resourceful when providing ways to design morphological filters [27,31]. Such filters compound a family of image operators which are increasing and idempotent. In particular, openings compound a very well known class of anti-extensive morphological filters which is the basis (with their dual counterparts, the closings - which are extensive) for the design of several morphological operators, such as the alternating sequential filters [31] and the openings and closings by reconstruction [27]. The last ones are also classified as connected filters.

Connected operators belong to a category of MM operators whose property is the reduction of regions of an image while not introducing new borders [20]. A common characteristic of some operators is the fact that a simplified image has a reduced number of flat zones (regions with a constant grayscale value). Such property is very useful in frameworks such as compression, segmentation or even in pattern recognition, in the reduction of the image statistics and in the simplification of the number of samples required in such techniques. Reconstruction operators, levelings [25] and area openings [34] are among well known connected operators.

© Springer International Publishing AG 2017
J. Angulo et al. (Eds.): ISMM 2017, LNCS 10225, pp. 249–260, 2017.
DOI: 10.1007/978-3-319-57240-6_20

By now, it is worth mentioning the area opening operator [34]. Let us consider, as a connected component, a flat zone valued one in a binary image. Given a binary image and a non-negative numerical thresholding value as input parameters, this operator removes from the binary image all connected components whose areas (defined by its number of pixels) are lower than the numerical parameter. As a result, well defined connected component are usually preserved while possible noisy components are discarded. This filter can be extended to grayscale case by the application of the thresholding decomposition [34]: After the decomposition of a grayscale image (with k graylevels) in k binary image slices - each slice given by a simple thresholding with the threshold value ranging from 0 to k - the binary case is applied to each slice and, then, the slices are recombined to compose the filtered image.

In the last example, we may consider area as a filtering criterion and the non-negative numerical parameter as a thresholding value assigned to that criterion. This concept is generalized by the attribute opening [6], a particular connected operator that removes connected components according to an increasing criterion (when the criterion is not increasing, the operator is an attribute thinning). Among the increasing criteria, we cite the area measurement [32,33]. Again, through thresholding decomposition, authors extend attribute operators from binary case to the grayscale one. Besides the cited generalization for filtering/processing images according to given criteria, attribute operators are applied to solve several problems such as image compression, segmentation of medical images and structural analysis of ore [6].

As stated above, the extension of attribute filtering from binary to grayscale images is straightforward. However, such extension to color images is not natural or straightforward, due the lacking of a natural complete lattice representation for such images [27]. Literature presents several strategies to extend MM to color images [2,4,7-9,15-18,21,27,29].

The goal of this paper is to propose an extension of attribute operators to color images. This extension is based on the thresholding decomposition of a color gradient of the input color image. The gradient valleys give a good representation, in grayscale, of the objects in the color image, preserving characteristics such as size and shape. In this way, color images may be analyzed by the application of classical attribute operators to the color gradient valleys.

A strong motivation to propose the extension to color images is, although attribute operators are well known for their application to the structural analysis of objects, to allow the use of a color feature or statistic as criteria for analysis. Two criteria based on unsupervised segmentation evaluation methods [37] are applied to improve the quality of a color segmentation of the input gradient.

This paper is organized as follows: Sect. 2 reviews some preliminary concepts. Section 3 introduces the extension of attribute operators to color images. Section 4 describes how average color error and entropy may be used as a homogeneity criterion for a color attribute operator. Section 5 demonstrate the application of average color error and entropy criterion to image segmentation. And, Sect. 6 concludes the paper with the final remarks.

2 Preliminary Concepts

Let $E \subset \mathbb{Z} \times \mathbb{Z}$ be a rectangular finite subset of points. A *binary image* may be denoted by a subset $X \subseteq E$. Let the power set $\mathcal{P}(E)$ be the set of all binary images.

Let $K = [0, k]$ be a totally ordered set. Denote by $Fun[E, K]$ the set of all functions $f : E \to K$. A *graylevel image* is one of these functions.

Let $Fun[E, \mathbf{C}]$ be the set of all functions $\mathbf{f} : E \to \mathbf{C}$, where $\mathbf{C} = \{\mathbf{c}_1, \mathbf{c}_2, \mathbf{c}_3\}$ and $\mathbf{c}_i \in \mathbb{Z} : 0 \leq \mathbf{c}_i \leq k$. $Fun[E, \mathbf{C}]$ denotes the set of all *color images*.

Let A and B be two image sets, as described in the last three paragraphs. An *image operator* (*operator*, for simplicity) is a mapping $\psi : A \to B$.

Let B_c be the structuring element that defines a connectivity [27]. A *connected component* of E is a subset $X \subset E$ such that, $\forall x, y \in X$, there is a path $P(x, y) = (p_0, p_1, ..., p_t)$, $p_i \in X$, such that $p_0 = x$, $p_t = y$ and $\forall i \in [0, t-1], \exists b \in B_c : p_i + b = p_{i+1}$.

2.1 Color Gradient

Literature presents several ways to compute color gradients [10,23]. They are usually designed taking into account the analysis of each color component image under a certain color space model. For instance, color gradients may be designed under RGB [10,13], HSL [29] or L*a*b* [30]. In this work, the color gradient is computed under the L*a*b* color space model. In this perceptual model, the Euclidean distance may be applied to evaluate the perceptual distance between two colors.

Let $\mathbf{f} \in Fun[E, \mathbf{C}]$ be a color image under the L*a*b* color space model [10]. Let $D(\mathbf{a}, \mathbf{b})$ be the Euclidean distance between two colors $\mathbf{a}, \mathbf{b} \in \mathbf{C}$, under the L*a*b* color space. The *color gradient* $\nabla_{B_{\mathrm{grad}}} : Fun[E, \mathbf{C}] \to Fun[E, \mathbb{R}_+]$ is given by, $\forall x \in E$,

$$\nabla_{B_c}(\mathbf{f})(x) = \bigvee_{b_1, b_2 \in B_c} D(\mathbf{f}(x + b_1), \mathbf{f}(x + b_2)). \tag{1}$$

In order to apply the thresholding decomposition, the color gradient needs to be converted to an image $\overline{\nabla}(\mathbf{f}) \in Fun[E, K]$, as follows:

1. Normalize $\nabla_{B_{\mathrm{grad}}}(\mathbf{f})$ from the interval $[\min\{\nabla_{B_{\mathrm{grad}}}(\mathbf{f})\}, \cdots, \max\{\nabla_{B_{\mathrm{grad}}}(\mathbf{f})\}]$ to $[0, \cdots, k]$ (rounding it down);
2. Image $\overline{\nabla}(\mathbf{f}) \in Fun[E, K]$ is given by the negation of the normalized gradient. This is done because objects are represented by valleys in the color gradient, and the negation of such valleys makes them sliceable by the thresholding decomposition.

2.2 Thresholding Decomposition

Let $\mathbf{thr} : Fun[E, K] \times K \to \mathcal{P}(E)$ be the *thresholding* function, given by, $\forall f \in Fun[E, K], \forall t \in K$,

$$\mathbf{thr}(f, t) = \{x \in E : f(x) \geq t\}. \tag{2}$$

Let $f_{bin} : \mathcal{P}(E) \rightarrow Fun[E, k]$ be the mapping that gives a numerical representation of a binary image $X \subseteq E$, such that, $\forall x \in E$,

$$f_{bin}(X)(x) = \begin{cases} 1, & \text{if } x \in X, \\ 0, & \text{otherwise.} \end{cases} \tag{3}$$

Let $f \in Fun[E, K]$. The *thresholding decomposition* is given by,

$$f = \sum_{t=1}^{k} f_{bin}(\mathbf{thr}(f, t)). \tag{4}$$

In other words, a grayscale image f may be decomposed as a stack of binary images, each one provided by the thresholding of f by a distinct graylevel $k \in K$. The "addition" of all binary images in that stack returns image f.

Thresholding decomposition is a way to extend some operators - designed for binary images - to the grayscale context [6]. Formally, the extension of $\psi : \mathcal{P}(E) \rightarrow \mathcal{P}(E)$ to $\Psi : Fun[E, K] \rightarrow Fun[E, K]$ is given by,

$$\Psi(f) = \sum_{t=1}^{k} f_{bin}(\psi(\mathbf{thr}(f, t))). \tag{5}$$

2.3   Attribute Operators

Attribute operators are applied to remove all structures that do not fit into a given criterion [6]. Such criterion is usually tied to a measurement and a comparison to a numerical parameter p. For instance, an image structure must be removed if its area is not greater than p - this is the criterion for area opening.

Let $X \in \mathcal{P}(E)$ be a binary image. Let $C \subseteq X$ be a connected component. The trivial operator $\Gamma_T : \mathcal{P}(E) \rightarrow \mathcal{P}(E)$ evaluates if C satisfies a criterion T [6]:

$$\Gamma_T(C) = \begin{cases} C, & \text{if } C \text{ satisfies criterion } T, \\ \varnothing, & \text{otherwise,} \end{cases} \tag{6}$$

where $\Gamma_T(\varnothing) = \varnothing$.

The attribute operator is given by, for all $X \in \mathcal{P}(E)$,

$$\Gamma^T(X) = \bigcup_{C \subseteq X} \Gamma_T(C), \tag{7}$$

where C is a connected component of X. Note that this is the binary case - the extension of Eq. 7 to grayscale case is given by thresholding decomposition (see Eq. 5).

For detailed information about attribute openings and thinnings, see [6]. Notice that only structural criteria are mentioned in that paper.

3   Color Attribute Operators

This section introduces an extension of the attribute operators to color images. The proposed framework allows the choice and application of criteria based on color information for attribute filtering of color images. Analysis of local morphological structures from a color gradient provides regions where information is collected from the color input image. And, such morphological structures are filtered in function of the collected color information.

Let $\overline{\Gamma}_T : \mathcal{P}(E) \times Fun[E, \mathbf{C}] \to \mathcal{P}(E)$ denote the *color trivial operator*. It is similar to Eq. 6, but $\overline{\Gamma}_T(C, \mathbf{f})$ may also take into account local color information given by $\{\mathbf{f}(x) : x \in C\}$, if a criterion T involves color features or statistics.

Connected component C is enough for assessment when criterion T is structural, like in graylevel version [6]. For criteria based on color features or statistics, we cite average color error [5,36], color harmony [28] and entropy [12]. In this paper, we apply the color entropy criterion and the average color error (see Sect. 4).

The computation of the *color attribute operator* $\overline{\Gamma}^T : Fun[E, \mathbf{C}] \to Fun[E, K]$ is described in Algorithm 1. Note that, except for the application of the color trivial operator $\overline{\Gamma}_T(C, \mathbf{f})$, the color attribute operator is computed like the grayscale version [6].

Algorithm 1. Color Attribute Operator

Input:
- image $\mathbf{f} \in Fun[E, \mathbf{C}]$;
- criterion T (a numerical parameter is usually assigned to T for purpose of analysis);

Output: image $\overline{\Gamma}^T(\mathbf{f}) \in Fun[E, K]$;

begin

1     $g = \overline{\nabla}(\mathbf{f})$;

2   decompose g into a set of slices $G_i = \mathbf{thr}(g, i), \forall i \in [1..k]$;

3   **foreach** G_i **do**

      $F_i = \bigcup_{C \subseteq G_i} \overline{\Gamma}_T(C, \mathbf{f})$;

    end

4     $\overline{\Gamma}^T(\mathbf{f}) = \sum_{i=1}^{k} f_{bin}(F_i)$;

end

The use of the filtered image $\overline{\Gamma}^T(\mathbf{f})$ depends on the application of the method. For instance, one can use the *residue* of $\overline{\Gamma}^T(\mathbf{f})$ [14]. Another approach is to apply the watershed operator [3,27] to the negation of $\overline{\Gamma}^T(\mathbf{f})$ in order to compute a hierarchical segmentation [26] of $\mathbf{f}$. This approach is adopted in this work.

4   Criteria: Unsupervised Segmentation Evaluation Methods

Haralick and Shapiro [19] affirmed that good segmentations provides homogeneous regions. Indeed, unsupervised evaluation methods for segmentation quality assessment usually have a component to measure this aspect [37]. We propose to use a homogeneity metric in a color attribute operator with the objective of obtaining a segmentation with homogeneous regions.

In order to evaluate the regions for segmentation we decided to modify the trivial operator to not just evaluate the elements of the connected component but also the elements of it's influence zone.

Let G_t be a set of connected components of $\overline{\nabla}(\mathbf{f})$ given a threshold t. Let W_t be a watershed of the inverse of $\overline{\nabla}(\mathbf{f})$ using G_t as markers. Let $C \in G_t$ be a connected component. Let R_w be the region generated by C in W_t.

We define a operator based on average color error as it has already been used to evaluate homogeneity of segmentations [5,22]. The trivial average color error operator is given by the sum of the euclidean distance between the color information of each element of the connected component and the mean value of the color information of the region:

$$e(R, f) = \sum_{x \in R} D(f(x), \mu(R, f)), \tag{8}$$

where μ returns the mean value of the color information in the region.

The average color error criterion T_e is defined by $T_e = (e(R, f) \geq t)$. In other words, *average color error of $e(R, f)$ must be greater than or equal to t.*

We also define a operator based on entropy which was used in the function E [36] as a component evaluating homogeneity of the regions of a segmentation. The trivial entropy operator is given by the entropy of the color information of the elements of the region:

$$S(R, f) = -\sum_{x \in R} p(f(x)) \log_2 p(f(x)), \tag{9}$$

where p returns the probability of the color information in the region.

The entropy criterion T_S is defined by $T_S = (S(R, f) \geq t)$. In other words, *entropy of $S(R, f)$ must be greater than or equal to t.*

5   Improvements for Color Segmentation

We use the watershed on the negation of $\overline{\Gamma}^T(\mathbf{f})$ to obtain a segmentation. Varying the threshold value results in a sequence of segmentation ranging from a image with just one region (when only a minimum remains) to the supersegmentation (when no minima is removed). This enables to control a trade-off between the number of regions and the attribute of the regions.

5.1   Experimental Results

In our experiment we used the Berkeley Segmentation Dataset and Benchmark 300 [24]. The experiments were implemented using SDC Morphological Toolbox for python [11] and scikit-image [35].

For each image we applied each operator described in the session 5 sliding the attribute value which resulted in a sequence of segmentations, varying from a coarse to a fine segmentation.

Then we confronted the color attributes with the structural ones (namely the area, height and volume) using the same framework and also with the Simple Linear Iterative Clustering (SLIC) [1]. For each resulting segmentation we obtained a comparison segmentation with the same number of regions.

Figure 1 illustrates some of the segmentations obtained during the experiment on sample "138078". Observe how this paper's operators are sensible to details in the same time of identifying large homogeneous regions.

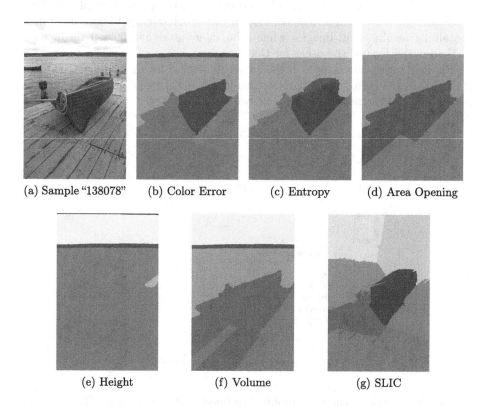

(a) Sample "138078"   (b) Color Error   (c) Entropy   (d) Area Opening

(e) Height   (f) Volume   (g) SLIC

Fig. 1. (b), (c), (d), (e), (f) and (g) are segmentations of (a) with 8 regions.

The analysis was made by measuring the segmentations using E evaluation method proposed by Zhang et al. [36]. It uses a combination of entropy to evaluate intra-region uniformity and a layout entropy to evaluate inter-region uniformity. The former decreases as the segmentation becomes finer while the latter increases. A good segmentation finds more homogeneous regions without increasing too much complexity.

For balancing the intra and inter components of Zhang's function, we used $22/S(f)$ as a weight for the first component. The division by the maximum entropy has a role of evening the weight for images with different color diversity. The constant was found empirically running the experiment several times. There was no need for a specific weight on the second component as the images have the same number of pixels.

A graphic was plotted for each image on the base. These graphics contain a curve for each method and show how it performed based on the number of regions. As we consider that is more important to analyze segmentation with fewer regions, we plotted the regions' dimension in a log scale.

To measure how our methods performed against another we calculated a score based on the area between the two curves as the proportion of this area where our method's curve was below. An ordered bar graphic was generated containing results of all images, when a bar is greater than 50% it means our method outperformed the other for that image.

Figure 2 shows an example of the curves generated for the sample "138078". It is relevant to highlight that the score can reach 100% even if it was just slightly better on all of the number of regions evaluated.

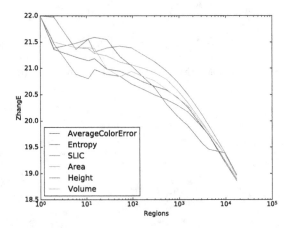

Fig. 2. Curves from sample "138078"

Figure 3 shows all the bar graphics obtained in the experiment. The average color error was better in 80%, 83%, 87% and 75% of the images with an average score of 74.33%, 81.10%, 75.23% and 72.04% in relation to the area, height, volume and SLIC, respectively. Also, the entropy was better in 77%, 86.33%, 71.67% and 78% of the images with a an average score of 69.78%, 82.83%, 66.07% and 72.34% in relation to the same methods, respectively.

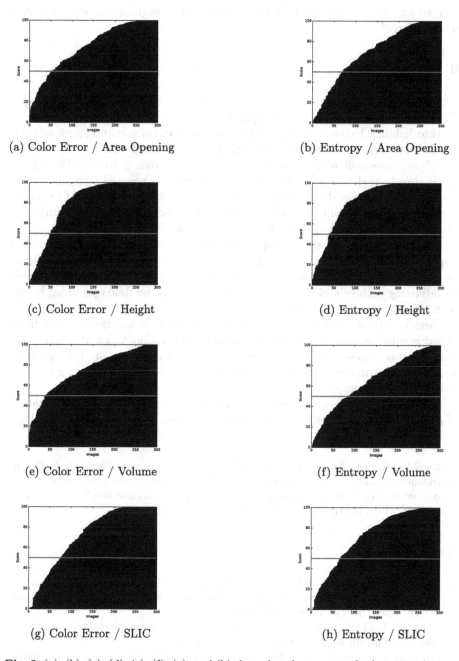

(a) Color Error / Area Opening

(b) Entropy / Area Opening

(c) Color Error / Height

(d) Entropy / Height

(e) Color Error / Volume

(f) Entropy / Volume

(g) Color Error / SLIC

(h) Entropy / SLIC

Fig. 3. (a), (b), (c), (d), (e), (f), (g), and (h) shows how better a method was in relation to another.

The results shows that, though not for every case, the color operators where able to find good segmentations and also in finding segmentations that looked quite different.

6 Conclusion

This paper proposes an extension of the attribute operators to color images. This extension is supported by two pillars: (i) the processing of a color gradient by thresholding decomposition; and (ii) the choice of color criteria such as color features or statistics, in addition to well known structural ones.

Objects are well represented by valleys in the color gradient. This gradient is thresholded in a set of binary slices and their connected components are measured and assessed according to the chosen operator criterion. Connected components that do not satisfy the criterion are discarded; the remaining ones are added together to compose the operator result.

This paper also shows the application of two color attribute operators to improve the image segmentation of color images. It allows the homogeneity of the resulting regions to be decreased in expense of increasing their numbers by adjusting a threshold value. In an experiment our methods provided better segmentation in most of the images.

Besides attribute operators were extended to color images, they became more versatile with the extension of criteria to color information. Color attribute operators have been also applied to solve image compression. In [14], entropy from local color information [12] were applied to improve entropy encoding of color images for data compression.

Future works include the proposal of improvements to the implementation of color attribute operators and the proposal of new color criteria for attribute color processing.

Acknowledgment. First author would like to thank Conselho Nacional de Desenvolvimento Científico e Tecnológico (CNPq), Brazil, for the master scholarship. Second author would like to thank Conselho Nacional de Desenvolvimento Científico e Tecnológico (CNPq), Brazil, for the post-doctoral scholarship.

References

1. Achanta, R., Shaji, A., Smith, K., Lucchi, A., Fua, P., Süsstrunk, S.: SLIC superpixels compared to state-of-the-art superpixel methods. IEEE Trans. Pattern Anal. Mach. Intell. **34**(11), 2274–2282 (2011)
2. Aptoula, E., Lefevre, S.: A comparative study on multivariate mathematical morphology. Pattern Recogn. **40**(11), 2914–2929 (2007)
3. Beucher, S., Meyer, F.: The Morphological Approach to Segmentation: The Watershed Transformation (chap. 12). In: Mathematical Morphology in Image Processing, pp. 433–481. Marcel Dekker (1992)

4. Boroujerdi, A.S., Breuß, M., Burgeth, B., Kleefeld, A.: PDE-based color morphology using matrix fields. In: Aujol, J.-F., Nikolova, M., Papadakis, N. (eds.) SSVM 2015. LNCS, vol. 9087, pp. 461–473. Springer, Cham (2015). doi:10.1007/978-3-319-18461-6_37

5. Borsotti, M., Campadelli, P., Schettini, R.: Quantitative evaluation of color image segmentation results. Pattern Recogn. Lett. **19**, 741–747 (1998)

6. Breen, E.J., Jones, R.: Attribute openings. Thinnings Granulometries **64**(3), 377–389 (1996)

7. Burgeth, B., Kleefeld, A.: Morphology for color images via Loewner order for matrix fields. In: Hendriks, C.L.L., Borgefors, G., Strand, R. (eds.) ISMM 2013. LNCS, vol. 7883, pp. 243–254. Springer, Heidelberg (2013). doi:10.1007/978-3-642-38294-9_21

8. Burgeth, B., Kleefeld, A.: An approach to color-morphology based on Einstein addition and Loewner order. Pattern Recogn. Lett. **47**, 29–39 (2014)

9. Burgeth, B., Kleefeld, A.: Order based morphology for color images via matrix fields. In: Westin, C.-F., Vilanova, A., Burgeth, B. (eds.) Visualization and Processing of Tensors and Higher Order Descriptors for Multi-Valued Data. MV, pp. 75–95. Springer, Heidelberg (2014). doi:10.1007/978-3-642-54301-2_4

10. Busin, L., Vandenbroucke, N., Ludovic, L.: Color spaces and image segmentation. Advances in Imaging and Electron Physics **151**, 65–168 (2008)

11. Dougherty, E.R., Lotufo, R.A.: Hands-on Morphological Image Processing, vol. 71. SPIE Optical Engineering Press, Washington (2003)

12. Duda, R.O., Hart, P.E., Stork, D.G.: Pattern Classification. Wiley, Hoboken (2001)

13. Evans, A.N., Liu, X.U.: A morphological gradient approach to color edge detection. IEEE Trans. Image Process. **15**(6), 1454–1463 (2006)

14. Flores, F.C., Evans, A.N.: Attribute operators for color images: the use of entropy from local color information to image compression (2017, submitted)

15. Flores, F.C., Polidório, A.M., Lotufo, R.A.: The weighted gradient: a color image gradient applied to morphological segmentation. J. Braz. Comput. Soc. **11**(3), 53–63 (2006)

16. van de Gronde, J.J., Roerdink, J.B.T.M.: Group-invariant frames for colour morphology. In: Hendriks, C.L.L., Borgefors, G., Strand, R. (eds.) ISMM 2013. LNCS, vol. 7883, pp. 267–278. Springer, Heidelberg (2013). doi:10.1007/978-3-642-38294-9_23

17. Hanbury, A., Serra, J.: Mathematical morphology in the HLS colour space. In: BMVC, pp. 1–10 (2001)

18. Hanbury, A., Serra, J.: Mathematical morphology in the cielab space. Image Anal. Ster. **21**(3), 201–206 (2002)

19. Haralick, R.M., Shapiro, L.G.: Image segmentation techniques. Comput. Vis. Graph. Image Process. **29**(1), 100–132 (1985)

20. Heijmans, H.J.A.M.: Introduction to connected operators. In: Dougherty, E.R., Astola, J.T. (eds.) Nonlinear Filters for Image Processing, pp. 207–235. SPIE-The International Society for Optical Engineering, Bellingham (1999)

21. Lambert, P., Chanussot, J.: Extending mathematical morphology to color image processing. In: Proceedings CGIP 2000 (2000)

22. Liu, J., Yang, Y.H., Member, S.: Multiresolution color image segmentation. IEEE Trans. Pattern Anal. Mach. Intell. **16**(7), 689–700 (1994)

23. Lucchese, L., Mitra, S.K.: Color image segmentation: a state-of-the-art survey. Proc. Indian Natl. Sci. Acad. (INSA-A) **67**(2), 207–221 (2001). Delhi, Indian

24. Martin, D., Fowlkes, C., Tal, D., Malik, J.: A Database of human segmented natural images and its application to evaluating segmentation algorithms and measuring ecological statistics. In: Proceedings of 8th International Conference Computer Vision. vol. 2, pp. 416–423, July 2001
25. Meyer, F.: From connected operators to levelings. In: Heijmans, H., Roerdink, J. (eds.) Proceedings of ISMM 1998 Mathematical Morphology and its Applications to Image and Signal Processing, pp. 191–198. Kluwer Academic Publishers (1998)
26. Meyer, F.: Hierarchies of partitions and morphological segmentation. In: Kerckhove, M. (ed.) Scale-Space 2001. LNCS, vol. 2106, pp. 161–182. Springer, Heidelberg (2001). doi:10.1007/3-540-47778-0_14
27. Najman, L., Talbot, H. (eds.): Mathematical Morphology: From Theory to Applications. Wiley, Hoboken (2013)
28. Ou, L.C., Luo, M.R.: A colour harmony model for two-colour combinations. Color Res. Appl. **31**(3), 191–204 (2006)
29. Rittner, L., Flores, F.C., Lotufo, R.A.: A tensorial framework for color images. Pattern Recogn. Lett. **31**(4), 277–296 (2010)
30. Ruzon, M.A., Tomasi, C.: Edge, junction, and corner detection using color distributions. IEEE Trans. Pattern Anal. Mach. Intell. **23**(11), 1281–1295 (2001)
31. Serra, J., Vincent, L.: An overview of morphological filtering. Circuits Syst. Signal Process. **11**(1), 47–108 (1992)
32. Vachier, C.: Extraction de caractéristiques, segmentation d'image et morphologie mathématique. Ph.D. thesis, Ecole des Mines de Paris (1995)
33. Vachier, C., Meyer, F.: Extinction value: a new measurement of persistence. In: Proceedings of 1995 IEEE Workshop on Nonlinear Signal and Image Processing. vol. I, pp. 254–257. IEEE (1995)
34. Vincent, L.: Grayscale area openings and closings, their efficient implementation and applications. In: First Workshop on Mathematical Morphology and its Applications to Signal Processing, pp. 22–27 (1993)
35. van der Walt, S., Schönberger, J.L., Nunez-Iglesias, J., Boulogne, F., Warner, J.D., Yager, N., Gouillart, E., Yu, T.: the scikit-image contributors: scikit-image: image processing in Python. PeerJ **2**, e453 (2014)
36. Zhang, H., Fritts, J.E., Goldman, S.A.: An entropy-based objective evaluation method for image segmentation. In: Storage and Retrieval Methods and Applications for Multimedia 2004. vol. 5307, pp. 38–49, December 2003
37. Zhang, H., Fritts, J.E., Goldman, S.A.: Image segmentation evaluation: a survey of unsupervised methods. Comput. Vis. Image Underst. **110**(2), 260–280 (2008)

Ultimate Opening Combined with Area Stability Applied to Urban Scenes

Beatriz Marcotegui[1(✉)], Andrés Serna[2], and Jorge Hernández[3]

[1] MINES ParisTech, PSL Research University,
CMM - Centre for Mathematical Morphology,
35 rue Saint Honoré, Fontainebleau, France
`beatriz.marcotegui@mines-paristech.fr`
[2] Trimble Inc., 3D Scanning Operating Site,
174 Av. du Maréchal de Lattre de Tassigny, Fontenay-sous-Bois, France
`andres_serna@trimble.com`
[3] SAFRAN Paris-Saclay, Rue Des Jeunes Bois, Châteaufort, France
`jorge.hernandez-londono@safrangroup.com`
`http://cmm.mines-paristech.fr`, `http://www.trimble.com/3d-laser-scanning/`

Abstract. This paper explores the use of ultimate opening in urban analysis context. It demonstrates the efficiency of this approach for street level elevation images, derived from 3D point clouds acquired by terrestrial mobile mapping systems. An area-stability term is introduced in the residual definition, reducing the over-segmentation of the vegetation while preserving small significant regions.

We compare two possible combinations of the Ultimate Opening and the Area Stability: first as a multiplicative factor, then as a subtractive term. On the one hand, multiplicative factor is very strict and many significant regions may be eliminated by the operator. On the other hand, a subtractive factor is more easily controlled according to the image dynamics. In our application, the latter provides the best results by preserving small meaningful objects such as poles and bollards while avoiding over-segmentation on more complex objects such as cars and trees.

Keywords: Residual approach · Ultimate opening · Urban scene analysis · Elevation images

1 Introduction

Ultimate opening (UO) is a morphological operator based on numerical residues. It produces relevant partial partitions of generic images. The operator successively applies a decreasing family of openings, γ_λ, $\lambda = 0, 1, ... N - 1$, that forms a size granulometry:

$$\forall i, j : 0 \leq i \leq j \leq N - 1 \Rightarrow \gamma_i \geq \gamma_j \tag{1}$$

Then, the residues between successive openings are computed: $r_i = \gamma_i - \gamma_{i+1}$ and the maximum residue is kept for each pixel. Then, this operator returns two

© Springer International Publishing AG 2017
J. Angulo et al. (Eds.): ISMM 2017, LNCS 10225, pp. 261–268, 2017.
DOI: 10.1007/978-3-319-57240-6_21

significant pieces of information for each pixel: $R(I)$, the maximal residue which carries contrast information and $q(I)$ which indicates the size of the opening leading to this residue, corresponding to the size of the structure containing the considered pixel. Thus, the ultimate opening definition is given by the following formula:

$$R(I) = \max(r_i(I)) = \max(\gamma_i(I) - \gamma_{i+1}(I)), \forall i \geq 1$$
$$q(I) = \begin{cases} max\,\{i+1 \mid r_i(I) = R(I)\} & \text{if } R(I) > 0 \\ 0 & \text{if } R(I) = 0 \end{cases} \qquad (2)$$

In this residual framework, other openings can be used: openings by reconstruction [15] or attribute openings [16]. These openings preserve the shape of the analyzed objects. Moreover attribute openings allow an efficient implementation based on a max-tree representation. Their use in real-time applications becomes possible for relatively large images.

Successful applications based on this operator have been developed such as rock analysis [13], automatic text localization [16], façade segmentation [9]. Morphological profiles are extensively used in the remote sensing field: they address classification approaches in urban areas [2,3,10], ground and building extraction from lidar data [12,14]. As far as we know, these techniques have not been used for classification purposes on elevation images at the street level, acquired by terrestrial mobile mapping systems equipped with lidar scanners.

In this paper we apply ultimate opening to street level elevation images. We analyze the results and improve them introducing an area stability term in the residual definition. The paper is organized as follows: Sect. 2 details UO computation on a synthetic image. Section 3 introduces the area-stability term in the residual framework. Then, Sect. 4 demonstrates the efficiency of this method to segment urban scenes into significant objects. Finally, Sect. 5 concludes the paper and gives some perspectives for future developments.

2   Ultimate Opening Computation

Figure 1 illustrates the intermediate steps of UO calculation. The attribute opening [6] used in this example is the height (y-extent) of the bounding box containing a connected component. The associated function q corresponds then to the height attribute of segmented structures. Figure 1(a) shows the original image I. Results of successive height attribute openings (of size $1, 2, 3$ and 4) are illustrated in the Fig. 1(a–d). Then residues are computed (Fig. 1(e–g)) and for each pixel, the maximum residue r_i is selected and stored in $R(I)$ (Fig. 1(h)) and the size of the opening leading to this residue is recorded in $q(I)$ (Fig. 1(i)).

As aforementioned, the ultimate opening operator can be applied for segmenting generic images. In this paper, we explore its performance on elevation images from urban scenarios. Consider the Fig. 2, showing a 3D point cloud from Assas street in Paris. Figure 2, first column, illustrates the ultimate height opening on the elevation image from that 3D point cloud.

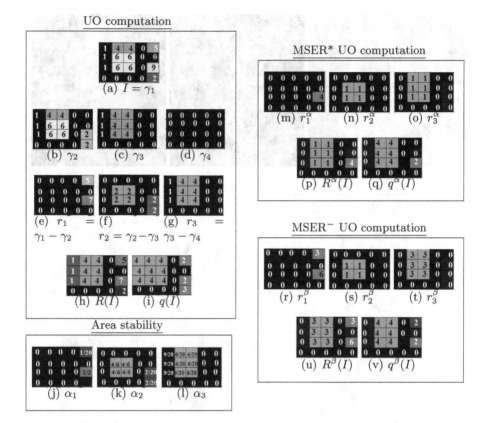

Fig. 1. Height UO computation. (a) Input image I. (b–d) height openings. (e–g) residues. (h–i) resulting $R(I)$ and $q(I)$ images. (j–l) area stability. (m–o) MSER* residues and (p–q) resulting images. (r–t) MSER⁻ residues and (u–v) resulting images.

3 UO Combined with Area Stability

Despite its performance, a drawback of the ultimate opening is the numerous spurious regions that are segmented in the background. Belhedi and Marcotegui [1] address this problem. These spurious regions are assumed to correspond to noisy regions with low area stability (inspired from MSER approach [11]). In order to reduce the influence of noise [1] propose to introduce an area stability weighting factor, α_i, in the residue definition:

$$r_i^\alpha = r_i * \alpha_i \ with \ \alpha_i = \frac{area_i}{area_{i+1}} \tag{3}$$

where $area_i$ is the area of the connected component of r_i containing the considered pixel, in the threshold decomposition framework.

Figure 1 illustrates the $MSER^*UO$ computation on the previously used synthetic image. First, the area stability weighting factors, α_i are computed in Fig. 1(j–l). Then r^α are obtained multiplying α_i by r_i in Fig. 1(m–o). Finally,

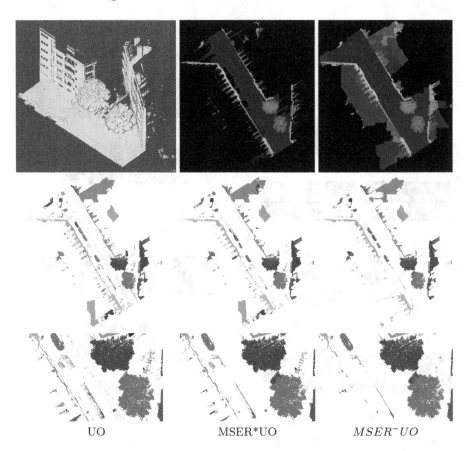

UO   MSER*UO $MSER^-UO$

Fig. 2. Ultimate opening results on Assas street, Paris (TerMob2_LAMB93_0021 point cloud acquired by IGN). First row: 3D point cloud, elevation image I and its interpolated version $\hat{I}$. Second row: $q(\hat{I})$. Third row: a zoom on second row for visualization purposes.

the maximal residue is computed for each pixel in R^α and the corresponding attribute size in q^α. Comparing Fig. 1(i) and (q), we can observe that q^α contains less noise than q

The result of $MSER^*UO$ on an elevation image from a urban environment is shown in Fig. 2, second column. Many noisy regions have faded. However bollards dismiss too, as they are small regions that merge directly with large regions without transition zones. They are then considered as noisy regions and are filtered out.

The multiplicative factor has a high influence in the final residual value and as we have seen it may delete significant small regions. We propose to introduce the area stability information as a subtractive term instead of multiplicative factor. Thus, the residue is defined as:

$$r_i^\beta = r_i - \beta_i \; with \; \beta_i = k * (1 - \alpha_i) \qquad (4)$$

where k is a maximal influence value, related to the image dynamics, given that the area stability is bounded between 0 and 1. In that way, the influence of the stability can be controlled by factor k. Figure 1(r–t) show the corresponding $MSER^-UO$ residues and Fig. 1(u–v) the resulting images. The region in the top right corner is preserved in spite of its low area stability.

Figure 2, third column shows the result of the UO with a subtractive MSER term $(MSER^-UO)$. In our experiments, the k parameter has been chosen to be one hundredth of the maximum elevation of the image ($k = max\{I\}/100$). Since the highest structures in the elevation image are trees and facades higher than 15 m, this selected k value allows to extract objects in the decimeter scale. We can observe that most noisy regions are not present anymore while preserving bollards and trees without over-segmentation. The third row shows a zoom of the second row in order to appreciate the differences between the three ultimate opening versions.

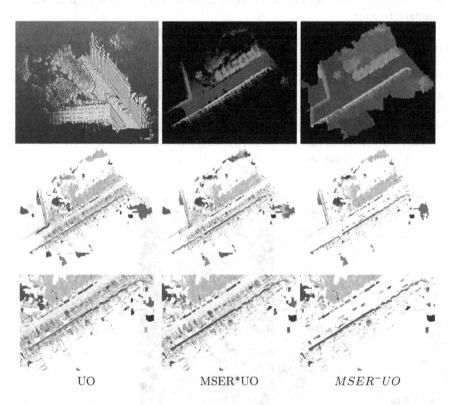

UO   MSER*UO $MSER^-UO$

Fig. 3. Ultimate opening results on Montparnasse street, Paris (acquired by CAOR-MINES ParisTech). First row: 3D point cloud, elevation image I and its interpolated version $\hat{I}$. Second row: q($\hat{I}$). Third row: a zoom on second row for visualization purposes.

4   Results

A urban scene semantization contest was organized by the French Mapping Agency (IGN) [5]. 3D point clouds were acquired in Paris by Stereopolis, the IGN terrestrial mobile mapping system equipped with lidar sensors. The method leading to the best results on TerraMobilita/iQmulus database [21] apply to the elevation image obtained by a vertical projection of the 3D point cloud. This method, presented in [18,20], is based on the hypothesis that each object contains only one significant maximum on the elevation image, which is true for objects such cars and poles. In that sense, the method relies on the $h - maxima$ operator [17] to filter out low contrasted maxima and use the significant maxima as input for a constrained watershed [4] in order to obtain segmented objects. The main drawback of this method is that it does not work on complex objects such as vegetation and motorcycles. Moreover, the h parameter is a key parameter: if it is too low it over-segments textured areas and if it is too high it removes significant regions.

Ultimate opening is a good candidate to deal with urban analysis on this type of images. Image interpolation based on the morphological fill-holes operator introduced in [8] and used also in [18,20] is adopted in this paper in order to fill occluded areas and assure connectivity. The result of ultimate height opening applied to the interpolated image is shown in Fig. 2 first column. We can observe that the trees are correctly segmented, without over-segmentation and the car next to the trees is also correctly handled. Compared with the competing approach [19] the trees are not over-segmented and the car is correctly handled, even if it is not a maximum.

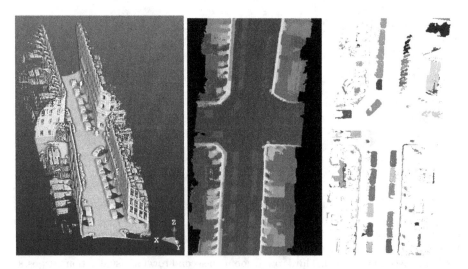

Fig. 4. Ultimate opening results on Madame street, Paris (point cloud acquired by CAOR MINES ParisTech). Left: 3D rendering. Middle: interpolated elevation image $\hat{I}$. Right: q($\hat{I}$).

The maximal residue image R produces many spurious regions in the background that do not correspond to regions of interest. A thresholding of R can help removing these regions but may also remove low contrasted regions. The use of area stability weight penalizes regions with important changes in area that are probably related to the presence of noise, artifacts and contrast variation in the background or unintended connection between components.

Figures 3 and 4 show additional results on elevation images derived from 3D point clouds acquired by the CAOR-MINES ParisTech acquisition system [7]. The same remarks on the quality of the result are valid for these images.

5   Conclusions and Perspectives

This paper demonstrates the interest of the ultimate opening for the analysis of urban scenes. It produces relevant partitions avoiding over-segmentation of trees as well as preserving the objects next to them, even if they do not correspond to regional maxima in the elevation images. A drawback of this operator is that it produces many noisy regions in the background. The combination of an area stability term in the residual definition highly reduces this problem. We propose to use a subtraction term instead of a multiplicative factor previously introduced in the literature. In our experiments, we show that this subtractive term better handles significant small regions while efficiently removing noisy ones. It amounts to applying an adaptive contrast threshold that depends on the area stability of the region. In future work, we will quantify the benefit of our approach in urban scene analysis.

References

1. Belhedi, A., Marcotegui, B.: Adaptive scene-text binarisation on images captured by smartphones. IET Image Process. **10**(7), 515–523 (2016)
2. Benediktsson, J.A., Arnason, K., Pesaresi, M.: The use of morphological profiles in classification of data from urban areas. In: IEEE/ISPRS Joint Workshop Remote Sensing and Data Fusion over Urban Areas, pp. 30–34. IEEE (2001)
3. Benediktsson, J.A., Pesaresi, M., Arnason, K.: Classification and feature extraction for remote sensing images from urban areas based on morphological transformations. IEEE Trans. Geosci. Remote Sens. **41**(9), 1940–1949 (2003)
4. Beucher, S., Lantuéjoul, C.: Use of watersheds in contour detection. In: International Workshop on Image Processing, Real-time Edge and Motion Detection (1979)
5. Brédif, M., Vallet, B., Serna, A., Marcotegui, B., Paparoditis, N.: TerraMobilita/IQmulus urban point cloud classification benchmark. In: Workshop on Processing Large Geospatial Data (2014)
6. Breen, E.J., Jones, R.: Attribute openings, thinnings, and granulometries. Comput. Vis. Image Underst. **64**(3), 377–389 (1996)
7. Goulette, F., Nashashibi, F., Ammoun, S., Laurgeau, C.: An integrated on-board laser range sensing system for on-the-way city, road modelling. In: The ISPRS International Archives of Photogrammetry, Remote Sensing, Spatial Information Sciences, vol. XXXVI-1, pp. 1–6 (2006)

8. Hernández, J., Marcotegui, B.: Segmentation et interprétation de nuages de points pour la modélisation d'environnements urbains. Revue Francaise de Photogrammetrie et de Teledetection **191**, 28 (2008)

9. Hernández, J., Marcotegui, B.: Ultimate attribute opening segmentation with shape information. In: Wilkinson, M.H.F., Roerdink, J.B.T.M. (eds.) ISMM 2009. LNCS, vol. 5720, pp. 205–214. Springer, Heidelberg (2009). doi:10.1007/978-3-642-03613-2_19

10. Lefevre, S., Weber, J., Sheeren, D.: Automatic building extraction in VHR images using advanced morphological operators. In: Proceedings of IEEE/ISPRS Joint Workshop Remote Sensing and Data Fusion over Urban Areas (2007)

11. Matas, J., Chum, O., Urban, M., Pajdla, T.: Robust wide baseline stereo from maximally stable extremal regions. In: Proceedings of British Machine Vision Conference, pp. 384–396 (2002)

12. Mongus, D., Lukac, N., Zalik, B.: Ground and building extraction from LiDAR data based on differential morphological profiles and locally fitted surfaces. ISPRS J. Photogram. Remote Sens. **93**, 145–156 (2014)

13. Outal, S., Beucher, S.: Controlling the ultimate openings residues for a robust delineation of fragmented rocks. In: The 10th European Congress of Stereology and Image Analysis, June 2009

14. Pedergnana, M., Marpu, P.R., Mura, M.D., Benediktsson, J.A., Bruzzone, L.: Classification of remote sensing optical and lidar data using extended attribute profiles. IEEE J. Sel. Top. Sig. Process. **6**(7), 856–865 (2012)

15. Pesaresi, M., Benediktsson, J.A.: A new approach for the morphological segmentation of high-resolution satellite imagery. IEEE Trans. Geosci. Remote Sens. **39**(2), 309–320 (2001)

16. Retornaz, T., Marcotegui, B.: Scene text localization based on the ultimate opening. In: Proceedings of the 8th International Symposium on Mathematical Morphology, vol. 1 of ISMM 2007, pp. 177–188, Rio de Janeiro, Brazil, October 2007

17. Schmitt, M., Prêteux, F.: Un nouvel algorithme en morphologie mathématique,: les rh maxima et rh minima. In: Proceedings 2ieme Semaine Internationale de l'Image Electronique, pp. 469–475 (1986)

18. Serna, A.: Semantic analysis of 3D point clouds from urban environments: ground, facades, urban objects and accessibility. Ph.D. thesis, Mines ParisTech - C.M.M., Fontainebleau - France, December 2014

19. Serna, A., Marcotegui, B.: Detection, segmentation and classification of 3d urban objects using mathematical morphology and supervised learning. ISPRS J. Photogram. Remote Sens. **93**, 243–255 (2014)

20. Serna, A., Marcotegui, B., Decenciere, E., Baldeweck, T., Pena, A.-M., Brizion, S.: Segmentation of elongated objects using attribute profiles, area stability: application to melanocyte segmentation in engineered skin. Pattern Recogn. Lett. **47**, 172–182 (2014). Advances in Mathematical Morphology

21. Vallet, B., Brédif, M., Serna, A., Marcotegui, B., Paparoditis, N.: Terramobilita/iqmulus urban point cloud analysis benchmark. Comput. Graph. **49**, 126–133 (2015)

PDE-Based Morphology

PDE for Bivariate Amoeba Median Filtering

Martin Welk[(⊠)]

Institute of Biomedical Image Analysis, Private University for Health Sciences,
Medical Informatics and Technology (UMIT), Eduard-Wallnöfer-Zentrum 1,
6060 Hall, Tyrol, Austria
martin.welk@umit.at

Abstract. Amoebas are image-adaptive structuring elements for morphological filters that have been introduced by Lerallut et al. in 2005.
Iterated amoeba median filtering on grey-scale images has been proven
to approximate asymptotically for vanishing structuring element radius
a partial differential equation (PDE) which is known in image filtering by the name of self-snakes. This approximation property helps to
understand the properties of both, morphological and PDE, image filter
classes. Recently, also the PDEs approximated by multivariate median filtering with non-adaptive structuring elements have been studied. Affine
equivariant multivariate medians turned out to yield more favourable
PDEs than the more popular L^1 median. We continue this work by considering amoeba median filtering of bivariate images using affine equivariant medians. We prove a PDE approximation result for this case. We validate the result by numerical experiments on example functions sampled
with high spatial resolution.

1 Introduction

Image processing methods based on superficially disparate paradigms often show
surprising similarities in their results. For example, discrete image filters designed
in a morphological frameworks can often be connected to partial differential
equations (PDEs). Dilation and erosion can be linked to Hamilton–Jacobi PDEs
[2]; the median filter is known to approximate in a continuous limit a curvature
motion PDE [10]. The important role of adaptive filters in all branches of image
processing has triggered interest in extending such connections also to adaptive
filters.

In mathematical morphology, adaptive filters can be based on adaptive structuring elements. Morphological amoebas which have been introduced by Lerallut,
Decencière and Meyer [14,15] are one class of such image-adaptive structuring
elements. The essence of the amoeba construction is the definition of a spatially
variant metric on the image domain that combines the spatial distance with local
image contrast. The structuring element of a pixel is then defined as a neighbourhood of prescribed radius with regard to this metric. Thereby, preferably
pixels with similar intensities are included. The shape of the structuring element
then adapts flexibly to image details, giving reason to the name amoebas.

© Springer International Publishing AG 2017
J. Angulo et al. (Eds.): ISMM 2017, LNCS 10225, pp. 271–283, 2017.
DOI: 10.1007/978-3-319-57240-6_22

Morphological amoebas are naturally linked to graph morphology [21]. They have been compared to further types of adaptive structuring elements [6,9,22]. They have also been employed to construct an active contour method for image segmentation [25]. Amoeba-based filters for multi-channel images, such as colour images or diffusion tensor data sets, have also been considered [8,13].

For univariate amoeba median filtering, it has been shown [29] that it approximates the so-called *self-snakes* PDE [18] in a space-continuous limit. In this paper, we want to investigate amoeba median filtering of multi-channel images under this aspect. In recent image processing literature, multivariate median filtering is mostly based on the so-called L^1 median [13,20,30]. For multivariate median filtering using the L^1 median, a PDE limit becomes fairly complicated already in the non-adaptive case [28]. However, in [27] it was shown that two affine equivariant median concepts, the Oja median [16] and the transformation–retransformation L^1 median [5,12,17], give rise to image filters that can be related to simpler and more manageable PDEs.

Our Contribution. Motivated by the previously mentioned results, we study amoeba filters based on the affine equivariant Oja and transformation–retransformation L^1 median. For the purpose of the present paper, we restrict ourselves to the bivariate case, i.e. two-channel images, and one specific amoeba metric. We derive a PDE that is approximated by the amoeba median filter under consideration. More precisely, one step of amoeba median filtering in a space-continuous setting asymptotically approximates a time step of an explicit scheme for the PDE when the radius of structuring elements tends to zero; the time step size goes to zero with the square of the radius.

The focus of our contribution is theoretic. Therefore we refrain from presenting image filtering experiments; apart from flow fields, there are not many meaningful application cases for bivariate images. Instead we will validate our approximation results by numerical experiments on example functions sampled with high spatial resolution.

Structure of the Paper. We recall the main facts on amoeba filtering procedures in Sect. 2. Section 3 is devoted to multivariate median concepts. Our analysis of bivariate amoeba median filtering is presented in Sect. 4, followed by the numerical validation in Sect. 5. A conclusion is given in Sect. 6.

2   Amoeba Filtering

Discrete Amoeba Filtering. The construction of amoeba structuring elements [14] relies on an *amoeba metric*. Given a discrete image $u = (u_i)_{i \in \mathcal{J}}$ over a discrete domain such as $\mathcal{J} = \{1, \ldots, M\} \times \{1, \ldots, N\}$, the amoeba metric measures the distance between two neighbouring pixels $i, j \in \mathcal{J}$ as $d(i, j) := \varphi(\|i - j\|, \beta |u_i - u_j|)$, where $\varphi : \mathbb{R}_+^2 \to \mathbb{R}_+$ is a 1-homogeneous continuous function. In [14] the L^1 sum $\varphi(s,t) \equiv \varphi_1(s,t) := s + t$ is used; an alternative is the L^2 sum $\varphi(s,t) \equiv \varphi_2(s,t) := \sqrt{s^2 + t^2}$. The neighbourhood relation may be defined by 4-neighbourhoods as in [14] or 8-neighbourhoods; other choices are possible. The contrast-scale parameter $\beta > 0$ balances the influence of spatial and intensity information.

Summation along paths $P := (i_0, i_1, \ldots, i_k)$ where i_j and i_{j+1} for $j = 0, \ldots, k-1$ are neighbours yields the path length $L(P) = \sum_{j=0}^{k-1} d(i_j, i_{j+1})$. For each pixel $i \in \mathcal{J}$ one then defines the amoeba structuring element of radius ϱ as the set of all pixels $j \in \mathcal{J}$ for which a path P from i to j exists that has $L(P) \leq \varrho$.

Amoebas can be used in a straightforward way as structuring elements for e.g. dilation, erosion, or median filtering.

Continuous Amoeba Filtering. [29] This procedure can easily be translated to a continuous-scale setting: Assume $u : \mathbb{R}^2 \supset \Omega \to \mathbb{R}$ is a smooth function over the compact image domain Ω. The construction is understood best by considering the (rescaled) *image graph* $\Gamma := \{(\boldsymbol{x}, \beta u(\boldsymbol{x})) \mid \boldsymbol{x} \in \Omega\} \subset \Omega \times \mathbb{R}$. Note that Γ is a section in the bundle $\Omega \times \mathbb{R}$. A continuous amoeba metric on Ω can then be defined by introducing the infinitesimal metric $\mathrm{d}s := \varphi(\|\mathrm{d}\boldsymbol{x}\|, \beta |\mathrm{d}u|)$ on Γ, with $\|\cdot\|$ denoting some norm on $\mathbb{R}^2$, and φ as above, and projecting it back to Ω. In general $\mathrm{d}s$ will be a Finsler metric. For the special choice where $\|\cdot\|$ is the Euclidean norm, and $\varphi \equiv \varphi_2$, the metric on Γ is induced from the Euclidean metric on $\Omega \times \mathbb{R}$, yielding a Riemannian metric $\mathrm{d}s$ on Ω. In any case, it can be integrated along continuous curves in Ω, and gives rise to distances $d(\boldsymbol{x}, \boldsymbol{y})$ between points $\boldsymbol{x}, \boldsymbol{y} \in \Omega$. The continuous-scale amoeba $\mathcal{A}_\varrho(\boldsymbol{x})$ of radius ϱ around $\boldsymbol{x} \in \Omega$ is then the set of all $\boldsymbol{y} \in \Omega$ for which $d(\boldsymbol{x}, \boldsymbol{y}) \leq \varrho$ holds.

Again, it is straightforward to conceive a continuous-scale amoeba filter: For each location $\boldsymbol{x} \in \Omega$, one selects the neighbourhood $\mathcal{A}_\varrho(\boldsymbol{x})$. Using the Borel measure on Ω, the intensities $u(\boldsymbol{y})$ for $\boldsymbol{y} \in \mathcal{A}_\varrho(\boldsymbol{x})$ give rise to a density on (a subset of) $\mathbb{R}$. The filtered value $v(\boldsymbol{x})$ is then obtained by applying an aggregation operator such as maximum (for dilation), minimum (for erosion), or median to the density. For example, the median of the density is the infimum μ of values $z \in \mathbb{R}$ for which $u(\boldsymbol{y}) \leq z$ on at most half of the area of $\mathcal{A}_\varrho(\boldsymbol{x})$; i.e. in generic cases the level line $u(\boldsymbol{y}) = \mu$ bisects the amoeba area.

Continuous Amoeba Filtering of Multivariate Images. [13] The amoeba construction can easily be transferred to images which take values in $\mathcal{R} = \mathbb{R}^m$, such as RGB colour images ($m = 3$), planar flow fields ($m = 2$), or symmetric 2×2 matrices ($m = 3$). Let a continuous-scale multivariate image $\boldsymbol{u} : \mathbb{R}^2 \supset \Omega \to \mathbb{R}^m$ be given. Its graph $\Gamma := \{(\boldsymbol{x}, \beta \boldsymbol{u}(\boldsymbol{x})) \mid \boldsymbol{x} \in \Omega\}$ is a section in the bundle $\Omega \times \mathbb{R}^m$. Choosing a norm $\|\cdot\|_\mathcal{R}$ on $\mathbb{R}^m$ one defines $\mathrm{d}s := \varphi(\|\mathrm{d}\boldsymbol{x}\|, \beta \|\mathrm{d}\boldsymbol{u}\|_\mathcal{R})$ on Γ and obtains the amoeba metric by projection to Ω. The construction of amoebas $\mathcal{A}_\varrho(\boldsymbol{x})$ then translates verbatim.

Multivariate amoeba filters can then be defined in the same way as grey-value filters, provided suitable aggregation operators are available on the density of values $\boldsymbol{u}(\boldsymbol{y}) \in \mathbb{R}^m$ for $\boldsymbol{y} \in \mathcal{A}_\varrho(\boldsymbol{x})$. Examples of such operators are supremum/infimum operators for symmetric matrices [3,4], multivariate medians [11,13,16,20,30] or quantiles [31]. We will discuss multivariate median concepts in more detail in the next section.

Specification of Amoeba Metric for this Paper. We will focus on continuous-scale amoeba filtering of bivariate images $u : \Omega \to \mathbb{R}^2$, where the amoeba metric is induced from the Euclidean metric in $\Omega \times \mathbb{R}^2 \subset \mathbb{R}^4$, i.e. $\|\cdot\|_{\mathcal{R}}$ is the Euclidean norm, and $\varphi \equiv \varphi_2$. From now on, we will refer to this setting as *Euclidean amoeba metric.*

3   Multi-channel Median

Attempts to generalise the concept of median from the univariate setting to multivariate data go back more than a century, see the work by Hayford from 1902 [11].

L^1 *Median.* One of the most popular multivariate median concepts, which is nowadays termed L^1 median, has been introduced as early as 1909 by Weber to location theory [23] and by Gini in 1929 to statistics [7]; it reappeared later in [1,7,24] and many other works. The L^1 median generalises the well-known property of the univariate median to minimise the sum of distances to given data values on the real line; thereby, the L^1 median of points $\boldsymbol{a}_1, \ldots, \boldsymbol{a}_n \in \mathbb{R}^m$ is defined as

$$\boldsymbol{\mu}_{L^1}(\boldsymbol{a}_1, \ldots, \boldsymbol{a}_n) := \operatorname*{argmin}_{\boldsymbol{x} \in \mathbb{R}^m} \sum_{i=1}^{n} \|\boldsymbol{x} - \boldsymbol{a}_i\| \tag{1}$$

where $\|\cdot\|$ is the Euclidean norm in $\mathbb{R}^m$. The L^1 median concept has been used in image processing for symmetric matrices [30] as well as for RGB images [13,20]. A first attempt to derive the PDE approximated by such a median filter with non-adaptive structuring elements has been made in [28], with a later correction in [27].

Affine Equivariance. The L^1 median is equivariant under similarity transformations of $\mathbb{R}^m$ (combinations of Euclidean transformations and scalings): Let T be such a transform; then we have $\boldsymbol{\mu}_{L^1}(T(\boldsymbol{a}_1), \ldots, T(\boldsymbol{a}_n)) = T(\boldsymbol{\mu}_{L^1}(\boldsymbol{a}_1, \ldots, \boldsymbol{a}_n))$. However, the univariate median is equivariant under much more general transformations, namely under arbitrary strictly monotonous mappings of $\mathbb{R}$. Given that a Euclidean structure is often unnatural to impose on data sets, interest in alternative multivariate median concepts that feature at least affine equivariance has arisen. From the various concepts developed for this purpose, see [19] for an overview, we mention two approaches.

Oja Median. The first concept was introduced in 1983 by Oja [16] and termed *simplex median;* nowadays also the name *Oja median* has gained popularity. It generalises the same property of the univariate median as mentioned above. However, instead of viewing $|x - a_i|$ on $\mathbb{R}$ as a *distance,* it views it as *measure of an interval* on the real line, thus a one-dimensional simplex. Replacing one-dimensional simplices with m-dimensional ones, the simplex median is of points $\boldsymbol{a}_1, \ldots, \boldsymbol{a}_n \in \mathbb{R}^m$ is defined as

$$\boldsymbol{\mu}_{\mathrm{Oja}}(\boldsymbol{a}_1, \ldots, \boldsymbol{a}_n) := \operatorname*{argmin}_{\boldsymbol{x} \in \mathbb{R}^m} \sum_{1 \leq i_1 < \ldots < i_m \leq n} |[\boldsymbol{x}, \boldsymbol{a}_1, \ldots, \boldsymbol{a}_m]| \tag{2}$$

where $|[x, a_1, \ldots, a_m]|$ is the m-dimensional volume of the simplex spanned by the $m + 1$ points $x, a_1, \ldots, a_m$.

Transformation–Retransformation L^1 Median. An alternative approach is to upgrade the L^1 median to affine equivariance by a transformation–retransformation procedure [5,12,17]. Here, an affine transformation is determined based on the covariance matrix of input data, and the data thereby transformed to a normalised form with an isotropic covariance matrix. Then the L^1 median is computed and transformed back to the original coordinates. This approach allows to combine the more favourable algorithmic complexity of the L^1 median, as compared to the Oja median, with affine equivariance.

In the context of iterated median filtering of multivariate images, the Oja median has been studied in [26] for bivariate images; in [27] both Oja and transformation–retransformation L^1 median were considered in bivariate and trivariate settings, and PDEs approximated by both types of median filters with non-adaptive structuring elements were derived. Compared to the PDE mentioned earlier for the standard L^1 median filter, these PDEs are more favourable as their coefficient functions are simpler, and coincide for both filter types. This means that, despite the fact that Oja and transformation–retransformation L^1 median of point sets are not the same, their corresponding image filters can be viewed as different discrete realisations of the same affine equivariant median filter concept [27].

Median Concepts Studied in this Paper. Based on the encouraging findings in [27], we focus here on affine equivariant median filters for bivariate images based on the Oja and transformation–retransformation L^1 median.

4   PDE for Affine Equivariant Bivariate Amoeba Median Filtering

To begin with amoeba median filtering with Oja median, we can state our first result.

Theorem 1. *Let a bivariate image $u : \mathbb{R}^2 \supset \Omega \to \mathbb{R}^2$, $(x, y) \mapsto (u, v)$ be given, for which the Jacobian Du of u has rank 2 throughout Ω. For fixed contrast scale $\beta > 0$, and amoeba radius $\varrho \to 0$, one step of amoeba median filtering of u with Oja median approximates a time step of size $\tau = \varrho^2/24$ of an explicit scheme for the PDE system*

$$\partial_t u = T_1(Du)\partial_{\eta\eta} u + T_2(Du)\partial_{\xi\xi} u + T_3(Du)\partial_{\eta\xi} u \tag{3}$$

where η is the major, and ξ the minor eigenvector of the structure tensor $J := J(Du) := \nabla u \nabla u^{\mathrm{T}} + \nabla v \nabla v^{\mathrm{T}} = Du^{\mathrm{T}} Du$. The coefficient matrices $T_i(Du)$, are given by

$$\boldsymbol{T}_i(\mathrm{D}\boldsymbol{u}) := \boldsymbol{R}\,\boldsymbol{\Theta}_i\big(|\partial_{\boldsymbol{\eta}}\boldsymbol{u}|, |\partial_{\boldsymbol{\xi}}\boldsymbol{u}|\big)\,\boldsymbol{R}^{\mathrm{T}}, \qquad i = 1, 2, 3, \tag{4}$$

$$\boldsymbol{\Theta}_1(r, s) := \mathrm{diag}\big(\vartheta_1(r), \vartheta_2(r, s)\big), \qquad \vartheta_1(z) := \frac{1 - 8\beta^2 z^2}{(1 + \beta^2 z^2)^2}, \tag{5}$$

$$\boldsymbol{\Theta}_2(r, s) := \mathrm{diag}\big(\vartheta_2(r, s), \vartheta_1(s)\big), \qquad \vartheta_2(w, z) := \frac{3}{(1 + \beta^2 w^2)(1 + \beta^2 z^2)}, \tag{6}$$

$$\boldsymbol{\Theta}_3(r, s) := -2\begin{pmatrix} 0 & \vartheta_3(r, s) \\ \vartheta_3(s, r) & 0 \end{pmatrix}, \qquad \vartheta_3(w, z) := \frac{w}{z}\,\frac{1 + 4\beta^2 z^2}{(1 + \beta^2 w^2)(1 + \beta^2 z^2)}, \tag{7}$$

where $\boldsymbol{R} = (\mathrm{D}\boldsymbol{u}^{-1})^{\mathrm{T}}\boldsymbol{P}\,\mathrm{diag}\big(|\partial_{\boldsymbol{\eta}}\boldsymbol{u}|, |\partial_{\boldsymbol{\xi}}\boldsymbol{u}|\big)$ *is a rotation matrix that aligns the principal components of the variation of* $\boldsymbol{u}$ *with the coordinate axes in both the* (x, y) *and* (u, v) *spaces;* $\boldsymbol{P} = (\boldsymbol{\eta} \mid \boldsymbol{\xi})$ *is the eigenvector matrix of* $\boldsymbol{J}$.

The proof of this theorem relies on two principles that allow to reduce the general situation of the theorem to a sequence of more specialised cases, which are successively treated in three lemmas. The first of these principles has been established in [25] where univariate amoeba median filtering was considered in a setting where a function u was filtered using amoebas computed from a different function f. Then u influences the filter result by the curvature of its level lines, whereas f exerts its influence via the asymmetric shape of the amoeba. The decomposition of the effect of the amoeba median filter into a curvature-related and an asymmetry-related part remains valid even when both functions are the same, and we will use it in the proof of Lemma 3. The second principle is to use invariances of the amoeba construction and the median filter to normalise the local geometry of a configuration. This was done in a non-adaptive setting for the L^1 median in [28] using Euclidean transformations, whereas [27] used affine transformations to normalise the geometry for affine equivariant medians. In our context, the amoeba construction as well as the median filters are Euclidean equivariant which allows to align the function by rotations to one with diagonal Jacobian, see Lemma 3. Deriving in this setting the amoeba shape from the bivariate function and treating further the function being filtered and the amoeba separately, according to the first principle above, allows to exploit the affine equivariance of the median filter for a further normalisation, making the Jacobian of the function a unit matrix, see Lemmas 1 and 2. On this level, the result for non-adaptive bivariate median filtering from [27] can be invoked for the curvature contribution, Lemma 1, whereas the asymmetry contribution is accessible to direct analysis via the definition of the Oja median, Lemma 2.

Lemma 1 (from [27], Lemma 2 there). *Let* $\boldsymbol{u}$ *be given as in Theorem 1 with the origin* $\boldsymbol{0} = (0, 0)$ *in the interior of* Ω. *Assume that the Jacobian* $\mathrm{D}\boldsymbol{u}(\boldsymbol{0})$ *is the* 2×2 *unit matrix* $\boldsymbol{I}$, *i.e.* $u_x = v_y = 1$, $u_y = v_x = 0$. *One step of Oja median filtering of* $\boldsymbol{u}$ *at* $\boldsymbol{0}$ *with the disc* D_ϱ *of radius* ϱ *as (non-adaptive) structuring element approximates an explicit time step of size* $\tau = \varrho^2/24$ *of the PDE system*

$$u_t = u_{xx} + 3u_{yy} - 2v_{xy}, \qquad v_t = 3v_{xx} + v_{yy} - 2u_{xy}. \tag{8}$$

The second lemma refers to the filtering of a linear bivariate function with amoeba structuring elements computed from another function.

Lemma 2. *Let u be given as in Theorem 1 with 0 in the interior of Ω, and $\mathbf{D}u(x) = I$ for all $x \in \Omega$. At 0, let an amoeba structuring element $\mathcal{A}(0)$ be given in polar coordinates (r, φ) with $x = r \cos \varphi$, $y = r \sin \varphi$ by its contour $r(\varphi) = \varrho - a(\varphi)$, $a(\varphi) := \frac{1}{2} \varrho^2 \beta^2 (\alpha_1 \cos^3 \varphi + \alpha_2 \cos^2 \varphi \sin \varphi + \alpha_3 \cos \varphi \sin^2 \varphi + \alpha_4 \sin^3 \varphi)$. Then one step of Oja median filtering of u at 0 approximates an explicit time step of size $\tau = \varrho^2/24$ of the PDE system*

$$u_t = -9\beta^2 \alpha_1 - 3\beta^2 \alpha_3, \qquad v_t = -3\beta^2 \alpha_2 - 9\beta^2 \alpha_4. \qquad (9)$$

Proof (of Lemma 2). Due to the linearity of u, the Oja median of u within the disc $D_\varrho(0)$ with contour $r(\varphi) = \varrho$ equals $u(0)$. We study how the perturbation of the amoeba contour $r(\varphi)$ by $-a(\varphi)$ changes the gradient g of the objective function of (2) at 0. Remembering that the Oja median is the minimiser for a sum of triangle areas in the u-v plane, this means that we are interested in the net contribution of those triangles which are added or removed when switching from the disc $D_\varrho(0)$ to the actual amoeba. Neglecting a higher order error, the added or removed areas can be projected to the circumference of the disc. We are therefore led to consider the gradients $g_{B,A;M}$ of the areas of triangles MAB w.r.t. M, where M is the candidate median point, $A \equiv A(\varphi) = (\varrho \cos \varphi, \varrho \sin \varphi)$ and B in the disc D_ϱ. Up to higher order terms, g will be the resultant of $g_{B,A;M}$ for all such triangles, each weighted with $-a(\varphi)$ and evaluated at $M = 0$. Note that the weights $-a$ can take either sign, depending on whether the amoeba boundary is inside or outside the disc near A.

For $M = 0$ the straight line MA bisects D_ϱ; thus, for each point $B \in D_\varrho$ also the point B' obtained by reflecting B on MA is in D_ϱ. Thus, the gradients of the triangle areas of MAB and MAB' (w.r.t. M) can be combined into the gradient $g_{B,A,B';M}$ of the area of the quadrangle $MB'AB$ (a kite, a.k.a. deltoid). The vector $g_{B,A,B';M}$ is perpendicular to BB' and proportional to $\frac{1}{2}|BB'|$, the height of B over the line MA. Thus, $g_{B,A,B';M} = -\frac{1}{2}|BB'|(\cos\varphi, \sin\varphi)$. Integration over B in a half-disc yields the aggregated gradient for all triangles MAB with the point A as $g_{A;M} = -\frac{2}{3}\varrho^3 (\cos \varphi, \sin \varphi)$. Integrating over directions φ with weights $-a(\varphi)$ yields

$$g = \int_0^{2\pi} -a_\varphi g_{A(\varphi);M} \, d\varphi = -\frac{\varrho^6 \beta^2 \pi}{12} \begin{pmatrix} 3\alpha_1 + \alpha_3 \\ \alpha_2 + 3\alpha_4 \end{pmatrix}. \qquad (10)$$

From the proof of Lemma 2 in [27] (i.e. Lemma 1 above) it can be read off that a gradient g for (2) is compensated by a shift of M by $g/(-\frac{2}{3}\pi\varrho^4)$ which means that the median of u within the amoeba $\mathcal{A}(0)$ amounts to $u(0) - \beta^2 \begin{pmatrix} 3\alpha_1 + \alpha_3 \\ \alpha_2 + 3\alpha_4 \end{pmatrix}$, which is exactly the time step for the PDE system (9) as claimed in the lemma.

Lemma 3. *Let u be given as in Theorem 1 with the origin $0 = (0,0)$ in the interior of Ω. Assume that the Jacobian $\mathbf{D}u(0)$ is diagonal, with $u_x \geq v_y > 0$, $u_y = v_x = 0$. Then one step of amoeba median filtering with the Oja median at 0 approximates for $\varrho \to 0$ an explicit time step of size $\tau = \varrho^2/24$ of the PDE system*

$$u_t = \vartheta_1(u_x)\, u_{xx} + \vartheta_2(u_x, v_y)\, u_{yy} - 2\vartheta_3(u_x, v_y)\, v_{xy}, \tag{11}$$

$$v_t = \vartheta_2(u_x, v_y)\, v_{xx} + \vartheta_1(v_y)\, v_{yy} - 2\vartheta_3(v_y, u_x)\, u_{xy}, \tag{12}$$

with $\vartheta_{1,2,3}$ as in Theorem 1.

Proof (of Lemma 3). In [28, Sect. 4.1.2] the effect of an univariate amoeba median filter step was analysed into two parts, a *curvature contribution* coming from the curvature of level lines of the function being filtered, and an *asymmetry contribution* reflecting the asymmetry of the amoeba. Interactions of the two manifest only in higher order terms w.r.t. ϱ that can safely be neglected in the PDE analysis. The same decomposition can also be applied here. We denote by $\boldsymbol{f}$ the function obtained from $\boldsymbol{u}$ by linearisation at $\boldsymbol{0}$. Then the curvature contribution at $\boldsymbol{0}$ is equivalent to a step of amoeba median filtering of the function $\boldsymbol{u}$ with the amoeba $\mathcal{A}_{\boldsymbol{f}}(\boldsymbol{0})$ computed from $\boldsymbol{f}$, whereas for the asymmetry contribution $\boldsymbol{f}$ is filtered using the amoeba $\mathcal{A}_{\boldsymbol{u}}(\boldsymbol{0})$ computed from $\boldsymbol{u}$.

To start with the curvature contribution, $\mathcal{A}_{\boldsymbol{f}}(\boldsymbol{0})$ is an ellipse with half-axes ϱ/U in x direction, and ϱ/V in y direction, with $U := \sqrt{1 + \beta^2 u_x^2}$, $V := \sqrt{1 + \beta^2 v_y^2}$. By the affine equivariance of the Oja median, we can apply first to the x-y plane an affine transform T_{xy} with the transformation matrix $\boldsymbol{M} := \operatorname{diag}(U, V)$, so the amoeba is turned into the disc D_ϱ, and the image into $\tilde{\boldsymbol{u}}$ with Jacobian $\mathrm{D}\tilde{\boldsymbol{u}} = \mathrm{D}\boldsymbol{u}\,\boldsymbol{M}^{-1}$. Second, the u-v plane is transformed by an affine transform T_{uv} with the transformation matrix $\mathrm{D}\tilde{\boldsymbol{u}}^{-1}$ which yields a bivariate image $\hat{\boldsymbol{u}}$ with unit Jacobian. Now the hypotheses of Lemma 1 are satisfied for $\hat{\boldsymbol{u}}$. Using (8) for $\hat{\boldsymbol{u}}$ and reverting T_{uv} and T_{xy} yields the curvature contribution

$$u_t^{\mathrm{curv}} = \frac{u_{xx}}{U^2} + \frac{3u_{yy}}{V^2} - \frac{2u_x v_{xy}}{v_y U^2}, \qquad v_t^{\mathrm{curv}} = \frac{3v_{xx}}{U^2} + \frac{v_{yy}}{V^2} - \frac{2v_y u_{xy}}{u_x V^2}. \tag{13}$$

Turning to the asymmetry contribution, we compute the contour of $\mathcal{A}_{\boldsymbol{u}}(\boldsymbol{0})$ in a way that after applying the same two affine transformations as before suits the hypothesis of Lemma 2. To this end, consider the vector $\boldsymbol{w}^* := (U \cos \varphi, V \sin \varphi)$, with U and V as above, and normalise it to $\boldsymbol{w} := \boldsymbol{w}^*/\|\boldsymbol{w}^*\|$. Then the directional derivatives of $\boldsymbol{u}$ become $\partial_{\boldsymbol{w}} \boldsymbol{u} = (u_x V \cos \varphi/G, v_y U \sin \varphi/G)$, where $G := \sqrt{V^2 \cos^2 \varphi + U^2 \sin^2 \varphi}$, and $\partial_{\boldsymbol{ww}} \boldsymbol{u} = \frac{V^2}{G^2} \cos^2 \varphi \partial_{xx} \boldsymbol{u} + \frac{2UV}{G^2} \cos \varphi \sin \varphi \partial_{xy} \boldsymbol{u} + \frac{U^2}{G^2} \sin^2 \varphi \partial_{yy} \boldsymbol{u}$. The point of the amoeba contour in direction $\boldsymbol{w}$ is given up to higher order terms by $r\boldsymbol{w}$ where $r \equiv r(\boldsymbol{w})$ satisfies the condition

$$\varrho = \int_0^r \sqrt{1 + \beta^2 |\partial_{\boldsymbol{w}} \boldsymbol{u} + s \partial_{\boldsymbol{ww}} \boldsymbol{u}|^2}\, \mathrm{d}s. \tag{14}$$

The r.h.s. of this equation is the length of a straight line in direction $\boldsymbol{w}$ under the Euclidean amoeba metric. Note that the actual shortest path from $\boldsymbol{0}$ to the amoeba contour under the amoeba metric can deviate from this line but as pointed out in [28, 4.1.2] this deviation only influences higher order terms. Again up to higher order terms (14) can be evaluated to

$$\varrho = r\sqrt{1 + \beta^2 |\partial_w u|^2} \left(1 + r\frac{\beta^2 \langle \partial_w u, \partial_{ww} u \rangle}{2(1 + \beta^2 |\partial_w u|^2)}\right), \tag{15}$$

$$r = \frac{\varrho}{\sqrt{1 + \beta^2 |\partial_w u|^2}} \left(1 - \frac{\varrho \beta^2 \langle \partial_w u, \partial_{ww} u \rangle}{2(1 + \beta^2 |\partial_w u|^2)^{3/2}}\right). \tag{16}$$

Applying the affine transform T_{xy} as above, $\boldsymbol{w}^*$ becomes the vector $(\cos\varphi, \sin\varphi)$, and $r(\boldsymbol{w})$ is transformed into $r(\varphi)$ as in the hypothesis of Lemma 2 with

$$\alpha_1 = \frac{u_x u_{xx}}{U^3}, \qquad\qquad \alpha_2 = \frac{2u_x u_{xy} + v_y v_{xx}}{U^2 V}, \tag{17}$$

$$\alpha_3 = \frac{u_x u_{yy} + 2v_y v_{xy}}{UV^2}, \qquad\qquad \alpha_4 = \frac{v_y v_{yy}}{V^3}. \tag{18}$$

The derivatives u_x etc. herein refer to the untransformed function $\boldsymbol{u}$. Applying further T_{uv} as above, the setting is fully adapted to the hypothesis of Lemma 2. Reverting T_{uv} and T_{xy} on (9), then inserting (17), (18) yields the desired asymmetry contribution as

$$u_t^{\text{asym}} = -9\beta^2 \frac{u_x^2 u_{xx}}{U^4} - 3\beta^2 \frac{u_x^2 u_{yy}}{U^2 V^2} - 6\beta^2 \frac{u_x v_y v_{xy}}{U^2 V^2}, \tag{19}$$

$$v_t^{\text{asym}} = -3\beta^2 \frac{v_y^2 v_{xx}}{U^2 V^2} - 9\beta^2 \frac{v_y^2 v_{yy}}{V^4} - 6\beta^2 \frac{u_x v_y u_{xy}}{U^2 V^2}. \tag{20}$$

Combining (13) and (19), (20) as $\boldsymbol{u}_t := \boldsymbol{u}_t^{\text{curv}} + \boldsymbol{u}_t^{\text{asym}}$ yields (11), (12).

Proof (of Theorem 1). Consider a bivariate image $\boldsymbol{u}$ with regular Jacobian. By translation invariance, it can be assumed that the filter is considered at location $\boldsymbol{0}$. As eigenvectors of the structure tensor, $\boldsymbol{\eta}$ and $\boldsymbol{\xi}$ are orthonormal. Moreover, the directional derivatives of $\boldsymbol{u}$ in these directions, i.e. $\mathrm{D}\boldsymbol{u} \cdot \boldsymbol{\eta}$ and $\mathrm{D}\boldsymbol{u} \cdot \boldsymbol{\xi}$, are orthogonal. Thus both the x-y plane and the u-v plane can be rotated in order to align these orthogonal pairs with the respective coordinate axes. For the x-y plane, this is achieved using the rotation matrix $\boldsymbol{P}$; for the u-v plane, taking into account the non-unit lengths of the two orthogonal vectors, one obtains $\boldsymbol{R}^{-1}$ as the appropriate rotation matrix. Applying both rotations yields a bivariate image that satisfies the conditions of Lemma 3. Applying the inverse rotations to (11), (12) yields the PDE system stated in (3)–(7).

We turn now to the second affine equivariant median filter under consideration, the transformation–retransformation L^1 median. As in the case of nonadaptive filtering [27], we find that it approximates the same PDE as the Oja median filter.

Theorem 2. *Let a bivariate function $\boldsymbol{u} : \Omega \to \mathbb{R}^2$ with $\mathrm{D}\boldsymbol{u}$ of rank 2 be given as in Theorem 1. For $\varrho \to 0$, one step of amoeba median filtering with the transformation–retransformation L^1 median approximates a time step of size $\tau = \varrho^2/24$ of the same PDE (3)–(7) as for the Oja median.*

Proof (of Theorem 2). We follow the same strategy as for Theorem 1. Euclidean and affine transformations can again be used to reduce the generic geometric setting to the hypotheses of Lemma 3 and further to the hypotheses of Lemmas 1 and 2. As pointed out in [27, Sect. 3.1.5], this reduction procedure reproduces the transformations of the transformation–retransformation L^1 median. Thus, it remains to secure statements analogous to Lemmas 1 and 2 for the standard L^1 median. As for Lemma 1, this is a special case of Lemma 1 from [27]. To prove the L^1 analogue of Lemma 2 one calculates the gradient of the objective function of (1), which is even easier than for the Oja median because, adopting notations from the proof of Lemma 2, said gradient is composed just of the vectors $MA = (\cos\varphi, \sin\varphi)$ with appropriate weighting.

Remark 1. In the univariate case an interesting relation holds between the PDEs associated with non-adaptive and adaptive median filters: The self-snakes PDE [18] $u_t = |\nabla u| \operatorname{div}\big(g(|\nabla u|)\nabla u/|\nabla u|\big)$ – the limit case of the amoeba filter – is obtained from the curvature motion PDE $u_t = |\nabla u| \operatorname{div}\big(\nabla u/|\nabla u|\big)$ – the limit case of the nonadaptive filter – by inserting the edge-stopping function g within the divergence expression.

   We have to defer a more detailed discussion of the PDE system (3)–(7) to future work. In this context, an interesting question to be answered will be whether a relation similar to that between univariate self-snakes and curvature motion can be established between (3)–(7) and the PDE system derived in [27] for the non-adaptive case.

5   Experiments

We validate our PDE approximation results in the axis-aligned setting of Lemma 3 for eight simple bivariate example functions $u = (u, v)$ given by $u(x, y) = u_x x + \frac{1}{2}u_{xx}x^2 + u_{xy}xy + \frac{1}{2}u_{yy}y^2$, $v(x, y) = v_y y + \frac{1}{2}v_{xx}x^2 + v_{xy}xy + \frac{1}{2}v_{yy}y^2$, where all derivatives refer to the location $(0, 0)$. The example functions are collected in Table 1.

   For the functions u and their linearisations f, we compute approximations of amoeba structuring elements $\mathcal{A}_\varrho^u(0)$ and $\mathcal{A}_\varrho^f(0)$ with contrast scale $\beta = 1$ and amoeba radius $\varrho = 1$ on a discrete Cartesian grid with sample distance $h = 0.015$. For good approximation of rotational invariance, we base our amoeba computation on neighbourhoods of radius $h\sqrt{10}$, which include 36 points instead of 4- or 8-neighbourhoods. (This makes sense for our fairly smooth example functions; in an image filtering application, however, the use of such large neighbourhoods would spoil the adaptation of amoebas to fine image details.) The sizes of these amoebas are also given in Table 1.

   The Oja and transformation–retransformation L^1 medians of u within $\mathcal{A}_\varrho^f(0)$ are then computed by gradient descent methods, see [27], and compared with the isolated curvature contribution (13). In the same way, the medians of f within $\mathcal{A}_\varrho^u(0)$ are computed and compared with the isolated asymmetry contribution (19), (20), and the medians of u within $\mathcal{A}_\varrho^u(0)$ with the full PDE system (11),

Table 1. Bivariate example functions u for the validation of PDE approximations, and sizes (in pixels) of the amoeba approximations.

Test case	$u(x,y)$	$v(x,y)$	$|\mathcal{A}_\varrho^u(0)|$	$|\mathcal{A}_\varrho^f(0)|$
a	$x + 0.15\,x^2$	y	6992	6977
b	$x + 0.3\,xy$	y	6995	6977
c	$x + 0.15\,y^2$	y	6973	6977
d	$2x + 0.15\,x^2$	$0.5\,y$	5584	5573
e	$2x + 0.3\,xy$	$0.5\,y$	5583	5573
f	$2x + 0.15\,y^2$	$0.5\,y$	5571	5573
g	$2x - 0.15\,x^2 + 0.1\,xy + 0.1\,y^2$	$0.5\,y + 0.05\,x^2 + 0.3\,xy - 0.1\,y^2$	5574	5573
h	$2x - 0.25\,x^2 + 0.2\,xy + 0.15\,y^2$	$0.5\,y + 0.05\,x^2 + 0.6\,xy - 0.15\,y^2$	5556	5573

Table 2. Amoeba median values and time steps of approximated PDEs for the example functions u from Table 1. All values are multiplied by 10^4.

Test case	Coordinate	Curvature contribution			Asymmetry contribution			Complete filter		
		TRL^1	Oja	PDE	TRL^1	Oja	PDE	TRL^1	Oja	PDE
a	u	62	63	63	-281	-282	-281	-216	-216	-219
	v	0	0	0	0	0	0	0	0	0
b	u	0	0	0	0	0	0	0	0	0
	v	-125	-125	-125	-187	-187	-188	-312	-311	-313
c	u	187	187	188	-90	-90	-94	94	97	94
	v	0	0	0	0	0	0	0	0	0
d	u	24	24	25	-183	-182	-180	-158	-155	-155
	v	0	0	0	0	0	0	0	0	0
e	u	0	0	0	0	0	0	0	0	0
	v	-50	-50	-50	-116	-116	-120	-166	-167	-170
f	u	300	299	300	-237	-237	-240	62	62	60
	v	0	0	0	0	0	0	0	0	0
g	u	-23	-24	-25	-95	-96	-100	-122	-121	-125
	v	-61	-59	-59	74	73	75	14	14	17
h	u	-128	-133	-142	-159	-159	-180	-304	-301	-322
	v	-118	-108	-108	75	75	95	-37	-35	-13

(12). In Table 2 the results of median computations and the corresponding time steps with size $\tau = \varrho^2/24$ are given.

As can be seen, the results of both median filters agree well with the PDE time steps for the first six example functions a–f in which only one second-order coefficient is nonzero, with deviations in the order of 10^{-4}. In the last two examples g, h where all six second-order coefficients are nonzero, there are deviations in the order of 10^{-3} for individual coordinates. However, these deviations are well within the range to be expected for the given grid discretisation.

6   Summary and Outlook

We have derived a PDE approximated by amoeba median filtering of bivariate images with the affine equivariant Oja median or transformation–retransformation L^1 median, for Euclidean amoeba metrics. Numerical computations on sampled test functions confirmed the approximation. A more detailed discussion of the PDE will be a subject of forthcoming work where also the degenerate case det $\mathbf{D}u = 0$ that we excluded here (as before in [27]) should be investigated.

Our hope is that the result presented will contribute to a deeper understanding of adaptive filtering and the relations between morphological and PDE-based image filters. We are optimistic that we will be able in our ongoing work to extend the result to practically more meaningful cases such as three-channel planar or volume images, such as done in [27] for non-adaptive median filters.

References

1. Austin, T.L.: An approximation to the point of minimum aggregate distance. Metron **19**, 10–21 (1959)
2. Brockett, R.W., Maragos, P.: Evolution equations for continuous-scale morphology. In: Proceedings of IEEE International Conference on Acoustics, Speech and Signal Processing, San Francisco, CA, vol. 3, pp. 125–128 (1992)
3. Burgeth, B., Kleefeld, A.: An approach to color-morphology based on Einstein addition and Loewner order. Pattern Recogn. Lett. **47**, 29–39 (2014)
4. Burgeth, B., Welk, M., Feddern, C., Weickert, J.: Mathematical morphology on tensor data using the Loewner ordering. In: Weickert, J., Hagen, H. (eds.) Visualization and Processing of Tensor Fields. Mathematics and Visualization, pp. 357–368. Springer, Berlin (2006)
5. Chakraborty, B., Chaudhuri, P.: On a transformation and re-transformation technique for constructing an affine equivariant multivariate median. Proc. AMS **124**, 2539–2547 (1996)
6. Ćurić, V., Hendriks, C.L., Borgefors, G.: Salience adaptive structuring elements. IEEE J. Sel. Top. Sig. Process. **6**, 809–819 (2012)
7. Gini, C., Galvani, L.: Di talune estensioni dei concetti di media ai caratteri qualitativi. Metron **8**, 3–209 (1929)
8. González-Castro, V., Debayle, J., Pinoli, J.C.: Color adaptive neighborhood mathematical morphology and its application to pixel-level classification. Pattern Recogn. Lett. **47**, 50–62 (2014)
9. Grazzini, J., Soille, P.: Edge-preserving smoothing using a similarity measure in adaptive geodesic neighbourhoods. Pattern Recogn. **42**, 2306–2316 (2009)
10. Guichard, F., Morel, J.M.: Partial differential equations and image iterative filtering. In: Duff, I.S., Watson, G.A. (eds.) The State of the Art in Numerical Analysis. Number 63 in IMA Conference Series (New Series), pp. 525–562. Clarendon Press, Oxford (1997)
11. Hayford, J.F.: What is the center of an area, or the center of a population? J. Am. Stat. Assoc. **8**, 47–58 (1902)
12. Hettmansperger, T.P., Randles, R.H.: A practical affine equivariant multivariate median. Biometrika **89**, 851–860 (2002)

13. Kleefeld, A., Breuß, M., Welk, M., Burgeth, B.: Adaptive filters for color images: median filtering and its extensions. In: Trémeau, A., Schettini, R., Tominaga, S. (eds.) Computational Color Imaging. LNCS, vol. 9016, pp. 149–158. Springer, Cham (2015)

14. Lerallut, R., Decencière, É., Meyer, F.: Image processing using morphological amoebas. In: Ronse, C., Najman, L., Decencière, E. (eds.) Mathematical Morphology: 40 Years On. Computational Imaging and Vision, vol. 30, pp. 13–22. Springer, Dordrecht (2005)

15. Lerallut, R., Decencière, É., Meyer, F.: Image filtering using morphological amoebas. Image Vis. Comput. **25**, 395–404 (2007)

16. Oja, H.: Descriptive statistics for multivariate distributions. Stat. Probab. Lett. **1**, 327–332 (1983)

17. Rao, C.R.: Methodology based on the l_1-norm in statistical inference. Sankhyā A **50**, 289–313 (1988)

18. Sapiro, G.: Vector (self) snakes: a geometric framework for color, texture and multiscale image segmentation. In: Proceedings of 1996 IEEE International Conference on Image Processing, Lausanne, Switzerland, vol. 1, pp. 817–820 (1996)

19. Small, C.G.: A survey of multidimensional medians. Int. Stat. Rev. **58**, 263–277 (1990)

20. Spence, C., Fancourt, C.: An iterative method for vector median filtering. In: Proceedings of 2007 IEEE International Conference on Image Processing, vol. 5, pp. 265–268 (2007)

21. Ta, V.T., Elmoataz, A., Lézoray, O.: Nonlocal PDE-based morphology on weighted graphs for image and data processing. IEEE Trans. Image Process. **20**, 1504–1516 (2012)

22. Verdú-Monedero, R., Angulo, J., Serra, J.: Anisotropic morphological filters with spatially-variant structuring elements based on image-dependent gradient fields. IEEE Trans. Image Process. **20**, 200–212 (2011)

23. Weber, A.: Über den Standort der Industrien. Mohr, Tübingen (1909)

24. Weiszfeld, E.: Sur le point pour lequel la somme des distances de n points donnés est minimum. Tôhoku Math. J. **43**, 355–386 (1937)

25. Welk, M.: Analysis of amoeba active contours. J. Math. Imaging Vis. **52**, 37–54 (2015)

26. Welk, M.: Partial differential equations for bivariate median filters. In: Aujol, J.F., Nikolova, M., Papadakis, N. (eds.) Scale Space and Variational Methods in Computer Vision. LNCS, vol. 9087, pp. 53–65. Springer, Cham (2015)

27. Welk, M.: Multivariate median filters and partial differential equations. J. Math. Imaging Vis. **56**, 320–351 (2016)

28. Welk, M., Breuß, M.: Morphological amoebas and partial differential equations. In: Hawkes, P.W. (ed.) Advances in Imaging and Electron Physics, vol. 185, pp. 139–212. Elsevier Academic Press (2014)

29. Welk, M., Breuß, M., Vogel, O.: Morphological amoebas are self-snakes. J. Math. Imaging Vis. **39**, 87–99 (2011)

30. Welk, M., Feddern, C., Burgeth, B., Weickert, J.: Median filtering of tensor-valued images. In: Michaelis, B., Krell, G. (eds.) Pattern Recognition. LNCS, vol. 2781, pp. 17–24. Springer, Berlin (2003)

31. Welk, M., Kleefeld, A., Breuß, M.: Non-adaptive and amoeba quantile filters for colour images. In: Benediktsson, J.A., Chanussot, J., Najman, L., Talbot, H. (eds.) Mathematical Morphology and Its Applications to Signal and Image Processing. LNCS, vol. 9082, pp. 398–409. Springer, Cham (2015)

A Unified Approach to PDE-Driven Morphology for Fields of Orthogonal and Generalized Doubly-Stochastic Matrices

Bernhard Burgeth[1][(✉)] and Andreas Kleefeld[2]

[1] Faculty of Mathematics and Computer Science, Saarland University,
66041 Saarbrücken, Germany
burgeth@math.uni-sb.de
[2] Institute for Advanced Simulation, Jülich Supercomputing Centre,
Forschungszentrum Jülich GmbH, Wilhelm-Johnen-Straße,
52425 Jülich, Germany
a.kleefeld@fz-juelich.de

Abstract. In continuous morphology two nonlinear partial differential equations (PDEs) together with specialized numerical solution schemes are employed to mimic the fundamental processes of dilation and erosion on a scalar valued image. Some attempts to tackle in a likewise manner the processing of higher order data, such as color images or even matrix valued images, so-called matrix fields, have been made. However, research has been focused almost exclusively on real symmetric matrices. Fields of non-symmetric matrices, for example rotation matrices, defy a unified approach. That is the goal of this article. First, the framework for symmetric matrices is extended to complex-valued Hermitian matrices. The later offer sufficient degrees of freedom within their structures such that, in principle, any class of real matrices may be mapped in a one-to-one manner onto a suitable subset of Hermitian matrices, where image processing may take place. Second, both the linear mapping and its inverse are provided. However, the non-linearity of dilation and erosion processes requires a backprojection onto the original class of matrices. Restricted by visualization shortcomings, the steps of this procedure are applied to the set of 3D-rotation matrices and the set of generalized doubly-stochastic matrices.

Keywords: Matrix field · Hermitian matrix · Symmetric matrix · Orthogonal matrix · Doubly-stochastic matrix · Loewner order · Procrustes problem · Continuous morphology · Erosion · Dilation · Rouy-Tourin scheme

1 Introduction

For almost half a century mathematical morphology has embraced the challenges brought on by the increasing complexity and difficulty of image processing tasks. Since its set- and lattice-theoretic beginnings with the fundamental works of

© Springer International Publishing AG 2017
J. Angulo et al. (Eds.): ISMM 2017, LNCS 10225, pp. 284–295, 2017.
DOI: 10.1007/978-3-319-57240-6_23

Matheron and Serra [12,18] in the late sixties mathematical morphology keeps pace with the fast-paced developments in image processing, as the large body of publications in journals and conference proceedings bears witness, see [19, 20,24], just to name a few. The range of data types that can be processed and analyzed with morphological methods has expanded vastly from scalar- to vector- or general multi-valued data such as tensor fields or frames [21,22]. In image processing usually the notion tensor field refers to a field of symmetric (even positive definite) matrices with real entries.

This article, however, is devoted to the development of PDE-driven morphological image processing tools for 3D fields of Hermitian and, henceforth, other special matrices. We will denote any mapping $H : \Omega \longmapsto M_{\mathbb{F}}(n)$ from a two- or three-dimensional image domain Ω into the set $M_{\mathbb{F}}(n)$ of $n \times n$- real ($\mathbb{F} = \mathbb{R}$) or complex ($\mathbb{F} = \mathbb{C}$) matrices as a matrix field. For the sake of brevity we will refer to them as $M_{\mathbb{F}}(n)$-valued images, or even shorter, as $M_{\mathbb{F}}(n)$-fields. A corresponding terminology is used when we are dealing with the vector spaces $\mathrm{Sym}(n)$, the space of symmetric $n \times n$-real matrices, $\mathrm{Skew}(n)$, the space of skew-symmetric $n \times n$-real matrices, and $\mathrm{Her}(n) \subset M_{\mathbb{C}}(n)$, the space of Hermitian $n \times n$-complex matrices. These Hermitian matrices will play a key role in all that follows. We will also be interested in subsets of matrices of $M_{\mathbb{R}}(n)$ such as $\mathrm{SO}(n)$, the set of rotations of $\mathbb{R}^n$, a group with respect to matrix multiplication, or $\mathrm{GDS}(n)$, the vector space of generalized doubly-stochastic matrices, which contains $\mathrm{DS}(n)$, the convex, compact set of doubly-stochastic $n \times n$ matrices. Although the subsequent reasoning is valid for $n \in \mathbb{N}$ we will most often focus on the case $n = 3$, in view of potential applications, and to $\Omega \subset \mathbb{R}^2$ for the sake of simplicity. The processing and analysis of $\mathrm{Sym}(3)$-fields has been at the center of numerous research efforts since they play an important role, for example, in diffusion tensor magnetic imaging, a technique in medical imaging capable to capture the overall nerve fibre structure in human tissue. Some of this research was aiming at creating a general framework that allows to transfer as many techniques as possible from scalar image processing to the setting of $\mathrm{Sym}(3)$-fields. This framework is made possible by the rich functional-algebraic calculus available for $\mathrm{Sym}(3)$-fields. However, $\mathrm{SO}(3)$- and $\mathrm{GDS}(3)$-fields do not possess structures rich enough to support the aforementioned transfer, and hence, it is not immediately clear how to establish image processing methods for this type of data.

The key idea to circumvent this obstacle is complexification. By this we mean the embedding of $M_{\mathbb{R}}(n)$ (and its subsets $\mathrm{SO}(3)$ and $\mathrm{GDS}(3)$ for $n = 3$) into $\mathrm{Her}(n)$ in a one-to-one manner. In this way we will be able to take advantage of the elementary fact that each Hermitian matrix is unitarily similar to a diagonal matrix with real-valued entries. In other words, $\mathrm{Her}(n)$ is just as convenient a space as $\mathrm{Sym}(n)$ is, and many techniques developed for $\mathrm{Sym}(n)$ will carry over directly to $\mathrm{Her}(n)$. Since subsets of $M_{\mathbb{R}}(n)$ are mapped to subsets of $\mathrm{Her}(n)$ such an embedding opens up paths for the treatment of subsets otherwise not accessible to image processing techniques. We are interested in continuous, PDE-driven morphology governed by the two well-known equations, see [16,17,23],

$$\partial_t h = \pm \|\nabla h\|_2 := \sqrt{(\partial_x h)^2 + (\partial_y h)^2} \quad \text{on} \quad \Omega \times (0, +\infty)$$
$$\partial_n h = 0 \quad \text{on} \quad \partial\Omega \times (0, +\infty) \tag{1}$$
$$h(x, y, 0) = g(x, y) \quad \text{for all} \quad (x, y) \in \Omega.$$

The evolution process governed by Eq. (1) is initialized with the original image g and yields transformed versions $h(\cdot, t)$ for any $t \in (0, +\infty)$. Here $\partial_n h$ denotes the outward normal derivative of h at the boundary $\partial\Omega$ of the image domain Ω. The plus sign $+$ realizes the dilation, while the minus sign $-$ corresponds to erosion. These equations may be solved by means of the schemes of Rouy-Tourin [15] (used in this article), Osher-Sethian [13,14], or by the more sophisticated FCT-scheme, [2]. It is our aim to establish Hermitian matrix field counterparts for both PDE and numerical solution scheme. However, since even a 2×2-Hermitian matrix has 4 degrees of freedom our capabilities of visualization of such matrix fields are very limited, at best. The same holds true for the aforementioned classes of matrices. Finding proper graphical representations of high-dimensional data such as matrix fields will be a challenge for the visualization and image processing community. Therefore, we will focus more on the theoretical aspects providing a proof-of-concept rather than the visual evaluation of experiments.

The article is structured as follows: a short summary of Hermitian matrices' properties are given in Sect. 2. In Sect. 3 the basic calculus such as the concept of function, partial derivatives, gradient, length, product, pseudo supremum, and pseudo infimum is illustrated for Hermitian matrix fields. These building blocks are used in Sect. 4 to define PDE-driven morphology for Hermitian fields, specifically for the dilation process using the Rouy-Tourin scheme. The next section extends the previous concept to an arbitrary matrix field containing real-valued square matrices. Section 6 deals with two different classes: orthogonal and doubly-stochastic matrices. To ensure that the results are in the right class, a backprojection is introduced as well. Numerical experiments are illustrated in Sect. 7 and a short summary and conclusion is given in Sect. 8.

2   The $\mathbb{R}$-Vector Space of Hermitian Matrices

The set Her(n) of Hermitian matrices consist of all complex $n \times n$-matrices H that satisfy $H = H^*$, where H^* stands for matrix transposition $H^\top$ concatenated with component-wise complex conjugation $\overline{H}$. Any Hermitian matrix H can be diagonalized with a suitable unitary matrix in the form $H = UDU^*$ and all the eigenvalues are real [9,25]. Here, U is unitary, that is, $U^*U = UU^* = I$, and $D = \text{diag}(d_1, \ldots, d_n)$ is a diagonal matrix with real entries in decreasing order, $d_1 \geq \ldots \geq d_n$. The real part, $\Re(H)$, of a Hermitian Matrix H is symmetric, $\Re(H) \in \text{Sym}(n)$, while the imaginary part, $\Im(H)$ is skew-symmetric, $\Im(H) \in$ Skew(n). Clearly, we have $H = \Re(H) + i\Im(H)$. H is called positive semidefinite if all of its eigenvalues are positive: $d_1 \geq \ldots \geq d_n \geq 0$. This is equivalent to the nonegativity of the quadratic form $z^*Hz \geq 0$, where z is a complex vector, $z \in \mathbb{C}^n$. If the inequalities are strict, the matrix is called positive definite. A matrix H is negative (semi-)definite if $-H$ is positive (semi-)definite. If the H

is none of the above, then the Hermitian matrix is called indefinite. This provides the set of Hermitian matrices with a partial order "$\geq$": $A \geq B$ if and only if $A - B$ is positive semidefinite. In the case of real symmetric matrices this order is often referred to as Loewner order (see [1]). However, Her(n) equipped with this order is not a lattice. As it is pointed out in [3], and in more detail in [4], a rich functional algebraic calculus can be set up for symmetric matrices. As a consequence numerous filtering and analysis methods can be derived for such fields of symmetric matrices in a rather straightforward way from their scalar counterparts (see [3,6]). Hermitian matrices as immediate generalizations of symmetric matrices possess an equally rich and easily manageable calculus. Processing a matrix field amounts to apply an operator $\mathcal{O}$ to the field. This operation might be a simple concatenation with a scalar function f, or it might be an application of a function of several variables, a matrix valued differential operator, or even a step in a numerical algorithm, such as a time step in an explicit scheme to solve partial differential equations. This setting requires that a certain amount of operations from matrix calculus is at our disposal. In order to make the article as self-contained as possible we present some basic notions from calculus for fields of Hermitian matrices in the next chapter albeit in a very abridged form.

3   Basic Calculus for Hermitian Matrix Fields

A very brief account of some basic definitions for the formulation of a differential calculus for Hermitian matrix fields is the subject of this section. Since the algebraic theory of Hermitian matrices parallels that of symmetric matrices, unitary matrices playing the role of orthogonal ones in the symmetric case, the material in [3,4] may be generalized to Hermitian matrices in a rather direct way. Therefore, we only juxtapose basic operations of this calculus with their counterparts in the scalar setting. We call the matrices psup(A, B) and pinf(A, B) as defined in Table 1 for $A, B \in$ Her(n) (pseudo-)supremum resp., (pseudo-)infimum. They are the upper, resp., lower matrix valued bounds of smallest, resp., largest trace. As such they serve as suitable replacements for supremum and infimum in the non-lattice setting of Her(n) equipped with the Loewner order (see [5,11]).

4   Basic PDE-Driven Morphology

The correspondence between real and matrix calculus has given rise to matrix valued partial differential equations (PDEs) together with matrix valued solution schemes gleaned from the real-valued counterparts in the case of symmetric matrices [4]. This is now possible for Hermitian matrices as well.

4.1   Continuous Morphology: Hermitian-Valued PDEs

The matrix-valued counterparts of the morphological PDEs $\partial_t h = \pm \|\nabla h\|_p$ for dilation ($+$) and erosion ($-$) proposed for symmetric matrices in [3] may now be interpreted in terms of Hermitian matrices $H(x, t) \in$ Her(3):

Table 1. Transfering elements of scalar valued calculus to the Hermitian setting.

setting	scalar valued	matrix-valued
function	$f : \begin{cases} \mathbb{R} \longrightarrow \mathbb{R} \\ x \mapsto f(x) \end{cases}$	$F : \begin{cases} \mathrm{Her}(n) \longrightarrow \mathrm{Her}(n) \\ H \mapsto U \mathrm{diag}(f(\lambda_1), \dots, f(\lambda_n)) U^* \end{cases}$
partial derivatives	$\partial_\omega h,$ $\omega \in \{t, x_1, \dots, x_d\}$	$\overline{\partial}_\omega H := (\partial_\omega h_{ij})_{ij},$ $\omega \in \{t, x_1, \dots, x_d\}$
gradient	$\nabla h(x) := (\partial_{x_1} h(x), \dots, \partial_{x_d} h(x))^*,$ $\nabla h(x) \in \mathbb{R}^d$	$\overline{\nabla} H(x) := (\overline{\partial}_{x_1} H(x), \dots, \overline{\partial}_{x_d} H(x))^*,$ $\overline{\nabla} H(x) \in (\mathrm{Her}(n))^d$
length	$\|w\|_p := \sqrt[p]{|w_1|^p + \cdots + |w_d|^p},$ $\|w\|_p \in [0, +\infty[$	$\|W\|_p := \sqrt[p]{|W_1|^p + \cdots + |W_d|^p},$ $\|W\|_p \in \mathrm{Her}^+(n)$
supremum	$\sup(a, b)$	$\mathrm{psup}(A, B) = \frac{1}{2}(A + B + |A - B|)$
infimum	$\inf(a, b)$	$\mathrm{pinf}(A, B) = \frac{1}{2}(A + B - |A - B|)$

$$\overline{\partial}_t H = \pm \||\overline{\nabla} H\||_p \tag{2}$$

with $(x, t) \in \Omega \times [0, \infty)$ due to the rich functional calculus encompassing the one for symmetric matrices.

4.2 Continuous Morphology: Hermitian-Valued Solution Scheme

For simplicity, we use $p = 2$ and choose a variant of the first-order upwind scheme of Rouy and Tourin [15,23] adapted to our matrix-valued setting. Precisely, it is given by

$$H_{i,j}^{n+1} = H_{i,j}^n$$
$$+ \tau \left(\mathrm{psup} \left(\frac{1}{h_x} \mathrm{psup} \left(-D_-^x H_{i,j}^n, 0 \right), \frac{1}{h_x} \mathrm{psup} \left(D_+^x H_{i,j}^n, 0 \right) \right)^2 \right.$$
$$\left. + \mathrm{psup} \left(\frac{1}{h_y} \mathrm{psup} \left(-D_-^y H_{i,j}^n, 0 \right), \frac{1}{h_y} \mathrm{psup} \left(D_+^y H_{i,j}^n, 0 \right) \right)^2 \right)^{1/2},$$

where

$$D_+^x H_{i,j}^n := H_{i+1,j}^n - H_{i,j}^n \quad \text{and} \quad D_-^x H_{i,j}^n := H_{i,j}^n - H_{i-1,j}^n.$$

In the above formulas the notation H_{ij}^n means the $\mathrm{Her}(n)$-matrix at the pixel centered at $(ih_x, jh_y) \in \Omega \subset \mathbb{R}^2$ at the time-level $n\tau$ of the evolution with time-step $\tau > 0$. Note that the erosion process can be described similarly just by incorporating a sign switch. It displays a similar numerical behavior to the first-order variant of Osher and Sethian's scheme, [13,14]. However, here we will not process fields of Hermitian matrices for their own sake. The set of Hermitian 2×2-matrices is already a 4D-manifold and hence eludes direct visualization. Instead, various classes of matrices with real entries are going to be embedded into $\mathrm{Her}(n)$ with suitable $n = 2, 3, \ldots$ and processed in this "detour space" before being transformed back. In the next section, the largest set of real $n \times n$-matrices is embedded into $\mathrm{Her}(n)$: $\mathrm{M}_{\mathbb{R}}(n)$ itself.

5   An Isomorphism Between $\mathrm{M}_{\mathbb{R}}(n)$ and $\mathrm{Her}(n)$

Every matrix $M \in \mathrm{M}_{\mathbb{R}}(n)$ can be decomposed into a symmetric and a skew-symmetric part, that is $M = S + A$. Now, we define a complex mapping $\varXi :$ $\mathrm{M}_{\mathbb{R}}(n) \to \mathrm{Her}(n)$ by

$$\varXi : M \longmapsto -S + \mathrm{i}A = -\frac{1}{2}(M + M^{\top}) + \frac{\mathrm{i}}{2}(M - M^{\top})$$

This is a linear and invertible mapping with

$$\varXi^{-1} : H \longmapsto -\frac{1}{2}(H + H^{\top}) - \frac{\mathrm{i}}{2}(H - H^{\top})$$

for any $H \in \mathrm{Her}(n)$. Despite its simplicity it is vital to the subsequent approach: instead of tackling a problem in $\mathrm{M}_{\mathbb{R}}(n)$ with its relatively poor algebraic structure we transfer the problem to $\mathrm{Her}(n)$, take advantage of the rich calculus there when applying an operation $\mathcal{O}$, and finally transform the (pre-)solution back to $\mathrm{M}_{\mathbb{R}}(n)$. In other words, the subsequent diagram displayed in (4) (with a slight abuse of notation) commutes. We define $\varXi(M) = -S + \mathrm{i}A$ for $M = S + A$ rather than the straightforward $\varXi(M) = S + \mathrm{i}A$ to ensure that, when $\varXi$ is employed in the $\mathrm{SO}(2)$-setting, the Loewner order in $\mathrm{Her}(2)$ corresponds in a direct monotone manner to the size of the rotation angle (the only free parameter in $\mathrm{SO}(2)$). Let us assume that an operation $\mathcal{O}$ is formulated in the language of the Hermitian-calculus sketched in Table 1. Then it is the concatenated operation $\varXi^{-1} \circ \mathcal{O} \circ \varXi$ that processes matrices from $\mathrm{M}_{\mathbb{R}}(n)$, as displayed in (4). Even more interestingly, subsets (classes) C of $\mathrm{M}_{\mathbb{R}}(n)$ may be processed in this manner as well, provided some type of additional class-preserving mapping is involved. This is detailed in the next section.

6   Application: Morphology for Classes of Matrices

In this section, we prepare the path to elementary morphology for certain subsets or classes of $\mathrm{M}_{\mathbb{R}}(n)$. We use the terminology "class" C to indicate that the subset has some additional structure of algebraic or topological nature. Exemplarily we consider the subgroup, with respect to matrix multiplication, of orthogonal 3×3-matrices, $C_1 = \mathrm{O}(3)$, and $C_2 = \mathrm{GDS}(3)$, the set of generalized doubly-stochastic

3×3 matrices S: its entries are not necessarily nonnegative real numbers while both row- and column-sums equal 1 (hence, the doubly-stochastic matrices form a convex, compact subset). The general strategy is a follows: First, we embed the matrix class C_i of $M_{\mathbb{R}}(n)$ into $\mathrm{Her}(n)$ by means of Ξ $(i = 1, 2)$. Then the (image-)processing step is performed exploiting the rich algebraic structure of $\mathrm{Her}(n)$, followed by the canonical mapping Ξ^{-1} back to $M_{\mathbb{R}}(n)$ as indicated in the diagram (4). However, setting $\tilde{\mathcal{O}} := \Xi^{-1} \circ \mathcal{O} \circ \Xi$ and choosing $R \in C_i$ we might encounter the difficulty that $\tilde{\mathcal{O}}(R) \notin C_i$. As a way out, we need to find the $\| \cdot \|_{\mathrm{F}}$-best approximation $\tilde{R} \in C_i$ to $\tilde{\mathcal{O}}(R)$, that is,

$$\| \tilde{\mathcal{O}}(R) - \tilde{R} \|_{\mathrm{F}}^2 \longrightarrow \min, \tag{3}$$

where $\| \cdot \|_{\mathrm{F}}$ stands for the Frobenius norm for matrices. This results in a projection Pr_{C_i} from $M_{\mathbb{R}}(n)$ back to C_i, see diagram (4).

$$
\begin{array}{ccc}
\mathrm{Her}(n) & \xrightarrow{\ \mathcal{O}\ } & \mathrm{Her}(n) \\
\Xi \uparrow & & \downarrow \Xi^{-1} \\
C_i \subset M_{\mathbb{R}}(n) & \underset{\underbrace{\Xi^{-1} \circ \mathcal{O} \circ \Xi}_{\tilde{\mathcal{O}}}}{\longrightarrow} M_{\mathbb{R}}(n) & \xrightarrow{\ Pr_{C_i}\ } C_i
\end{array} \tag{4}
$$

Therefore, for instance, taking the (pseudo-)supremum $\mathrm{psup}^{C_i}(M_1, M_2)$ of two matrices $M_1, M_2 \in C_i$ amounts to calculate

$$\mathrm{psup}^{C_i}(M_1, M_2) = Pr_{C_i}\left(\Xi^{-1}\left(\mathrm{psup}^{H(3)}(\Xi(M_1), \Xi(M_2)) \right) \right).$$

The notation psup^{C_i} is employed to indicate that the supremum is an element of C_i due to the applied backprojection Pr_{C_i}. The notation pinf^{C_i} is used likewise. This reckoning extends naturally to more elaborate operations that might even depend on several matrix arguments. The appearance of a projection Pr_{C_i} makes any operation, be that averaging, taking the (pseudo-)supremum/infimum, dilation/erosion or a time step in an explicit numerical PDE-solution scheme, an approximate operation. It is worth mentioning that $\Xi(QRQ^{\top}) = Q\Xi(R)Q^{\top}$ holds for any orthogonal matrix Q, a property sometimes referred to as "rotational invariance". If the projection Pr_{C_i} is uniquely determined, it is rotationally invariant as well, due to the invariance of the Frobenius norm in (3). Hence, if $\mathcal{O}$ has this property, it is passed on to the whole processing line indicated in diagram (4). This is the case for the aforementioned (pseudo-)supremum/infimum operation since the definiteness of a Hermitian matrix is unitarily, hence especially, rotationally invariant.

In the next subsection, we address the problem of finding a suitable backprojection P_{C_i} onto one of the aforementioned matrix classes C_1 and C_2, in other words, a solutions to (3). These problems are instances of so-called matrix-nearness or Procrustes problems, see [7,8].

6.1   Backprojection Onto $C_1 = O(3)$

In the case of the group $O(n)$ the solution to this so-called orthogonal Procrustes problem is the orthogonal factor in the polar decomposition of $\tilde{\mathcal{O}}(R)$. The polar decomposition as needed in this context states that every real square invertible matrix A can be uniquely written as $A = OP$ with positive definite $P = \sqrt{A^\top A}$ and orthogonal $O = AP^{-1}$, see [9]. Consequently, our closed form solution for invertible $\tilde{\mathcal{O}}(R)$ with $n = 3$ reads:

$$\tilde{R} = Pr_{O(3)}(\tilde{\mathcal{O}}(R)) = \tilde{\mathcal{O}}(R)\left(\tilde{\mathcal{O}}(R)^\top \tilde{\mathcal{O}}(R)\right)^{-1/2}.$$

Indeed, this formula establishes a projection $Pr_{O(3)}$ from $M_\mathbb{R}(n)$ back to $O(3)$. Further, we see that $\tilde{R} = Pr_{O(3)}(\tilde{\mathcal{O}}(R)) \in SO(3)$ whenever $\det(\tilde{\mathcal{O}}(R)) > 0$.

6.2   Backprojection Onto $C_2 = GDS(3)$

Although by far more elaborate, there exists an explicit solution of the nearest matrix problem (3) in case of the generalized doubly-stochastic matrices $C_2 = GDS(3)$ as well. According to [10] the optimal GDS-matrix is given explicitly by $\tilde{A} = (I_n - J_n)A(I_n - J_n) + J_n$, where I_n denotes the $n \times n$-unit matrix and in J_n all entries equal $1/n$, $J_n = (1/n)_{k,l \in \{1,\dots,n\}}$. This boils down in the case $n = 3$ to

$$\tilde{R} = Pr_{GDS(3)}(\tilde{\mathcal{O}}(R)) = (I_3 - J_3)\tilde{\mathcal{O}}(R)(I_3 - J_3) + J_3.$$

7   Experiments

A visualization of highdimensional data, e.g. 3×3-Hermitian or doubly-stochastic matrices $M = (m_{i,j})_{i,j=1,2,3}$, that conveys the information captured in these arrays in a intuitive and enlightning but at the same time practical manner is a very challenging task and as such beyond the scope of this article. Therefore, we depict the nine different matrix channels $m_{i,j}$, after an appropriate re-scaling, as nine separate gray-value images, albeit arranged in a 3-by-3 pattern as a tiled image. We employ the scheme of Rouy-Tourin with time step size $\tau = 0.01$ to mimic the process of dilation with evolution times $T = 1$ and $T = 2$. Figure 1(a) shows a tiled version, i.e. a channelwise view of the original 15×15-field of rotations. The left half of the field consists of rotations around the x-axis with angle $\frac{\pi}{2}$, the right side of rotations around the y-axis by the same angle, while the center 5×5-square is made up by rotations around the z-axis again with angle $\frac{\pi}{4}$. However, for the center pixel this angle is reduced to $\frac{\pi}{2}$. The field represented in Fig. 1(b) has the following structure: the 3×3-unit matrix (center), its "flipped version" (the ring), and two non-symmetric permutation matrices (left and right). They represent four of the six extreme points of the convex compact set of doubly-stochastic matrices and belong to the affine space of generalized doubly-stochastic matrices.

7.1   Dilation of a SO(3)-Field

The field given in Fig. 1(a) is processed in two different ways: in method 1 the evolution via PDE takes place in Her(3) and the backprojection $P_{O(3)}$ (effectively $P_{SO(3)}$) is done once as a final step at the end. In method 2 the backprojection onto the (non-linear) manifold SO(3) is performed after each time step, leading, theoretically, to a more accurate result.

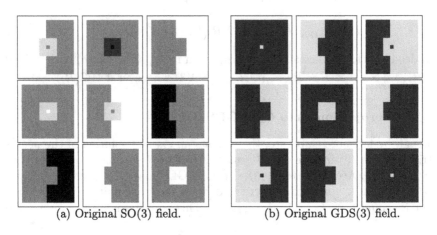

(a) Original SO(3) field.   (b) Original GDS(3) field.

Fig. 1. Tiled view of original 15×15-fields of SO(3)- and GDS(3)-matrices.

As expected we observe in general a blurring of the field due to the known dissipative effects of the numerical scheme. What we see in Fig. 2(a),(b) and (c) and (d) is a dilation (visually areas with large values are expanding while areas with smaller values are shrinking in each channel); pixels P1 $(8, 10)$ and P2 $(8, 11)$ originally carrying a z-axis-rotation with angle $\frac{\pi}{4} \approx 0.7854$ (P1) and a y-axis-rotation with angle $\frac{\pi}{2} \approx 1.5708$ (P2) are altered by an evolution lasting $T = 1$ to a rotation around $(-0.0036, 0.3420, 0.9397)^{\top}$ with angle 0.4092 (P1) and a $(-0.0011, 0.6864, 0.7272)^{\top}$-rotation axis with angle 0.4345 (P2), if method 1 is utilized. Instead, method 2 leads to a rotation around $(0.0005, 0.5217, 0.8531)^{\top}$ with angle 0.4145 (P1) and a $(0.0000, 0.8770, 0.4804)^{\top}$-rotation axis with 0.5108. Hence, method 2 (with repeated backprojections) is preferable over method 1, although visually a difference is barely noticeable.

7.2   Dilation of a Field of Generalized Doubly-Stochastic Matrices

Due to the linearity of the backprojection onto GDS(3), the two variants of the numerical procedure described above coincide.

The dilation of Fig. 1(b) with evolution times $T = 1$ and $T = 2$ leads to results that seem to be in accordance with the findings in Sect. 7.1. However, some fine details, such as the cross-like structure in the center or ridges (e.g.,

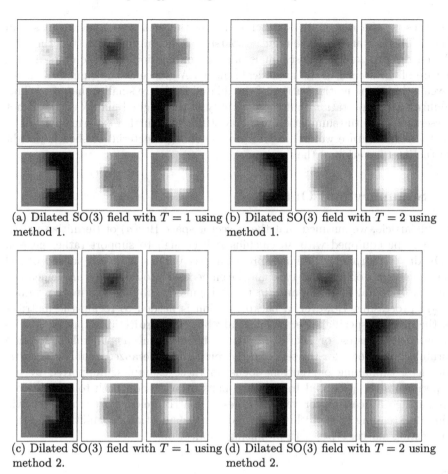

(a) Dilated SO(3) field with $T = 1$ using method 1.

(b) Dilated SO(3) field with $T = 2$ using method 1.

(c) Dilated SO(3) field with $T = 1$ using method 2.

(d) Dilated SO(3) field with $T = 2$ using method 2.

Fig. 2. Experiment 1.

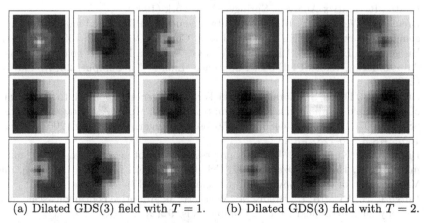

(a) Dilated GDS(3) field with $T = 1$.

(b) Dilated GDS(3) field with $T = 2$.

Fig. 3. Experiment 2.

the tiles on the diagonal) do emerge during the evolution as one can see in Fig. 3. This is caused by the relative sparsity of the original data (two thirds of the entries are zero, the rest equals one) in conjunction with the cross-shaped nature of the stencil in the numerical scheme. A glance at the numerical values reveals that some matrix entries of the dilated fields are smaller than zero. This confirms our expectation that the positivity of the original matrix entries is not preserved, while the summation property still holds. Further experiments will be postponed to future work that will provide enough material to shed new light on the interpretation of the results.

8   Summary and Outlook

In this article, we outlined that the $\mathbb{R}$-vector space Her(n) of Hermitian matrices may be equipped with an calculus rich enough to support rather general PDE-driven morphological image processing techniques. Now, any subsets of real square matrices of a specific type may be embedded into Her(n) by an elementary but fundamental mapping and, in this "disguise", subjected to matrix valued image processing. However, a backprojection is needed to regain the matrices of the original specific type. In linear algebra such projections appear as solutions to Procrustes- or nearest-matrix-problems. Convenient explicit solution formulae are known for the sets of orthogonal and generalized doubly-stochastic matrices, explaining our current focus on these examples. Supported by the experiments the proposed framework opens up a general path to (morphological) image processing for various types of matrix fields. Future research will aim at extending these techniques (and our visualization capabilities) to other interesting matrix types with promising applications.

References

1. Bhatia, R.: Matrix Analysis. Springer, New York (1996)
2. Breuß, M., Weickert, J.: Highly accurate schemes for PDE-based morphology with general convex structuring elements. Int. J. Comput. Vis. **92**(2), 132–145 (2011)
3. Burgeth, B., Bruhn, A., Didas, S., Weickert, J., Welk, M.: Morphology for tensor data: ordering versus PDE-based approach. Image Vis. Comput. **25**(4), 496–511 (2007)
4. Burgeth, B., Didas, S., Florack, L., Weickert, J.: A generic approach to the filtering of matrix fields with singular PDEs. In: Sgallari, F., Murli, A., Paragios, N. (eds.) SSVM 2007. LNCS, vol. 4485, pp. 556–567. Springer, Heidelberg (2007). doi:10.1007/978-3-540-72823-8_48
5. Burgeth, B., Kleefeld, A.: Towards processing fields of general real-valued square matrices. In: Hotz, I., Özarslan, E., Schultz, T. (eds.) Modeling, Analysis, and Visualization of Anisotropy. Springer, Heidelberg (2017)
6. Burgeth, B., Pizarro, L., Breuß, M., Weickert, J.: Adaptive continuous-scale morphology for matrix fields. Int. J. Comput. Vis. **92**(2), 146–161 (2011)
7. Higham, N.: Matrix procrustes problems. Ph.D. thesis, The University of Manchester (1993)

8. Higham, N.J.: Matrix Nearness Problems and Applications. Oxford University Press, Oxford (1989)
9. Horn, R.A., Johnson, C.R.: Matrix Analysis. Cambridge University Press, Cambridge (1990)
10. Khoury, R.N.: Closest matrices in the space of generalized doubly stochastic matrices. J. Math. Anal. Appl. **222**(2), 562–568 (1998)
11. Kleefeld, A., Meyer-Baese, A., Burgeth, B.: Elementary morphology for SO(2)- and SO(3)-orientation fields. In: Benediktsson, A.J., Chanussot, J., Najman, L., Talbot, H. (eds.) Mathematical Morphology and Its Applications to Signal and Image Processing, pp. 458–469. Springer International Publishing, Heidelberg (2015)
12. Matheron, G.: Eléments pour une théorie des milieux poreux. Masson, Paris (1967)
13. Osher, S., Fedkiw, R.P.: Level Set Methods and Dynamic Implicit Surfaces, Applied Mathematical Sciences, vol. 153. Springer, New York (2002)
14. Osher, S., Sethian, J.A.: Fronts propagating with curvature-dependent speed: algorithms based on Hamilton-Jacobi formulations. J. Comput. Phys. **79**, 12–49 (1988)
15. Rouy, E., Tourin, A.: A viscosity solutions approach to shape-from-shading. SIAM J. Numer. Anal. **29**, 867–884 (1992)
16. Sapiro, G.: Geometric Partial Differential Equations and Image Analysis. Cambridge University Press, Cambridge (2001)
17. Sapiro, G., Kimmel, R., Shaked, D., Kimia, B.B., Bruckstein, A.M.: Implementing continuous-scale morphology via curve evolution. Pattern Recogn. **26**, 1363–1372 (1993)
18. Serra, J.: Echantillonnage et estimation des phénomènes de transition minier. Ph.D. thesis, University of Nancy, France (1967)
19. Serra, J.: Image Analysis and Mathematical Morphology, vol. 1. Academic Press, London (1982)
20. Serra, J.: Anamorphoses and function lattices (multivalued morphology). In: Dougherty, E.R. (ed.) Mathematical Morphology in Image Processing, pp. 483–523. Marcel Dekker, New York (1993)
21. van de Gronde, J.J., Lysenko, M., Roerdink, J.B.T.M.: Path-based mathematical morphology on tensor fields. In: Hotz, I., Schultz, T. (eds.) Visualization and Processing of Higher Order Descriptors for Multi-Valued Data, pp. 109–127. Springer, Berlin (2015)
22. van de Gronde, J.J., Roerdink, J.B.T.M.: Sponges for generalized morphology. In: Benediktsson, J.A., Chanussot, J., Najman, L., Talbot, H. (eds.) ISMM 2015. LNCS, vol. 9082, pp. 351–362. Springer, Cham (2015). doi:10.1007/978-3-319-18720-4_30
23. van den Boomgaard, R.: Numerical solution schemes for continuous-scale morphology. In: Nielsen, M., Johansen, P., Olsen, O.F., Weickert, J. (eds.) Scale-Space 1999. LNCS, vol. 1682, pp. 199–210. Springer, Heidelberg (1999). doi:10.1007/3-540-48236-9_18
24. Wilkinson, M.H.F., Roerdink, J. (eds.): Mathematical Morphology and Its Application to Signal and Image Processing. LNCS. Springer, Heidelberg (2009)
25. Zhang, F.: Matrix Theory: Basic Results and Techniques Universitext. Springer, New York (1999)

Matrix-Valued Levelings for Colour Images

Michael Breuß[1(✉)], Laurent Hoeltgen[1], and Andreas Kleefeld[2]

[1] Institute for Mathematics, Brandenburg Technical University,
Platz der Deutschen Einheit 1, 03046 Cottbus, Germany
{breuss,hoeltgen}@b-tu.de
[2] Forschungszentrum Jülich, Institute for Advanced Simulation,
Jülich Supercomputing Centre, Wilhelm-Johnen-Straße, 52425 Jülich, Germany
a.kleefeld@fz-juelich.de

Abstract. Morphological levelings represent a useful tool for the decomposition of an image into cartoon and texture components. Moreover, they can be used to construct a morphological scale space. However, the classic construction of levelings is limited to the use of grey scale images, since an ordering of pixel values is required.

In this paper we propose an extension of morphological levelings to colour images. To this end, we consider the formulation of colour images as matrix fields and explore techniques based on the Loewner order for formulating morphological levelings in this setting. Using the matrix-valued colours we study realisations of levelings relying on both the completely discrete construction and the formulation using a partial differential equation. Experimental results confirm the potential of our matrix-based approaches for analysing texture in colour images and for extending the range of applications of levelings in a convenient way to colour image processing.

Keywords: Mathematical morphology · Dilation · Erosion · Loewner order · Leveling

1 Introduction and Motivation

The analysis of texture in an image is an important task in image processing. It is for instance fundamental for performing the decomposition of an image into cartoon and texture components, and gives useful information for many potential applications like inpainting or segmentation. Among the interesting tools for filtering textures are the morphological levelings. While the concept behind levelings can be traced back to the works [6] (see also [5]) the notion itself was to our knowledge coined by Meyer in [14]. A distinct feature of the morphological levelings is their ability to preserve object boundaries during the filtering, similar to morphological amoebas cf. [11,18]. This in particular makes them a potentially useful building block for many basic image processing tasks such as image segmentation. The appearance of the levelings triggered further research expanding the knowledge about their theoretical properties and possible applications. Mentioning some milestones, in [15] the corresponding scale space

© Springer International Publishing AG 2017
J. Angulo et al. (Eds.): ISMM 2017, LNCS 10225, pp. 296–308, 2017.
DOI: 10.1007/978-3-319-57240-6_24

has been studied, and the potential for performing image decompositions and texture analysis tasks has been explored in [13].

Since levelings rely in their original construction on a lattice structure between scales, they inherently depend on the total order relation of grey scale values. When considering the processing of colour images, this is a significant issue since such vector-valued images do not incorporate a natural ordering of the colour information. We refer to [2] for a comprehensive overview on the efforts to rank colour information for morphological processing purposes. Let us also mention the more recent work [7] where frame-theoretical methods are applied to obtain morphological processes for colours, and the paper of Zanoguera and Meyer [20] where non-separable vector levelings are considered.

In this paper we propose and study the extension of morphological levelings to colour images based on their recent formulation in terms of matrix fields introduced in [4,10]. In this setting, each pixel bears colour information in terms of a symmetric 2×2 matrix. For such matrix fields the Loewner order can be used to define an ordering of matrices. This can be explored for defining matrix-valued counterparts of morphological filters. Both the fully discrete formulation of dilation/erosion [4,10] as well as formulations based on partial differential equations (PDEs) of those processes [3] have been proposed in this framework, whereas the median filter and its extensions have recently been the subject of even more extensive investigations [9,18,19].

Our Contributions. Relying on the matrix-valued framework for dilation/erosion as developed in [3,10] we present colour-valued formulations of the morphological levelings. Thereby we especially rely on the discrete leveling formulation via the triphase operator Λ as defined in [12], and the PDE-based formulation used for studying scale-space properties in [15]. Both possible approaches are carried over within our framework to colour image processing and are evaluated experimentally. On the technical side, let us note in this context that the definition of the PDE-based colour levelings involves some intricate choices in the discretisation process. We study here a technically different discretisation of matrix-valued derivatives as in [3] as well as two possible implementations of the mechanism that steers application of dilation/erosion in that formulation. Our experiments demonstrate the potential of our approach for texture-related colour image processing tasks.

2   Levelings for Grey-Valued Images

In this section, we briefly review how to define a leveling for a given grey-valued image f and a marker image M. We describe both the fully discrete, complete lattice based construction as well as the continuous-scale PDE-based formulation, see e.g. [13] for a detailed discussion of both approaches.

2.1   Discrete Construction of Levelings

A leveling $\Lambda(M|f)$ is a fixed point of the operator $\lambda(M|f)$ defined by

$$\lambda(M|f) = [f \wedge \delta_B(M)] \vee \epsilon_B(M) \tag{1}$$

where we have employed the usual notations $\vee$ for the supremum, $\wedge$ the infimum, δ_B the dilation, and ϵ_B for erosion, respectively (see for instance [12]). The fixed point is obtained in the limit of the iteration of λ, i.e. more precisely, by calculating the iteration

$$u_{k+1} = \lambda(u_k|f) \tag{2}$$

for $k = 0, 1, 2, \ldots$ with $u_0 = M$. Hence, $\Lambda(M|f)$ is given by u_∞.

Recalling the definitions of the fundamental operations used in (1), let us note that one needs to consider a structuring element B. In the discrete setting, this is effectively the shape of a window that slides over the pixels in an image and defines the grey values accumulated in the indicated operation. Making use of the structuring element, one can define dilation and erosion in general as

$$\delta_B(f)(x) := (f \oplus B)(x) := \sup_{a \in B} f(x - a)$$
$$\epsilon_B(f)(x) := (f \ominus B)(x) := \inf_{a \in B} f(x - a)$$

The supremum $\vee$ and infimum $\wedge$ of two images f_1 and f_2 are defined in a pointwise sense as

$$(f_1 \vee f_2)(x) := \max\{f_1(x), f_2(x)\}$$
$$(f_1 \wedge f_2)(x) := \min\{f_1(x), f_2(x)\}$$

2.2   PDE-Based Formulation of Levelings

The PDE-based formulation of levelings in a grey valued setting amounts to the following problem formulation over an image domain Ω, cf. [12]:

$$\begin{aligned} \partial_t u(t, x) &= \operatorname{sgn}(u - f) \|\nabla u\|, & \forall x \in \Omega \\ u(0, x) &= (K_\sigma * f)(x), & \forall x \in \Omega \\ \partial_n u(t, x) &= 0, & \forall x \in \partial\Omega, \ \forall t > 0 \end{aligned} \tag{3}$$

where $(K_\sigma * f)(x)$ represents a Gaussian convolution of f with mean $\mu = 0$ and standard deviation σ and where $\partial_n u(t, x)$ denotes the derivative in outer normal direction along the boundary of our domain.

Let us note that the time t can be understood as the scale parameter in the corresponding scale space. The leveling is obtained as the steady state of the time evolution i.e. formally for $t \to \infty$. However, let us already note here that in practice integration over relatively small time intervals is sufficient to obtain reasonable results.

For the suitable interpretation of (3) let us remark that the underlying PDEs

$$\partial_t u(t, x) = \pm\|\nabla u\| \tag{4}$$

describe scalar-valued dilation (+) respectively erosion (−). Thus one can observe that the PDE in (3) includes in addition to (4) the signum function, working as a switch between the dilation/erosion processes.

3   Processing Colour Images

To define a leveling for a colour image consisting of three channels (r, g, b) is not a trivial task. As indicated beforehand this is due to the non-existing total order of colour values. In order to define suitable operations for colour images, we largely rely on the framework developed in several recent works on matrix-valued colour image processing, for details we refer to [10,19]. We now give a brief account of its building blocks.

3.1   Colour Images and Matrix Fields

We now briefly recall the conversion of (r, g, b) values to matrices, summarising the presentation in [3]. Via this conversion, a given RGB image is transformed in two steps into a matrix field of equal dimensions in which each pixel is assigned a symmetric 2×2 matrix.

To this end, the RGB colours are first transformed to the HCL colour space, assuming that red, green and blue intensities are normalised to $[0, 1]$. This is a standard procedure, cf. [1]. For a pixel with intensities r, g, b, we obtain its hue h, chroma c and luminance l via $M = \max\{r, g, b\}$, $m = \min\{r, g, b\}$, $c = M - m$, $l = \frac{1}{2}(M + m)$, and $h = \frac{1}{6}(g - b)/M$ modulo 1 if $M = r$, $h = \frac{1}{6}(b - r)/M + \frac{1}{3}$ if $M = g$, $h = \frac{1}{6}(r - g)/M + \frac{2}{3}$ if $M = b$.

After modifying the luminance l as by $\tilde{l} := 2l - 1$, and interpreting c, $2\pi h$, and $\tilde{l}$ as radial, angular and axial coordinates of a cylindrical coordinate system, we have a bijection from the (r, g, b) unit cube mapping onto a solid bi-cone, see Fig. 1.

The bi-cone is then transformed into Cartesian coordinates via $x = c\cos(2\pi h)$, $y = c\sin(2\pi h)$, and $z = \tilde{l}$. In a final step, the coordinates (x, y, z) are put as entries into a symmetric matrix $A \in \mathrm{Sym}(2)$ via

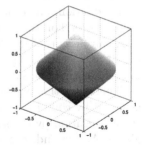

Fig. 1. Colour bi-cone (Color figure online)

$$A := \frac{\sqrt{2}}{2} \begin{pmatrix} z - y & x \\ x & z + y \end{pmatrix} \qquad (5)$$

We have thus obtained by (5) a bijection between the RGB colour space and the set of all matrices that correspond to points of the defined solid bi-cone.

3.2   Functions of Matrix-Valued Arguments

For constructing the levelings we need to give a meaning to matrix-valued counterparts of scalar-valued functions $\varphi : \mathbb{R} \to \mathbb{R}$, such as the signum function. The question arises how such functions operating in a scalar setting can be applied in the matrix-valued framework.

To tackle the issue, let us recall that any matrix $A \in \mathrm{Sym}(2)$ can be written using the decomposition $A = V\mathrm{diag}\,(\lambda_1, \lambda_2)\,V^\top$, where $V := (v_1, v_2)$ accumulates the eigenvectors v_1, v_2 of A as column vectors and $\lambda_{1,2}$ are the corresponding eigenvalues. Making use of this decomposition, one may define a function φ of a matrix A via

$$\varphi(A) := V \operatorname{diag}(\varphi(\lambda_1), \varphi(\lambda_2)) V^\top \qquad (6)$$

in terms of its standard scalar representation. For instance with $\varphi(\cdot) := \operatorname{sgn}(\cdot)$ we thus obtain the signum function of a symmetric matrix.

Let us note that this construction is in accordance to the extension of exponential and logarithm functions to matrices available in the literature. There the extension is based on the Taylor series expansion and boils down to applying the scalar functions at the eigenvalues. For details on the existing techniques see [8].

3.3   The Loewner Order Approach to Mathematical Morphology

As indicated we propose to use the concept developed for morphological colour processing in [4] that relies on the Loewner order. This is a half-order $\preccurlyeq$ for symmetric matrices in which, for two symmetric matrices A and B, it holds $A \preccurlyeq B$ iff $B - A$ is positive semidefinite.

As discussed for instance in [18] in detail, this concept of comparing two matrices can be extended to tuples of matrices. For any such tuple $\mathcal{A} := (A_1, A_2, \ldots, A_k)$, all matrices that are upper bounds for $\mathcal{A}$ with respect to the Loewner order form a convex set, which we denote here as $\mathcal{U}$. To distinguish within this set for instance a unique supremum as needed for defining a matrix-valued dilation, an additional ordering relation needs to be applied.

To this end one may compare matrices by their trace tr, so that the supremum of $\mathcal{A}$ can be obtained as the minimal element C of $\mathcal{U}$ such that $\operatorname{tr}(C) \leq \operatorname{tr}(X)$ for all matrices $X \in \mathcal{U}$. Note that this is in analogy to the standard definition in calculus that the supremum is the smallest upper bound of a set.

Therefore, the analogon of discrete lattice-based dilation or erosion can be achieved by combining the selection of pixels using a structuring element with the aggregation by the supremum or respective infimum operation as discussed above. At the end, the resulting matrix field is transformed back to a colour image.

4   Levelings for Colour Images

We now describe the setups for colour levelings that we study in this paper. In doing this there is a certain emphasis on the scheme construction for PDE-based levelings as this is technically the most challenging task.

4.1   Component-Wise Approach

The by far easiest approach to realise colour levelings is to apply the previous explanations of grey-valued dilation/erosion and the leveling filtering for each channel separately, and to combine the resulting three filtered channels at the end to obtain a new colour image. We also discuss this possibility for clarifying the advantages of our constructions.

4.2   Construction Based on Discrete Colour Morphology

An important point of this work is to investigate if the approach from [10] can be carried over to work for levelings by applying the matrix-valued colour framework described above within the construction given in (1) and (2).

To this end, for one filtering step with dilation/erosion, the agglomeration is performed in this paper pixelwise either over cross- or block-shaped structuring elements of size 3×3 as indicated, so that the corresponding supremum/infimum of matrices is computed as mentioned in Sect. 3.3. Let us also note that, as written in (2), the marker image is updated during the iterative procedure.

4.3   Construction of Levelings Based on Colour-Valued PDEs

For our numerical solution strategy to solve the PDE from (3) we need to discretise matrix-valued differential expressions in time and space, and we also need to take care of numerical boundary conditions.

For approximating the temporal derivative we make use of a standard forward difference scheme. In addition we use an adapted Rouy-Tourin discretisation for the spatial gradient. Rouy and Tourin [17] suggested to approximate the derivative u_z in a direction z by

$$u_z \approx \max \left(\max(D_z^- u, 0), - \min(D_z^+, 0) \right) \tag{7}$$

where D_z^+ is the forward difference operator and D_z^- the backward difference operator in z direction. In our implementation these finite differences are computed component wise on our symmetric matrices.

Let us note that our means to realise the Rouy-Tourin discretisation as in (7) differs from the one given in [3] where a formulation based on computing the supremum/infimum of three matrices is employed. In the scalar setting the formulas are equivalent, but in the matrix-valued setting our approach above is easier to realise than the one from [3].

Similarly the Neumann boundary conditions are enforced by mirroring our data along the boundary in each channel. Following Pizarro et al. [16] we compute these maxima and minima of our matrix-valued data as

$$\max(A, B) = \frac{1}{2}(A + B + |A - B|) \tag{8}$$

$$\min(A, B) = \frac{1}{2}(A + B - |A - B|) \tag{9}$$

where the absolute value function operates on the eigenvalues of the matrix $A - B$, compare Sect. 3.2. The so obtained spatial derivatives u_x and u_y will again be symmetric 2×2 matrices. The corresponding computation of the gradient norm is obtained through

$$\|\nabla u\| = \sqrt{u_x^2 + u_y^2} \tag{10}$$

where the square root applies again to the eigenvalues. Let us remark that $u_x^2 + u_y^2$ is a symmetric matrix if u_x and u_y are symmetric matrices.

The sign of $u - f$ can be computed in several possible ways. A straightforward approach would be to apply the signum function to the eigenvalues of the matrices $u - f$ in the style of Sect. 3.2. In that case we must take precautions that the matrix-matrix product between $\mathrm{sgn}(u - f)$ and $\|\nabla u\|$ remains a symmetric matrix. This could for example be achieved by considering the Jordan matrix product [16]

$$A \bullet B = \frac{1}{2}(AB + BA) \tag{11}$$

Alternatively, one may consider approaches that yield a scalar value for the sign of a 2×2 matrix. In this latter case the symmetry of the matrices is always preserved. Such an approach could for example be based on the Loewner order. In that case we say that the sign of $u - f$ is 1 if $u \succ f$, -1 if $u \prec f$ and 0 else.

5   Experimental Results

In this section, we present several results for constructing levelings for colour images. As a common point of our proceeding, we generally employ test images of relatively low resolution in order to make the effects of the filtering processes obvious.

In what follows, we first demonstrate that the channel-wise approach gives inferior results. After that we aim to clarify differences between the fully discrete and the PDE-based approaches. Furthermore we illustrate the potential of our methods at hand of filtering image patches featuring different types of textures.

5.1   The Channel-Wise Method: False Colours

In this first experiment we apply the component-wise and for comparison the Loewner order approach to the classic test image *Peppers*, downsampled to resolution 32×32 which is shown in Fig. 2(a). These two approaches are in some sense comparable directly since they are both different means to realise the iterative leveling construction from (1) and (2).

We employ here a cross-shaped structuring element of size 3×3. The marker images M_i are constructed via $2D$ Gaussian convolution G_{σ_i} of the original image f with standard deviation $\sigma_i = 2^{i-1}$, that is $M_i = f * G_{\sigma_i}$ as shown in Fig. 2(b) and (f). The levelings $u_i = \Lambda(M_i | u_{i-1})$ with $u_0 = f$ are shown in Fig. 2(c) and (g). Additionally we depict the residuals $r_i = f - u_i$ in Fig. 2(d) and (h). For each channel of the colour image r_i, we add 0.5 to avoid negative numbers. In Fig. 2(e) we show $u_1 - u_2$. Here, we also add 0.5 to each channel.

As we can see, we are easily able to identify levelings by means of the channel-wise approach. One may also observe the desired segmentation-like effect in the output images, and that edges of the original input image stay sharp after leveling construction. This effect is also observable in the (translated) difference image $u_1 - u_2$ displayed in Fig. 2(e). Turning to the observable colours, however, we can see even without any comparison to other approaches that the green banana pepper on the left gets an orange-red colour as seen in Fig. 2(c) and (g). This

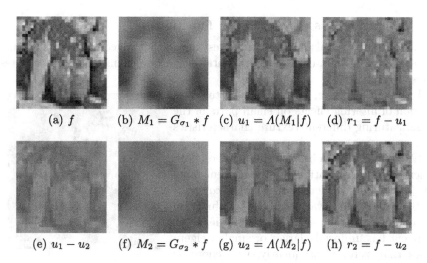

Fig. 2. Levelings for the component-wise approach. Notice the false colours appearing at the banana pepper in the left half of u_1, u_2, see especially Fig. 2(g). (Color figure online)

effect is also observable by the relatively prominent green colour in the residual images in Fig. 2(d) and (h). This is an undesirable false colour arising by the missing coupling of channels in this approach.

Let us now evaluate if the Loewner order approach gives in comparison better results. Using the same computational parameters, we obtain the results depicted in Fig. 3. While we obtain qualitatively comparable results, we do not obtain a false colour.

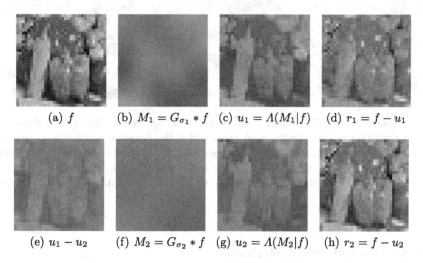

Fig. 3. Levelings for the discrete Loewner order approach. No false colour appears. (Color figure online)

5.2   Discretisations of the Signum Function

Let us introduce at this point a classic test image that we consider in this paper for the following part of the experimental evaluation, namely the *Baboon* test image, see Fig. 4.

To be more precise, we focus for better visualisation of filtering effects on small parts of the
image featuring different texture characteristics
(in a loose sense).

We test at this point the influence of both
presented implementations of the signum function. To this end, we denote as *Loewner comparison* the scalar-valued variant of the sign function
where we test at hand of the Loewner order if the
argument has uniformly larger or smaller eigenvalues than zero. Consequently, we denote as
Jordan product the method following strictly the
proceeding of Sect. 3.2 yielding a matrix-valued
sign function.

Fig. 4. *Baboon* test image

A visualisation of the obtained results for both choices is given in Fig. 5. As
computational parameters we employ here a stopping time of $t = 1$ as this will
turn out to suffice for efficient texture discrimination, and $\sigma = 2$. As observable, the Loewner comparison method incorporates a mechanism that tends to
sharpen edges at the expense of slight artefacts along them.

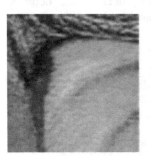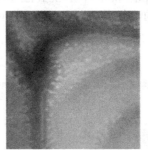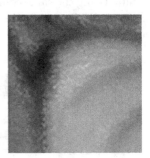

Fig. 5. Results of our PDE implementation for the indicated two different implementations of the sign function. **Left:** Input image. **Middle:** Leveling result computed with
signum function based on Loewner comparison. **Right:** Corresponding leveling by Jordan product. As observable, both possible implementations give reasonable results.
However, in the Loewner comparison result we observe an enhancement of the strong
edges in the image.

5.3   Comparison of All Methods

At hand of three distinct parts of the Baboon test image, see Fig. 6, we now give a
detailed comparison of all the introduced methods – channel-wise, fully discrete,

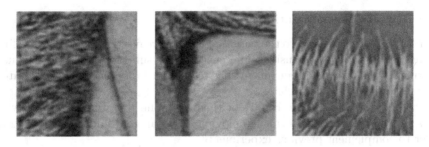

Fig. 6. Zooms into the baboon test image used for the detailed comparison.

PDE-based using Loewner comparison and Jordan product – that complements and extends the previous tests.

Let us comment on the leveling results. Here in this first test we stay with our marker images relatively close to the original ones, using $\sigma = 1$. For the component-wise approach there are no highly prominent false colours in this test. The fully discrete approach works basically as expected, yet one may notice a slight darkening of some colours. For the PDE-based methods we take here the stopping time $t = 1$ as this yields already reasonable results. We observe in the PDE-based levelings a more prominent smoothing effect that might be due to a numerical artefact introduced by the Rouy-Tourin discretisation (which is inherently of first order), but the luminance of colours appears to be preserved (Fig. 7).

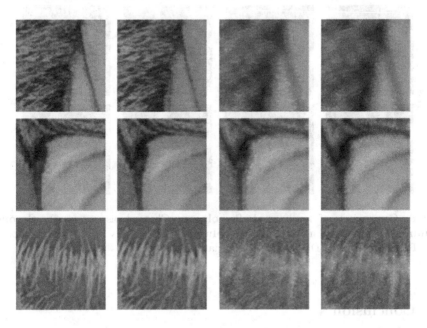

Fig. 7. Comparison of levelings. **Left column:** Channel-wise approach. **Second column:** Fully discrete method. **Third column:** PDE-based with Loewner comparison. **Right column:** PDE-based with Jordan product.

5.4   Texture Discrimination

We now turn our attention to the potential for texture filtering of the new leveling methods. To this end we consider difference images, subtracting the RGB valued filtering result from the input image. The so-obtained difference image is shifted pixelwise by 127.5 in each channel.

In Fig. 8 we present the results of this step. The test was performed with the same input images, this time employing Gaussian convolution with $\sigma = 2$ in order to complement previous experiments.

We observe that already the channel-wise approach seems to be able to filter the texture. Due to the near absence of false colours in this experiment the result is actually better here than with the fully discrete, Loewner-based method, where the observed change in luminance accounts for areas with large contributions in the difference images. This may be a problem for automatic processing of the results. The, to our impression, most convincing results are delivered by the PDE-based methods, as structures related to texture can clearly be identified.

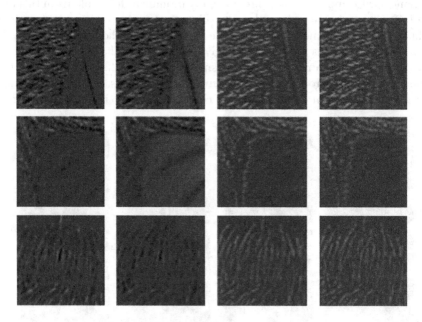

Fig. 8. Comparison of levelings. **Left column:** Channel-wise approach. **Second column:** Fully discrete method. **Third column:** PDE-based with Loewner comparison. **Right column:** PDE-based with Jordan product.

6   Conclusion

Constructing levelings for colour images is not an easy task. The simplest approach to process each channel of the colour image separately may lead to false colours. The discrete approach based on the method in [10] may introduce a change in the luminance of colours which may be an undesirable property in an application. The PDE-based approach based on Jordan product seems to yield the best results among the tested methods and deserves a more detailed analysis in future work.

References

1. Agoston, M.: Computer Graphics and Geometric Modeling: Implementation and Algorithms. Springer, London (2005)
2. Aptoula, E., Lefévre, S.: A comparative study on multivariate mathematical morphology. Pattern Recogn. **40**(11), 2914–2929 (2007)
3. Boroujerdi, A.S., Breuß, M., Burgeth, B., Kleefeld, A.: PDE-based color morphology using matrix fields. In: Aujol, J.-F., Nikolova, M., Papadakis, N. (eds.) SSVM 2015. LNCS, vol. 9087, pp. 461–473. Springer, Cham (2015). doi:10.1007/978-3-319-18461-6_37
4. Burgeth, B., Kleefeld, A.: Morphology for color images via Loewner order for matrix fields. In: Hendriks, C.L.L., Borgefors, G., Strand, R. (eds.) ISMM 2013. LNCS, vol. 7883, pp. 243–254. Springer, Heidelberg (2013). doi:10.1007/978-3-642-38294-9_21
5. Crespo, J.: Adjacency stable connected operators and set levelings. Image Vis. Comput. **28**(10), 1483–1490 (2010)
6. Crespo, J., Serra, J., Schafer, R.: Image segmentation using connected filters. In: Serra, J., Salembier, P. (eds.) Workshop on Mathematical Morphology, pp. 52–57 (1993)
7. van de Gronde, J., Roerdink, J.: Group-invariant colour morphology based on frames. IEEE Trans. Image Process. **23**, 1276–1288 (2014)
8. Higham, N.: Functions of Matrices. SIAM, Philadelphia (2008)
9. Kleefeld, A., Breuß, M., Welk, M., Burgeth, B.: Adaptive filters for color images: median filtering and its extensions. In: Trémeau, A., Schettini, R., Tominaga, S. (eds.) CCIW 2015. LNCS, vol. 9016, pp. 149–158. Springer, Cham (2015). doi:10.1007/978-3-319-15979-9_15
10. Kleefeld, A., Burgeth, B.: An approach to color-morphology based on Einstein addition and Loewner order. Pattern Recogn. Lett. **47**, 29–39 (2014)
11. Lerallut, R., Decenciére, E., Meyer, F.: Image filtering using morphological amoebas. Image Vis. Comput. **25**(4), 395–404 (2007)
12. Maragos, P.: Algebraic and PDE approaches for lattice scale-spaces with global constraints. Int. J. Comput. Vision **52**(2–3), 121–137 (2003)
13. Maragos, P., Evangelopoulos, G.: Leveling cartoons, texture energy markers, and image decomposition. In: Proceedings of ISMM, pp. 125–138. MCT/INPE (2007)
14. Meyer, F.: The levelings. In: Proceedings of the ISMM. Kluwer Academic Publishers (1998)
15. Meyer, F., Maragos, P.: Nonlinear scale-space representation with morphological levelings. J. Vis. Commun. Image Represent. **11**(2), 245–265 (2000)
16. Pizarro, L., Burgeth, B., Breuß, M., Weickert, J.: A directional Rouy-Tourin scheme for adaptive matrix-valued morphology. In: Wilkinson, M.H.F., Roerdink, J.B.T.M. (eds.) ISMM 2009. LNCS, vol. 5720, pp. 250–260. Springer, Heidelberg (2009). doi:10.1007/978-3-642-03613-2_23

17. Rouy, E., Tourin, A.: A viscosity solutions approach to shape-from-shading. SIAM J. Numer. Anal. **29**(3), 867–884 (1992)
18. Welk, M., Kleefeld, A., Breuß, M.: Quantile filtering of colour images via symmetric matrices. Math. Morphol. - Theory Appl. **1**, 136–174 (2016)
19. Welk, M., Kleefeld, A., Breuß, M.: Non-adaptive and amoeba quantile filters for colour images. In: Benediktsson, J.A., Chanussot, J., Najman, L., Talbot, H. (eds.) ISMM 2015. LNCS, vol. 9082, pp. 398–409. Springer, Cham (2015). doi:10.1007/978-3-319-18720-4_34
20. Zanoguera, F., Meyer, F.: On the implementation of non-separable vector levelings. In: Talbot, H., Beare, R. (eds.) Proc. ISMM. CSIRO Publishing (2002)

Nonlocal Difference Operators on Graphs for Interpolation on Point Clouds

François Lozes$^{(\boxtimes)}$ and Abderrahim Elmoataz

Université de Caen Basse-Normandie and the ENSICAEN in the
GREYC UMR CNRS 6072 Laboratory, Image Team,
6 Boulevard Maréchal Juin, 14050 Caen Cedex, France
{francois.lozes,abderrahim.elmoataz-billah}@unicaen.fr

Abstract. In this paper we introduce a new general class of partial difference operators on graphs, which interpolate between the nonlocal ∞-Laplacian, the Laplacian, and a family of discrete gradient operators. In this context we investigate an associated Dirichlet problem for this general class of operators and prove the existence and uniqueness of respective solutions. We propose to use this class of operators as general framework to solve many interpolation problems in a unified manner as arising, e.g., in image and point cloud processing. (AE is supported by the European FEDER Grant (PLANUCA Project) and the project ANR GRAPHSIP.)

1 Introduction

Partial differential equations (PDEs) involving the p-Laplace and ∞-Laplace operators still generate a lot of interest both in the setting of Euclidean domains as well as on discrete graphs. These operators in their different forms, i.e., continuous, discrete, local, and nonlocal, are at the interface of many scientific fields as they are used to model many interesting phenomena, e.g., in mathematics, physics, engineering, biology, and economy. Some closely related applications can be found in image processing, computer vision, machine learning, and game theory, see e.g., [1,4,7] and references therein.

In this paper, we introduce and study a novel adaptive general class of Partial difference Equations (PdEs) on weighted graphs. These equations are based on finite difference operators which adaptively interpolate between two discrete upwind gradients and the p-Laplacian operator on graphs. The advantage of the involved family of operators is its adaptivity with respect to potential applications, i.e., to handle many local and nonlocal interpolation problems in image and data processing within the same unified framework, e.g., inpainting, colorization, and semi-supervised clustering.

Furthermore, in order to solve the associated Dirichlet problem in the setting of discrete graphs we propose an algorithm which can be used to unify interpolation tasks on both conventional images and point cloud data.

© Springer International Publishing AG 2017
J. Angulo et al. (Eds.): ISMM 2017, LNCS 10225, pp. 309–316, 2017.
DOI: 10.1007/978-3-319-57240-6_25

The rest of this work is organized as follows. In Sect. 2 we provide basic definitions and notations, which are used throughout this work. Furthermore, we recall our previous works on PdEs on graphs and the p-Laplacian on graphs. In Sect. 3 we derive a novel partial difference operator which interpolates between the graph p-Laplacian, the graph ∞-Laplacian, and discrete gradient operators. Then, we study an associated Dirichlet problem and prove the existence and uniqueness of respective solutions. Section 4 presents several applications and interpolation problems on real world images and point clouds. Finally, a short discussion in Sect. 5 concludes this paper.

2 Partial Difference Operators on Graphs

In this section we introduce the basic notations used throughout this paper. Additionally, we recall various definitions of difference operators on weighted graphs from previous works in order to define in this context derivatives, the p-Laplace operator, and some morphological operators on graphs. More details on these operators can be found in [5,11,18].

2.1 Notations and Preliminaries

A weighted graph $G = (V, E, w)$ consists of a finite set $V = \{v_1, \ldots, v_N\}$ of N vertices and a finite set $E \subset V \times V$ of weighted edges. We assume G to be undirected, with no self-loops and no multiple edges. Let (u, v) be the edge of E that connects two vertices u and v of V. Its weight, denoted by $w(u, v)$, represents the similarity between its vertices. Similarities are usually computed by using a positive symmetric function $w : V \times V \to \mathbb{R}^+$ satisfying $w(u, v) = 0$ if $(u, v) \notin E$. The notation $u \sim v$ is also used to denote two adjacent vertices. The degree of a vertex u is defined as $\delta_w(u) = \sum_{v \sim u} w(u, v)$. A function $f : V \to \mathbb{R}$ of $H(V)$ assigns a real value $f(u)$ to each vertex $u \in V$.

The two *upwind gradient norm operators* $\|\nabla_w^\pm f\|_\infty : \mathcal{H}(V) \to \mathcal{H}(V)$ for a function $f \in \mathcal{H}(V)$ can be defined as:

$$\|(\nabla_w^\pm f)(u)\|_\infty = \max_{v \sim u}\left(\sqrt{w(u,v)}\big(f(v) - f(u)\big)^\pm\right), \tag{1}$$

where $(x)^+ = \max(x, 0)$ and $(x)^- = -\min(x, 0)$.

2.2 Morphological Nonlocal Dilation, Erosion and Mean on Graphs

These gradients were also used to approximate certain continuous Hamilton-Jacobi equations on a discrete domain [10,18]. For example, given two functions $f, \mu : \Omega \subset \mathbb{R}^n \to \mathbb{R}$, then any continuous equation of the form:

$$\frac{\partial f(x, t)}{\partial t} = \mu(x)\|\nabla f(x, t)\|_p, \tag{2}$$

can be numerically approximated in a discrete setting as:

$$\frac{\partial f(u,t)}{\partial t} = \mu^+(u)\|(\nabla_w^+ f)(u,t)\|_p - \mu^-(u)\|(\nabla_w^- f)(u,t)\|_p, \tag{3}$$

where $\mu^+(u) = \max(\mu(u),0)$ and $\mu(u)^- = -\min(\mu(u),0)$.

In particular, if $\mu \equiv 1$, $p = \infty$, and if we employ a forward Euler time discretization with $\Delta t = 1$ (for stability we have $\Delta t \leq 1$), this equation can be rewritten as:

$$f^{k+1}(u) = f^k(u) + \|(\nabla_w^+ f^k)(u)\|_\infty, \tag{4}$$

with $f^k(u) = f(u, k\Delta t)$. This can be interpreted as a single iteration of the following *nonlocal dilation* type operator:

$$f^{k+1}(u) = NLD(f^k)(u), \tag{5}$$

where $NLD : \mathcal{H}(V) \to \mathcal{H}(V)$ is defined as:

$$NLD(f)(u) = f(u) + \max_{u \sim v}\left(\sqrt{w(u,v)}(f(v) - f(u))^+\right). \tag{6}$$

Similarly, for the case $\mu \equiv -1$ and $p = \infty$ we have $NLE : \mathcal{H}(V) \to \mathcal{H}(V)$ defined as:

$$NLE(f)(u) = f(u) - \max_{u \sim v}\left(\sqrt{w(u,v)}(f(v) - f(u))^-\right). \tag{7}$$

Likewise, for the case of the continuous Laplacian, the discretization leads to the operator $NLM : \mathcal{H}(V) \to \mathcal{H}(V)$, which is the well-known nonlocal mean filter [6], defined as

$$NLM(f)(u) = \frac{\sum_{v \sim u} w(u,v)f(v)}{\delta_w(u)}. \tag{8}$$

3   A New Family of Graph Adaptive Operators

In this section we propose a new family of discrete operators on weighted graphs which corresponds to a graph operators with gradients terms and we investigated an associated Dirichlet problem.

3.1   Definition

Based on the discussed PdE framework on graphs in Sect. 2, we are now able to propose a novel family of operators denoted by $\Delta_{\alpha,\beta,\gamma} : \mathcal{H}(V) \to \mathcal{H}(V)$ for a function $f \in \mathcal{H}(V)$ by:

$$\Delta_{\alpha,\beta,\gamma} f(u) = \alpha(u)\|\nabla_w^+ f(u)\|_\infty - \beta(u)\|\nabla_w^- f(u)\|_\infty + \gamma(u)\Delta_{w,2} f(u), \tag{9}$$

with $u \in V$, $\alpha(u), \beta(u), \gamma(u) : V \to \mathbb{R}$ and $\alpha(u) + \beta(u) + \gamma(u) = 1$. By a simple factorization of the ∞-Laplacian this new family of operators can be rewritten as:

$$\Delta_{\alpha,\beta,\gamma}f = 2\min(\alpha(u),\beta(u))\Delta_{w,\infty}f(u) + (\alpha(u) - \beta(u))^{+}\|\nabla_w^{+}f(u)\|_\infty$$
$$- (\alpha(u) - \beta(u))^{-}\|\nabla_w^{-}f(u)\|_\infty + \gamma\Delta_{w,2}f(u). \tag{10}$$

With $\alpha(u), \beta(u), \gamma(u)$ constants, we retrieve formulation presented in [12]. We propose to defined $\alpha(u), \beta(u), \gamma(u)$ as:

$$\alpha(u) = \frac{\sum_{f(v)-f(u)>\epsilon} w(u,v)}{\delta_w(u)} \qquad \beta(u) = \frac{\sum_{f(v)-f(u)<\epsilon} w(u,v)}{\delta_w(u)}, \tag{11}$$

and $\gamma(u) = 1 - \alpha(u) - \beta(u)$. Note that this family of operators is directly related to the nonlocal average operator: $\Delta_{\alpha,\beta,\gamma}f = NLA(f) - f$, for which we refer to the operator $NLA : \mathcal{H}(V) \to \mathcal{H}(V)$ as 'Nonlocal Average' with

$$NLA(f)(u) = \alpha(u)NLD(f)(u) + \beta(u)NLE(f)(u) + \gamma(u)NLM(f)(u), \tag{12}$$

and the operators NLD, NLE, and NLM as introduced in Sect. 2.

3.2   Dirichlet Problem

In the following we focus on a PdE related to the proposed family of graph operators with gradient terms. In particular, we investigate an associated Dirichlet problem. Let $G = (V, E, w)$ be an undirected, weighted, and connected graph, $A \subset V$ a subset of vertices, the boundary of A defined as $\partial A = V \backslash A$ and $g : \partial A \to \mathbb{R}$. We consider the PdE as:

$$\begin{cases} (\Delta_{\alpha,\beta,\gamma}f)(u) = 0, & u \in A, \\ f(u) = g(u), & u \in \partial A, \end{cases} \tag{13}$$

for the general case $\gamma \neq 0$. We could demonstrate as in [12] that the problem (13) has a unique solution.

4   Unified Interpolation for Inverse Problems on Images and Point Clouds

Many tasks in computer vision and image processing can be formulated as interpolation problems. Image and video colorization [15], inpainting [2,17], and semisupervised segmentation [13,19] are examples of these interpolation problems. In general, interpolation consists of estimating appropriate values in regions of missing data while staying coherent with respect to the given data. Until today many methods have been developed and proposed for image interpolation [6,11,13,18]. Among them, a significant amount of methods is based on local or nonlocal PDEs or variational methods.

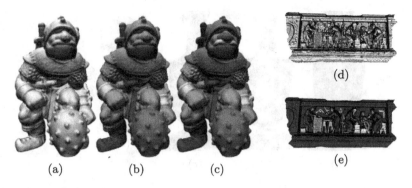

Fig. 1. Colorization of points clouds. (a) Uncolored dwarf, (b) half-colored dwarf, (c) full colored dwarf, (d) original point cloud, (e) colorized point cloud. (Color figure online)

In this work we propose to use the in Sect. 3 introduced family of graph operators as a unified framework. Among other tasks, this framework can be used to solve semi-supervised segmentation or clustering, image inpainting, as well as colorization of point clouds. To perform this task we propose to solve the discussed Dirichlet problem from (13), for which $A \subset V$ is the subset of vertices associated to the missing information. Note that the initial value function g is application-dependent and will be defined for each application in the sequel.

To solve (13) we make use of the following associated evolution equation problem:

$$\begin{cases} \frac{\partial}{\partial t} f(u,t) = \Delta_{\alpha,\beta,\gamma} f(u,t), & u \in A, \\ f(u,t) = g(u), & u \in \partial A, \\ f(u,t=0) = f_0(u), & u \in A, \end{cases} \tag{14}$$

for which f_0 is an initial function that is also application-dependent. To solve (14) iteratively we use an explicit forward Euler time discretization. Using $\Delta_{\alpha,\beta,\gamma} = NLA(f) - f$ and setting $\Delta t = 1$, we get the following nonlocal average filter:

$$\begin{cases} f^{n+1}(u) = NLA(f^n)(u) & u \in A, \\ f^{n+1}(u) = g(u), & u \in \partial A, \\ f^0(u) = f_0(u), & u \in A. \end{cases} \tag{15}$$

Graph Construction: The first step in the graph construction, consists in defining the sets V and E from a given dataset. Let us consider a dataset P as a set of data points $\{p_1, \ldots, p_n\} \in \mathbb{R}^n$. To each data point we first associate a vertex of a proximity graph G to define a set of vertices V. Then, we determine the edge set E from the neighbors of each vertex v_i. We consider the k Nearest Neighbors Graph (k-NNG): $v_j \sim v_i$ if the distance between p_i and p_j is among the k-th smallest distances from p_i to all the other data points. To speed up the k-NN algorithm, a kD-tree can also be used [3].

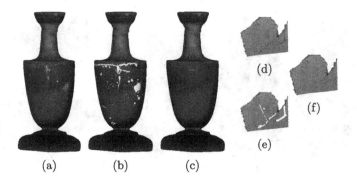

Fig. 2. Restoration of antique objects. (a) Original vasis, (b) vasis to inpaint, (c) restored vasis, (d) original wall, (e) wall to inpaint, (f) restored wall. (Color figure online)

Once the graph has been created, it has to be weighted. If one does not want to take care of the vertices similarities, the weight function w can be set to $w = 1$. A better one can be obtained using patches [6]. For images, a patch $\boldsymbol{P}(v_i)$ centered at a vertex $v_i \in V$ is a vector of values (e.g., coordinates, intensities) defined by $\boldsymbol{P}(v_i) = \left(f^0(v_j) : v_j \in B(v_i, n) \right)^T$ where $B(v_i, n)$ is a square of size n^2 centered at v_i. Using patches, $w : V \times V \to \mathbb{R}$ is defined by: $w(v_i, v_j) = exp\left(-\frac{\|\boldsymbol{P}(v_i) - \boldsymbol{P}(v_j)\|_2^2}{\sigma^2} \right)$. We have proposed a novel definition of patches to three-dimensional point cloud that can be used for any graph representation associated to meshes or 3D point clouds, see [16] for more details.

3D Colorization: Image colorization is the process of adding colors to monochromatic images. To colorize monochrome images the luminance channel is used to determine pixels similarities which enable color diffusion from scribbles. In the case of 3D data however, the intensity channel is missing and similarities between points have to be determined in a different way. To the best of our knowledge Leifman and Tal [14] are the only researchers which have proposed a method for *mesh colorization* up to now. The colorization is then performed by solving a constrained quadratic optimization problem (as in [15]). Let $f^0 : V \to \mathbb{R}^3$ be a function that assigns RGB colors to vertices. Let $A \subset V$ be the subset of vertices with unknown colors and ∂A the subset of vertices for which $g : \partial A \to \mathbb{R}^3$ gives the user-specified color scribbles. Then, we are able to use the iteration scheme (15) to perform 3D colorization of point cloud data. Figure 1 shows results of the method to colorize several 3D point clouds.

Nonlocal Inpainting: Digital inpainting can be simply formulated as reconstructing a damaged or incomplete image by filling the missing information in certain regions. In recent years many methods have been developed for interpolating geometry [8], texture [9], or both geometry and texture [2]. Among the proposed interpolation methods a significant number of algorithms are based on PDEs or variational methods, see e.g., [2] and references therein. Recent works tend to unify local and nonlocal interpolation approaches [13]. With respect

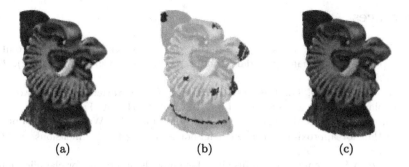

Fig. 3. Illustration of segmentation on a colored 3D point cloud. See text for more details. (a) Original, (b) label, (c) segmentation result. (Color figure online)

to (13) we propose to formulate the inpainting problem as follows: A is the set of pixels with missing information, $g : \partial A \to \mathbb{R}^c$, represents the known information (for which c is the number of color channels of the image), and $f : A \to \mathbb{R}^c$ represents the image to be reconstructed. Using this notation we are able to use the iteration scheme (15) to perform nonlocal inpainting. We illustrate this approach in Fig. 2 to inpaint the texture reconstruction on colored 3D point cloud data.

Semi-supervised Segmentation: We propose to consider the semi-supervised segmentation task as an interpolation problem, for which the function to be interpolated is the label function specifying the partition. Considering a partition into two classes A and B, with $N = 2$ the number of classes to segment. A multiphase segmentation can be performed by applying the iteration scheme (13) N times and considering the label A as a class and B as the other classes. In this case, the label function L, associating a class to each vertex, defined as $L : V \to \{C_i\}_{i=1,\dots,N}$ with $\{C_i\}$ the set of class labels is computed as: $L(u) = C_i | f_i(u) = \max\limits_{j=1,\dots,N} f_j(u)$. Figure 3 shows exemplary results of the method to segment a 3D colored point cloud. The graph is built in a similar way as in the Sect. 4.

5   Conclusion

In this paper we have introduced a novel family of graph operators with gradient terms. These partial difference operators interpolate between nonlocal ∞-Laplacian, nonlocal Laplacian, and gradient terms on graphs. We considered an associated Dirichlet problem for this class of operators and have proven the existence and uniqueness of respective solutions. Finally, we have demonstrated the applicability of these operators in terms of a unified framework to solve many inverse problems in image processing, 3D point cloud processing.

References

1. Andreu, F., Mazón, J., Rossi, J., Toledo, J.: A nonlocal p-Laplacian evolution equation with Neumann boundary conditions. J. Math. Pures Appl. **90**(2), 201–227 (2008)
2. Arias, P., Facciolo, G., Caselles, V., Sapiro, G.: A variational framework for exemplar-based image inpainting. IJCV **93**(3), 319–347 (2011)
3. Arya, S., Mount, D.M., Netanyahu, N.S., Silverman, R., Wu, A.Y.: An optimal algorithm for approximate nearest neighbor searching fixed dimensions. J. ACM **45**(6), 891–923 (1998)
4. Bertozzi, A.L., Flenner, A.: Diffuse interface models on graphs for classification of high dimensional data. Multiscale Model. Simul. **10**(3), 1090–1118 (2012)
5. Bougleux, S., Elmoataz, A., Melkemi, M.: Local and nonlocal discrete regularization on weighted graphs for image and mesh processing. Int. J. Comput. Vis. **84**(2), 220–236 (2009)
6. Buades, A., Coll, B., Morel, J.M.: Nonlocal image and movie denoising. IJCV **76**(2), 123–139 (2008)
7. Bühler, T., Hein, M.: Spectral clustering based on the graph p-Laplacian. In: Proceedings of the 26th Annual ICML, pp. 81–88. ACM (2009)
8. Chan, T.F., Kang, S.H., Shen, J.: Euler's elastica and curvature-based inpainting. SIAM J. Appl. Math. **63**, 564–592 (2002)
9. Criminisi, A., Pérez, P., Toyama, K.: Region filling and object removal by exemplar-based image inpainting. IEEE Trans. Image Process. **13**(9), 1200–1212 (2004)
10. Desquesnes, X., Elmoataz, A., Lézoray, O.: Eikonal equation adaptation on weighted graphs: fast geometric diffusion process for local and non-local image and data processing. JMIV **46**(2), 238–257 (2013)
11. Elmoataz, A., Lezoray, O., Bougleux, S.: Nonlocal discrete regularization on weighted graphs: a framework for image and manifold processing. IEEE Trans. Image Process. **17**(7), 1047–1060 (2008)
12. Elmoataz, A., Lozes, F., Toutain, M.: Nonlocal PDEs on graphs: from tug-of-war games to unified interpolation on images and point clouds. JMIV **57**, 1–21 (2016)
13. Gilboa, G., Osher, S.: Nonlocal linear image regularization and supervised segmentation. Multiscale Model. Simul. **6**(2), 595–630 (2007)
14. Leifman, G., Tal, A.: Mesh colorization. Comput. Graph. Forum **31**(2), 421–430 (2012)
15. Levin, A., Lischinski, D., Weiss, Y.: Colorization using optimization. ACM Trans. Graph. **23**(3), 689–694 (2004)
16. Lozes, F., Elmoataz, A., Lezoray, O.: Partial difference operators on weighted graphs for image processing on surfaces and point clouds. IEEE Trans. Image Process. **23**(9), 3896–3909 (2014)
17. Schönlieb, C.B., Bertozzi, A.: Unconditionally stable schemes for higher order inpainting. Commun. Math. Sci. **9**(2), 413–457 (2011)
18. Ta, V.T., Elmoataz, A., Lézoray, O.: Nonlocal PDEs-based morphology on weighted graphs for image and data processing. IEEE Trans. Image Process. **20**(6), 1504–1516 (2011)
19. Zhou, D., Schölkopf, B.: Discrete regularization. In: Semi-supervised Learning, pp. 221–232. MIT Press (2006)

Scale-Space Representations and Nonlinear Decompositions

Function Decomposition in Main and Lesser Peaks

Robin Alais[✉], Petr Dokládal, Etienne Decencière, and Bruno Figliuzzi

CMM, Center for Mathematical Morphology,
PSL Research University - MINES ParisTech,
35 rue Saint Honoré, Fontainebleau, France
robin.alais@mines-paristech.fr

Abstract. This article shows how the *dynamics* extinction value can be used to compute the decomposition of a function as a sum of simpler components. We show that this decomposition induces a hierarchical segmentation of the domain of definition, and a new partial ordering on nonnegative functions. Removing some of the components according to different criteria leads to new morphological operators. Their properties are discussed and illustrated in the last section. In particular, we see that thresholding on the supports' areas simplifies textured zones, while retaining perceptually salient elements of the image.

Keywords: Mathematical morphology · Extinction values · Tree representation · Dynamics · Gray-scale thinnings

1 Introduction

Since their introduction by Vachier and Meyer [1], *extinction values* have been used as a measure of importance for the extrema of a signal [2,3]. When introduced, they were defined with respect to a granulometry; the extinction value of a maximum is the parameter of the smallest opening to completely erase it. The concept was then generalized [4]; for instance, volumic extinction values can be defined, although there is no corresponding granulometry.

While extinction values are typically used to select relevant markers, prior to watershed segmentation, this article introduces a new function decomposition associated to extinction values. In this article, we focus on a specific one, *dynamics* [5], but the idea can be easily extended to other ones.

The input signal can be written as a sum of simpler components that we will denote as main and secondary *peaks*, following a mountain climbing analogy. Moreover, the inclusions of the peaks' supports define a forest structure. This representation is different from state-of-the-art techniques such as the *max-tree* [6] or the *tree of shapes* [7].

Peaks are then attributed various measures of importance; removing or altering the least important peaks (in the sense of those criteria) leads to new operators, some of which we detail in this work. These operators differ from classic attribute openings or thinnings [8]. In particular, they do not obey the threshold

© Springer International Publishing AG 2017
J. Angulo et al. (Eds.): ISMM 2017, LNCS 10225, pp. 319–330, 2017.
DOI: 10.1007/978-3-319-57240-6_26

decomposition principle [9]: we extend binary connected operators to gray-level operators by considering the supports of our peaks, instead of considering threshold sets.

All of the methodology presented in this article is applicable on signals defined on any finite simple graph; however, in the context of image processing, signals are usually defined on a regular grid in a two- or three-dimensional space. The decomposition that we introduce and all related operators have the additional attractive property of being invariant by isometries (rotations, translations, symmetries and combinations thereof).

2   Notations

In the following, E denotes a finite, connected, simple graph. In most image processing applications, typically E would be a subset of $\mathbb{Z}^2$ with a rectangular or hexagonal grid. For 3-D images, it could as well be a subset of $\mathbb{Z}^3$ with any arbitrary connectivity.

For a node $x \in E$, $N(x)$ denotes the neighborhood of x, *i.e.* all nodes $y \in E$ that share an edge with x.

The *support* of a function f is the set of vertices on which it takes non-zero values: $\mathrm{supp}(f) = \{x \in E | f(x) \neq 0\}$.

3   Decomposition in Main and Lesser Peaks

In this section, we introduce a new function decomposition, that we call the *decomposition in main and lesser peaks*, by two different, equivalent ways of computing it. The first approach is based on successive *geodesic reconstructions* [10] from the *global* maxima of a signal - as this transformation in a way 'ignores' maxima of lesser values, which is in our case the desired behavior. The second approach is much more closely related to the theory of extinction values [1,4], and although we focus here solely on *dynamics* [5], it can easily be extended to other extinction values.

3.1   Decomposition via Successive Geodesic Reconstructions

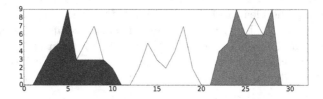

Fig. 1. Example of the elementary operation: the maximum of f is $M = 9$, reached at $X_M = \{5, 24, 28\}$. There are two connected components in the support of the reconstruction $\Gamma(f)$, so we obtain two functions f_9^1 and f_9^2. One is shown in blue, the other in green. (Color figure online)

Definition 1 (Geodesic reconstruction under the global maxima).
  Let $f \in \mathrm{Fun}(E, \mathbb{R}^+)$, $f \neq 0$.
  Let $M = \max\limits_{x \in E} f(x)$, and $X_M = \{x \in E \mid f(x) = M\}$ the set of points where it is reached.
  We define $\Gamma(f)$ as the geodesic reconstruction of f with markers X_M.
  Let C^1, C^2, ... C^K be the K connected components of $\mathrm{supp}(\Gamma(f))$, and for $1 \leqslant k \leqslant K$, let f_M^k be the restriction of $\Gamma(f)$ on C^k: $f_M^k = \Gamma(f) \times \mathbb{1}_{C^k}$
  In particular, $\Gamma(f) = \sum_{k=1}^{K} f_M^k$.
  If f is identically zero, we take $\Gamma(0) = 0$ and the above summation is empty.
  If f is not identically zero, we call the f_M^ks the main peaks *of function f.*

Note that by construction, Γ is an anti-extensive operator: $\Gamma(f) \leqslant f$, so the residue $f - \Gamma(f)$ is a nonnegative function, to which we can apply operator Γ again. By iterating the process until the residue is zero, we obtain a first way of computing our decomposition (see Algorithm 1).

Algorithm 1. Reconstruction-based algorithm

Given a function $f \in \mathrm{Fun}(E, \mathbb{R}^+)$: initialize with $i = 0$, $R_0 = f$.
  repeat
    Let $d_i = \max(R_i)$
    Compute the main peaks of R_i: $\Gamma(R_i) = \sum\limits_{k=1}^{K_i} f_{d_i}^k$
    $R_{i+1} \leftarrow R_i - \Gamma(R_i)$
    $i \leftarrow i + 1$
  until $R_i = 0$

Let I be the greatest integer such that $R_I \neq 0$; the previous algorithm yields our decomposition:

$$f = \sum_{i=0}^{I} \sum_{k=1}^{K_i} f_{d_i}^k$$

where, by construction, $d_0 = M = \max f$.

Definition 2. *We call the $f_{d_0}^k$ the* main peaks *of f, while for $i \geqslant 1$, we call the $f_{d_i}^k$* lesser peaks *of f.*

Figure 1 illustrates operator Γ and the notion of *main peaks*.

It is easy to show that the previous algorithm converges in a finite number of steps: keeping the same notations as above, we have:

$$\forall x \in X_M, \quad f(x) = \Gamma(f)(x) = M \tag{1}$$

If $f \neq 0$, $X_M \neq \emptyset$ and the support of $f - \Gamma(f)$ is strictly included in the support of f. Since we chose E to be a finite graph, the number of steps required for the algorithm to converge is upper-bounded by the cardinal of E.

Equation 1 also implies that the sequence $(d_i)_{1 \leqslant i \leqslant I}$ is strictly decreasing: $d_0 > d_1 > \ldots > d_I$.

3.2   Decomposition via Razings

The second way of computing the decomposition is quite similar to the watershed algorithm by flooding [11], except we raze the relief instead of flooding it. The idea is that we start from the regional maxima and gradually expand labels by following descending edges. When two or more regions meet, we stop propagating the labels that started from the lowest maxima, subtract the razed peak from the original function and propagate the remaining label. If two or more regions meet that started at the same - highest - altitude, we merge the regions and replace their labels by a single one. This way of handling equality cases can lead to 'twin peaks', like the green component in Fig. 1, and it is debatable; many image processing algorithms would arbitrarily break such ties, considering for instance the lexicographical order on the coordinates when working on a rectangular subset of $\mathbb{Z}^2$, or the area extinction value, as was proposed by Grimaud [5]. The former approach is not invariant by rotation (swapping the coordinates will break ties in a different way) and is not applicable to the general case when the domain of definition is a finite simple graph; the latter is more attractive, but does not handle the case where the area extinction values are equal, too. Our way of handling ties ensures invariance by rotation when working on a regular grid, is well-defined, and as we shall see, yields interesting properties, in particular regarding the supports of the peaks.

Let us consider again a function $f \in \mathrm{Fun}(E, \mathbb{R}^+)$. If $f = 0$ its decomposition in peaks is just an empty summation; in the following, let us assume f is not identically zero and that its support is connected (if it is not, we simply apply the following procedure on each connected component).

Let us consider the set X_{RM} of all regional maxima of f. A *regional maximum* of f is a connected set $X \subset E$ such that f takes the same value on each point of X and stricly lower values on neighboring points of X.

Let $X_{RM}^1, \ldots X_{RM}^K$ be the K connected components of X_{RM}, and let $L = \{l_1, ..., l_L\} = f(E) \setminus \{0\}$ the set of values ('levels') taken by f on its support, sorted in increasing order: $0 < l_1 < l_2 < ... < l_L$.

The algorithm then goes as follows: we start by defining K functions $f^1, \ldots, f^K$ which are simply the restrictions of f on the connected components $X_{RM}^1, \ldots, X_{RM}^K$:

$$\forall k \in \{1, \ldots, K\}, f^k = f \times \mathbb{1}_{X_{RM}^k}$$

We will then consider the connected components of the upper level sets of f, and progressively assign new points to one of the f^ks until we have $f = \sum_{k=1}^K f^k$.

Let us denote by $F^l = \{x \in E | f(x) \geqslant l\}$ the upper set of function f at level l; for each i from L to 1, for each connected component C of F^{l_i}, let us consider all k's such that $\mathrm{supp}(f^k) \subset C$ (there is at least one such k).

There are two cases to consider: if only one of the f^k has a maximum strictly greater than all others ($\max(f^k) > \max_{k' \neq k} \max(f^{k'})$), we consider it a potential main peak and assign to the others, which are necessarily secondary peaks, their final values:

$$\forall k' \neq k : \begin{cases} f^{k'} \leftarrow f^{k'} - l_i \\ \text{Finalize} f^{k'} \end{cases}$$
$$f \leftarrow f - \sum_{k' \neq k} f^{k'}$$
$$f^k \leftarrow f \times \mathbb{1}_C$$

If there is only one f^k whose support is a subset of C, obviously we would just assign it the value $f \times \mathbb{1}_C$. Else, if there is a tie between two or more peaks $(\max(f^{k_1}) = \max(f^{k_2}) = \ldots = \max(f^{k_m}) = \max_k \max(f^k))$, we first merge all these higher peaks together:

$$f^{k_1} \leftarrow f^{k_1} \vee f^{k_2} \vee \ldots \vee f^{k_m}$$
$$f^{k_2} \leftarrow 0$$
$$f^{k_3} \leftarrow 0$$
$$\ldots$$
$$f^{k_m} \leftarrow 0$$

then as before:

$$\forall k' \notin \{k_1, k_2, \ldots, k_m\} : \begin{cases} f^{k'} \leftarrow f^{k'} - l_i \\ \text{Finalize} f^{k'} \end{cases}$$
$$f \leftarrow f - \sum_{k' \neq k_1} f^{k'}$$
$$f^{k_1} \leftarrow f \times \mathbb{1}_C$$

In the previous instructions, 'finalize' $f^{k'}$ means that when a peak f^k is overtaken by a higher peak f^k at level l_i, we subtract l_i from it at this iteration, but then we will not change its value at any subsequent iteration.

After the last iteration, we have $f = \sum_{k=1}^{K} f^k$; this does not mean there are necessarily K peaks, as some terms of the summation can be zero. After removing those, regrouping the terms with the same maximum value d_i and reindexing, we can rewrite the summation as

$$f = \sum_{i=0}^{I} \sum_{k=1}^{K_i} f_{d_i}^k$$

as we had before. When we are not concerned with the values d_i, we may skip the double summation and write $f = \sum_{j=1}^{J} f^j$ for ease of reading.

Note that although this algorithm is mathematically valid for any finite simple graph E and for any nonnegative real-valued function f, it can be more easily rewritten in the more restrictive - but more often encountered - case where E is a regular grid and f takes values in a finite set, typically $\{0, \ldots, 255\}$ for 8-bit images. Efficient algorithms exist for simulating floodings or razings, based on hierarchical queues [10]. Although this paper does not aim at providing a thorough algorithmic study, this remark means that computing the decomposition using this algorithm is roughly as time-expensive as computing the watershed transformation of an image (Fig. 2).

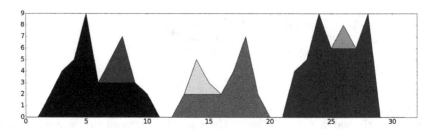

Fig. 2. Example of the decomposition

3.3 Properties of the Decomposition

Let $f = \sum_{i=0}^{I} \sum_{k=1}^{K_i} f_{d_i}^k$, as obtained by one of the previous algorithms, for a function $f : E \to \mathbb{R}^+$.

Proposition 1. *Nature of the peaks*
  Every peak p obtained by our decomposition ($p = f_{d_i}^k$ for some i and k) satisfies the properties that:

1. $\mathrm{supp}(p)$ *is connected*
2. $\Gamma(p) = p$

  Conversely, each function q satisfying these properties is by itself its own peak decomposition.

Proof. The first point is obvious by construction of the second algorithm. The fact that $\Gamma(p) = p$ comes from the fact that a function f^k obtained by the second algorithm cannot have local maxima that are not global maxima, else they would have been assigned another component $f^{k'}$.
  The reciprocal sense is basically the definition given in the first algorithm. $\square$

We denote by $\mathbb{P}$ the set of functions from E to $\mathbb{R}^+$ satisfying those properties. In particular, peaks have no local maxima that are not global maxima.
  A function $c : \mathbb{P} \to \{0, 1\}$ will be referred to as an *intrinsic criterion*. In particular, any binary criterion [8] T (defined not on the space of functions but on the space of subsets of E) can be associated a criterion c_T on $\mathbb{P}$ by considering the support of a peak p:

$$c_T(p) = T(\mathrm{supp}(p)).$$

Definition 3. *Dynamics of a maximum*
  The dynamics [5] of a maximum is the difference in altitude between this maximum and the maximal lowest altitude of a point on a path joining it to another, strictly higher, maximum.

Proposition 2. *If x_m is the index of a local maximum of f with dynamics d, then x_m is the index of a global maximum of f_d^k for some k, and $f_d^k(x_m) = d$.*

The proof can be found in the original article defining the notion of dynamics [5], as the original proposed algorithm is similar to the second algorithm we propose; the originality of ours lies in the decomposition it computes along with the dynamics.

A noticeable difference is how equality cases are handled: in [5], the author considers that two close maxima of the same value being assigned the same dynamics is undesirable behavior, since dynamics are often used as a way to select markers prior to segmentation, typically using the watershed algorithm. In this context, selecting many similar close extrema (usually minima, actually) is indeed undesirable behavior.

The dynamics of the global maximum (or global maxima) of a function f is ill-defined; the most common convention is to assign it the value $\max(f) - \min(f)$; other authors assign it an infinite value; if dynamics are to be used for ranking extrema by order of importance, the essential point is just to assign it a higher value than all local ones. In order to be consistent with the definitions of our algorithms, we will adopt the convention that the dynamics of the global maximum is simply its value. This is consistent with the fact that the value 0 plays a specific role in our decomposition. For instance, if the support of f is not connected, f and $f + \lambda$, where $\lambda > 0$ may not have the same main peaks (f may have several, while $f + \lambda$ necessarily has only one). Following our mountain-climbing analogy, the value 0 can be thought of as the sea level; even if the support of f is equal to the entire domain E, meaning that for all x in E, $f(x) > 0$, our reference level is not the minimum of f, but 0.

Proposition 3. *For any two components $f_{d_i}^k$ and $f_{d_j}^{k'}$, $\mathrm{supp}(f_{d_i}^k)$ and $\mathrm{supp}(f_{d_j}^{k'})$ are either nested or disjoint.*

In particular, if $\mathrm{supp}(f)$ is connected, then the number of main peaks $K_0 = 1$, $\Gamma(f) = f_{d_0}$ and:

$$\forall i > 1, \forall k \in \{1, \ldots, K_i\} \quad \mathrm{supp}(f_{d_i}^k) \subset \mathrm{supp}(f_{d_0})$$

This can be proved by induction; consider the razing-based algorithm: at the beginning, all supports are disjoint. Supports can only increase with iterations, and when peaks meet at level l_i, they are either merged together (if there is a tie) or some are finalized and the greater is assigned the value $l_i > 0$ on the supports of the other.

Proposition 4. *For any two components f_{d_i} and f_{d_j}:*

$$\mathrm{supp}(f_{d_i}) \subset \mathrm{supp}(f_{d_j}) \Rightarrow d_i < d_j$$

Proposition 5. *Stability by truncation*

Let us consider a function $f \in \mathrm{Fun}(E, \mathbb{R}^+)$ and its decomposition in main and lesser peaks $f = \sum_{i=0}^{I} \sum_{k=1}^{K_i} f_{d_i}^k$.

Let us now consider a function $g = \sum_{i=0}^{I} \sum_{k=1}^{K_i} \delta_i^k f_{d_i}^k$, where $\delta_i^k \in \{0, 1\}$, that is, g is a truncated summation of the components of f.

The peaks of g, as defined in the first section, are simply the peaks of f that appear in the summation; the decomposition of g can be written:

$$g = \sum_{i=0}^{I} \sum_{\substack{1 \leqslant k \leqslant K_i \\ \delta_i^k = 1}} f_{d_i}^k$$

In Sect. 4, we present several operators based on the idea of keeping only certain peaks, depending on certain criteria. The previous proposition asserts that, provided the criteria only depend on intrinsic properties (that is, the choice of δ_i^k only depends on $f_{d_i}^k$), these operators are idempotent.

3.4   Induced Tree Structure

Definition 4. *If* $\mathrm{supp}(f)$ *is connected, then* $\mathrm{supp}(f_1) = \mathrm{supp}(f)$, *so Proposition 3 naturally defines a tree structure on the components. Let* f_{d_j} *be a child node of* f_{d_i} *(we drop the superscripts for ease of reading) if:*

- $\mathrm{supp}(f_{d_j}) \subset \mathrm{supp}(f_{d_i})$
- *There is no k such that* $\mathrm{supp}(f_{d_j}) \subset \mathrm{supp}(f_{d_k}) \subset \mathrm{supp}(f_{d_i})$

If $\mathrm{supp}(f)$ *is not connected, the previous relationship yields a forest structure, with a tree for each connected component (Fig. 3).*

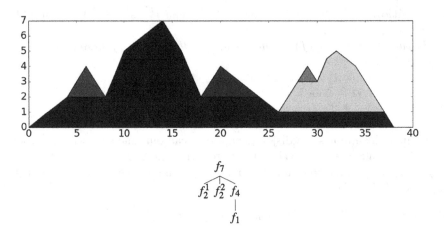

Fig. 3. Example of the decomposition and its associated tree: we have f_7 in blue, f_4 in yellow, f_2^1 and f_2^2 in green and red, f_1 in cyan. (Color figure online)

Proposition 6. *Stability by anamorphosis*

The forest decomposition is stable by any nonnegative anamorphosis that goes through the origin; if u is an increasing, invertible function from $\mathbb{R}^+$ *to* $\mathbb{R}^+$ *with* $u(0) = 0$, *then f and* $u(f)$ *yield the same forest structure.*

3.5 Ordering Induced by the Decomposition

Let us consider f and g two functions from E into $\mathbb{R}^+$, and their decompositions in main and lesser peaks, with the peaks labeled respectively from 1 to J_f and from 1 to J_g:

$$f = \sum_{j=1}^{J_f} f^j, \quad g = \sum_{j=1}^{J_g} g^j$$

We write $f \preceq g$ if there exists $\alpha : \{1, \ldots, J_f\} \to \{1, \ldots, J_g\}$ such that:

$$\forall j \in \{1, \ldots, J_g\}, \sum_{i \mid \alpha(i)=j} f^i \leqslant g^j \tag{2}$$

It can be easily seen that $f \preceq g \Rightarrow f \leqslant g$, but the reverse is not true in general.

Unfortunately, the couple $(\mathrm{Fun}(E, \mathbb{R}^+), \preceq)$ is not a lattice, nor even a semilattice: neither the infimum nor supremum of two functions are guaranteed to exist.

4 Morphological Operators on $(\mathrm{Fun}(E, \mathbb{R}^+), \preceq)$

In this section, we present some operators based on intrinsic criteria. All these operators are anti-extensive and idempotent; however, they lack the property of being increasing when considering the usual ordering on functions, so they are not morphological filters on the lattice $(\mathrm{Fun}(E, \mathbb{R}^+), \leqslant)$ (they are only thinnings).

They are increasing, though, if considering the order $\preceq$ introduced in the previous section. It must be kept in mind that since $(\mathrm{Fun}(E, \mathbb{R}^+), \preceq)$ is not a lattice, they stand outside the usual framework of mathematical morphology, but they can be thought of as morphological filters on a poset.

Definition 5. *General form for criterion-defined operators*
  Given a function f, its peak decomposition

$$f = \sum_{i=0}^{I} \sum_{k=1}^{K_i} f_{d_i}^k$$

and an intrinsic criterion c, we define:

$$\gamma_c : \begin{cases} \mathrm{Fun}(E, \mathbb{R}^+) \to \mathrm{Fun}(E, \mathbb{R}^+) \\ f \mapsto \sum_{i=0}^{I} \sum_{k=1}^{K_i} c(f_{d_i}^k) f_{d_i}^k \end{cases}$$

It is straightforward that any such operator is anti-extensive and idempotent; it is not, in general, increasing, be it for the usual order $\leqslant$ or the peak-induced order $\preceq$; we must add the constraint that the intrinsic criterion c is itself increasing (for either order, as they are the same on P). Fortunately, this requirement is easily met.

Definition 6. *Peak dynamics thresholding*

This tranformation has a parameter δ, which is the minimum dynamics of peaks to be kept; the intrinsic criterion is a threshold above this value. Formally:

$$c_\delta : P \rightarrow \{0,1\}$$
$$p \mapsto \begin{cases} 1 \ if & \max(p) \geqslant \delta \\ 0 \ else \end{cases}$$

Definition 7. *Peak area thresholding*

This tranformation has a parameter A, which is the minimum area (cardinal of the support) of peaks to be kept; the intrinsic criterion is simply a threshold above this area. Formally:

$$c_A : P \rightarrow \{0,1\}$$
$$p \mapsto \begin{cases} 1 \ if & \#\mathrm{supp}(p) \geqslant A \\ 0 \ else \end{cases}$$

Definition 8. *Peak volume thresholding*

This time, the parameter V is the minimal volume (sum of the values) of peaks to be kept; the intrinsic criterion is again a simple threshold above this parameter. Formally:

$$c_A : P \rightarrow \{0,1\}$$
$$p \mapsto \begin{cases} 1 \ if & \sum_{x \in E} p(x) \geqslant V \\ 0 \ else \end{cases}$$

These operators are illustrated in Fig. 4, in one dimension. Figure 5 illustrates peak area thresholding in two dimensions, and compares it to area opening, as

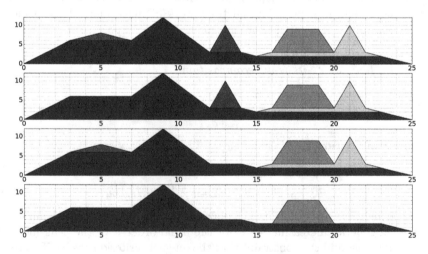

Fig. 4. From top to bottom: original signal, peak dynamics thresholding with $\delta = 3$, peak area thresholding with $A = 3$, peak volume thresholding with $V = 15$.

defined by Vincent [12]. Gray-scale area opening razes all maxima until they become large enough flat zones, reducing the overall contrast of the image. Peak area thresholding, on the other hand, does not affect high or isolated maxima, thus preserving the most salient bright structures.

Fig. 5. Comparison between classical area opening and peak area thresholding; the original image is shown on the top left, the second column shows the result of a gray-scale area opening of size 350 and its residue; the rightmost column shows the result of a peak area thresholding of size 350 and its residue (the contrast of the residues has been enhanced for better visualization). We can see that peak area thresholding flattens textured zones, such as the hair, forehead and right cheek, but does not alter the most salient ones, such as the right eye or the reflection on the hat.

5 Conclusions and Perspectives

In this article, we presented the methodology and mathematical foundations for obtaining a function decomposition based on a particular extinction value: dynamics. Instead of assigning a scalar value to each maximum, we assign it a function, that we call a *peak*, associated to this maximum. Those peaks can in turn be assigned different measures: maximum value, area, volume; binary criteria can also be computed on their supports. We exhibit some properties of our decomposition, and present a new order on functions induced by it.

Although we focused here on a *dynamics*-driven decomposition, similar decompositions can as easily be obtained by considering other extinction values in the definition of Sect. 3.2. Area or volume extinction values will lead to different decompositions, and may be more sensible than dynamics, depending on the application and on the expected type of noise.

In order to introduce the concepts, we considered nonnegative functions and the value zero as our reference level; however, it is easily generalized to real-valued functions by considering the positive and negative part and applying the decomposition to each, yielding self-dual operators.

References

1. Vachier, C., Meyer, F.: Extinction value: a new measurement of persistence. In: IEEE Workshop on Nonlinear Signal and Image Processing, vol. 1, pp. 254–257 (1995)
2. Angulo, J., Serra, J.: Automatic analysis of DNA microarray images using mathematical morphology. Bioinformatics **19**(5), 553–562 (2003)
3. Meyer, F., Vachier, C.: On the regularization of the watershed transform. Adv. Imaging Electron Phys. **148**, 193–249 (2007)
4. Vachier, C.: Extraction de caractéristiques, segmentation d'image et morphologie mathématique. Ph.D. thesis. École Nationale Supérieure des Mines de Paris, December 1995
5. Grimaud, M.: New measure of contrast: the dynamics, vol. 1769, pp. 292–305. International Society for Optics and Photonics (1992)
6. Salembier, P., Oliveras, A., Garrido, L.: Antiextensive connected operators for image and sequence processing. IEEE Trans. Image Process. **7**(4), 555–570 (1998)
7. Ballester, C., Caselles, V., Monasse, P.: The tree of shapes of an image. In: ESAIM: Control, Optimisation and Calculus of Variations, vol. 9, pp. 1–18, January 2003
8. Breen, E.J., Jones, R.: Attribute openings, thinnings, and granulometries. Comput. Vis. Image Underst. **64**(3), 377–389 (1996)
9. Heijmans, H.J.A.M.: Theoretical aspects of gray-level morphology. IEEE Trans. Pattern Anal. Mach. Intell. **13**(6), 568–582 (1991)
10. Beucher, S.: Segmentation d'Images et Morphologie Mathématique. Ph.D. thesis. Ecole Nationale Supérieure des Mines de Paris (1990)
11. Beucher, S., Meyer, F.: The morphological approach to segmentation: the watershed transformation. Opt. Eng.-New York-Marcel Dekker Inc. **34**, 433 (1992)
12. Vincent, L.: Morphological area openings, closings for grey-scale images. In: O, Y.L., Toet, A., Foster, D., Heijmans, A.M., Meer, P. (eds.) Shape in Picture, pp. 197–208. Springer, Heidelberg (1994)

Morphological Processing of Gaussian Residuals for Edge-Preserving Smoothing

Marcin Iwanowski[✉]

Institute of Control and Industrial Electronics, Warsaw University of Technology,
ul.Koszykowa 75, 00-662 Warszawa, Poland
`iwanowski@ee.pw.edu.pl`

Abstract. The Gaussian filter, thanks to its low-pass filtering properties, removes noise and smooths the image surface. On the other hand, smoothing influences remarkably contours of image objects because it does not preserve contrast at edges. In this paper, a method is described, that allows for correction of the result of Gaussian filtering, thanks to which the original sharpness of edges is restored. This correction is based on the morphological reconstruction-based processing of the residual of the Gaussian filter that is added back to the smoothed image. The final result of filtering using the proposed approach is an image consisting of smooth flat surfaces and sharp boundaries.

1 Introduction

The Gaussian filter is the most widely used linear filtering tool. It is a low-pass filter that removes the image noise, makes the image surfaces smooth, but, on the other hand, it introduces also a blurring of image edges. There exist however a relatively wide area of its applications, where one requires smoothing of flat image surfaces while preserving the meaningful edges, related to e.g. contours of image objects. The digital processing of photographs belongs to such applications. In this paper, a novel approach to solving this problem is proposed. It is based on the Gaussian residuals, which are defined as the differences between the original image and the image filtered using the Gaussian filter. Such residuals are characterized by high response within all regions were the Gaussian filter have modified the image content – in particular on edges and noise. The Gaussian residual is, in the proposed method, filtered using the morphological approach based on the reconstruction. It removes, from Gaussian residuals, regions that refer to noise on the original image, while keeping untouched regions related to meaningful edges. The latter are preserved thanks to known properties of morphological reconstruction. The morphologically processed residuals are added back to the Gaussian-filtered image. Such an addition restores the original edges without however reducing the smoothing over the flat image regions.

The paper is organized as follows. In Sect. 2, previous works are shortly presented. In Sect. 3, the Gaussian residuals are introduced. Section 4 contains the description of the proposed concept – a way in which residuals are morphologically processed. Section 5 shows some results and finally, Sect. 6 concludes the paper.

© Springer International Publishing AG 2017
J. Angulo et al. (Eds.): ISMM 2017, LNCS 10225, pp. 331–341, 2017.
DOI: 10.1007/978-3-319-57240-6_27

2   Previous Works

The problem of edge-preservation in smoothing filters has already been addressed in the literature. One of the first filters of this type was the Kuwahara filter [1, 2, 9], that was based on the division of the pixel neighborhood into four overlapping regions, where the mean and standard deviation of pixel values are computed. Based on the standard deviation, the appropriate mean is chosen as the filter response.

The approach described in [3] is a representative of a wider branch of research on edge-preserving smoothing that is based on the adaptive approach. According to this approach, the way of using the blurring mask is modified in such a way that neighbors are considered in the linear combination only if there is no edge in the underlying image.

One of the most popular filters, based on the similar idea, is a bilateral filter [6, 7]. In this filter, typical for the linear filters, weighting has been modified. The special factor that depends on the pixel values differences has been added, what allowed for reducing the blurring in the edge regions.

In [5] residuals are used to preserve edges in smoothing filter. It investigates an idea that is similar to one presented in the current paper, but in that case, the Gaussian residuals are linearly processed in an iterative way to get better preservation of edges in the final image.

An application of morphological processing to address the edge-preservation problem was proposed in [10]. The method makes use of the adaptive scheme combined with the geodesic time propagation.

3   Gaussian Residuals

The linear low-pass filter is defined as a convolution of the image and a filter mask:

$$I' = I * F \Leftrightarrow I'(p) = \sum_{q \in F} I(p+q) \cdot F(q), \tag{1}$$

where I, I' are input and output images, respectively; p, q are image pixels and F is a mask of the filter such that $\sum_{q \in F} F(q) = 1$. In case of the Gaussian filter, the mask is defined as:

$$G(p) = \frac{1}{2\pi\sigma^2} e^{-\frac{\|p\|^2}{2\sigma^2}}, \tag{2}$$

where σ stands for the standard deviation, that in this case is a parameter defining the strength of a filter. By subtracting the result of filtering from the original image, the Gaussian residual is obtained:

$$\mathcal{R}_G(I) = I - I * G. \tag{3}$$

In order to further process the residual using the morphological approach, the residual is split into two fractions: positive and negative:

$$\mathcal{R}_G^+(I) = 0.5\left(\mathcal{R}_G(I) + |\mathcal{R}_G(I)|\right) \; ; \; \mathcal{R}_G^-(I) = 0.5\left(|\mathcal{R}_G(I)| - \mathcal{R}_G(I)\right). \tag{4}$$

The fractions of residual fulfills the following obvious relations:

$$\mathcal{R}_G(I) = \mathcal{R}_G^+(I) - \mathcal{R}_G^-(I) \; ; \; I = I * G + \mathcal{R}_G^+(I) - \mathcal{R}_G^-(I). \tag{5}$$

The second of above equation describes the restoration of the original image from filtered one using residuals. An example image ('Lena'), its Gaussian-filtered version and both residuals[1] are shown in Fig. 1. When considering results of filtering from the point of view the edge-preserving properties of a filter, the complete set of all image pixels may be split into two subsets. To the first one belong all pixels that are located within the relatively flat image regions (low intensity variance). The result of the Gaussian filtering within such regions should be preserved, which means that the intensity is smoothed. To the second subset of pixels belong all meaningful edges i.e. regions with high intensity variance. These regions should not be smoothed since it would reduce the sharpness of edges. When looking at residuals, each of the above subsets is characterized by different properties. See Fig. 1 and its enlarged part shown in Fig. 2 (the region that is enlarged is outlined in Fig. 1(a)). The regions of small intensity variance on the original image are visible at the residual image as speckle noise – small groups of pixels of relatively low intensity. The meaningful edge regions (high variance) are, in turn, represented by thick paths of relatively high magnitude. To get the final filtering result with preserved edges, the former should be removed from the residuals, while the latter should be kept.

4   Morphological Processing of Residuals

The principal idea of the proposed method consists of morphological processing of both residuals, to remove low-variance regions while preserving high-variance ones. Finally, the morphologically processed residuals are used to correct the result of the Gaussian filtering according to the following rule:

$$I_{out} = I_{in} * G + \mathcal{M}\left(\mathcal{R}_G^+(I_{in})\right) - \mathcal{M}\left(\mathcal{R}_G^-(I_{in})\right), \tag{6}$$

where $\mathcal{M}(I)$ stands for the result of morphological processing. The morphological operator $\mathcal{M}$ is, in fact, a *selection* operator that removes from the residual image regions that are characterized by low-variance on the input image I_{in}. Since the low-variance regions are characterized by low values of pixels within the residual image, the thresholding of the latter:

$$\mathcal{R}_G^+(I_{in}) < t \; ; \; \mathcal{R}_G^-(I_{in}) < t, \tag{7}$$

---

[1] To increase the visibility of details, the complement of residual is displayed in all figures in this paper.

(a)   (b)

(c)   (d)

Fig. 1. The original image (a), the result of Gaussian filtering (b), Gaussian residuals (c), (d).

Fig. 2. Enlarged views of a region of the image, from left to right: input image, Gaussian filer, two resuduals.

results in binary masks indicating the low-variance regions. The threshold t is a sensitivity parameter that allow for adjusting the visibility of edges. The complement of binary masks defined by the Eq. (7) is a high-variance regions marker. These regions, contrary to the low-variance ones should be preserved in the final image, i.e. values of pixels belonging to these regions should be restored from the original image. In order to obtain the fine restoration, the morphological reconstruction by dilation [4,8] is used. An application of the reconstruction guarantees that the complete patches referring to image edges are fully restored. The reconstruction mask image is equal to the residual image while reconstruction markers are produced by mapping of above binary mask into extremal values of the resudial image (minimum value assigned to binary 1's, while maximum – to 0's). The morphological operator $\mathcal{M}$ from the Eq. (6) is thus defined as:

$$\mathcal{M}(I) = \rho_I\left(\min\left(I, (I \geq t)|_{\{min\{I\}, max\{I\}\}}\right)\right), \qquad (8)$$

where $\rho_I(J)$ stand for the morphological reconstruction (by dilation) of the mask I from markers J, $\min(I, J)$ stands for the point-wise minimum of two argument images, and $min\{I\}, max\{I\}$ – for the global minimum and maximum pixel value of the image I.

Detailed results of processing of residuals are shown in Figs. 3 and 4. They show the cross-sections of images when applying the morphological processing of residuals with threshold $t = 30$ (processed images are 8-bit, i.e. the maximum pixel value equals 255). The final, filtered image is shown[2] in Fig. 5(b). In Fig. 3 two functions are given, both are cross-section of the input image along the horizontal line shown in Fig. 1(a). The first one, indicated by dashed line is the cross-section of the residual $\mathcal{R}_G(I_{in}) = \mathcal{R}_G^+(I_{in}) - \mathcal{R}_G^-(I_{in})$. It contains both low- and high-variance regions. The continuous line represents the morphologically processed residuals i.e. $\mathcal{M}\left(\mathcal{R}_G^+(I_{in})\right) - \mathcal{M}\left(R_G^-(I_{in})\right)$. The latter consists of flattened low-variance regions and untouched high-variance ones. The cross-section along the same line on the original I_{in}, Gaussian-filtered $I_{in} * G$ and final I_{out} images are shown in Fig. 4. In the cross-section of the final image, the result of Gaussian blur has been preserved only within the low-variance regions while within the high-variance ones the output image follows the original one.

5   Results

Apart from the parameters of the Gaussian blur – size of the mask and standard deviation, the proposed approach requires a single parameter. This parameter – threshold t allows for setting up the sensitivity of the method to local variations of pixel intensities. Images shown in Fig. 5 presents results of processing using the proposed approach with various t values of the input Lena image (shown in Fig. 1(a)). The selected (see the rectangle on the original image) region has been enlarged and is presented (for the same sequence of t values) in Fig. 6.

---

[2] The paper printing process may reduce the quality of figures. To see filtering results in their original form, it is advised to look at all figures in this paper in its digital version displayed at the computer monitor.

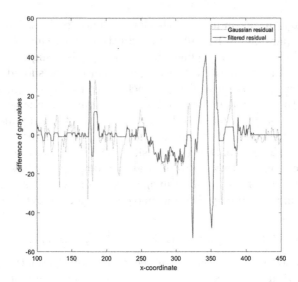

Fig. 3. The cross section of residuals before and after morphological processing.

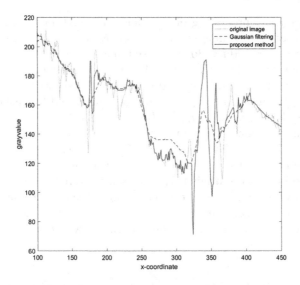

Fig. 4. The cross-section of the images: original, blurred and produced by the proposed method.

Enlarged fragments of images that have been obtained are showing the absolute difference between the original image and the filtering result. They show regions that have been removed in the result of filtering.

The choice of the parameter t influences the number of details that are restored by adding the result of morphological processing to the result of the Gaussian blur. The lower its value has been set-up, the higher number of details

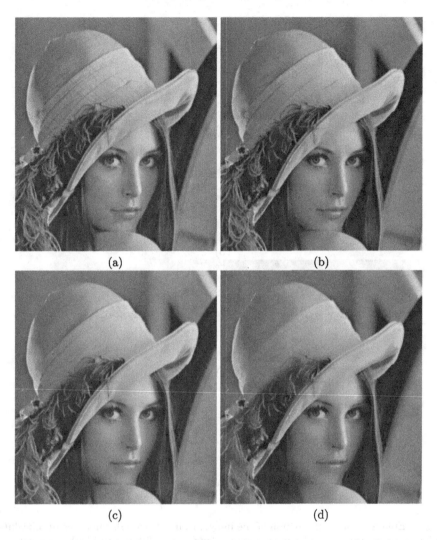

(a)   (b)

(c)   (d)

Fig. 5. The results of the proposed method with various sensitivity thresholds: $t = 15$ (a), $t = 30$ (b), $t = 45$ (c) and $t = 60$ (d).

is restored and reversely. Its value may be set-up directly as the gray level value (as in the case shown if Figs. 5 and 6) or as a fraction α of the maximal value of the residue: $t = \alpha \cdot max\{\mathcal{R}_G^+(I_{in})\}$ and the same for $\mathcal{R}^-$. In such a case, due to the fact that maximal values of both residuals may differ one to another, absolute values of both thresholds may be different for the same α.

The proposed method may be applied to remove the artifacts that are introduced by lossy compression methods. Figure 7 shows at position (a) the Lena image compressed using the *jpeg* method with highest possible compression level. The typical for this type of compression blocky appearance is present. The image

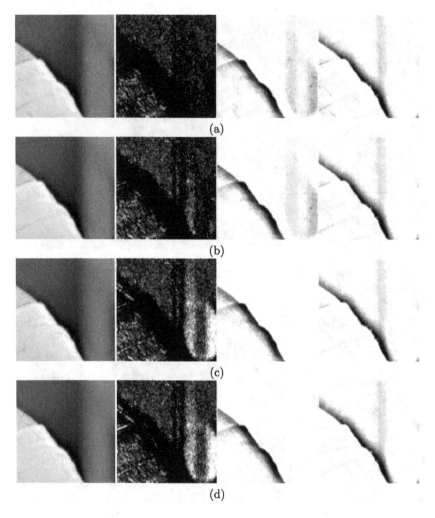

(a)

(b)

(c)

(d)

Fig. 6. Enlarged views of a region of the image, from left to right: final result, absolute difference with the original image, complements of two filtered residuals – operations with various sensitivity thresholds: $t = 15$ (a), $t = 30$ (b), $t = 45$ (c) and $t = 60$ (d).

filtered using the proposed method (see Fig. 7 for images and Fig. 8 for cross-sections) is free from these artifacts on flat regions, some remainders are only present around edges.

Some other results of filtering using the proposed approach are shown in Figs. 9 and 10. They present two well-known test images: 'clown' and a part of 'building'. In the first case ('clown') initial image is corrupted by relatively high level of noise that may be efficiently removed by the Gaussian filtering with, however, strong edge blurring. An application of the proposed approach allowed for keeping the meaningful edges in their original form. In the second case (part

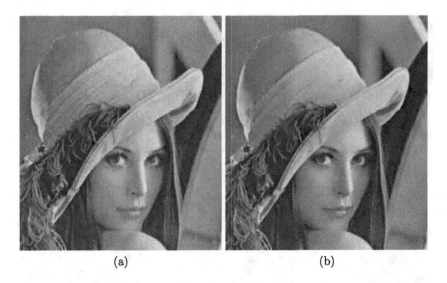

(a)   (b)

Fig. 7. An application of the method to remove the jpeg compression artifacts: jpeg-coded with high compression rate (corrupted) image (a), filtered image (b).

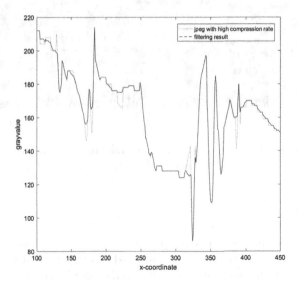

Fig. 8. The cross-sections of the images from Fig. 7.

of the 'building' image) the walls and roof of the exhibited building are covered by a texture. An application of the proposed method allowed for smoothing the textured regions while keeping the edges of the building originally sharp.

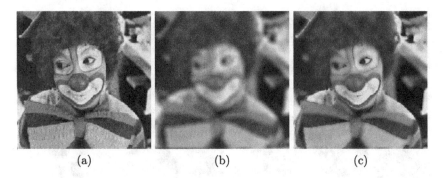

(a)   (b)   (c)

Fig. 9. The original image – 'clown' test image (a), the result of the Gaussian filtering (b), and of the proposed method (c).

(a)   (b)   (c)

Fig. 10. The original image – selected region from the 'building' test image (a), the result of the Gaussian filtering (b), and of the proposed method (c).

6   Conclusions

In the paper, the method for edge-preserving smoothing based on Gaussian filtering and morphological processing of residuals was presented. The method is based on the morphological processing makes use of the reconstruction by dilation operator performed on the residue of the Gaussian filter. The proposed method allows for separating the low-variance regions of the input image from the high-variance ones. By adding morphologically processed residual to result of the Gaussian filtering the method introduces the correction of the edge-regions by restoring the original contrast of edges.

The method produces promising results of processing. The result of filtering consists of smooth low-variance (flat) image regions where existing pixel varia-tions are nicely smoothed and sharp edges. The sharpness of edges is restored from the original image. The method requires the sensitivity threshold parameter

t that allows for tuning the cut-off level of local variance. The proposed method may be applied to improve the quality of various types of images, e.q. digital photographs. It can also be used to remove artifacts caused by lossy compression methods.

References

1. Kuwahara, M., Hachimura, K., Eiho, S., Kinoshita, M.: Processing of RI-angiocardiographic images. In: Preston Jr., K., Onoe, M. (eds.) Digital Processing of Biomedical Images, pp. 187–202. Springer, New York (1976)
2. Nagao, M., Matsuyama, T.: Edge preserving smoothing. Comput. Graph. Image Process. **9**, 394–407 (1979)
3. Saint-Marc, P., Chem, J.-S., Medoni, G.: Adaptive smoothing: a general tool for early vision. IEEE PAMI **13**(6), 514–529 (1991)
4. Vincent, L.: Morphological grayscale reconstruction in image analysis: applications and efficient algorithms. IEEE Trans. Image Process. **2**(2), 265–272 (1993)
5. Wheeler, M.D., Ikeuchi, K.: Iterative smoothed residuals: a low-pass filter form smoothing with controlled shrinkage. IEEE PAMI **18**(3), 334–337 (1996)
6. Tomasi, C., Manduchi, R.: Bilateral filtering for gray and color images. In: Proceedings of the IEEE ICCV (International Conference on Computer Vision), Bombay, India (1998)
7. Barash, D.: A fundamental relationship between bilateral filtering, adaptive smoothing, and the non-linear diffusion equation. IEEE PAMI **24**(6), 844–847 (2002)
8. Soille, P.: Morphological Image Analysis - Principles and Applications. Springer, Heidelberg (2004)
9. Hong, V., Palus, H., Paulus, D.: Edge preserving filters on color images. In: Bubak, M., Albada, G.D., Sloot, P.M.A., Dongarra, J. (eds.) ICCS 2004. LNCS, vol. 3039, pp. 34–40. Springer, Heidelberg (2004). doi:10.1007/978-3-540-25944-2_5
10. Grazzini, J., Soille, P.: Edge-preserving smoothing using a similarity measure in adaptive geodesic neighbourhoods. Pattern Recogn. **42**, 2306–2316 (2009)

Quasi-Flat Zones for Angular Data Simplification

Erchan Aptoula[1], Minh-Tan Pham[2(✉)], and Sébastien Lefèvre[2]

[1] Institute of Information Technologies - Gebze Technical University,
41400 Kocaeli, Turkey
[2] IRISA - Université Bretagne Sud, UMR 6074, 56000 Vannes, France
`minh-tan.pham@irisa.fr`

Abstract. Quasi-flat zones are based on the constrained connectivity paradigm and they have proved to be effective tools in the context of image simplification and super-pixel creation. When stacked, they form successive levels of the α- or ω-tree powerful representations. In this paper we elaborate on their extension to angular data, whose periodicity prevents the direct application of grayscale quasi-flat zone definitions. Specifically we study two approaches in this regard, respectively based on reference angles and angular distance computations. The proposed methods are tested both qualitatively and quantitatively on a variety of angular data, such as hue images, texture orientation fields and optical flow images. The results indicate that quasi-flat zones constitute an effective means of simplifying angular data, and support future work on angular tree-based representations.

Keywords: Quasi-flat zones · Image partition · Image segmentation · Connectivity · Orientation field · Hue · Optical flow

1 Introduction

Although flat zones [1], i.e. connected image regions of constant pixel intensity, represent semantically homogeneous image areas and are thus invaluable for segmentation purposes, they almost always produce oversegmented or too fine image partitions. That is why, there have been multiple attempts at relaxing the pixel connectivity criterion from as early as the 1970s, thus leading to a variety of solutions such as *jump connections* [2] and *quasi-flat zones* [3], capable of hierarchical image partitioning and multiscale image representation [4]. Given their potential in terms of image simplification and super-pixel creation, quasi-flat zones in particular have been studied extensively and multiple definitions have been elaborated with varying degrees of flexibility and efficiency [5–8]. Specifically, Soille [7] has provided a solution based on both local and global pixel intensity variation criteria leading to unique image partitions. Angulo and Serra

This work was supported by the BAGEP Award of the Science Academy and by the Tubitak Grant 115E857.

J. Angulo et al. (Eds.): ISMM 2017, LNCS 10225, pp. 342–354, 2017.
DOI: 10.1007/978-3-319-57240-6_28

have explored the application of quasi-flat zones in color image segmentation [9], while Aptoula et al. [8] focused on color image simplification. Crespo et al. [10] and Weber et al. [11] have investigated the use of quasi-flat zones in the context of region merging and interactive video segmentation, respectively. Furthermore, Aptoula [12] has recently explored the impact of multivariate quasi-flats zones on the simplification of hyperspectral remote sensing images.

In the light of the success of quasi-flat zones with the simplification and segmentation of grayscale and color images, here we focus on their definition for angular data. In particular, angular images can be encountered in the image processing community in various forms, such as the hue channel of color images in polar color spaces [13], oriented textures [14] as well as optical flow datasets [15]. However, the periodicity of angular values most often prevents the direct application of graylevel solutions and instead demands custom definitions. That is why, in this paper we concentrate specifically on the definition of quasi-flat zones for angular data. As a matter of fact, we investigate two approaches, respectively based on angular distances and reference angles. Besides elaborating on their theoretical and practical properties, we also put them to qualitative and quantitative test with the hue channels from the Berkeley dataset, texture orientation fields from Outex database, as well as with optical flow images.

The rest of the paper is organized as follows. Section 2 provides background information on quasi-flat zones, while Sect. 3 explains the proposed quasi-flat zone definitions for angular data. Experiment results are provided in Sect. 4, and Sect. 5 is devoted to concluding remarks.

2   Background

Let $f : E \to T$ be a digital image, where E is its definition domain, the discrete coordinate grid (usually $\mathbb{N}^2$ for 2D images) and T the set of possible pixel values (for instance a subset of $\mathbb{R}$ or $\mathbb{Z}$). A path $\pi(p, q)$ between two pixels $p, q \in E$ in this case, is a sequence of $n > 1$ pixels $(p = p_1, \ldots, p_n = q)$ such that any two successive pixels of the said sequence are adjacent (for instance w.r.t. 4- or 8-adjacency). Moreover, any two pixels are said to be connected if there exists a path between them; while the connected component associated with a pixel p is the set of pixels $C(p)$ containing p and all those connected to p. Once one starts taking into account the intensity $f(p) \in T$ associated with each pixel p, various custom connectivity relations can be defined, based for instance on pixel-wise intensity dissimilarity (or based on any other pixel attribute, e.g. color purity, texture orientation etc.):

$$\forall p, q \in E, \; d(p, q) = \|f(p) - f(q)\| \tag{2.1}$$

where $\|\cdot\|$ is a norm. In this case, a couple of pixels p, q are said to be α-connected if there exists a path $\pi(p, q)$ between them where the maximal dissimilarity between any couple of successive pixels on the said path is below a certain threshold α, thus leading to the definition of α-connected components:

$$C^\alpha(p) = \{p\} \cup \{q \mid \widehat{d}(p,q) \le \alpha\} \tag{2.2}$$

$$\widehat{d}(p,q) = \bigwedge_{\pi \in \Pi} \left\{ \bigvee_{i \in [1,\dots,N_\pi - 1]} \{d(p_i, p_{i+1}) \mid \langle p_i, p_{i+1} \rangle \in \Pi\} \right\} \tag{2.3}$$

with Π being the set of all possible paths between p and q and N_π the length of path π. Note that flat zones are a particular case of C^α where $\alpha = 0$. Moreover, a crucial property of α-connected components concerns their hierarchy w.r.t. the value of α:

$$\forall \alpha' \le \alpha, \; C^{\alpha'}(p) \subseteq C^\alpha(p) \tag{2.4}$$

in other words, for increasing values of α, the α-connected component associated with a pixel p is guaranteed to contain the previous ones, leading to the so-called α-tree representations [16,17]. This principle has formed the basis of several extensions. As a matter of fact, early examples employing Eq. (2.2) go as back as the late 70s [5,6], followed by various significant extensions and modifications, aiming to remedy inconveniences such as the chaining effect and the non-uniqueness of the resulting image partitions. Both of these inconveniences have been resolved by Soille's [7] $C^{\alpha,\omega}$ quasi-flat zones (Fig. 1), where besides α, the so-called local variation criterion, an additional global variation criterion ω is employed:

$$C^{\alpha,\omega}(p) = \max\{C^{\alpha'}(p) \mid \alpha' \le \alpha \text{ and } R(C^{\alpha'}(p)) \le \omega\} \tag{2.5}$$

where $R(C^\alpha)$ is the maximal dissimilarity between the intensities (or some other alternative attribute) of any two pixels of C^α. Consequently, $C^{\alpha,\omega}$ of a pixel p is the widest $C^{\alpha'}$ (i.e. built with the highest $\alpha' \le \alpha$ thanks to property (2.4)) where the maximal inter-pixel dissimilarity is less than or equal to ω. Examples of $C^{\alpha,\omega}$ are shown in Fig. 1. Furthermore, a framework unifying the various quasi-flat zones definitions under the concept of logical predicates has also been proposed by Soille in [18].

Considering its positive theoretical and practical properties, in the rest of the paper we concentrate exclusively on $C^{\alpha,\omega}$ in order to realize its extension to angular images.

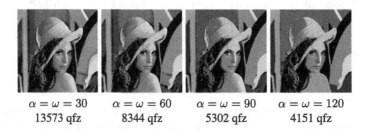

| $\alpha = \omega = 30$ | $\alpha = \omega = 60$ | $\alpha = \omega = 90$ | $\alpha = \omega = 120$ |
| 13573 qfz | 8344 qfz | 5302 qfz | 4151 qfz |

Fig. 1. Examples of $C^{\alpha,\omega}$ on 256×256 Lenna (8-bit) for various local and global criteria.

3   Angular Quasi-Flat Zones

The constant diversification of image acquisition means has led to various distinct pixel data types, ranging from scalars (e.g. grayscale images), to vectors (e.g. color, multispectral, hyperspectral images) and even tensors (e.g. diffusion tensor MRI images). Angular images in particular or in other words images where each pixel p represents an angle $f(p) \in [0, 2\pi]$, are often encountered in practice as auxiliary sources of information (Fig. 2). For instance, in the case of color images, polar (or phenomenal) color spaces such as HSV, HLS and their derivatives, describe color in terms of luminance, saturation and hue triplets; the last component is indeed an angular value (Fig. 2a). Other sources of angular content include oriented textures (Fig. 2b), where pixel values represent the local orientation and optical flow images, where each pixel is characterized by a flow vector possessing both a magnitude and an orientation (Fig. 2c).

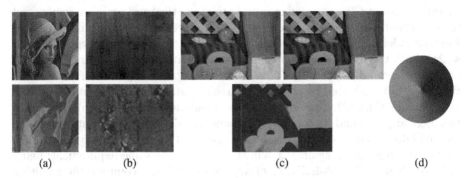

(a)   (b)   (c)   (d)

Fig. 2. Examples of angular images: (a) Lenna color image and (colored) hue channel, (b) wood texture and orientation field, (c) 2 successive frames and the related optical flow orientation, and (d) the color coding used to represent angular data. (Color figure online)

Although they constitute a rich source of information, the processing of angular images on the other hand is a not a straightforward issue. More precisely, the inherent periodicity of angular data leads to a discontinuity at the origin which very often prevents the direct application of image analysis techniques, that otherwise work perfectly with standard grayscale images, thus rendering it imperative to develop custom solutions adapted to this type of data.

Quasi-flat zones are no exception to this situation. Since any two pixels of similar angular value $f(p) = 2\pi - \varepsilon$ and $f(q) = 2\pi + \varepsilon$ located at opposing sides of the origin are bound to be placed in distinct α-connected components, thus leading to severe discontinuities. This can be observed in Fig. 3 with the Lenna image, which possesses a mostly reddish hue content (Fig. 3b). Hence, applying $C^{\alpha,\omega}$ directly on its hue channel leads to visually very poor quasi-flat zone results, see Fig. 3c and d.

Given these inconveniences we present in this section two different approaches of computing quasi-flat zones for angular images, that avoid effectively the aforementioned discontinuity problem.

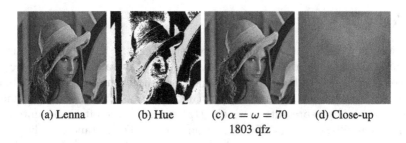

(a) Lenna   (b) Hue   (c) $\alpha = \omega = 70$ (d) Close-up
                                        1803 qfz

Fig. 3. Example of applying $C^{\alpha,\omega}$ directly on the hue channel of a color image, followed by setting each quasi-flat zone to its mean hue. (Color figure online)

3.1   Angular Distance Based Approach

In order to adapt $C^{\alpha,\omega}$ to angular data, both theoretical and practical issues need to be resolved. Let us focus first on the theoretical requirements of this extension. Plus, considering that we now deal with angular images of the type $f : E \to [0, 2\pi]$, both the local α and global ω variation criteria represent arc lengths. If one studies carefully Eq. (2.5) of $C^{\alpha,\omega}$, two data dependent parts can be observed that need to be adapted to processing angular data: (i) the computation of C^{α} by means of Eq. (2.2), that requires the calculation of the dissimilarity d_θ of two angular pixels and its comparison against α; (ii) and the computation of the maximal dissimilarity $R(C^{\alpha})$ between all pixels of C^{α}, which once again requires d_θ, and its comparison against ω. These are in fact the same requirement: being able to calculate the dissimilarity of any two angles and compare the resulting distance against a predefined distance such as α and ω. However, since the $d_\theta : [0, 2\pi]^2 \to [0, \pi]$ of any two angles is in fact handled as an arc length, its comparison against α and ω is trivial; which leaves as sole requirement the definition of d_θ. To this end we adopt the solution in Refs. [13,14], where angular distances have been employed as means for establishing a lattice structure on the hue circle of color images:

$$\forall\, h, h' \in [0, 2\pi],\ \ d_\theta(h, h') = \begin{cases} |h - h'| & \text{if} \quad |h - h'| < \pi \\ 2\pi - |h - h'| & \text{if} \quad |h - h'| \geq \pi \end{cases} \tag{3.1}$$

Consequently, we can proceed to define α-connected components for angular data (C_θ^{α}) merely by replacing the distance expression of C^{α} with its angular counterpart:

$$C_\theta^{\alpha}(p) = \{p\} \cup \{q \mid \widehat{d_\theta}(p, q) \leq \alpha\} \tag{3.2}$$

Thus, we reach the angular $C_\theta^{\alpha,\omega}$:

$$C_\theta^{\alpha,\omega}(p) = \max\{C_\theta^{\alpha'}(p) \mid \alpha' \leq \alpha \text{ and } R_\theta(C_\theta^{\alpha'}(p)) \leq \omega\} \tag{3.3}$$

where R_θ represents the maximal distance d_θ between any two pixels contained in C_θ^{α}. Nevertheless, although $C_\theta^{\alpha,\omega}$ is theoretically sound, there are additionally two practical issues that require consideration: (i) The amount $\alpha - \alpha'$ by which α is going to be decreased each time the global variation criterion is not satisfied, (ii) and the computation efficiency of R_θ. To explain, when dealing with grayscale images, if the $\alpha \in [0, 255]$ argument of $C^{\alpha, w}$ leads to an α-connected component C^α that does not verify the global variation criterion w, it is the immediately smaller value $\alpha' = \alpha - 1$ that is employed in its place. With real values however, representing either angular or some other form of data, it is no longer possible to simply select the next smaller value[1]. Consequently, when the initial α value in Eq. (3.3) leads to an α-connected component C_θ^α that does not verify the global variation criterion w, a fixed decrementation step $\beta \ll \alpha$ becomes necessary. Thus, if α fails to verify the global criterion, we try next $\alpha' = \alpha - \beta$. Naturally, a large β value will result in faster quasi-flat zone computations, since less attempts will be made in order to determine the α' value that verifies the global criterion, while on the other hand the said α' value will be a poorer approximation w.r.t. using a smaller β value. A more effective solution for this problem could be to set α' to the smallest dissimilarity $d_{\theta, min}$ between the pixels of C_θ^α, since the α' values such that $d_{\theta, min} \leq \alpha' < \alpha$ will not modify the content of C_θ^α and thus have no effect on the global variation criterion.

The second practical issue concerns the computation of $R_\theta(C_\theta^\alpha)$. To explain, when dealing with standard grayscale images, the calculation of $R(C^\alpha)$ is realized with a linear complexity w.r.t. the number of pixels in C^α. Each time a pixel is added into C^α, it suffices to compare it only against the pre-calculated maximum and minimum values, i.e. an operation of constant complexity. However, when it comes to angular images this strategy is no longer applicable, since the new maximal angular distance might be in fact between any angle couple and not necessarily between an angle and one of the previous extrema (Fig. 4). The naive approach of calculating $R_\theta(C_\theta^\alpha)$ would consist in cross-comparing all pixels of C_θ^α every time a new pixel is added, which implies a complexity $O(n^3)$ in terms of pixels within C_θ^α and is practically unacceptable. An improvement would be to preserve the maximal dissimilarity of C_θ^α (a trivial computation for 1 and 2 pixels) and each time a new pixel q arrives into C_θ^α, one would need to compare only q against the pixels of C_θ^α and update the maximal dissimilarity if q possesses a greater distance to C_θ^α, thus reducing complexity to $O(n^2)$.

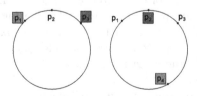

Fig. 4. The most distant angles are illustrated in red & blue. On the right, it is shown that the most distant couple does not have to contain one of the previous extrema p_1 and p_3. (Color figure online)

---

[1] Let us observe however that this issue can be overcome with algorithms specifically designed to build tree-based representations for high-depth data.

Further acceleration can be achieved if we search for the most distant pixel to q within C_θ^α with a binary search. However, in order to avoid the discontinuity at the origin, which prevents the classic application of the binary search algorithm, one can divide the interval $[0, 2\pi]$ into two bins $[0, \pi]$ and $]\pi, 2\pi]$ and place all pixels of C_θ^α into their respective bins, where in each bin they are kept within a sorted data structure. A scalar sort is feasible since the individual bins will not contain the discontinuity at the origin. Next, each time a new pixel q arrives into C_θ^α, we locate the pixel of maximal distance to q with a binary search within the opposing bin to that of q. If the opposing is empty, it suffices to compare against the extreme pixels (that are known thanks to the sorted data structure) of the remaining bin. Of course this is by no means an optimal solution. Besides, as this is the first paper on angular quasi-flat zones, it focuses rather on their feasibility, with computation efficiency being a future work topic.

Figure 5 shows the quasi-flat zone results obtained for Lenna using various local and global variation criteria. Differences w.r.t. a mere grayscale application (Fig. 3) are emphasized at hue discontinuity regions.

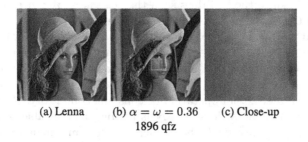

(a) Lenna   (b) $\alpha = \omega = 0.36$ (c) Close-up

1896 qfz

Fig. 5. Example of applying $C_\theta^{\alpha,\omega}$ (with $\beta = 10^{-4}$) on Lenna's hue channel by setting each quasi-flat zone to its mean hue. Spatial inconsistencies from Fig. 3d have been removed.

3.2 Reference Angle Based Approach

A relatively simpler and more efficient way of computing quasi-flat zones from angular data is to first convert the input into a grayscale image and then apply the efficient algorithm of $C^{\alpha,\omega}$ from Eq. (2.5), as already done with color images [8]. The conversion process can be realized using the angular distance of Eq. (3.1). More precisely, we associate each angular pixel value $f(q) \in [0, 2\pi]$ with its distance $d_\theta(\theta_{ref}, f(q))$ to a reference angle $\theta_{ref} \in [0, 2\pi]$, which leads to a grayscale image of distances. Distance images of Lenna for various reference hues are shown in Fig. 6.

Once the grayscale distance image is obtained, the computation of quasi-flat zones by means of $C^{\alpha,\omega}$ is straightforward. Figure 7 shows the quasi-flat zones obtained for $\theta_{ref} = 0.0$. As the results appear visually very similar, if not identical to the previously presented angular distance based approach, one is tempted to adopt the reference based approach for simplicity's sake. Yet, there are serious differences between the two methods. For instance, this one

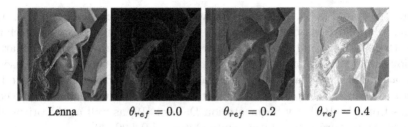

| Lenna | $\theta_{ref} = 0.0$ | $\theta_{ref} = 0.2$ | $\theta_{ref} = 0.4$ |

Fig. 6. Angular distance images of Lenna's hue channel for various reference values.

requires not only a reference angle value, but if we are to employ the efficient implementations of $C^{\alpha,\omega}$ then it also subquantizes the resulting pixel values into a range of integers as well, instead of dealing with real-valued angles.

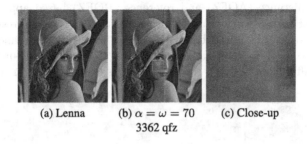

(a) Lenna   (b) $\alpha = \omega = 70$ (c) Close-up
3362 qfz

Fig. 7. Example of applying $C^{\alpha,\omega}$ on the distance image of Lenna's hue channel using $\theta_{ref} = 0.0$, followed by setting each quasi-flat zone to its mean hue.

Moreover, how can one choose the reference angle? A similar question is encountered during the morphological processing of hue and has been explored in details in [13]. Multiple strategies are available. Since setting it at some arbitrary value can be deemed as a last resort, an alternative is to employ as reference the most dominant angle of the input image, or even better, if multiple dominant angles are present, take them all into account. This is however a direction for future research and we will not focus on it here. In short, concerning the reference angle based approach, the following questions arise: What is the effect of the reference angle choice on the quality of the resulting image partitions? Does the performance of the proposed two approaches depend on data type (e.g. hue, orientation field, etc.)? And of course, which one is better suited for image simplification? We now proceed to a series of experiments that will help us answering these questions.

4   Experiments

We present here a series of experiments conducted with the main goal of comparing the proposed angular quasi-flat zone approaches against each other and more precisely in order to answer the questions raised at the end of the previous section. As stated in [7], image simplification and super-pixel creation constitute the main applications of quasi-flat zones; that is why they have been chosen for both qualitative and quantitative performance comparison. In particular, we employ three datasets of angular content; for qualitative comparison we use the orientation of the optical flow images made available by Baker et al. [15] and for quantitative image simplification evaluation we use the hue channel of the color images from the Berkeley Segmentation Dataset [19] as well as the orientation fields of the texture segmentation suite Outex-SS-00000 [20].

4.1 Orientation of Optical Flow

The dataset provided in [15] contains the optical flow ground truth of eight images, out of which we picked three for qualitative comparison purposes. We first computed the orientation images of the provided ground-truth and then both angular distance (AQFZ) and reference (RQFZ) based approaches have been computed with distinct α and ω arguments, so as the final results possess comparable quasi-flat zone numbers. We set $\beta = 10^{-4}$ for AQFZ and $\theta_{ref} = 0.0$ for RQFZ. The colored results are shown in Fig. 8, as well as zoomed versions to highlight differences between the two methods occuring at a finer scale.

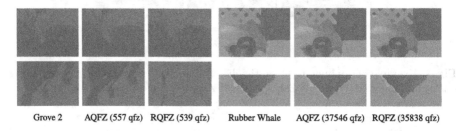

Grove 2   AQFZ (557 qfz)   RQFZ (539 qfz)   Rubber Whale   AQFZ (37546 qfz)   RQFZ (35838 qfz)

Fig. 8. Colored results (w.r.t. color coding in Fig. 2) of computing angular distance (AQFZ) and reference (RQFZ) based quasi-flat zones from the orientation of optical flow data (top line), and related close-ups (bottom line). (Color figure online)

Based on Fig. 8, both AQFZ and RQFZ appear at first to provide results of similar visual quality. However, upon a closer inspection of Grove 2 image, it becomes evident that for a comparable number of quasi-flat zones, RQFZ has a less reliable behavior than AQFZ, since it leads to serious under-segmentation. On the hand, we observe RQFZ to preserve a higher level of detail for Rubber Whale image at Fig. 8. As far as AQFZ is concerned, its results appear to be overall spatially and spectrally consistent. We consider the next datasets for a more healthy comparison.

4.2 Hue Channels

Following the preliminary qualitative evaluation with the orientation of optical flow images, we now focus on a quantitative evaluation of AQFZ and RQFZ through the segmentation of the hue channel of the images contained in the

Berkeley dataset [19]. In detail, it contains 300 color images of size 481×321 pixels and 3269 reference segmentations, realized by 28 distinct experts. The relatively high number of reference segmentation maps per image enables a more objective comparison of our results against real-world practical requirements.

Moreover, in order to evaluate the image partitions produced by either method, two criteria will be employed: the over-segmentation ratio (OSR) [21] and maximal precision (MP) [22]. The former is defined as:

$$OSR = \frac{\text{number of quasi-flat zones}}{\text{number of reference regions}} \tag{4.1}$$

This ratio expresses directly the degree of over-segmentation. In fact, in a way it provides us with the merging degree required for achieving the closest possible segmentation to the reference.

MP or maximal precision on the other hand, focuses on pixel-based distances between the reference segmentation and the quasi-flat zones. More precisely, each quasi-flat zone is associated with the reference region it shares the highest number of pixels. Consequently, one can compute for each quasi-flat zone the degree with which it overlaps with the desired reference region. Thus, one can calculate a confusion matrix C, where C_{ij} represents the number of pixels affected to reference region i, while in fact those pixels belong to reference region j. Formally:

$$MP = \frac{\Sigma_{i=1}^{\text{number of reference regions}} C_{ii}}{\text{number of pixels}} \tag{4.2}$$

For instance $MP = 0.8$ means that we have achieved 80% pixel-based accuracy w.r.t. the ground-truth. Hence, by using both MP and OSR one can effectively evaluate a segmentation result, with the ultimate goal being the minimization of OSR and the maximization of MP. In order to achieve this goal, we computed both AQFZ ($\beta = 10^{-4}$) and RQFZ for all the available images, using a wide interval of values for both α and ω and computed both MP and OSR for all reference segmentation maps, that were then averaged. As far as the reference angle value of RQFZ is concerned, three distinct hues have been employed (red, yellow and green). The resulting plot is shown in Fig. 9a.

Judging from the results, one can assert the overall superiority of AQFZ with respect to RQFZ, both in terms of MP and OSR. This difference can be theoretically explained by the fact that AQFZ deals with the actual real-valued angles, thus having access to a higher resolution of α' values when searching for the greatest $\alpha' \leq \alpha$ that verifies the global variation criterion. While RQFZ on the other hand, transforms its input first into grayscale distance images and thus loses precision prior to realizing grayscale quasi-flat zone computations. Moreover, according to Fig. 9a one can also observe that the choice of reference angle has indeed a non-negligible effect on segmentation performance. The choice of an optimal reference angle however is beyond the scope of this paper.

Furthermore, despite its shortcomings RQFZ has a distinct advantage over AQFZ in terms of execution speed. The average computation time of AQFZ and RQFZ across all images of the Berkeley dataset is 5552 ms and 400 ms,

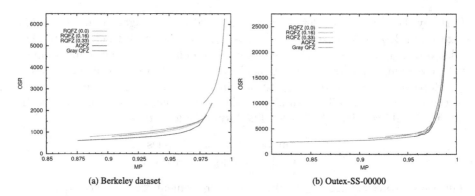

(a) Berkeley dataset   (b) Outex-SS-00000

Fig. 9. OSR-MP plot for the QFZ results of the Berkeley and Outex-SS-00000 dataset. The reference angle in use is indicated between parentheses. (Color figure online)

respectively. Experiments were carried out on a 2.3 GHz system with 3GB of memory. Apparently, the relative superiority of AQFZ comes at a very steep computational cost. All the same, as these are the first results on angular quasi-flat zone computation, we are confident that AQFZ can be rendered more efficient, especially if embedded within a tree framework.

4.3   Texture Orientation Fields

As a further source of angular images we have used additionally the orientation field of textures. In particular, the Outex-SS-00000 [20] supervised segmentation test suite contains 100 images, such as the one shown in Fig. 10a, each composed of five distinct textures acquired at various illumination and rotation conditions. Given this grayscale data, for each texture we computed its orientation field according to [23], and subsequently AQFZ and RQFZ similarly as previous experiment. The ground-truth of the segmentation suite being available, the MP and OSR plot that has been obtained is shown in Fig. 9b.

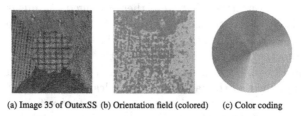

(a) Image 35 of OutexSS (b) Orientation field (colored)   (c) Color coding

Fig. 10. A texture example of OutexSS and its orientation field.

Although the differences are not as emphasized as in the case of hue images, the relative performances are reproduced. Namely, AQFZ is once again superior to RQFZ especially in terms of MP regardless of RQFZ's reference angle, while RQFZ appears once more to be sensitive to the choice of reference angle. The reproduction of the relative performances with two distinct datasets and additionally with distinct image types (hues and orientation fields) constitute a strong indication in favor of AQFZ, even though computationally speaking it is significantly disadvantaged w.r.t. RQFZ.

5   Conclusion

In this paper, we have adapted the existing definitions of quasi-flat zones to angular data. Both theoretical and practical issues have been addressed. Two distinct approaches have been introduced: angular distance based and reference angle based methods. The former relies on arc lengths for local and global variation comparisons while the latter is based on converting the angular images into grayscale. Following the experimental study using three angular image types, it has been shown that the strong potential of quasi-flat zones as image simplification tools, encompasses angular data as well.

Nevertheless, although the proposed methods can both resolve the discontinuity issues arising when using directly the grayscale methods on angular data, they have significant differences. In detail, the angular distance based approach has been observed to outperform its counterpart with respect to both segmentation criteria that have been employed, while its computational cost and implementation complexity remain much higher. The reference angle based approach on the other hand is simple to implement and efficient, yet it requires setting a reference angle, which is unclear how to realize optimally at this point.

Having established the groundwork for computing quasi-flat zones from angular images, future work will continue on three directions. First and foremost, the angular distance based approach has a serious efficiency problem, which we believe can be resolved with more advanced data structures. Moreover, although the reference angle based method is a fast and relatively effective solution, the need for a reference angle is another issue that needs attention. And last, given the work on color (vectorial) quasi-flat zones, the combination of both orientation and magnitude of optical flow pixels during quasi-flat zone computation is another direction worth exploring.

Let us also observe that having brought some first definitions of constrained connectivity on angular data, we can now process such data using multiscale tree based representations, i.e. hierarchies of partitions such as α-tree and ω-tree (but also binary partition tree). The availability of efficient algorithms to build such representations from scalar values [24] will certainly ease tackling computational issues raised in this study, and demonstrate the potential of multiscale representations for angular data.

References

1. Salembier, P., Serra, J.: Flat zones filtering, connected operators, and filters by reconstruction. IEEE Trans. Image Proc. **4**, 1153–1160 (1995)
2. Serra, J.: Connectivity for sets and functions. Fundamenta Informaticae **41**, 147–186 (2000)
3. Meyer, F.: The levelings. In: Proceedings of ISMM (1998)

4. Cousty, J., Laurent, N., Yukiko, K., Silvio, G.: Hierarchical segmentations with graphs: quasi-flat zones, minimum spanning trees, and saliency maps. Technical report, LIGM (2016)
5. Nagao, M., Matsuyama, T., Ikeda, Y.: Region extraction and shape analysis in aerial photographs. Comp. Graph. Image Proc. **10**, 195–223 (1979)
6. Hambrusch, S., He, X., Miller, R.: Parallel algorithms for gray-scale digitized picture component labeling on a mesh-connected computer. J. Parallel Distrib. Comput. **20**, 56–58 (1994)
7. Soille, P.: Constrained connectivity for hierarchical image partitioning and simplification. IEEE Trans. PAMI **30**, 1132–1145 (2008)
8. Aptoula, E., Weber, J., Lefèvre, S.: Vectorial quasi-flat zones for color image simplification. In: Hendriks, C.L.L., Borgefors, G., Strand, R. (eds.) ISMM 2013. LNCS, vol. 7883, pp. 231–242. Springer, Heidelberg (2013). doi:10.1007/978-3-642-38294-9_20
9. Angulo, J., Serra, J.: Color segmentation by ordered mergings. In: Proceedings of ICIP (2003)
10. Crespo, J., Schafer, R., Serra, J., Gratin, C., Meyer, F.: The flat zone approach: a general low-level region merging segmentation method. Signal Process. **62**, 37–60 (1997)
11. Weber, J., Lefèvre, S., Gançarski, P.: Interactive video segmentation based on quasi-flat zones. In: Proceedings of ISPA (2011)
12. Aptoula, E.: The impact of multivariate quasi-flat zones on the morphological description of hyperspectral images. Int. J. Remote Sens. **35**(10), 3482–3498 (2014)
13. Aptoula, E., Lefèvre, S.: On the morphological processing of hue. Image Vis. Comput. **27**, 1394–1401 (2009)
14. Hanbury, A., Serra, J.: Morphological operators on the unit circle. IEEE Trans. Image Process. **10**, 1842–1850 (2001)
15. Baker, S., Lewis, D.S.J.P., Roth, S., Black, M., Szeliski, R.: A database and evaluation methodology for optical flow. Int. J. Comp. Vis. **92**, 1–31 (2011)
16. Ouzounis, G.K., Soille, P.: Pattern spectra from partition pyramids and hierarchies. In: Soille, P., Pesaresi, M., Ouzounis, G.K. (eds.) ISMM 2011. LNCS, vol. 6671, pp. 108–119. Springer, Heidelberg (2011). doi:10.1007/978-3-642-21569-8_10
17. Ouzounis, G.K., Soille, P.: The alpha-tree algorithm. JRC Scientific and Policy Report (2012)
18. Soille, P.: On genuine connectivity relations based on logical predicates. In: Proceedings of ICIAP (2007)
19. Martin, D., Fowlkes, C., Tal, D., Malik, J.: A database of human segmented natural images and its application to evaluating segmentation algorithms and measuring ecological statistics. In: Proceedings of ICCV (2001)
20. Ojala, T., Mäenpää, T., Pietikäinen, M., Viertola, J., Kyllonen, J., Huovinen, S.: Outex: new framework for empirical evaluation of texture analysis algorithms. In: Proceedings of ICPR (2002)
21. Carleer, A., Debeir, O., Wolff, E.: Assessment of very high spatial resolution satellite image segmentations. Photogram. Eng. Remote Sens. **71**, 1285–1294 (2005)
22. Derivaux, S., Forestier, G., Wemmert, C., Lefèvre, S.: Supervised image segmentation using watershed transform, fuzzy classification and evolutionary computation. Pattern Recogn. Lett. **31**, 2364–2374 (2010)
23. Hanbury, A., Serra, J.: Analysis of oriented textures using mathematical morphology. In: Proceedings of Workshop Austrian Assian for Pattern Recognition (2002)
24. Havel, J., Merciol, F., Lefèvre, S.: Efficient tree construction for multiscale image representation and processing. J. Real Time Image Process. 1–18 (2016)

Computational Morphology

Connected Morphological Attribute Filters on Distributed Memory Parallel Machines

Jan J. Kazemier[1], Georgios K. Ouzounis[2], and Michael H.F. Wilkinson[1]([✉])

[1] Johann Bernoulli Institute, University of Groningen, P.O. Box 407, 9700
Groningen, AK, The Netherlands
jan.kazemier@tno.nl, m.h.f.wilkinson@rug.nl
[2] DigitalGlobe, Inc., 1300 W 120th Ave, Westminster, CO 80234, USA
gouzouni@digitalglobe.com

Abstract. We present a new algorithm for attribute filtering of
extremely large images, using a forest of modified max-trees, suitable
for distributed memory parallel machines. First, max-trees of tiles of the
image are computed, after which messages are exchanged to modify the
topology of the trees and update attribute data, such that filtering the
modified trees on each tile gives exactly the same results as filtering a regular max-tree of the entire image. On a cluster, a speed-up of up to $53\times$
is obtained on 64, and up to $100\times$ on 128 single CPU nodes. On a shared
memory machine a peak speed-up of $50\times$ on 64 cores was obtained.

1 Introduction

Attribute filters are powerful tools for filtering, analysis and segmentation [1,2].
In the binary case attribute filters [1] remove connected components if some
property (or attribute) of the latter, such as area, fails some threshold. This can
be generalised to grey scale by performing the following steps:

- Compute all threshold sets.
- Organize all threshold sets into sets of connected components.
- Compute attribute values for each connected component of each set.
- Apply the binary attribute filter to each connected component set.
- Compute the combined filtered results of all threshold sets.

Components of threshold sets at higher levels can only be subsets of, or equal to
those at lower threshold sets. Using this nesting property along the grey-scale,
the set of all components can be organised into a tree structure referred to as a
max-tree (or min tree if we want to filter background components) [3], as shown
in Fig. 1. Max-trees and min-trees are collectively referred to as component trees
[4]. Once a component tree has been constructed and the attributes for each
node have been computed, the attribute filter computed by removing unwanted
nodes. Component trees are also used for multi-scale analysis, e.g., in the form
of pattern spectra and morphological profiles [5–7].

The most time consuming part in attribute filtering and analysis using connected operators is building the max-tree. Many algorithms for computing these

© Springer International Publishing AG 2017
J. Angulo et al. (Eds.): ISMM 2017, LNCS 10225, pp. 357–368, 2017.
DOI: 10.1007/978-3-319-57240-6_29

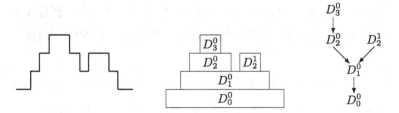

Fig. 1. A 1-D signal f (left), the corresponding peak components (middle) and the Max-Tree (right). Figure from [8].

trees exist [3,9,10]. A shared-memory, parallel algorithm for creating max-trees has been proposed [8]. It relies in building separate max-trees for disjoint parts of the image, using a variant of the algorithm from [3]. These trees are subsequently merged into a single max-tree. This approach was extended in [11] to computing differential morphological profiles. The key problem limiting the algorithm to shared-memory machines, and limiting the size of images that can be processed is the fact that the entire max-tree needs to be stored in memory. Though efficient, it cannot handle images of tens of gigapixels, or even terapixels, such as those collected in remote sensing, or ultramicroscopy.

These very high-resolution images require a different approach of building the max-tree, which does not require the entire tree to be stored in memory. In this paper we develop an algorithm that computes a forest of max-trees for disjoint tiles of the input image, and then modifies these trees in such a way that filtering them results in the same output as filtering the whole tree. To be able to use distributed memory or hybrid memory model machines, an implementation of this new algorithm using MPI is given. A key part of the proposed solution is a structure called a *boundary tree*, which allows efficient communication between the processes; passing all required information, whilst keeping communication to a minimum.

The rest of this paper is organized as follows: Sect. 2 discusses our novel algorithm, and results are discussed in Sect. 3. The conclusion and an overview of future work are given in Sect. 4.

2   The Proposed Algorithm

In this section the design of the distributed memory parallel solution for filtering connected components based on their attribute(s) is described. In the following discussion we will focus on max-trees, but min-trees can be built equivalently.

The key idea is the following: Assuming we have a very large image stored as a series of disjoint tiles, we can of course compute a max-tree for each tile using any suitable algorithm, and merge them using the merging algorithm in [8]. However, rather than passing the entire max-tree to a neighbouring tile for merger, we only select the necessary nodes, i.e. those that correspond to connected components that touch the boundary of the tile. Any component that does not touch the boundary cannot possibly be changed as a result of a merger. Therefore we create a subtree of the max-tree called the boundary-tree, effectively pruning away all branches of the full max-tree that do no touch the tile boundary. We can then merge the two boundary-trees of neighbouring tiles, by passing the boundary-tree of one processor to its neighbour. As a boundary-tree is essentially a pruned max-tree, we can use the same merging algorithm as in [8], where we use an extra flag to indicate which nodes have changed during merger. Only these nodes in the two involved max-trees (not boundary-trees!) are then updated, in terms of changed attributes and parent relationships. For a two-tile situation, the entire process can be summarised as:

1. Load the grey scale image tiles from disk.
2. Compute local max-trees for both parts.
3. Extract boundary-trees from max-trees
4. Merge the boundary-trees
5. Use the changed nodes in the merged boundary-tree to correct each of the local max-trees.

For more than two tiles, we assume the image is cut into a $P = a \times b$ grid, with P the number of processes, and a and b powers of 2. As before, each processor performs the single-processor algorithm on its tile. We end up with P max-trees. The boundary-tree is sent to the appropriate adjacent processor, which merges the boundary-trees, whilst updating the attribute(s). After merging, the corrected attributes and parent-child relations are propagated back to the corresponding processors. Once this update is completed, a boundary-tree of the merged region is created. This process repeats until the boundary-tree of the entire images has been formed. After this, the same filtering step as used in the shared memory program can be used, per processor, on its own tile, using the updated attribute(s) and topology.

Note that the order of the message size is $\mathcal{O}(G\sqrt{N})$ in this scheme, since we only need to send the border plus ancestors, rather than $\mathcal{O}(N)$ if we send full max-trees. The number the message steps is $\mathcal{O}(\log P)$. For an 8 bit-per-pixel $40,000^2$ tile this is a savings of at least a factor of 78 in communication and memory overhead.

The following sub-sections describe each of the steps in detail.

2.1 Building the Max-Tree

For each tile the max-tree is built using the same algorithm as used in the shared memory parallel solution from [8], albeit some differences in how the data is structured. In our solution all data is confined in MaxNodes as much as possible.

The max-tree is represented as a one-dimensional array of MaxNodes, which is a mapping to the 2D image, i.e. every node represents a pixel in the image (plus extra information) and vice versa. A node in the max-tree, called a MaxNode consist of the following data:

- `index`, its index in the array.
- `parent`, the index of its parent.
- `area`, the area of all child-nodes pointing to this node. This is the attribute we filter on.
- `gval`, the grey value, or intensity, of the pixel.
- `filter`, the grey value of the pixel after filtering.
- `flags`, boolean flags that are set during execution.
- `borderindex`, after being put in the boundary-tree, this is the index thereof.
- `process`, the rank of the processor the initial max-tree is built on. This is used when sending back the updated attributes.

Only MaxNodes with a parent at a lower grey level represent a node in the max-tree. Such nodes are referred to as *level-roots*. Because we store the tree in an array, pointers are replaced by indices. The root of a max-tree (or boundary tree) has a parent field with a special index denoted ⊥.

The flags field is a series of bits, of which just three are used:

- `Reached`, whether the flood tree algorithm has already reached this node
- `Added`, whether this node is already in the border; to avoid duplicates
- `Changed`, whether this node is changed in a round of merging; to prevent overhead

The algorithm used to build the tree is exactly the same as that in [8], so for space considerations it is not reproduced here.

2.2   The Boundary-Tree

Once all local max-trees have been computed, a *boundary-tree* is created. This structure consists of a one-dimensional array, containing all max-tree nodes in the border, and all of their ancestors, i.e. parents through to the root, with added metadata. The boundary-tree is structured as follows:

- array; the array of all max-tree nodes in the border
- offset; an array of five offsets: for north, east, south, west and ancestors respectively; north is at always 0, just for convenience
- size; the number of elements in the boundary
- borderparent; an array with the index of this node in the boundary-tree and a pointer to which boundary-tree of the parent of this node belongs
- merged; an array of booleans of size P, indicating which processors are merged into this boundary-tree

Building the Boundary-Tree. The algorithm goes as follows: add each side: north, east, south and west. When the width or height of the tile is x, each side contains $x - 1$ nodes, as the final node of that side is contained in the next side (in the first side for the last side). We administer the offsets of the sides in the array when building the tree. The mapping to the one-dimensional array is shown in Fig. 2.

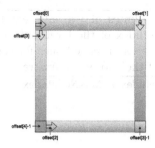

Fig. 2. Mapping of boundary to the 1D array

After the four sides are added, all ancestors of the nodes in the sides are added. The theoretical upper limit of the boundary including ancestors is: $(G - 1) * (L/2)$ where G is the number of grey levels, and L is the length of the border. The worst case occurs only when half the boundary pixels are local maxima, and are at grey level $G - 1$. We allocate the upper limit and shrink afterwards. For each node, we traverse the tree from boundary nodes to root. To make sure we add every node only once, we flag the nodes when adding them. For each ancestor we distinguish three cases:

1. The parent is ⊥, continue with the next node
2. The parent is new, add it to the border, point from the current node to its parent, and continue with the parent
3. The parent is already in the border node, point from the current node to its parent, and continue with the next node

After all ancestors are added we shrink the array to its final size.

2.3   Merging Tiles

The process for merging two tiles starts by sending the boundary-tree of a tile to its neighbour. Merging the nodes is done in a similar fashion to the sequential memory algorithm; but needs some more administration on the place of the nodes: is the node in the boundary-tree or in the tile itself. The process is as follows: for each node in the border, traverse the tree until you are at the bottom. The algorithm is shown in Algorithm 1, which is near identical to that in [8].

While merging two nodes in two trees, there basically are three options.

1. The parent of the current node to merge is ⊥; in this case we are done with the traversal for this branch; continue with the next node in the border.
2. The parent of the current node to merge is not in the merged tree yet; add this node by accumulating the area of this node, and point to its parent; then continue with the parent.
3. The parent of the current node to merge is already in the merged tree. Add the accumulated area of this node to that node, and point to it. Then traverse its parents to add the accumulated area.

Algorithm 1. Merging attributes of two adjacent boundary-trees. Function LEVROOT(x) returns the index of the level-root of x, PAR(x) the level-root of the parent of x. Function FUSE is called once for each pair of adjacent nodes along the common border of the two boundary-trees.

```
procedure FUSE(MaxNodeIndex x, MaxNodeIndex y )
    a ← 0
    b ← 0
    x ← LEVROOT(x)
    y ← LEVROOT(y)
    if GVAL(x) < GVAL(y) then
        SWAP(x,y)
    while x ≠ y ∧ y ≠⊥ do
        z ← PAR(x)
        if z ≠⊥ ∧ GVAL(z) >= GVAL(y) then
            Add a to x
            x ← z
        else
            b ← a+AREA(x)
            a ← AREA(x)
            AREA(x) ← b
            PARENT(x) ← y
            x ← y
            y ← z
    if y =⊥ then                                    ▷ Traverse down x
        while x ≠⊥ do
            AREA(x) ←AREA(x) +a
            x ← PAR(x)
```

We are done when all nodes in the border have been merged with the tree in the tile. We then need to propagate the retrieved information about the area of each merged node back to the tile that corresponds to the boundary-tree.

Alternating Tree Merges. As a connected component can theoretically span multiple tiles, or even all tiles, we have to merge all tiles recursively to obtain the correct attributes. In order to keep the boundary-tree as small as possible, to avoid overhead, we merge the longest edge of the grid first. This process is repeated until all merges have been done.

Synchronizing Attributes. After merging the attributes of the two trees, only the process that has performed the merging has the merged information. We therefore have to synchronise the information to the processes that the information belongs to. We chose to store the process number of each node of the boundary in an array in the boundary-tree. This way, after merging the attributes of two trees, but before combining the two trees to one, we can update all tiles by sending them the nodes. To keep the overhead small, we only send the nodes that have changed.

Combining Two Trees. In order to minimize the overhead when sending the boundary-tree, we grow the boundary-tree to a larger one by combining the nodes from both trees. Note that the nodes in the *middle* of the new tree, as shown in Fig. 3, labeled x and y can be omitted. We combine the trees after merging and syncing their attributes. There is a difference in combining the trees horizontally vs. vertically; as the offsets of the nodes are different.

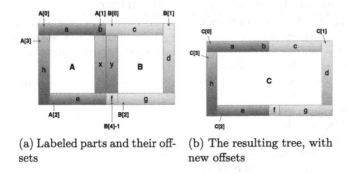

(a) Labeled parts and their off-   (b) The resulting tree, with
sets   new offsets

Fig. 3. Combining two boundary-trees horizontally

When combining two boundary-trees horizontally, we distinguish different parts. The labels are shown in Fig. 3a. An overview of how the nodes are placed in memory, using the labels from Fig. 3a is shown in Fig. 4. The vertical merge is done in an equivalent manner.

Fig. 4. Memory layout of the two boundary-trees A and B, combined in C

Adding Ancestors. After the nodes on the border of the area spanned by two boundary-trees has been extracted and put into a new boundary-tree C, we need to add all relevant ancestors from the two original boundary-trees A and B. This process is very similar to adding ancestors when creating a boundary-tree out of a local max-tree, but now we do not retrieve the nodes from the local max-tree, but from the boundary-trees A and B. The algorithm is shown in Algorithm 2.

For each node in the border of the new boundary-tree we check the original trees A and B for ancestors. We start by checking the origin of each node (that is: the node in A or B, which has been copied to C) for parents. If there is no parent of the current node, we can continue with the next node in the border of C. If there is a parent, which is already in C, we can point there from our current node and we can continue with the next node in the border of C. Otherwise the parent is new, we add it to C and point there from the current node. Then we continue with this parent, and repeat the process.

Algorithm 2. Adding ancestors. ORIGIN(n) indicates for each boundary node in C with which node in input boundary-trees A or B it corresponds; PARENT(n), denotes the parent of n; INDEX(n) returns the index of n.

---

procedure ADDANCESTORS(BoundaryTree A, BoundaryTree B, BoundaryTree C)
    s ← size of boundary of C ▷ Store the initial size of the boundary
    offset[4] ← s   ▷ Store the starting point of the ancestors
    Reserve upper limit of memory
    for all node n in border of C **do**   ▷ Process all nodes
        p ← PARENT(ORIGIN(n))
        curnode ← INDEX(n)
        while true **do**
            if $p = \perp$ **then**
                BORDERPARENT(curnode) ← $\perp$
                Continue with next
            if INBORDER(p) **then**   ▷ Of C
                BORDERPARENT(curnode) ← INDEX(p)
                Continue with next
            else   ▷ Add parent
                BORDERPARENT(curnode) ← INDEX(p)
                n ← p
    Shrink array to required size

Obtaining Attributes of the Full Tree. Combining the algorithms described in this chapter we can obtain the attributes for each local max-tree as if it was built in one large max-tree:

- Build a local max-tree for each process
- Build a boundary-tree for each local max-tree
- Recursively follow a pattern where vertical and horizontal merges/combinations are done based on the size of the grid of tiles
  - Send/receive the boundary-tree following that pattern
  - The receiving process merges the attributes and updates parent relations of two trees
  - Then sends back the nodes with changed attributes and parent pointers
  - Both the sending and receiving process apply the changes on their local max-tree
  - The receiving process combines the two boundary-trees
- Continue until all boundary-trees (and therefore their attributes) are merged

After merging and combining all boundary-trees, all local max-trees contain nodes with their updated attributes. Therefore we can use the same filtering process as in the shared memory solution [8] and apply it to each local max-tree. We end up with the filtered tiles and we can write them to files. Merging the images is considered a trivial post-processing step.

3   Results

In this section we show the results of executing the application. As a first essential test we verified that the output of the algorithm remained the same, regardless of the number of MPI processes used (data not shown).

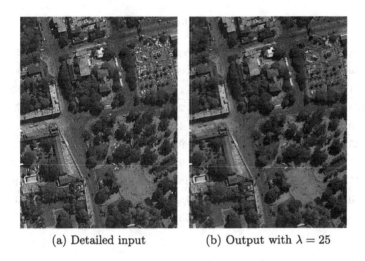

(a) Detailed input   (b) Output with $\lambda = 25$

Fig. 5. Example of execution with real life data

Initial tests were performed on the Zeus compute server: a 64-core AMD Opteron Processor 6276 with 512 GB of RAM. The main tests were on up to 128 standard nodes of the Peregrine cluster of the Center for Information Technology of the University of Groningen. The standard nodes have 24 Intel Xeon 2.5 GHz cores and 128 GB RAM, and are connected by a 56 Gb/s Infiniband network. We used only a single core per node, to simulate worst case-message passing conditions. We used a 30000 × 40000 pixel grey value VHR aerial image of the Haiti earthquake area. A small cutout is shown in Fig. 5.

We performed two experiments with this image. For Experiment 1, we split up the image into up to 128 tiles. Because the image is small enough that the max-tree to fits into memory of a single node of the cluster, this allowed us to measure speed-up $S(P)$ in the usual way as $t(1)/t(P)$, with P the number of processes, and $t(1)$ is the wall-clock time for one process, and $t(P)$ the wall-clock time for P processes. The disadvantage of this approach is that we can only work on comparatively small images. We therefore used multiple copies of the 30000×40000 image to generate a series of synthetic images of sizes 2, 4, 8, 16, and 32 times bigger than the original. Here we estimate speed-up as

$$S(P) = Pt(1)/t(P)$$

because the images scale in size with the number of processes in Experiment 2. In all measurements we took the minimum wall-clock time from ten runs, to minimise the effect of other programs running on the cluster. Given that the Peregrine is a "production" HPC cluster, multiple jobs may share a node, so timing may vary.

Table 1. Timings for the Haiti image

Tiles	Experiment 1			Experiment 2			
	Time (s)	$S(P)$	Speed (Mp/s)	Size (GB)	Time (s)	$S(P)$	Speed (Mp/s)
1×1	331.09	1.0	3.62	1.2	331.09	1.0	3.62
1×2	163.08	2.03	7.36	2.4	357.65	1.85	6.71
2×2	80.25	4.13	14.95	4.8	362.64	3.65	13.24
2×4	45.06	7.35	26.63	9.6	369.30	7.17	26.00
4×4	23.54	14.06	50.98	19.2	381.16	13.90	45.33
4×8	15.04	22.01	79.79	38.4	393.99	26.89	97.46
8×8	10.39	31.87	115.50	76.8	395.27	53.67	194.3
8×16	8.39	39.65	143.71	153.6	409.87	103.52	374.75

Looking at the speedup in Table 1 we can see that the algorithm performs rather well in terms of scaling, achieving roughly 75% efficiency on 32 processes. For Experiment 1, we see a near linear speed-up up to 8 processes, after which some slowdown is observed, dropping to about 50% at 64 cores, and some 25% at 128 cores. Note that at these large values of P the size of the max-trees of each tile is quite small compared to the boundary-tree size. This increases the message-passing overhead. As a variant, we ran the 32-process runs of Experiment 1 on 16 cores of 2 nodes to reduce message-passing overhead, but we could not notice any significantly different timings. In Experiment 2, speed-up is clearly better for large P on the cluster: at $P = 128$ we get no less than $103\times$ speed-up. The better results in Experiment 2 may be due to the fact that for large tiles the communication load is small compared to the compute load. In a final experiment on the Peregrine cluster, we combined experiment 1 and 2, using a 4×4 mosaic of the 1.2GB image, and subdividing the resulting 19.2GB image into 256 tiles. Using 16 cores on 16 nodes each, a performance of 302 Mpixel/s was obtained, or about 83 × speed-up.

The timings on the Zeus server, which could only run Experiment 1, were similar, and achieved up to 50.6× speed-up on 64 cores on a 3.88 Gpixel satellite image. Several short tests where run on a single, 48-core, 2.2 GHz Opteron node of a cluster at DigitalGlobe, using three panchromatic WorldView 2 and WorldView 3 satellite images: a 31888 × 31860 pixel image of the city of New York, USA at 0.5 m spatial resolution; a 35840 × 35840 pixel image of Madrid, Spain, at 0.5 m spatial resolution, and a 37711 × 54327 pixels image of Cairo, Egypt at 0.3 m spatial resolution. Wall-clock times on these images were 36.8 s, 51.6 s, and 87.0 s respectively, including I/O, very much comparable to the results obtained on the 64-core machine with similarly sized images. This indicates that the results were not due to some special configuration of the image used.

4   Conclusion

In this paper we showed how to approach a distributed memory algorithm for connected filters, building on the shared-memory version. The key notion is to build a forest of modified max-trees, rather than a single max-tree. This is possible through the use of the boundary-tree, which allows selective, recursive exchange of just the necessary data to modify the max-trees of each tile to yield exactly the same result as would be obtained by filtering the max-tree of the entire image. With minor adaptation, this approach also allows processing images with max-trees larger than local memory. We have shown that the algorithm scales well on different architectures, certainly up to 128 processes for large data sets.

Several improvements or extensions of the current work can be made. First of all the algorithm could be extended to 3-D volumes or videos, where you would have to redefine the neighbour function. This does have consequences for the complexity, as the size of the boundary-tree of a 3-D volume is $\mathcal{O}(GN^{\frac{2}{3}})$.

A key issue is the reliance on a limited grey-level range to keep the communication load manageable. Matas et al. [10] suggest using a fast linear max-tree algorithm that is efficient regardless of the bit depth of the image, but it uses the same merging algorithm as [8]. The problem is that the mergers of different trees scale linearly with the number of grey levels in the worst case, so bit depths beyond 16 bits require a different approach. Indeed, first merging lines in an image rather than using tiles or larger blocks maximizes the number of merges needed, exacerbating the problem.

Even for the current implementation, all the conclusions must of course be more thoroughly validated on a much more extensive, real-world data set. To that end, we are currently improving the I/O structure to accommodate more file formats, and easier integration within remote-sensing tool chains. Extensions for pattern spectra, and differential morphological and area profiles, and the use of other attributes will also be developed. The source code will be made available.

Acknowledgment. We would like to thank the Center for Information Technology of the University of Groningen for their support and for providing access to the Peregrine high performance computing cluster. The 64 core Opteron machine was obtained by funding for the HyperGAMMA project from the Netherlands Organisation for Scientific Research (NWO) under project number 612.001.110.

References

1. Breen, E.J., Jones, R.: Attribute openings, thinnings, and granulometries. Comput. Vis. Image Underst. **64**(3), 377–389 (1996)
2. Salembier, P., Wilkinson, M.H.F.: Connected operators. IEEE Sig. Process. Mag. **26**(6), 136–157 (2009)
3. Salembier, P., Oliveras, A., Garrido, L.: Antiextensive connected operators for image and sequence processing. IEEE Trans. Image Process. **7**(4), 555–570 (1998)
4. Najman, L., Couprie, M.: Building the component tree in quasi-linear time. IEEE Trans. Image Process. **15**(11), 3531–3539 (2006)
5. Pesaresi, M., Benediktsson, J.: Image segmentation based on the derivative of the morphological profile. In: Goutsias, J., Vincent, L., Bloomberg, D.S. (eds.) Mathematical Morphology and Its Applications to Image and Signal Processing, vol. 18, pp. 179–188. Springer, Heidelberg (2002)
6. Urbach, E.R., Roerdink, J.B.T.M., Wilkinson, M.H.F.: Connected shape-size pattern spectra for rotation and scale-invariant classification of gray-scale images. IEEE Trans. Pattern Anal. Mach. Intell. **29**(2), 272–285 (2007)
7. Ouzounis, G.K., Soille, P.: Differential area profiles. In: Proceedings of the 2010 20th International Conference on Pattern Recognition, pp. 4085–4088. IEEE Computer Society(2010)
8. Wilkinson, M.H.F., Gao, H., Hesselink, W.H., Jonker, J.E., Meijster, A.: Concurrent computation of attribute filters on shared memory parallel machines. IEEE Trans. Pattern Anal. Mac. Intell. **30**(10), 1800–1813 (2008)
9. Carlinet, E., Géraud, T.: A comparative review of component tree computation algorithms. IEEE Trans. Image Process. **23**(9), 3885–3895 (2014)
10. Matas, P., Dokládalová, E., Akil, M., Grandpierre, T., Najman, L., Poupa, M., Georgiev, V.: Parallel algorithm for concurrent computation of connected component tree. In: Blanc-Talon, J., Bourennane, S., Philips, W., Popescu, D., Scheunders, P. (eds.) ACIVS 2008. LNCS, vol. 5259, pp. 230–241. Springer, Heidelberg (2008). doi:10.1007/978-3-540-88458-3_21
11. Wilkinson, M.H.F., Soille, P., Pesaresi, M., Ouzounis, G.K.: Concurrent computation of differential morphological profiles on giga-pixel images. In: Soille, P., Pesaresi, M., Ouzounis, G.K. (eds.) ISMM 2011. LNCS, vol. 6671, pp. 331–342. Springer, Heidelberg (2011). doi:10.1007/978-3-642-21569-8_29

Statistical Threshold Selection for Path Openings to Detect Cracks

Petr Dokládal[(✉)]

CMM - Centre for Mathematical Morphology, MINES ParisTech,
PSL - Research University, 35, Rue Saint Honoré, 77300 Fontainebleau, France
petr.dokladal@mines-paristech.fr

Abstract. Inspired by the a contrario approach this paper proposes a way of setting the threshold when using parsimonious path filters to detect thin curvilinear structures in images.

The a contrario approach, instead of modeling the structures to detect, models the noise to detect structures deviating from the model. In this scope, we assume noise composed of pixels that are independent random variables. Henceforth, cracks that are curvilinear sequences of bright pixels (not necessarily connected) are detected as abnormal sequences of bright pixels.

In the second part, a fast approximation of the solution based on parsimonious path openings is shown.

Keywords: Mathematical morphology · Parsimonious path filters · Rank-max opening · A contrario model

1 Introduction

The detection of thin, curvilinear elements is a frequent task in automated visual inspection to detect cracks. Similar applications can be found in other image processing domains to detect the blood vessels (medicine), facial wrinkles (cosmetology), neurites or DNA molecule chain (biology), roads in satellite images, or a few others. An accurate detection of such structures that are thin, curved, discontinuous and often rare and sometimes absent (in which case we want to avoid false detection) from the images is a difficult problem. In mathematical morphology the operator for enhancing curvilinear, thin objects is the path opening.

However, the main difficulty to produce a binary result is to find a convenient value for the subsequent thresholding. One solution, based on the same hypothesis formulated using the a contrario model, and combined with the percolation, has been investigated by Aldea and Le Hégarat-Mascle [1]. Long, thin and dark cracks are detected as deviations from the model of the background by minimizing their probability to occur by chance. The minimization is done through a reconstruction from initial seeds preformed as long as dark pixels can be added the crack to make it longer and consequently less probable.

© Springer International Publishing AG 2017
J. Angulo et al. (Eds.): ISMM 2017, LNCS 10225, pp. 369–380, 2017.
DOI: 10.1007/978-3-319-57240-6_30

The same hypothesis of sparsity of defects is used throughout this paper. The principal contribution here is the proposition of a method of identifying parameter values such that the detected structures are perceptually significant. The principle is inspired from the idea of meaningful alignments proposed by Desolneux *et al.* [2], detecting perceptually significant, straight segments in images. The perceptual significance stems from the Helmholtz principle stating that a perceptually significant element is improbable to occur by chance. Any perceptual significant segment is hence a deviation from randomness. The model proposed in [2] is based on the hypothesis of random distribution of the orientation of the isophote[1] in noise. A perceptually significant segment is defined as a sequence of isophotes aligned with the orientation of the segment. Notice that a perceptually significant segment does not need to be contiguous.

In this work we extend the original model based on the orientation of isophotes to that of the distribution of intensity in images, and from straight segments to curvilinear structures. As well as in the case of the isophotes, the points do not necessarily need to be connected to form a perceptually significant structure, but only sufficiently densely follow each another in a sequence.

This idea can be efficiently implemented using mathematical morphology that indeed possesses an operator conceived for the detection of such structures. It is derived from the original concept of the path opening, due to Buckley and Talbot [3,4], and formalized in Heijmans *et al.* [5]. Later, Talbot and Appleton [6] propose a more efficient implementation and introduce incomplete path openings robust to noise due to the capability to tolerate missing pixels. The originally exponential complexity has been reduced later to a constant by Morard *et al.* [7] by limiting the search of paths to only a relevant subset of all paths in an image and using a constant-time opening algorithm for the filtering alongside these paths or a closing-opening to increase the robustness to noise. Using closing-opening instead of opening alone does not increase the complexity, and is indeed similar to another version proposed by Cokelaer *et al.* [8] that tolerates gaps up to some maximal admissible width.

The main contribution of the present paper is to set correct parameter values for parsimonious path filters to detect perceptually meaningful structures longer than some chosen minimal length. The morphological tool allowing to implement efficiently the above ideas is parsimonious incomplete path opening applying a 1-D rank-max-opening alongside these paths.

2   The Meaningfulness

According to the Helmholtz principle, an object is perceived as meaningful provided it is unlikely for it to occur by chance. The complement of these structures is the background that we assume random with a priory unknown distribution.

Definition 1 (ε-meaningful event (Desolneux *et al.* [2])). *An event of type "such configuration of points has such property" is ε-meaningful if the expectation of the number of occurrences in an image of this event is less than ε.*

---

[1] Normal to the gradient direction.

For an event to be meaningful one needs to consider $\varepsilon \ll 1$, in which case our perception is likely to see it. Notice that the ε-meaningfulness is related to the statistical p-significance.

In the following we develop a framework to detect thin, curvilinear structures in gray-scale images. These structures are perceived due to the deviation of their intensity from the normality. They are either brighter or darker. We develop the framework for bright structures; dark structures can be detected after inverting the image. In the simulated experiment Fig. 1 the sinusoidal curve - composed of closely grouped points somewhat brighter than the surrounding - is perceived as one object even though not necessarily contiguous. The difficulty to extract this structure is due to that we do not know the parameters: (i) the distribution of the noise, (ii) how much the curve points are brighter than the noise, (iii) how densely these points populate this curve.

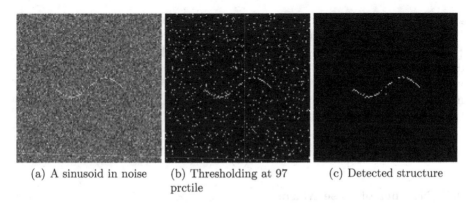

(a) A sinusoid in noise   (b) Thresholding at 97   (c) Detected structure
                                   prctile

Fig. 1. Detection of a meaningful structure in gaussian noise $\mathcal{N}(\mu = 0, \sigma^2 = 1)$; image size: 128×128. (a) A meaningful structure simulated as one sine period with intensity of 2.25 containing 56 randomly drawn points out of 70 (one period length). (b) thresholding at the 97 percentile, (c) the detected structure with $L = 20$, $k = 16$ (see text below for parameters setting).

In the remainder of this text, the ideas from Desolneux *et al.* [2] originally applied to isophote orientation, are developed and applied to the distribution of gray values. We use identical notation wherever possible and refer the reader to [2] for additional details.

In [2] the distribution of orientation of the isophotes in noise is uniformly distributed on $[0, 2\pi]$. It is independent of the values of the pixels, provided the pixels are independent, random variables. The detection of curvilinear structures can not use alignment of isophotes since the local orientation is not constant. We therefore extend the ideas from [2] to the distribution of pixel values in the image. Compared to the distribution of isophotes, the distribution of pixel values not only does not obey the uniform law but is even often unknown. Also an isophote is either aligned or misaligned with the segment. It is a boolean property. Analogously, a "pixel is brighter than" is also boolean but requires some threshold. On the other hand, the statement that "a pixel is bright" does not require a threshold and is not boolean any longer since it can be true to various degrees of truth. For example, a pixel can be "very bright", "extremely bright" or "exceptionally bright". Consequently, a curvilinear bright structure will be perceptually meaningful depending of two properties, its brightness and its length. For the same perception stimulus a shorter structure must be brighter than a longer one. Consequently, even a single, isolated pixel in noise can be perceptually meaningful provided it is exceptionally bright. Compared to that, one isophote in noise can never become meaningful.

Let f denote a realization, i.e. a gray-scale image, on a domain $M \times N \subset \mathbb{Z}^2$, of a random variable following the law F, assumed unknown. The complementary cumulative distribution function

$$\overline{F}(\alpha) = P[f(x) > f_\alpha] = p \qquad (1)$$

gives the probability p that the value of pixel x exceeds the quantile α.

Let C be a curvilinear structure formed by a sequence of l connected[2] points $\{x_1, x_2, \ldots, x_l\}$. Let X_i be a random variable equal to 1 if $f(x_i) > f_\alpha$ and 0 otherwise. Let $S_l = X_1 + X_2 + \ldots + X_l$, with $0 \leq S_l \leq l$, be a random variable counting in C the number of points brighter than f_α. Provided X_i are independent we have the probability that exactly k pixels in the curve C exceed the intensity f_α given by the binomial distribution

$$P[S_l = k] = \binom{l}{k} p^k (1 - p)^{l-k}$$

2.1   Number of False Alarms

A path is generated by a connectivity class given by a graph. The vertices of the graph represent the image pixels, and the edges connect pixels that are connected. The graph is directed and acyclic. A path in an image is therefore a path in the graph. The connectivity is assumed translation invariant, except on image borders.

Suppose every vertex is connected to n neighbors (with $n = 3$ in the example Fig. 4). There are therefore n^{l-1} l-pixel-long paths starting in any pixel in $\mathbb{Z}^2$. In a $N \times N$ image, the number of possible starting points for a l-pixel-long path (generated by either of these graphs) is $(N - l)^2$. Hence, the number of l-pixel-long paths generated by either of the connectivity graphs in Fig. 4 can be approximated by

$$\Pi(l; N) = (N - l)^2 n^{l-1}$$

Given the exponential rule we drop the multiplicative factor given by the number of possible rotations of the graph.

Following the idea from [2] we are interested in detecting structures "longer than" rather than "exactly as long as" since the exact length is not known.

---

[2] We define connectivity below.

We set up the detection threshold by decreasing the number of false alarms to below an acceptable (arbitrary but small enough) value.

In statistical terms the presence of a meaningful structure is a test and the number of possible paths along which it can occur is the number of trials. The probability of occurrence multiplied by the number of tests gives the expectation of false alarms NFA

$$NFA[k,l] = \Pi(N)P[S_l \geq k] = \sum_{i=k}^{l} \Pi(i,N)\binom{l}{i}p^i(1-p)^{l-i} \tag{2}$$

The upper length limit l in the sum being unknown we let it equal to the length of the diagonal of the image $\sqrt{2}N$. Comparing NFA to ε

$$NFA[k,l] \leq \varepsilon \tag{3}$$

is equivalent to saying whether the structure is ε meaningful.

2.2 Intensity Threshold

The first thing one can do is to set $k = l$ in Eq. 2 and isolate p as a function of the length l

$$p \leq \sqrt[l]{\frac{\varepsilon}{\Pi(l)e^{l-1}}} \tag{4}$$

Illustrated in Fig. 2a as a function of the length it shows that shorter structures, to be meaningful, must contain points that are less probable, that is more deviating from the normality. If the law of the intensity distribution in an image is known and invertible Eq. 4 can be used to determine the intensity threshold to detect meaningful structures. If the law is unknown or not invertible, a rank-order filter ξ_{1-p} in a sliding window with the rank $1 - p$ will select points with intensity above the quantile level $1 - p$, that is those with probability less or equal p.

Example: The intensity threshold corresponding to the probability p for a normal law $\mathcal{N}(\mu, \sigma^2)$, with $\mu = 0$, $\sigma = 1$, is given by Fig. 2b. It says that even an isolated point becomes meaningful provided its intensity exceeds 3.5σ. Long, connected structures are meaningful as long as their intensity remains above 0.5σ of the noise.

Isolating k, with $k \leq l$, from Eq. 2 allows saying in how many pixels out of a sequence of l connected pixels the intensity must exceed the p-th quantile value for the structure to become meaningful. This means that we search for structures that are composed of bright, not necessarily connected, sequences of points.

$$k(l) = \min\{k \in \mathbf{N}, NFA[k,l] \leq \varepsilon\} \tag{5}$$

Given some l, Eq. 5 can be directly solved for k. In a typical image, e.g. 512×512 pixels, the number of paths Π_l is a large number, given that it increases exponentially with l, itself bounded by the length of the diagonal of the image.

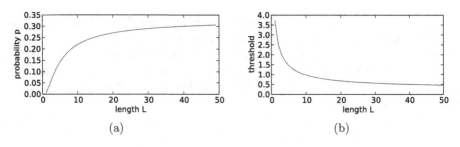

Fig. 2. (a) The probability of point as function of the length, (b) The threshold obtained as the quantile level corresponding to p for law $\mathcal{N}(\mu, \sigma^2)$, for $\mu = 0$, $\sigma = 1$.

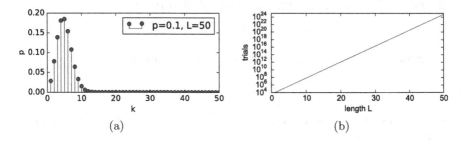

Fig. 3. (a) Probability mass function of the binomial law, (b) Number of 3-connected paths on $\mathbb{Z}^2$ for $L \leq 50$.

Consider a collection of L independent, random points. Let p denote the probability that the intensity of a point exceeds some threshold. The probability that in k points out of L the intensity exceeds some fixed threshold follows the binomial law, see illustration Fig. 3a for $p = 0.1$. If these L points form a path in a $\mathbb{Z}^2$ grid there are many trials to this test. There are 3^{L-1} paths generated by a three-connected graph (either of the graphs in Fig. 4) that start in every point of this grid. This gives a huge number on unbounded (or large) supports, see Fig. 3b.

The expectation of false alarms *NFA* is given Fig. 5a. We search for a configuration where *NFA* in a given image remains low (condition Eq. 3) that is the smallest k where the tail of the distribution remains under ε. See Fig. 5b for k expressed as a function of L, for $p = 0.05$.

Solving Eq. 3 requires counting the number of paths $\Pi(i, N)$ in Eq. 2. Exact counting not being easy on bounded supports, we provide a convenient approximation. Recall that a point in $\mathbb{Z}^2$ is an origin for 3^{L-1} L-pixel-long paths. However, restricting this estimation to bounded supports by multiplying this number by the number of points in a bounded support gives an unacceptably strong overestimation. We propose another, more precise approximation.

A pair of points (a, b) in $\mathbb{Z}^2$ is connected by a path which is either (i) a line segment running straight from a to b or (ii) a path oriented from a to b and containing as many right turns as left turns. The count of such paths is given by

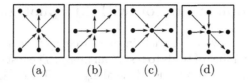

Fig. 4. An example of a commonly used connectivity graph (a), and its rotations (b–d)

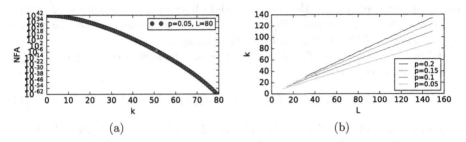

(a)   (b)

Fig. 5. (a) NFA as a function of k. In red: Choosing k for $\epsilon = 0.01$. (b) Minimum k ensuring meaningfulness as a function of the length L and parametrized by p. (Color figure online)

$$\sum_{k=1}^{L} \binom{L}{k}\binom{L-k}{k}$$

This number multiplied by the number of possible points L segments distant each from the other. This approximation provides a sufficiently precise estimation of paths on bounded supports.

Next, we fix some fixed fraction $r = const.$, $r \in \mathbb{R}^+$, $0 < r \le 1$, for having $k \in \mathcal{N}^+$ bright pixels in a sequence $k = \|rl_0\|$, with $\|.\|$ meaning rounding to nearest integer. We solve Eq. 2 for the lowest p satisfying the condition on the meaningfulness Eq. 3. Observe in Fig. 6 the values of p that we obtain for various r. Notice that the curve for $r = 1$ coincides with the one given by Fig. 2a (that of a complete path). Observe that for longer curves and higher r lower threshold suffices to ensure meaningfulness. These probabilities allow obtaining correct threshold values for the rank filter ξ_r, as explained below.

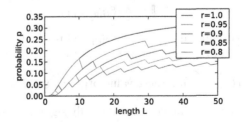

Fig. 6. The pixel probability p as a function of the sequence length L for various fixed fill fraction levels r.

3   Efficient Implementation with Morphological Tools

We have seen that an obvious obstacle in the search of efficient implementation is the overwhelming number of trials to test in an image. The test "intensity in k out of l points in a path exceeds a value" done alongside all paths is computationally very intensive; the number of paths in an image increases linearly with the image size and exponentially with the length of the paths. Testing even a much smaller number of trials in the case of straight lines in the algorithm in [2], takes seconds on a recent computer [9]. In what follows we present a convenient fast approximation of the solution using path openings.

Given some function $f : D \rightarrow V$, the morphological opening of f by a structuring element $B \subset D$ defined by

$$\gamma_B f = \sup\{B(u) + v \le f; (u, v) \in D \times V\} \tag{6}$$

is a supremum of all translates of B that do not exceed f, where $B(u) + v$ denotes B translated horizontally by u and vertically by v. That is, for some $x \in D$, there is a translate $B(u)$ such that for any $y \in B(u)$, $f(y) \ge \gamma_B f(x)$. Defining $g = \{x \mid \gamma_B f(x) > th\}$, then $x \in g$ means that there is some $B(u)$, $x \in B(u)$ that $f(y) > th$, for all $y \in B(u)$. For a connected B with $|B| = k$, g indicates where $f > th$ larger than k. In 1-D these subdomains are intervals.

In presence of noise the strict inclusion of B is likely to fail. This motivated the introduction of the rank-max opening, proposed by Ronse [10] and later described in Ronse and Heijmans [11], defined as the supremum of openings by all subsets of cardinal k of a chosen structuring element B

$$\gamma_{B,k} = \bigvee_i \{\gamma_{B_i} \mid B_i \subseteq B, \operatorname{card}(B_i) = k\} \tag{7}$$

where $1 \le k \le n$ with $n = \operatorname{card}(B)$. It acts as an opening tolerating $n - k$ missing pixels.

Note that r, in $r = k/l$, is to be understood as the rate of pixels that are required to be intensive, and inversely $1 - r$ the rate of tolerated missing pixels.

A naive implementation of Eq. 7 requires computing $\binom{n}{k}$ openings which is prohibitive for most applications. Instead of the naive implementation, Heijmans [12] shows that Eq. 7 can be efficiently implemented by using this identity

$$\gamma_{B,k} = \mathbb{1} \wedge (\delta_{\check{B}} \xi_{B,n-k+1}) \tag{8}$$

where $\mathbb{1}$ denotes the original image $\gamma_{B,k}$ is applied to, $\delta_{\check{B}}$ the conjugate dilation with $\check{B}(x) = B(-x)$ and $\xi_{B,r}$ a rank order filter

$$\xi_{B,r} f(x) = \text{r-th largest value of} \{f(y), y \in B(x)\}$$

Isolating in some original image f perceptually significant sequences of pixels longer than L is obtained using the following thresholding

$$\gamma_{L,k} > \xi_r$$

that is the main result of this paper. L is a length parameter that plays the role of the structuring element B in Eq. (8) meaning that the structuring element here is at least l-pixel long path.

A fast, approximate solution can be obtained using the parsimonious path selection as in [7] allowing to test only a fraction of all paths. The idea summarizes to the following principle. Let $f : D \rightarrow V$, with e.g. $V = R^+$ be an image such that D is a rectangular subset of Z^2. Suppose D equipped with a translation-invariant, acyclic connectivity graph $G : D \rightarrow \mathscr{P}(D)$, with $\mathscr{P}$ denoting the powerset. We say a sequence $\pi = (x_1, x_2, \ldots, x_n)$, $n \in \mathbb{N}$, of points is a path if $x_{i+1} \in G(x_i)$ for all $1 \leq i \leq n - 1$. The path length is n. Points x_1 and x_n are its starting and end points. The starting points are on the edges of the image and each path goes to the facing image edge so that the mean intensity is maximized alongside. The set of the paths generated by G in this way is denoted by Π_G. The opening is applied alongside all paths in Π_G. The complement of Π_G in D (the support of f) is set to 0.

Once the paths are isolated, similarly to applying a closing-opening as in [7], we apply a conjugate dilation and a rank filter alongside each of these paths. Both dilation and rank filter are reasonably fast on 1-D data even when implemented by definition. For time-critical applications one can use fast, $\mathcal{O}(1)$ algorithms for both, e.g. [13] for dilation and [14] for the rank order filter. We apply the morphological 1-D rank-max opening alongside the paths with statistically obtained parameters to detect cracks as statistically meaningful structures in images.

4   Results

We illustrate the method on the detection of road cracks. Prior to the detection of cracks, the original image - where cracks appear dark - is inverted using a morphological closing, so that an opening can be used to detect the cracks, and shadows are eliminated. Notice that even though the background texture is non-stationary no other preprocessing is needed as long as the rank order filter is applied in a sliding window. See Fig. 7 for the results.

We choose to detect cracks at least $L = 30$ pixels long. Notice that discontinuous cracks are still detected but appear as composed of segments that are not necessarily all longer than the parameter L. This occurs due to the non-increasingness of the opening (ensured by $\wedge$ in Eq. 8). The parameters have been estimated automatically from the pdf of the noise and L provided by user. The result can be compared with the result obtained using Parsimonious Incomplete Path Opening (PIPO) [7]. To obtain comparable settings we use the same set up: minimal length $L = 30$ pixels, and the same 30×30 sliding window threshold ξ_r for both. The tolerance to gaps up to 6 pixels has been chosen manually as to obtain a visually comparable result.

Regarding the timing: the processing time of a 1280×512 image is $85 \, \text{ms}$ for the rank filter, $180 \, \text{ms}$ the rank-max path opening, whereas it is only $51 \, \text{ms}$ for PIPO, on a $2.70 \, \text{GHz}$ Intel i7 CPU (without using the CPU's parallelism capabilities). Notice that even though a longer running time is needed to detect

(a) Original (inverted)

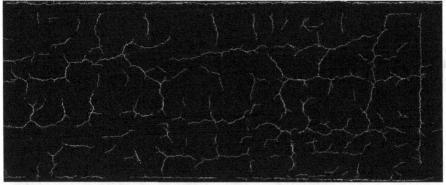

(b) Perceptually significant cracks $L \geq 30$

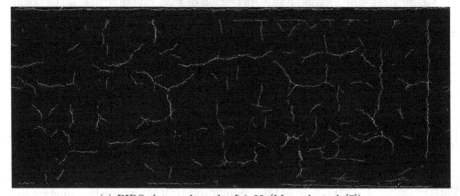

(c) PIPO detected cracks $L \geq 30$ (Morard *et al.* [7])

Fig. 7. Detection of road cracks. (top) original 1280×512 image, (middle) Perceptually significant cracks obtained for $L \geq 30$, $k = 24$, $r = 0.85$. (bottom) for comparison, cracks obtained with Parsimonious Incomplete Path Opening for $L = 30$, tolerated gaps 6 pixels. The sliding window size used for rank filter is a 30×30 rectangle for both.

perceptually significant cracks than that needed for PIPO the advantage is that all parameters can be estimated automatically to obtain a result of a comparable quality. The following section discusses other properties and limitations of this method.

5   Properties and Discussion

The result of rank-max path opening is not necessarily composed of connected paths the length of which is equal or greater than L as for (complete) path openings. Indeed, the tolerance to noise of rank-max openings makes the result disconnected if the object in the input image is disconnected (as in the simulation Fig. 1 and the road cracks Fig. 7). The rank-max opening may also create small spurious branches due to local random noise arrangements. As a direct consequence, the results obtained with a sequence of increasing lengths are not strictly decreasing (in the sense "being included") though global, non-strict decreasing is observed.

In presence of dense grouping of more cracks the threshold at some quantile is biased upwards which counterbalances the sensitivity. Nonetheless, this comes somewhat in relation with the perception law saying that isolated structures are perceived individually whereas groupings are perceived as a whole first.

One, well known, limitation comes from the path openings. Their underlying acyclic graph puts a limit on the tortuosity of the structures. The length of excessively tortuous structures becomes underestimated. This drawback is however widely accepted since counter-balanced by the linear computational complexity of this algorithm and rapidity even on large images.

Another limitation comes from the assumption of unstructured background noise. A geometrical pattern possibly present in a noise (making the noise be indeed a texture) is independent of its pdf but violates the assumption of uncorrelated pixel values. A possible solution to this problem might be to estimate the geometrical properties of the background clutter and their distributions and turn it into a model allowing to detect abnormal local pattern configurations.

References

1. Aldea, E., Le Hégarat-Mascle, S.: Robust crack detection for unmanned aerial vehicles inspection in an a-contrario decision framework. J. Electron. Imaging **24**(6), 061119 (2015)
2. Desolneux, A., Moisan, L., Morel, J.-M.: Meaningful alignments. Int. J. Comput. Vis. **40**(1), 7–23 (2000)
3. Buckley, M., Talbot, H.: Flexible linear openings and closings. Comput. Imaging Vis. **18**, 109–118 (2000)
4. Heijmans, H., Buckley, M., Talbot, H.: Path openings and closings. Probability, Netw. Algorithms, no. E 0403, 1–21 (2004)
5. Heijmans, H., Buckley, M., Talbot, H.: Path openings and closings. J. Math. Imaging Vis. **22**(2), 107–119 (2005)

6. Talbot, H., Appleton, B.: Efficient complete and incomplete path openings and closings. Image Vis. Comput. **25**(4), 416–425 (2007)
7. Morard, V., Dokládal, P., Decencière, E.: Parsimonious path openings and closings. IEEE Trans. Image Process. **23**(4), 1543–1555 (2014)
8. Cokelaer, F., Talbot, H., Chanussot, J.: Efficient robust d-dimensional path operators. IEEE J. Sel. Topics Sig. Process. **6**(7), 830–839 (2012)
9. Grompone von Gioi, R., Jakubowicz, J., Morel, J.-M., Randall, G.: LSD: a line segment detector. Image Process. Line **2**, 35–55 (2012)
10. Ronse, C.: Erosion of narrow image features by combination of local low rank and max filters. In: Proceedings of 2nd IEEE International Conference on Image Processing and its Applications, pp. 77–81 (1986)
11. Ronse, C., Heijmans, H.: The algebraic basis of mathematical morphology: II. openings and closings. Comput. Vis. Graph. Image Process. **54**(1), 74–97 (1991)
12. Heijmans, H.: Morphological image operators. Adv. Electron., Electron Phys. **1**(Suppl.) (1994). Academic Press, Boston
13. Dokládal, P., Dokládalová, E.: Computationally efficient, one-pass algorithm for morphological filters. J. Vis. Commun. Image Represent. **22**, 411–420 (2011)
14. Perreault, S., Hébert, P.: Median filtering in constant time. IEEE Trans. Image Process. **16**(9), 2389–2394 (2007)

Attribute Profiles from Partitioning Trees

Petra Bosilj[1,4]([✉]), Bharath Bhushan Damodaran[1], Erchan Aptoula[2],
Mauro Dalla Mura[3], and Sébastien Lefèvre[1]

[1] Univ. Bretagne Sud - IRISA, Vannes, France
{petra.bosilj,bharathbhushan.damodaran,sebastien.lefevre}@irisa.fr
[2] Institute of Information Technologies, Gebze Technical University, Gebze, Turkey
eaptoula@gtu.edu.tr
[3] GIPSA Laboratory, Department Image and Signal, Grenoble-INP,
Saint Martin d'Heres Cedex, France
mauro.dalla-mura@gipsa-lab.grenoble-inp.fr
[4] University of Lincoln, Lincoln, UK
pbosilj@lincoln.ac.uk

Abstract. Morphological attribute profiles are among the most prominent spatial-spectral pixel description tools. They can be calculated efficiently from tree based representations of an image. Although widely and successfully used with various inclusion trees (i.e., component trees and tree of shape), in this paper, we investigate their implementation through partitioning trees, and specifically α- and (ω)-trees. Our preliminary findings show that they are capable of comparable results to the state-of-the-art, while possessing additional properties rendering them suitable for the analysis of multivariate images.

Keywords: Attribute profiles · Partitioning trees · α-tree · (ω)-tree · Hyperspectral images

1 Introduction

Mathematical Morphology has offered effective ways to perform spatial analysis in many application domains of digital image processing. In the context of Earth Observation through satellite or airborne remote sensing, morphological tools have been popular in the last decades, especially due to their intrinsic ability to model spatial information within a multiscale framework. Following (Differential) Morphological Profiles (DMP) in the early 2000s [1], Attribute Profiles (AP) [2] and more recently Self-Dual Attribute Profile (SDAP) [3] are recognized as an efficient solution to provide multilevel spatial-spectral description of image pixels. Such a description can then be used as a pixelwise feature in a subsequent image interpretation task such as land cover classification.

P. Bosilj—The contributions made by author were done during the post-doc position at [1].

© Springer International Publishing AG 2017
J. Angulo et al. (Eds.): ISMM 2017, LNCS 10225, pp. 381–392, 2017.
DOI: 10.1007/978-3-319-57240-6_31

Profiles such as AP or SDAP rely on a tree-based representation of an image (min- and max-tree for AP, Tree of Shapes for SDAP). All these trees are indeed inclusion trees, which rely on an ordering relation of the image pixels. When dealing with multivariate images such as multi- or hyperspectral images, defining such an ordering is not straightforward and neither is the computation of AP/SDAP from multivariate images [4].

Conversely, partitioning trees have received much less attention in the context of multiscale characterizations of images. However, such trees are appealing in this context since they do not require the definition of a vectorial ordering, but only the selection of an appropriate distance measure. In this paper, we present some preliminary work aiming to demonstrate the relevance of building attribute profiles from partitioning trees.

The paper is organized as follows. In Sect. 2, we recall the attribute profiles and the tree models we rely on. We discuss the different steps to compute attribute profiles from partitioning trees in Sect. 3, before providing experimental results in Sect. 4. Finally, Sect. 5 concludes the paper and gives future research directions.

2   Related Work

In this section, we recall the principles of attribute profiles, attribute filtering and the different inclusion and partitioning trees compared in this work.

2.1   Attribute Profiles

APs are multiscale image/pixel description tools, constructed similarly to MPs, by successively applying a morphological operator with progressively increasing filter parameters (leading to a sequence of increasingly coarser filters). APs rely on morphological attribute filters (AFs). APs belong to the class of connected morphological filters, revolving around the core concept of connectivity, and deal directly with connected components (CCs) instead of pixels.

More formally, given a grayscale image $f : E \rightarrow \mathbb{Z}, E \subseteq \mathbb{Z}^2$, its upper-level sets are defined as $\mathcal{L}^t = \{f \geq t\}$ with $t \in \mathbb{Z}$ (resp. lower-level sets $\mathcal{L}_t = \{f \leq t\}$), i.e. the set of images obtained by thresholding an image at all possible values of their pixels. They are composed of the connected components ($CC \subseteq E$), typically based on 4 or 8-connectivity, which are referred to as peak components. AFs are applied to these peak components, using a predefined logical predicate T_κ^α, in general consisting of comparing the attribute α computed on CC against a threshold κ; e.g. T_{300}^{area}: "is the area of CC larger than 300 pixels?". Depending on the outcome of $T_{300}^{area}(CC)$, the connected component is either preserved or removed from the image. An output of the AF is computed by evaluating the predicate on all the connected components present in the input image.

Subsequently, an AP can be straightforwardly constructed using a sequence of AFs (often attribute thinnings and thickenings), that are applied to the input image using a set of ordered logical predicates. More precisely, given a predicate

T and a collection of L thresholds $\{\kappa_i\}_{1 \le i \le L}$ let γ^{κ_i} and ϕ^{κ_i} denote respectively the attribute thinnings and thickenings employing them. In this case the AP of a grayscale image f would become:

$$AP(f) = \{\phi^{\kappa_L}(f), \phi^{\kappa_{L-1}}(f) \ldots, \phi^{\kappa_1}(f), f, \gamma^{\kappa_1}(f), \ldots, \gamma^{\kappa_{L-1}}(f), \gamma^{\kappa_L}(f)\}. \quad (1)$$

Thus a pixel p of an image f can be characterized using the values it obtains across this sequential filtering process.

As far as the extension of AP to a multivariate image $\mathbf{f} : E \to \mathbb{Z}^r, r > 1$ is concerned, the widely (and mostly exclusively) encountered marginal strategy consists in first reducing the number of spectral dimensions (from r to n, $n \ll r$) through a variety of methods, and then in computing independently the AP of each resulting image band, that are finally concatenated in order to form the so-called extended attribute profile (EAP) [5]:

$$EAP(\mathbf{f}) = \{AP(band_1), AP(band_2), \ldots, AP(band_n)\}. \quad (2)$$

Here $band_i$ refers to either an image component after dimensionality reduction, but it might equivalently denote an actual spectral band of the input image if no reduction were performed, still leading to an EAP constructed marginally.

Since their initial introduction by Dalla Mura et al. [2], various aspects of APs have been intensively studied, e.g. combining them with alternative dimension reduction methods [6], multi-dimensional attributes [7], techniques for automatic threshold production [8], multivariate tree representations [9], histogram based AP representations [10], as well as their combination with deep learning [11]. For a recent survey on the topic the reader is referred to [12].

2.2   Trees

Hierarchical image representations used in contemporary mathematical morphology to manipulate the image connected components can be divided into inclusion and partitioning hierarchies. On one hand, the *inclusion* hierarchies comprise partial partitions of the image with nested supports and their components are formed by creating, inflating and merging image blocks [13] (cf. Fig. 1). On the other hand, the hierarchies from the class of *partitioning* trees comprise full partitions of the image, starting from the finest one in the the leaf nodes, iteratively merged until a single root node is formed (examples given in Fig. 2). Contrary to the inclusion trees which require a total order between the pixel values, partitioning trees require only a dissimilarity metric between neighboring pixels (i.e. a total ordering on the image edges instead of the pixels). They are also not extrema-oriented, capturing information about objects at intermediate gray levels, making them a strong candidate to to process multivariate data.

The seminal hierarchies **min and the max-trees** [14], belong to the class of inclusion trees. They are dual hierarchies structuring the inclusion relations between the peak components of the lower (resp. upper) level sets of the image, thus well suited for representing bright (resp. dark) structures in the image. The peak components of lower level sets $\mathcal{L}_t$ (resp. upper level sets $\mathcal{L}^t$) are nested

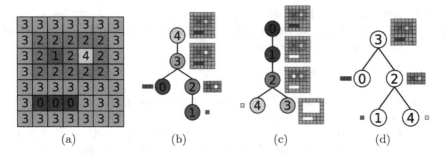

Fig. 1. The three different inclusion trees of a toy image (a). The min-tree is displayed in (b), while its dual max-tree is shown in (c). The self-dual tree of shapes is shown in (d). The (gray) levels of the nodes are displayed in the nodes, and the corresponding regions are shown beside the nodes, while the representation emphasizes the structure (region inclusion) of the image.

for increasing (resp. decreasing) values of t and form the nodes of the min-tree (resp. max-tree), thus fully representing a grayscale image. A min and a max-tree hierarchy for the image given in Fig. 1(a) are shown in Fig. 1(b) and (c).

The **Tree of Shapes** (ToS) [15] was introduced to unify a representation of bright and dark image structures, which are treated equivalently based on their contrast with the background. The *shapes* which compose the ToS are formed by filling the holes in the peak components of the image (used to form the min/max-tree), and do not intersect and are either nested or disjoint [16,17]. This makes the hierarchy (cf. Fig. 1(d)) an inclusion tree which is additionally self-dual and contrast invariant, while remaining a full representation of the image.

The α-tree is a partitioning tree and is formed based on the local range of its components. It comprises α-connected components [18,19], with α corresponding to their local range. Flat zones form the 0-CCs (for $\alpha = 0$) and correspond to the connected components of maximal size containing pixels at the same graylevel ($\mathcal{F}_t = \{f = t\}$) For $\alpha > 0$, an α-CC is defined as the CC of the maximal size such that only the neighboring pixels with gray level difference less or equal to α are considered connected. The α-CCs are nested for increasing values of α, forming a complete, self-dual hierarchy which can represent dark and bright but also regions at intermediate image levels. An example of this hierarchy can be seen on Fig. 2(b) for the image in Fig. 2(a). Due to the locality of the metric used, gray level variations within regions can be much higher than α, which is called the chaining effect [19] and is most prominent when the region gray levels gradually increase and decrease (e.g. $\alpha = 2$ in Fig. 2(b)).

To deal with this, different constrained hierarchies [19–21] are constructed from the α-tree, most notably the (ω)-**tree** [18], constraining the α-CCs by their global range. The (ω)-CCs are defined as the largest α-CCs in the image with the global range (i.e. the maximal dissimilarity between any two pixels belonging to that component) less or equal to some ω. Some of the α-CCs are removed from the hierarchy, but the maximal global range and thus the maximal level in the tree will typically be higher in the (ω)-tree. The hierarchy remains

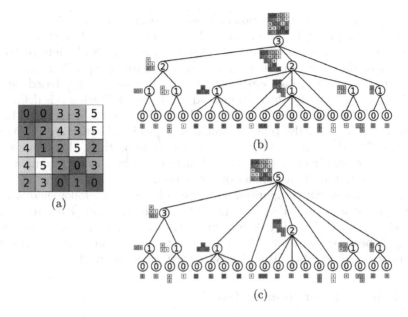

Fig. 2. For the toy image in (a), the α-tree is displayed in (b), while the constrained hierarchy (ω)-tree is shown in (c). The α (resp. (ω)) levels are displayed in the nodes and indicated by their height, with regions displayed besides the nodes. The (ω)-tree contains a subset of the α-tree nodes, but arranged through a larger span of levels.

self-dual, complete and capable of capturing regions of low, intermediate and high gray levels, but global range provides better grouping per level than just a local measure. An example of this can be seen in Fig. 2(c).

α and (ω)-trees were chosen in this paper over other partitioning hierarchies, such as Binary Partitioning Tree [22] as they do not require an external similarity parameter and are defined directly by the image pixel values.

3   Proposed Method

Although AFs have been part of the morphological toolbox for almost two decades, efficient implementations through max-tree (alternatively min-tree) based image representation [14] are the main reasons for their recent popularity, in addition to their flexibility. The advantage of these representations stems from the implementation possibility of attribute filtering in the form of node or branch removal from the tree representing the input image. This becomes especially interesting for the computation of AP since each tree needs to be computed only once and then multiple filtering outputs can be derived easily from it. Following their calculation using min and max-trees, the AP have also been calculated on the Tree of Shapes (ToS) [23]. Extending the idea of using different trees to calculate the AP, we examine the α and (ω)-trees as a basis for profile calculation.

To construct APs from partitioning trees, we consider an equivalent definition of APs through pixels. For some threshold κ_i, a pixel p would be characterized by the value it obtains in γ^{κ_i} and ϕ^{κ_i}. These values can also be interpreted as the levels of the nodes closest to a leaf in a max-tree (respectively min-tree) containing the pixel p, and satisfying the logical predicate $T^{\alpha}_{\kappa_i}$ based on the threshold κ_i employed by the thinning and thickening. Following this definition, when using an α-tree, a pixel value at level i of an AP is also determined based on the lowest node in the tree containing that pixel and satisfying the logical predicate $T^{\alpha}_{\kappa_i}$. Similarly to AP calculation for inclusion trees, this process also allows us to re-use the hierarchy in the calculation of subsequent AP components, and is elaborated in the remainder of the section.

For the sake of fair comparison with state-of-the-art, we follow here a marginal strategy as used in EAP and ESDAP. But let us recall that partitioning trees offer a greater flexibility than inclusion trees to deal with multivariate data such as hyperspectral images [24]. Exploring various dissimilarity metrics and constrained connectivity criteria is however left for future work.

3.1   Filtering a Partitioning Tree

Filtering partitioning trees is more restrictive than inclusion trees, since removing certain nodes would invalidate the hierarchy (e.g. all the children of a single parent need to be processed in the same manner, either preserved or removed). The AP is calculated following the interpretation that the lowest node containing a pixel and satisfying the logical predicate $T^{\kappa_i}_{\alpha}$ determines the value of that pixel in the ith component of the AP. It is possible that a region $R \subseteq E$ satisfying the logical predicate T^{κ}_{α} has both child nodes which do and do not satisfy T^{κ}_{α}. Thus, the region R will determine the value of a subset of pixels belonging to it, while the values of the other pixels will be determined based on the values of the descendant regions. From the implementation point of view, equal processing of all "sibling" regions is achieving by "collapsing" the child regions not satisfying the logical predicate T^{κ}_{α} into their parents by assigning them the parent node representation (cf. Sect. 3.2) and filtering their children from the tree. This can be seen in Fig. 3(a). Similarly to AP calculation of inclusion trees, this enables reusing the filtered hierarchy from one component of the AP to the next.

3.2   Node Representation

Every pixel at a certain level of the AP is assigned a value based on the node in the hierarchy selected to represent it. With inclusion trees (min/max-tree and the extension to ToS), the choice of representation is straightforward, with the AP corresponding precisely to applying a sequence of AF to the image. A node which is selected for representing one of its pixels will represent all the corresponding region pixels. An AP image is reconstructed from the filtered hierarchy by assigning to every pixel the node level in the hierarchy closest to the leaf which contains that pixel. On the other hand, no universal solution exists for node representation with partitioning trees, even for well-defined filtering.

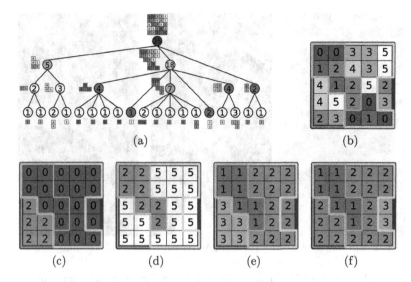

Fig. 3. α-tree filtering using different node representation strategies, filtered with the area attribute and threshold 5. In the filtered tree, the green nodes represent all of their pixels, orange and blue a part of their pixels, the red ones are represented by their parents model while the white ones are removed from the representation, which can be seen in (a). Corresponding areas on the image are shown in (b), with complete regions enclosed in green and partially represented regions in orange and blue. The image corresponding to filtering with the min, max, average and level strategies is shown in in (c)–(f). (Color figure online)

Additional complexity in the case of AP comes from the fact that a single tree node might not represent all of the pixels belonging to it. We explore in this paper three different node representation strategies, illustrated in Fig. 3(c)–(f).

Firstly, we attempted **min and max** representation (Fig. 3(c) and (d)) motivated by the fact that the original APs are a concatenation between two complementary series. In order to construct an anti-extensive (resp. extensive) series, the nodes are represented with the minimal (resp. maximal) pixel value they contain. However, both of these series are based on a different valuation of the same filtered hierarchy, thus introducing redundancy to the profile.

To avoid this redundancy, we experimented with the **average** representation (cf. Fig. 3(e)), where the filtered hierarchy for each threshold is used only once while the regions are represented by their average pixel value. This representation is typical when using the partitioning trees for image simplification.

Finally, we considered a strategy based on the **level** representation of the nodes (shown in Fig. 3(f)), inspired by considering that the values assigned to the pixels when calculating an AP from inclusion trees are their assigned levels in the tree. With our choices of partitioning trees, this means we are assigning the α (resp. (ω)) levels to the nodes from the α-tree (resp. (ω)-tree) hierarchy, which are what characterizes these nodes in the structure of the tree. This approach also always assigns a single scalar value as a node representation, which becomes

(a) Pavia University   (b) Ground Truth   (c) Training Set

Fig. 4. The Pavia University dataset (false colors) and its corresponding ground truth; its thematic classes (training set size/ground truth size) are: ▬ Asphalt (548/6631), ▬ Trees (524/3064), ▬ Bitumen (375/1330), ▬ Meadows (540/18649), ▬ Metal sheets (265/1345), ▬ Shadows (231/947), ▬ Gravel (392/2099), ▬ Bare soil (532/5029) and ▬ Self-blocking bricks (514/3682). (Color figure online)

significant when dealing with multivariate data (i.e. when the values of the pixels belonging to a single α-CC or (ω)-CC are not scalar but rather come from multiple image channels).

The selected strategy will also determine the length of the obtained descriptor. Given a collection of L thresholds for an attribute α, the length of the pixel-descriptor using a min and max representation will be $2L+1$, same as for the AP calculated on the min/max-trees and ToS. Using the average or level representation shortens the descriptor to the length of $L+1$ per pixel.

4   Experiments

The main goal of the experiments presented here is to compare the proposed tree based image representations against established alternatives, as a basis for computing attribute profiles with the end goal of pixel classification. More precisely, we compare α and (ω)-trees against max/min trees and the tree of shapes.

The dataset under consideration is an urban area of size 340×610 pixels and 9 thematic classes, acquired using the ROSIS-03 sensor with a 1.3 m spatial resolution over the city of Pavia, Italy. The ROSIS-03 sensor has 115 data channels with a spectral coverage ranging from 0.43 to 0.86 μm. After the elimination of 12 noisy bands, 103 bands have been left for processing (Fig. 4).

Classification has been realized using a Random Forest classifier composed of 100 trees. The number of variables involved in the training of the classifier was set to the square root of the feature vector length, as suggested by [25]. Classification performance has been measured by means of the κ statistic.

Table 1. The classification performance (κ statistic) of α- and (ω)-tree based AP for different node representations and for both attributes as well as their combination.

	α-tree			(ω)-tree		
	Area	Moment	Comb.	Area	Moment	Comb.
Max/min	0.9327	0.8521	0.9080	0.9264	0.8546	0.9429
Average	0.9240	0.8468	0.9379	0.8876	0.8449	0.9302
Level	0.8662	0.8477	0.8898	**0.9482**	0.8489	0.9479

For the Pavia University dataset, the learning step has been carried out using the standard training set widely used in the literature [2].

We are considering two of the most popular attributes: area and moment of inertia. We use the automatic settings according to [26] for area thresholds (λ_a) and manual settings for the moment of inertia (λ_m) according to [2]: $\lambda_a = \{770, 1538, 2307, 3076, 3846, 4615, 5384, 6153, 6923, 7692, 8461, 9230, 10000, 10769\}$ and $\lambda_m = \{0.2, 0.3, 0.4, 0.5\}$. However the aforementioned settings, whether automatic or manual, have been empirically determined by always the min/max-trees in mind. As the choice of threshold can have a significant effect on the performance of AP [2], for the sake of fairness we scale them with various multipliers $\mu = \{0.5, 0.6, 0.7, 0.8, 0.9, 1, 1.1, 1.2, 1.3, 1.4, 1.5, 1.6, 1.7, 1.8, 1.9, 2, 3, 4, 5, 6, 7, 8\}$. This allows us to calculate our profiles for a set of different area and moment thresholds, corresponding to $(\Lambda_a, \Lambda_m) = \{(\lambda_a^i, \lambda_m^i) = (\mu_i\lambda_a, \mu_i\lambda_m)\}$, and make a choice of thresholds most suitable for each hierarchy.

The performance of different profiles at $\mu = 1$ is compared to examine the effectiveness of different node representations for partitioning trees, shown in Table 1. Even though the max/min node representation effectively uses the same regions twice, producing the descriptor of twice the size and containing redundancy, it is more effective than the average representation which attempts to represent each region only once by its average gray level. The level representation performs much better with the (ω)-tree than for the α-tree, as the global range is a much better indicator of region complexity. The level representation also has the advantage of being half the size of the max/min representation (60 vs. 116 for the area attribute and 20 vs. 36 for the moments). As is additionally corresponds most closely to the pixel value interpretation for the original AP, we chose the level representation as our preferred one. A further advantage of this representation is its extendability to multivariate data: while the descriptor size for the min/max and the average representations will be multiplied by the number of bands in case of multivariate data representations, the length does not change when using the level representation.

We have also tried combining the moments and the area attribute, but have noticed that for the peak values of the area AP, addition of the moments AP does not improve performance. For this reason, the remainder of the experiments were carried out separately for the area and for the moments AP.

As far as their performance across a range of threshold values is concerned, in the case of the area attribute, the performances of all four tested trees are

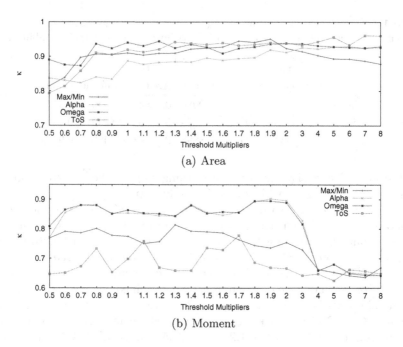

Fig. 5. Classifications scores (κ statistic) for the Pavia University dataset for the area and moment of inertia attributes using a variety of threshold settings.

relatively similar and consistent, underlining the reliability of the attribute under consideration (Fig. 5(a)). All the same, one can still observe that the (ω)-tree systematically outperforms the α-tree. Even though (ω)-tree leads to the best scores for lower thresholds, it is eventually the tree of shapes, that achieves the overall best classification performance level with the area attribute.

As to the moment of inertia attribute (Fig. 5(b)), both (ω)-tree and (α)-tree lead to very similar performances, surpassing their inclusion based counterparts for this shape attribute. The ToS, which was previously shown to outperform the min/max-tree based AP is performing the worst when moment of inertia is used, and with a higher level of variance across thresholds. While the performance of all examined hierarchies examined similar behavior for the area attribute, for the moments of inertia attribute the relative scores vary at a greater level, with the partitioning trees exhibiting both higher scores as well as stability across a larger interval of thresholds.

5   Conclusion

Attribute profiles have been one of the most successful morphological tools in remote sensing recently. They allow for multiscale image description through successive attribute filterings, and can be efficiently computed using inclusion trees such as min- and max-tree (AP) and tree of shapes (SDAP). In this paper, we propose to compute these profiles using another class of tree representations, namely partitioning trees.

We consider in this first attempt α-tree and (ω)-tree, applied with a marginal strategy on each band of a hyperspectral image (similarly to the standard computation of extended AP). We discuss the specific challenges raised when computing attribute profiles from partitioning trees, in particular the way to represent tree nodes and build the AP feature vectors. Our preliminary results, even though they are based only on a single hyperspectral dataset combined with a single classifier, still show the relevance of such an approach, and call for further exploration.

Indeed, while we have used here a marginal strategy to ensure a fair comparison with existing AP definitions, using partitioning trees allows building a single tree from a whole dataset (either the original image or its first principal components). Future work will thus aim to evaluate different local dissimilarity metrics as well as constrained connectivity criteria (e.g. global range) defined in a multivariate space.

Acknowledgments. This work was supported by the French Agence Nationale de la Recherche (ANR) under reference ANR-13-JS02-0005-01 (Asterix project), by the BAGEP Award of the Science Academy and by the Turkish TUBITAK Grant 115E857.

References

1. Benediktsson, J., Pesaresi, M., Arnason, K.: Classification and feature extraction for remote sensing images from urban areas based on morphological transformations. IEEE Trans. Geosci. Remote Sens. **41**(9), 1940–1949 (2003)
2. Dalla Mura, M., Benediktsson, J.A., Waske, B., Bruzzone, L.: Morphological attribute profiles for the analysis of very high resolution images. IEEE Trans. Geosci. Remote Sens. **48**(10), 3747–3762 (2010)
3. Dalla Mura, M., Benediktsson, J.A., Bruzzone, L.: Self-dual attribute profiles for the analysis of remote sensing images. In: Soille, P., Pesaresi, M., Ouzounis, G.K. (eds.) ISMM 2011. LNCS, vol. 6671, pp. 320–330. Springer, Heidelberg (2011). doi:10.1007/978-3-642-21569-8_28
4. Carlinet, E., Geraud, T.: MToS: a tree of shapes for multivariate images. IEEE Trans. Image Process. **24**(12), 5330–5342 (2015)
5. Dalla Mura, M., Atli Benediktsson, J., Waske, B., Bruzzone, L.: Extended profiles with morphological attribute filters for the analysis of hyperspectral data. Int. J. Remote Sens. **31**(22), 5975–5991 (2010)
6. Dalla Mura, M., Villa, A., Benediktsson, J.A., Chanussot, J., Bruzzone, L.: Classification of hyperspectral images by using extended morphological attribute profiles and independent component analysis. IEEE Geosci. Remote Sens. Lett. **8**(3), 542–546 (2011)
7. Aptoula, E.: Hyperspectral image classification with multi-dimensional attribute profiles. IEEE Geosci. Remote Sens. Lett. **12**(10), 2031–2035 (2015)
8. Marpu, P., Pedergnana, M., Dalla Mura, M., Benediktsson, J.A., Bruzzone, L.: Automatic generation of standard deviation attribute profiles for spectral spatial classification of remote sensing data. IEEE Geosci. Remote Sens. Lett. **10**(2), 293–297 (2013)

9. Aptoula, E., Dalla Mura, M., Lefèvre, S.: Vector attribute profiles for hyperspectral image classification. IEEE Trans. Geosci. Remote Sens. **54**(6), 3208–3220 (2016)

10. Demir, B., Bruzzone, L.: Histogram-based attribute profiles for classification of very high resolution remote sensing images. IEEE Trans. Geosci. Remote Sens. **54**(4), 2096–2107 (2016)

11. Aptoula, E., Ozdemir, M.C., Yanikoglu, B.: Deep learning with attribute profiles for hyperspectral image classification. IEEE Geosci. Remote Sens. Lett. **13**(12), 1970–1974 (2016)

12. Ghamisi, P., Dalla Mura, M., Benediktsson, J.A.: A survey on spectral-spatial classification techniques based on attribute profiles. IEEE Trans. Geosci. Remote Sens. **53**(5), 2335–2353 (2015)

13. Ronse, C.: Ordering partial partitions for image segmentation and filtering: merging, creating and inflating blocks. J. Math. Imaging Vis. **49**(2), 202–233 (2014)

14. Salembier, P., Oliveras, A., Garrido, L.: Anti-extensive connected operators for image and sequence processing. IEEE Trans. Image Process. **7**(4), 555–570 (1998)

15. Monasse, P., Guichard, F.: Scale-space from a level lines tree. J. Vis. Commun. Image Represent. **11**(2), 224–236 (2000)

16. Song, Y., Zhang, A.: Monotonic tree. In: Braquelaire, A., Lachaud, J.-O., Vialard, A. (eds.) DGCI 2002. LNCS, vol. 2301, pp. 114–123. Springer, Heidelberg (2002). doi:10.1007/3-540-45986-3_10

17. Ballester, C., Caselles, V., Monasse, P.: The tree of shapes of an image. ESAIM: Control Optim. Calc. Var. **9**, 1–18 (2003)

18. Soille, P.: On genuine connectivity relations based on logical predicates. In: International Conference on Image Analysis and Processing, pp. 487–492 (2007)

19. Soille, P.: Constrained connectivity for hierarchical image partitioning and simplification. IEEE Trans. Pattern Anal. Mach. Intell. **30**(7), 1132–1145 (2008)

20. Soille, P.: Preventing chaining through transitions while favouring it within homogeneous regions. In: Soille, P., Pesaresi, M., Ouzounis, G.K. (eds.) ISMM 2011. LNCS, vol. 6671, pp. 96–107. Springer, Heidelberg (2011). doi:10.1007/978-3-642-21569-8_9

21. Soille, P., Najman, L.: On morphological hierarchical representations for image processing and spatial data clustering. In: Köthe, U., Montanvert, A., Soille, P. (eds.) WADGMM 2010. LNCS, vol. 7346, pp. 43–67. Springer, Heidelberg (2012). doi:10.1007/978-3-642-32313-3_4

22. Salembier, P., Garrido, L.: Binary partition tree as an efficient representation for image processing, segmentation, and information retrieval. IEEE Trans. Image Process. **9**(4), 561–576 (2000)

23. Cavallaro, G., Dalla Mura, M., Benediktsson, J., Plaza, A.: Remote sensing image classification using attribute filters defined over the tree of shapes. IEEE Trans. Geosci. Remote Sens. **54**(7), 3899–3911 (2016)

24. Lefèvre, S., Chapel, L., Merciol, F.: Hyperspectral image classification from multiscale description with constrained connectivity and metric learning. In: IEEE Workshop on Hyperspectral Image and Signal Processing: Evolution in Remote Sensing (WHISPERS) (2014)

25. Breiman, L.: Random forests. Mach. Learn. **45**(1), 5–32 (2001)

26. Ghamisi, P., Benediktsson, J.A., Sveinsson, J.R.: Automatic spectral-spatial classification framework based on attribute profiles and supervised feature extraction. IEEE Trans. Geosci. Remote Sens. **52**(9), 5771–5782 (2014)

Object Detection

Distance Between Vector-Valued Fuzzy Sets Based on Intersection Decomposition with Applications in Object Detection

Johan Öfverstedt[1(✉)], Nataša Sladoje[1,2], and Joakim Lindblad[1,2]

[1] Centre for Image Analysis, Department of IT, Uppsala University,
Uppsala, Sweden
johan.ofverstedt@it.uu.se, joakim@cb.uu.se
[2] Mathematical Institute of Serbian Academy of Sciences and Arts,
Belgrade, Serbia

Abstract. We present a novel approach to measuring distance between multi-channel images, suitably represented by vector-valued fuzzy sets. We first apply *the intersection decomposition transformation*, based on fuzzy set operations, to vector-valued fuzzy representations to enable preservation of joint multi-channel properties represented in each pixel of the original image. Distance between two vector-valued fuzzy sets is then expressed as a (weighted) sum of distances between scalar-valued fuzzy components of the transformation. Applications to object detection and classification on multi-channel images and heterogeneous object representations are discussed and evaluated subject to several important performance metrics. It is confirmed that the proposed approach outperforms several alternative single- and multi-channel distance measures between information-rich image/object representations.

1 Introduction

Image analysis based on fuzzy object representations has been shown to exhibit very good performance under uncertainty and imprecision, inherently present in images [1]. Previous studies have demonstrated that the precision of various shape descriptors is significantly increased if fuzzy representations are used instead of traditional crisp ones [2,3]. A number of different image properties (such as intensity, gradient, texture) can be represented by utilizing fuzzy sets defined on the integer grid. The degree of membership of each pixel to a particular fuzzy set represents the degree of fulfilment of the corresponding represented property at the observed pixel. A classic fuzzy set, having a scalar-valued membership function, can model one property. It is, however, often beneficial to observe several properties simultaneously. Images with multiple spectral components, as well as information-rich combinations of scalar-valued object representations, have the potential to contribute to improved accuracy for detection, recognition and classification, if the used image analysis tools have a capability for utilizing the full information content. We argue that object representations

© Springer International Publishing AG 2017
J. Angulo et al. (Eds.): ISMM 2017, LNCS 10225, pp. 395–407, 2017.
DOI: 10.1007/978-3-319-57240-6_32

utilizing vector-valued fuzzy sets (VVFS), and tools based on them, are useful to capture this type of information.

In this study we extend the family of fuzzy set distances based on integration over α-cuts, proposed in [4], to VVFS, used for multi-channel image object representations. Distance measures are fundamental image processing tools which are used in a wide range of higher level tools and processing pipelines. A number of distance measures have been proposed for fuzzy sets, some of which compare memberships element-wise, ignoring spatial shape information, whereas others combine shape and membership in a single measure, usually exhibiting better performance [1,4]. We suggest an extension of distance measures applicable to scalar-valued fuzzy sets, based on utilization of a particular transform of a VVFS. The transform results in a new VVFS where each channel corresponds to a particular combination of inclusion/exclusion of the original fuzzy components, allowing to express object properties characterized by simultaneous presence of different combinations of channels (individual fuzzy components).

We show that this novel approach exhibits excellent performance in template matching on color images. The new measure is evaluated on a real-life task of detection and classification of cilia in transmission electron microscopy images, where heterogeneous object properties are combined in a vector-valued object representation. Advantages of the proposed approach are clearly demonstrated.

2   Preliminaries

2.1   Fuzzy Sets

A *fuzzy set* S [5] on a reference set X is a set of ordered pairs, $S = \{(x, \mu_S(x)) : x \in X\}$, where $\mu_S : X \to [0,1]$ is the *membership function* of S.

A crisp set $C \subset X$ is a special case of a fuzzy set, with its characteristic function as membership function

$$\mu_C(x) = \begin{cases} 1, & \text{for } x \in C \\ 0, & \text{for } x \in \overline{C}. \end{cases} \tag{1}$$

The *height* of a fuzzy set S on X is $h(S) = \max_{x \in X} \mu_S(x)$.

Let p be an element of the reference set X. A *fuzzy point* p defined at p in X (also called a fuzzy *singleton*) with height $h(p)$, is defined by a membership function

$$\mu_p(x) = \begin{cases} h(p), & \text{for } x = p \\ 0, & \text{for } x \neq p. \end{cases} \tag{2}$$

An α-*cut* of a fuzzy set S is a crisp set defined as $^{\alpha}S = \{x \in X : \mu_S(x) \geq \alpha\}$

The *support* of S is $\text{supp}(S) = \{x \in X : \mu_S(x) > 0\}$

The *complement* $\overline{S}$ of a fuzzy set S is $\overline{S} = \{(x, 1 - \mu_S(x)) : x \in X\}$

For fuzzy sets $\mathcal{A}$ and $\mathcal{B}$ on the same ref. set X the (standard) *intersection* is $\mathcal{A} \cap \mathcal{B} = \{(x, \min\{\mu_{\mathcal{A}}(x), \mu_{\mathcal{B}}(x)\}) : x \in X\}$ and the (standard) union is $\mathcal{A} \cup \mathcal{B} = \{(x, \max\{\mu_{\mathcal{A}}(x), \mu_{\mathcal{B}}(x)\}) : x \in X\}$.

Intersection and union over sets of fuzzy sets on set X are denoted $\bigcap\limits_{i \in \{1...n\}} \mathcal{S}_i$ and $\bigcup\limits_{i \in \{1...n\}} \mathcal{S}_i$, where $\bigcap\limits_{i \in \emptyset} \mathcal{S}_i = \{(x, 1) : x \in X\}$ and $\bigcup\limits_{i \in \emptyset} \mathcal{S}_i = \{(x, 0) : x \in X\}$

Functions f which are defined on crisp sets can be generalized to fuzzy sets by integration over α-cuts (*fuzzification principle*), assuming convergence,

$$f(\mathcal{S}) = \int_0^1 f(^\alpha \mathcal{S}) \, d\alpha. \tag{3}$$

2.2   Distances

A *distance* is a function $d \colon X \times Y \to \mathbb{R}_+ \cup \{0\}$. It can be computed between different entities. We consider distances between two points (in an integer grid), between a point and a set of points, and between two sets.

The most often used point-to-point distance measure is the Euclidean distance. A commonly used family of crisp point-to-crisp set distances is, for any point-to-point distance d, the closest point distance

$$d(x, Y) = \min_{y \in Y} d(x, y). \tag{4}$$

The (internal) Distance Transform (DT) of a (crisp) set $A \subset X$ is, for $x \in X$,

$$\mathrm{DT}[A](x) = \min_{y \in \overline{A}} d(x, y) = d(x, \overline{A}). \tag{5}$$

A family of set-to-set distances which exhibit excellent performance in practice for image processing tasks are the *Sum of Minimal Distances* [6] (SMD):

Definition 1. *For an arbitrary point-to-set distance d, Symmetric SMD is*

$$d_{SMD}(A, B) = \frac{1}{2}\left(\sum_{a \in A} d(a, B) + \sum_{b \in B} d(b, A)\right). \tag{6}$$

In Asymmetric SMD, only the point-to-set distances from A to B are considered:

Definition 2. *For an arbitrary point-to-set distance d, Asymmetric SMD is*

$$d_{ASMD}(A, B) = \sum_{a \in A} d(a, B). \tag{7}$$

For details about the SMD measures (6–7) we refer the reader to [4,6].

2.3   Distances Between Fuzzy Sets

The *fuzzy point-to-set inwards distance* d^α, based on integration over α-cuts [4], between a fuzzy point p and a fuzzy set S, is defined as

$$d^\alpha(p, S) = \int_0^{h(p)} d(p, {}^\alpha S)\, d\alpha, \tag{8}$$

where d is a point-to-set distance defined on crisp sets.

The *complement distance* of a fuzzy set distance d is defined as

$$\overline{d}(p, S) = d(\overline{p}, \overline{S}). \tag{9}$$

The *fuzzy point-to-set bidirectional distance* $\overleftrightarrow{d}^\alpha$ is defined as

$$\overleftrightarrow{d}^\alpha(p, S) = d^\alpha(p, S) + \overline{d}^\alpha(p, S). \tag{10}$$

The inwards distance between a fuzzy point p and a set S is equal to zero for all fuzzy points contained in the set, $h(p) \le \mu_S(p)$. The bidirectional distance offers increased differentiation between such points. Inserting (8) or (10) in Definitions 1 and 2 provides extensions of the SMD family of distances to fuzzy sets [4].

2.4   Fuzzy Domain Decomposition

Sets are commonly decomposed into parts with homogeneous properties, such that the sum (union) of the parts is the whole set. One such partitioning, adapted to fuzzy sets, is the *fuzzy domain decomposition* [7]: the membership function of a crisp reference set X (equal to 1 for $x \in X$) is decomposed to a set of fuzzy sets defined on X, each describing the degree of a particular property for elements of X, such that the memberships over all the component sets sum up to one.

Definition 3 [7]. *A fuzzy domain decomposition (of a domain X) is given by fuzzy sets $\mathcal{A}_i, i = 1, \ldots, n$, defined by their membership functions μ_i such that*

$$\sum_{i=1}^n \mu_i(x) = 1, \forall x \in X. \tag{11}$$

3   Novel Point-to-Set Distance Measures by Fuzzy Intersection Decomposition

The state-of-the-art point-to-set distances d^α, (8), and $\overleftrightarrow{d}^\alpha$, (10), being defined for classic scalar-valued fuzzy sets, are not directly applicable to vector-valued object representations. Various extension approaches can be considered, differing mainly in at which level of the distance measure definition the contributions from the different components are aggregated.

One possible extension approach is to aggregate the membership functions of the components and to apply a single channel point-to-set distance measure to the resulting fuzzy set. This approach risks losing valuable information when reducing the vector-valued fuzzy sets into a scalar-valued one.

Another approach is to apply the distance measure to each component in isolation and then to aggregate the distance values (e.g., as a weighted sum). This approach keeps more of the information contained in the channels, but loses essential information contained in the combination of different channels.

We here propose to consider a transform based on the notion of domain decomposition, such that regions with the same joint memberships are separated into individual fuzzy components. Following such a separation, where elements of each fuzzy set of the decomposition share, and preserve, multi-channel properties, the components are separately treated in the distance measure.

3.1   Vector-Valued Fuzzy Sets

Let us observe n fuzzy sets, $S_1, S_2, \ldots, S_n$, defined on a reference set X. Let $\mu_{S_i} : X \to [0,1]$ be the membership function of the fuzzy set S_i, for $i = 1, 2, \ldots, n$.

A *vector-valued fuzzy set* (VVFS) S on X is a set of ordered $(n+1)$-tuples

$$S = \{(x, \mu_{S_1}(x), \mu_{S_2}(x), \ldots, \mu_{S_n}(x)) : x \in X\}.$$

We denote with $\mu_S = (\mu_{S_1}, \mu_{S_2}, \ldots, \mu_{S_n})$ the membership function of a VVFS S.

A *vector-valued fuzzy point* p at a point $p \in X$ (also called a vector-valued fuzzy *singleton*), with the (vector-valued) height $h(p) = (h_1(p), h_2(p), \ldots, h_n(p))$, w.r.t. the components of a VVFS, is defined by a membership function

$$\mu_p(x) = \begin{cases} h(p) = (h_1(p), \ldots, h_n(p)), & \text{for} \quad x = p \\ 0, & \text{for} \quad x \neq p. \end{cases}$$

3.2   The Fuzzy Intersection Decomposition Transform

We present a *fuzzy domain decomposition* transform based on fuzzy set intersections and unions. This transform results in a VVFS which, in its fuzzy components, captures joint multi-channel properties of the initial VVFS.

Let $\ominus : S \times S \to S$ be the bounded difference [8] of fuzzy sets on the same reference set X, defined as $A \ominus B = \{(x, \max\{0, \mu_A(x) - \mu_B(x)\}) : x \in X\}$

Let U denote the index set $U = \{1, \ldots, n\}$ Let $\mathcal{P}(U)$ denote the power set of U. For a set $K \in \mathcal{P}(U)$, let $\overline{K} = U \setminus K$.

Definition 4 (Fuzzy Intersection Decomposition Transform (FIDT)).
The fuzzy intersection decomposition *of an n-component VVFS S is a transform from S to another VVFS $\hat{S}$ with 2^n components, such that one decomposition component $\hat{S}_K$ is defined for each $K \in \mathcal{P}(U)$ as a scalar-valued fuzzy set*

$$\hat{S}_K = \left(\bigcap_{k \in K} S_k \right) \ominus \left(\bigcup_{i \in \overline{K}} S_i \right). \tag{12}$$

The FIDT is a fuzzy domain decomposition according to Definition 3.

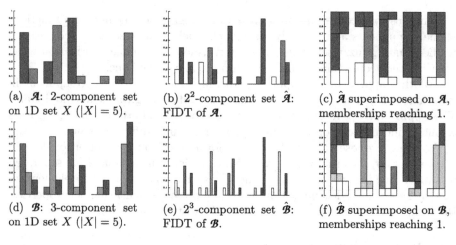

(a) $\mathcal{A}$: 2-component set on 1D set X ($|X| = 5$).

(b) 2^2-component set $\hat{\mathcal{A}}$: FIDT of $\mathcal{A}$.

(c) $\hat{\mathcal{A}}$ superimposed on $\mathcal{A}$, memberships reaching 1.

(d) $\mathcal{B}$: 3-component set on 1D set X ($|X| = 5$).

(e) 2^3-component set $\hat{\mathcal{B}}$: FIDT of $\mathcal{B}$.

(f) $\hat{\mathcal{B}}$ superimposed on $\mathcal{B}$, memberships reaching 1.

Fig. 1. FIDT applied to vector-valued fuzzy sets $\mathcal{A}$, $n = 2$ (shown in (a)) and $\mathcal{B}$, $n = 3$ (shown in (d)). Set elements are shown on the x-axis and their memberships on the y-axis. The 2^n transform component values (memberships of each of the 2^n components for each of the 5 elements) are displayed in (b) and (e). Decomposition components superimposed on the original sets are shown in (c) and (f). The transform component order (left-to-right) with corresponding color code used for index-sets K is ($\{1,2\}$: White, $\{1\}$: Red, $\{2\}$: Green, $\emptyset$: Black) for the 2-component set and ($\{1,2,3\}$: White, $\{1,2\}$: Yellow, $\{1,3\}$: Magenta, $\{2,3\}$: Cyan, $\{1\}$: Red, $\{2\}$: Green, $\{3\}$: Blue, $\emptyset$: Black) for the 3-component set. (Color figure online)

Figure 1 shows two examples of the FIDT applied to small synthetic VVFSs, defined on a discrete reference set X with $|X| = 5$. Figure 2 illustrates the decomposition of a color image, Mandrill. Here, a 3-component VVFS is used to represent the RGB image. The FIDT provides 2^3 fuzzy sets, shown in Fig. 2(b–i).

3.3   Point-to-Set Distances Using Fuzzy Intersection Decomposition

Having the decomposition set $\hat{\mathcal{S}}$, we can apply any point-to-set distance defined for scalar-valued fuzzy points and sets. The advantage from utilizing the suggested decomposition, instead of trivial decomposition into original channels, comes from preservation of information about properties resulting from simultaneous presence or absence of the original components in each pixel.

Definition 5 (Fuzzy Intersection Decomposition Point-to-Set Distance). *For a given weight function $\hat{w}$: $\mathcal{P}(U) \rightarrow \mathbb{R}_+ \cup \{0\}$ a family of point-to-set distance measures $d^\times$ between a vector-valued fuzzy point $\boldsymbol{p}$ and a vector-valued fuzzy set $\mathcal{S}$ is defined as*

$$d^\times(\boldsymbol{p}, \mathcal{S}, \hat{\boldsymbol{w}}) = \sum_{K \in \mathcal{P}(U)} \hat{\boldsymbol{w}}_K d(\hat{\boldsymbol{p}}_K, \hat{\mathcal{S}}_K). \tag{13}$$

$\hat{\boldsymbol{p}}_K$ and $\hat{\mathcal{S}}_K$ are components of decomposition $\hat{p}$ and $\hat{\mathcal{S}}$ as given by Definition 4, and d denotes any point-to-set distance for scalar-valued fuzzy points and sets.

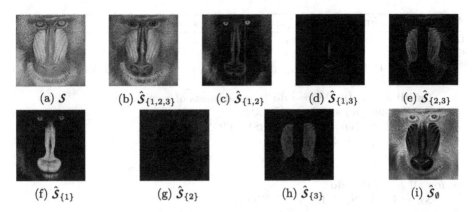

(a) S (b) $\hat{S}_{\{1,2,3\}}$ (c) $\hat{S}_{\{1,2\}}$ (d) $\hat{S}_{\{1,3\}}$ (e) $\hat{S}_{\{2,3\}}$

(f) $\hat{S}_{\{1\}}$ (g) $\hat{S}_{\{2\}}$ (h) $\hat{S}_{\{3\}}$ (i) $\hat{S}_{\emptyset}$

Fig. 2. The FIDT applied to the 3-component VVFS representation of the standard test RGB image *Mandrill*. (a) Original RGB image, S; (b) Degree of simultaneous presence of all three channels; (c–e) degree of simultaneous presence of two, in absence of the third channel; (f–h) degree of presence of one channel, in simultaneous absence of the remaining two; (i) degree of simultaneous absence of all three channels. (Color figure online)

We introduce the following notation: $d_\alpha^\times$ denotes the distance resulting from inserting the inwards α-cut distance d^α (8) as distance d in Definition 5 and $\bar{d}_\alpha^\times$ the resulting distance when the bidirectional α-cut distance $\bar{d}^\alpha$ (10) is used.

4   Distances Between Vector-Valued Fuzzy Sets

The point-to-set distances $d^\times$ given by (13) can be extended to distances between VVFSs, if inserted into any set distance based on point-to-set distances. We consider here the *SMD* set-to-set distances (1), which leads to the following:

Definition 6 (Symmetric Weighted SMD for VVFS). *Given weight functions* $w_{\mathcal{A}}, w_{\mathcal{B}} \colon X \to \mathbb{R}_+$ *and* $\hat{w} \colon \mathcal{P}(U) \to \mathbb{R}_+ \cup \{0\}$, *the* α-*cut based weighted symmetric sum of minimal distances, wSMD, between VVFS* $\mathcal{A}$ *and VVFS* $\mathcal{B}$, *is*

$$d_{wSMD}^\times(\mathcal{A}, \mathcal{B}, \hat{w}, w_{\mathcal{A}}, w_{\mathcal{B}}) = \frac{1}{2}\Big(\sum_{a \in \mathcal{A}} w_{\mathcal{A}}(a)d_\alpha^\times(a, \mathcal{B}, \hat{w}) + \sum_{b \in \mathcal{B}} w_{\mathcal{B}}(b)d_\alpha^\times(b, \mathcal{A}, \hat{w})\Big).$$
(14)

The bidirectional version $\bar{d}_{wSMD}^\times$ *is defined correspondingly, with* $\bar{d}_\alpha^\times$ *used as the point-to-set distance.*

In applications such as template matching the compared images may be of different sizes, hence an asymmetric treatment of the sets may be appropriate.

Algorithm 1. Fuzzy intersection decomposition algorithm

Input: $\mathcal{S}$ with $n \in \mathbb{N}_+$ components
Output: $\hat{\mathcal{S}}$
1: Init. trivial INTERSECTION_K and UNION_K, for K s.t. $|K| \leq 1$, based on $\mathcal{S}$
2: **for** $i \leftarrow 2$ **to** n **do**
3:     **for all** $K \in \mathcal{P}(\text{U}) : |K| = i$ **do** {For index-sets of cardinality i}
4:         $M \leftarrow \max(K)$ and $L \leftarrow K \setminus \{M\}$
5:         $\text{INTERSECTION}_K \leftarrow \text{INTERSECTION}_L \cap \mathcal{S}_M$
6:         $\text{UNION}_K \leftarrow \text{UNION}_L \cup \mathcal{S}_M$
7:     **end for**
8: **end for**
9: **for all** $K \in \mathcal{P}(\text{U})$ **do**
10:     $\hat{\mathcal{S}}_K \leftarrow \text{INTERSECTION}_K \ominus \text{UNION}_{\bar{K}}$
11: **end for**

Definition 7 (Asymmetric Weighted SMD for VVFS). *Given weight functions* $\boldsymbol{w_A}\colon X \to \mathbb{R}_+$ *and* $\hat{\boldsymbol{w}}\colon \mathcal{P}(U) \to \mathbb{R}_+ \cup \{0\}$, *the* α-*cut based weighted asymmetric sum of minimal distances, wASMD, between VVFS* $\boldsymbol{A}$ *and VVFS* $\boldsymbol{B}$, *is*

$$d^{\times}_{wASMD}(\boldsymbol{A}, \boldsymbol{B}, \hat{\boldsymbol{w}}, \boldsymbol{w_A}) = \sum_{a \in \boldsymbol{A}} \boldsymbol{w_A}(a) d^{\times}_{\alpha}(\boldsymbol{a}, \boldsymbol{B}, \hat{\boldsymbol{w}}). \tag{15}$$

$\bar{d}^{\times}_{wASMD}$ *is defined correspondingly, with* $\bar{d}^{\times}_{\alpha}$ *used as the point-to-set distance.*

5   Implementation and Complexity Analysis

Efficient computation of FIDT can be performed by following Algorithm 1. For each index-set K, at most three set operations are performed, each with time complexity $O(N)$, where N is the cardinality of the ref. set (i.e. number of pixels). Since there are 2^n index-sets, the time complexity of the algorithm is $O(2^n N)$.

The set distances are most efficiently computed by pre-computing all the FIDT components, and component-wise distance transforms corresponding to all considered membership levels, which are subsequently used as lookup tables from which each point-to-set distance is obtainable by one or two memory accesses.

All of the novel distances, $d^{\times}_{\alpha}$, $\bar{d}^{\times}_{\alpha}$, $d^{\times}_{wSMD}$ and $d^{\times}_{wASMD}$, have time complexity $O(2^n MN)$ (for M membership levels), given that a linear time DT algorithm exists for the chosen point-to-point measure.

6   Performance Evaluation

6.1   Template Matching

We first evaluate performance of the proposed distance measures on template matching in color images. Clearly, each channel may contain important information required for a successful match; we argue that improved performance is

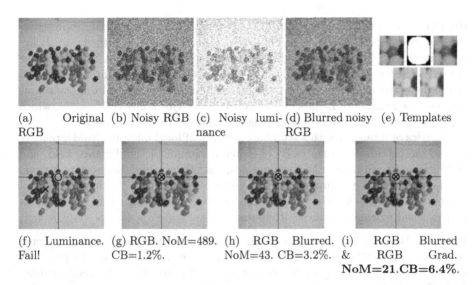

(a)   Original (b) Noisy RGB (c) Noisy lumi- (d) Blurred noisy (e) Templates
RGB   nance   RGB

(f)   Luminance. (g) RGB. NoM=489. (h) RGB Blurred. (i)   RGB   Blurred
Fail!   CB=1.2%.   NoM=43. CB=3.2%. &   RGB   Grad.
                                                        NoM=21.CB=6.4%.

Fig. 3. Template matching on the color image "Jelly beans". (a–d) Original image and versions of the distorted image. Templates in (e) are in the following order: RGB, weight mask, Blurred RGB, Luminance, Blurred Luminance. (f–i) Results of $d^{\times}_{\text{wASMD}}$ in ascending order of performance. Black cross "×" indicates the template position found by the algorithm; B/W circle shows the true template location (black cross in the middle of the circle indicates a match); CB of the global optimum/found template location is shown in cyan, overlaid on the original image. (Color figure online)

achieved if the joint multi-channel information is utilized, compared to utilization of each of the channels in separation.

We compare the proposed distance measure with the following alternative approaches: (i) considering a single-channel representation of a color image and applying a scalar-valued distance function; (ii) applying a scalar-valued distance function to each of the multiple channels separately and aggregating by summation; (iii) utilizing a morphological approach operating on a multi-channel representation. In the group (i) we observe sum of squared differences (SSD) and d_{ASMD} with d^{α} [4], applied on the luminance channel of the color image. In the group (ii) we observe weighted normalized cross-correlation (wNCC), SSD, and d_{ASMD} with d^{α}. The group (iii) is represented by a recent method based on the hit-and-miss transform (HMT), referred to as MHMT, [10]. MHMT takes a set of manually constructed structuring elements (SEs) each operating on a single component with its own spatial structure and threshold. Joint fit of all SEs is then evaluated in distance computation. For this study, 9 disks of various sizes and colors were used as SEs.

We observe common data-set USC-SIPI "Jelly beans". We add Gaussian noise to each channel of the source image to create heavy chromatic noise, reaching SNR = 5.40. The luminance image with, and without, smoothing ($\sigma = 1$), the RGB image with smoothing ($\sigma = 1$) and gradient magnitude representation [9] for each color component independently (smoothed with $\sigma = 2$) are computed along with corresponding versions of the template. A rounded mask is used as weighting function. Figure 3(a)–(e) shows the input image, examples of the distorted images, and the undistorted templates selected from the source images. The number of membership levels used for this study is $M = 31$, which captures most of the important details while suppressing some noise.

Metrics of interest for this study are number of local minima (NoM), relative size of the catchment basin (CB) of global optimum in the distance field, and the success rate in finding the original position of the template cut-out. A small NoM and large CB are important for local search approaches relying on gradient descent. Exhaustive search of the distance field corresponding to the test data-set is performed to compute these attributes.

Results. Table 1 presents the results of the template matching for all considered methods. Figure 3(f)–(i) visualizes the result of a test instance for the proposed distance measure $d^{\times}_{\text{wASMD}}$ on various object representations. The SSD method, MHMT on RGB, and d_{ASMD} applied to the Blurred Luminance representation were omitted from Table 1 since they failed to produce a single match.

Table 1. Template matching results averaged over 10 different noise realisations. The best obtained value for each metric is presented as bold.

Method	Object representation	n	NoM	CB	Match rate
wNCC	RGB	3	1426.9	1.0%	1
wNCC	RGB Blurred	3	286.8	1.4%	1
wNCC	RGB Blurred and RGB Grad.	6	370.8	1.3%	1
MHMT	RGB Blurred	3	**16.0**	0.0%	0.7
d^{α}_{ASMD}	Luminance	1	326.7	1.2%	0.3
d^{α}_{ASMD}	RGB	3	802.7	1.7%	1
d^{α}_{ASMD}	RGB Blurred	3	47.8	2.3%	1
d^{α}_{ASMD}	RGB Blurred and RGB grad.	6	49.3	2.4%	1
$d^{\times}_{\text{wASMD}}$	RGB	3	506.0	1.5%	1
$d^{\times}_{\text{wASMD}}$	RGB Blurred	3	32.5	4.7%	1
$d^{\times}_{\text{wASMD}}$	RGB Blurred and RGB Grad.	6	19.5	**7.4%**	1

Both the wNCC and the proposed measure produced a successful match in each of the 10 tests, while MHMT failed on three occasions. MHMT exhibits the best NoM, likely due to the fitting step, which retains only a few positions, thereby also causing a vanishingly small CB. The proposed measure, used with a 6-comp. object representation, exhibits the largest CB while keeping a low NoM.

6.2   Cilia Detection and Classification

We apply the proposed distance measure to automated detection of cilia objects in low magnification Transmission Microscopy (TEM) images. Cilia are hair-like organelles protruding from cells; their dysfunction causes a number of serious health problems. For setting a pathological diagnosis, regions of high cilia-density should be detected efficiently at a very low image resolution where each cilium is approximately 20 pixels in diameter and most of the characteristic structure is non-visible. An accurate classification is difficult under these conditions, but it is important (and possible) to detect regions of interest which subsequently can be explored at higher resolutions. The state-of-the-art method for this problem, [11], is to use wNCC with an optimised synthetic template. It is of interest to (i) avoid need for an optimized template, and (ii) reduce the number of falsely detected objects (false positives - FP).

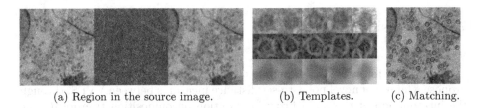

(a) Region in the source image.   (b) Templates.   (c) Matching.

Fig. 4. (a) and (b) A part of the source image and the used templates in their I,G and B representations. (c) The resulting matching in the IG representation of the same region: green circles mark true positives, red circles mark false positives and blue circles, considered neutral by the pathologist, are discarded from the statistical analysis. Images are best seen enlarged in electronic version. (Color figure online)

For evaluation we observe a Cilia source image, for which ground truth is available through annotation by a pathologist. We extract 5 templates of size 27×27 pixels, illustrated in Fig. 4. We consider the following (normalized) representations: (Orig.) Intensity (I), Gradient (G) and Blurred image (B). Gaussian smoothing with $\sigma = 1$ is used to generate G and B. To increase the small gradient values, $f(g) = \sqrt{g}$ is applied to G. $M = 15$ membership levels are used.

The distance value is computed for every position in the source image for each of the five templates and the minimum obtained distance is assigned to each position as the final value. Each local minimum is located in the distance map and ranked according to distance value. The distance map is thresholded at the level which, based on ground truth, maximizes $F_1 = 2 \cdot \frac{precision \cdot recall}{precision + recall}$.

Results. The performance of the proposed distance measure $\bar{d}^{\times}_{\mathrm{wASMD}}$ utilized with different representations (I,G,B and their combinations, IG, IB, GB, and IGB) is presented in Table 2 (and in Fig. 4(c)), together with the performance of wNCC on I. In all cases the same image cut-outs are used as templates. The proposed distance measure outperforms wNCC when cut-outs are used as templates. Using a vector-valued representation yields an improvement over using

Table 2. Cilia detection and classification results. Considered representations are Intensity (I), Gradient (G), Blurred image (B) and their 4 combinations. wNCC is applied on I. Max F_1-score, over all thresholded distances, with a known ground truth, is shown. TP (100) and FP (100) denote the number of true positives and false positives for the 100 best matches.

	I	G	B	IG	IB	GB	IGB	wNCC
Max F_1-score	0.274	0.258	0.185	0.369	0.250	0.329	**0.375**	0.189
TP (100)	12	**16**	9	14	11	15	15	13
FP (100)	69	73	79	66	71	62	**61**	85

only a single component. As an additional performance measure, the number of true positives (TP) and false positives (FP) are listed for the threshold set to the best 100 matches. Again, a combination of channels (in particular IGB, but also GB) provides smallest number of FP, while keeping a high number of TP.

7   Conclusion

We suggest to use vector-valued fuzzy set (VVFS) representations to capture image information provided simultaneously by multiple features of interest. We propose a family of novel point-to-set distance measures between VVFS, through the introduction of the novel fuzzy intersection decomposition transform. The transform components consist of scalar-valued fuzzy sets representing the combinations of the original components of VVFS to which we can apply existing state-of-the-art distance measures. We can conclude that the proposed transform and distance measure exhibit very good discriminatory performance in important image analysis tasks, such as template matching and object detection. Our tests, conducted on both synthetic and real image examples, confirm advantages of multi-channel representations, when image analysis tools are properly designed to handle the information-rich content.

Acknowledgement. A. Suveer, A. Dragomir and I.-M. Sintorn are acknowledged for acquisition and annotation of cilia images. Ministry of Science of the Republic of Serbia is acknowledged for support through Projects ON174008 and III44006 of MI-SANU.

References

1. Bloch, I.: On fuzzy distances and their use in image processing under imprecision. Pattern Recogn. **32**, 1873–1895 (1999)
2. Sladoje, N., Nyström, I., Saha, P.K.: Measurements of digitized objects with fuzzy borders in 2D and 3D. Image Vis. Comput. **23**, 123–132 (2005)
3. Sladoje, N., Lindblad, J.: Estimation of moments of digitized objects with fuzzy borders. In: Roli, F., Vitulano, S. (eds.) ICIAP 2005. LNCS, vol. 3617, pp. 188–195. Springer, Heidelberg (2005). doi:10.1007/11553595_23
4. Lindblad, J., Sladoje, N.: Linear time distances between fuzzy sets with applications to pattern matching and classification. IEEE Trans. Image Proces. **23**(1), 126–136 (2014)

5. Zadeh, L.A.: Fuzzy sets. Inf. Control **8**, 338–363 (1965)
6. Eiter, T., Mannila, H.: Distance measures for point sets and their computation. Acta Informatica **34**(2), 103–133 (1997)
7. Gander, M.J., Michaud, J.: Fuzzy domain decomposition: a new perspective on heterogeneous DD methods. In: Erhel, J., Gander, M.J., Halpern, L., Pichot, G., Sassi, T., Widlund, O. (eds.) Domain Decomposition Methods in Science and Engineering XXI. LNCSE, vol. 98, pp. 265–273. Springer, Cham (2014). doi:10.1007/978-3-319-05789-7_23
8. Zadeh, L.A.: Calculus of fuzzy restrictions. In: Fuzzy Sets and their Applications to Cognitive and Decision Processes, pp. 1–39 (1975)
9. Lopez-Molina, C., De Baets, B., Bustince, H.: Generating fuzzy edge images from gradient magnitudes. Comput. Vis. Image Underst. **115**(11), 1571–1580 (2011)
10. Weber, J., Lefévre, S.: Spatial and spectral morphological template matching. Image Vis. Comput. **30**, 934–945 (2012)
11. Suveer, A., Sladoje, N., Lindblad, J., Dragomir, A., Sintorn, I.M.: Automated detection of cilia in low magnification transmission electron microscopy images using template matching. In: IEEE Internaional Symposium on Biomedical Imaging, pp. 386–390 (2016)

Double-Sided Probing by Map of Asplund's Distances Using Logarithmic Image Processing in the Framework of Mathematical Morphology

Guillaume Noyel[1(✉)] and Michel Jourlin[1,2]

[1] International Prevention Research Institute,
95 cours Lafayette, 69006 Lyon, France
guillaume.noyel@i-pri.org
[2] Lab. H. Curien, UMR CNRS 5516,
18 rue Pr. B. Lauras, 42000 St-Etienne, France
http://www.i-pri.org

Abstract. We establish the link between Mathematical Morphology and the map of Asplund's distances between a probe and a grey scale function, using the Logarithmic Image Processing scalar multiplication. We demonstrate that the map is the logarithm of the ratio between a dilation and an erosion of the function by a structuring function: the probe. The dilations and erosions are mappings from the lattice of the images into the lattice of the positive functions. Using a flat structuring element, the expression of the map of Asplund's distances can be simplified with a dilation and an erosion of the image; these mappings stays in the lattice of the images. We illustrate our approach by an example of pattern matching with a non-flat structuring function.

Keywords: Map of Asplund's distances · Mathematical Morphology · Dilation · Erosion · Logarithmic Image Processing · Asplund's metric · Double-sided probing · Pattern recognition

1 Introduction

Asplund's metric is a useful method of pattern matching based on a double-sided probing, i.e. a probing by a greatest lower bound probe and a least upper bound probe. It was originally defined for binary shapes, or sets [1,7], by using the smallest homothetic shape (probe) containing the shape to be analysed and the greatest homothetic probe contained by the shape. Jourlin et al. [13,14] have extended this metric to functions and to grey-level images in the framework of the Logarithmic Image Processing (LIP) [15,16] using a multiplicative or an additive LIP law [11]. Then, Asplund's metric has been extended to colour and multivariate images by a marginal approach in [11,23] or by a spatio-colour (i.e. vectorial [21,22]) approach in [24].

Other approaches of double-sided probing have been previously defined in the framework of Mathematical Morphology [19,27]. The well-known hit-or-miss

© Springer International Publishing AG 2017
J. Angulo et al. (Eds.): ISMM 2017, LNCS 10225, pp. 408–420, 2017.
DOI: 10.1007/978-3-319-57240-6_33

transform [27] allows to extract all the pixels such that the first set of a structuring element fits the object while the second set misses it (i.e. fits its background). An extension based on two operations of dilation (for grey level images) has been proposed in [18]. It consists of a unique structuring element, which is used in the two dilations in order to match the signal from above and from below.

Banon and Faria [2] use two structuring elements obtained by two translations of a unique template along the grey level axis. They use an erosion and an anti-dilation to count the pixels whose values are in between the two structuring elements.

Odone et al. [25] use an approach inspired by the computation of the Hausdorff distance. They consider a grey level image as a tridimensional (3D) graph. They dilate by a 3D ball a template in order to compute a 3D "interval". Then, for any point of the image, they translate vertically the "interval" in order to contain the maximum number of points of the function and they count this number.

Barat et al. [4] present a unified framework for these last three methods. They show that they correspond to a neighbourhood of functions (i.e. a tolerance tube) with a different metric for each method. Their topological approach is named virtual double-sided image probing (VDIP) and they defined it as a difference between a grey-scale dilation and an erosion. For pattern matching, only the patterns which are in the tolerance tube are selected. It is a metric defined on the equivalence class of functions according to an additive grey level shift.

In [13], Jourlin et al. have introduced the logarithmic homothetic defined according to the LIP multiplication. This makes a compensation of the lighting variation due to a multiplicative effect, i.e. a thickening or a thinning of the object crossed by the light.

In the current paper, the important novelty introduced is the link between the map of Asplunds' metrics defined in the LIP multiplicative framework and the operations of Mathematical Morphology. We will show that the map of Asplunds' distances in the LIP multiplicative framework is the logarithm of the ratio between an anti-dilation and an anti-erosion.

This gives access to many other notions well defined in the corpus of Mathematical Morphology.

The paper is organised as follows: (1) a reminder of the main notions (LIP, Asplund's metrics, fundamental operations and framework of Mathematical Morphology), (2) the demonstration of the link between the map of Asplund's distances and Mathematical Morphology for flat structuring element (se) and for non-flat ones and (3) an illustration of pattern matching with Asplund's metric.

2   Prerequisites

In the current section, we remind the different mathematical notions and frameworks to be used: LIP model, Asplund's metric and the basis of Mathematical Morphology.

2.1   Logarithmic Image Processing (LIP)

The LIP model, created by Jourlin et al. [10,15–17], is a mathematical framework for image processing based on the physical law of transmittance. It is perfectly suited to process images acquired with transmitted light (when the object is located between the source and the sensor) but also with reflected light, due to the consistency of the model with human vision [6]. The mathematical operations performed using the LIP model are consistent with the physical principles of image formation. Therefore the values of an image defined in $[0, M[$ stay in this bounded domain. For 8 bits images $M = 256$ and the 256 grey levels are in the range of integers $[0, ..., 255]$.

A grey scale image f is a function defined on a domain $D \subset \mathbb{R}^N$ with values in $\mathcal{T} = [0, M[$, $M \in \mathbb{R}$. f is a member of the space $\mathcal{I} = \mathcal{T}^D$.

Due to the link with the transmittance law: $T_f = 1 - f/M$, the grey-scale is inverted in the LIP framework, 0 corresponds to the white extremity of the grey scale, when no obstacle is located between the source and the sensor, while the other extremity M corresponds to the black value, when the source cannot be transmitted through the obstacle.

In the LIP sense, the addition of two images corresponds to the superposition of two obstacles (objects) generating f and g:

$$f \,\triangle\, g = f + g - \frac{f.g}{M} \tag{1}$$

From this law, we deduce the LIP multiplication of f by a scalar $\lambda \in \mathbb{R}$:

$$\lambda \,\triangle\, f = M - M \left(1 - \frac{f}{M} \right)^{\lambda} \tag{2}$$

It corresponds to a thickness change of the observed object in the ratio λ. If $\lambda > 1$, the thickness is increased and the image becomes darker than f, while if $\lambda \in [0, 1[$, the thickness is decreased and the image becomes brighter than f.

The LIP laws satisfy strong mathematical properties. Let $\mathcal{F}(D, [-\infty, M[)$ be the set of functions defined on D with values in $]-\infty, M[$. We equipped it with the two logarithmic laws and $(\mathcal{F}(D, [-\infty, M[), \triangle, \triangle)$ becomes a real vector space. $(\mathcal{I}, \triangle, \triangle)$ is the positive cone of this vector space [16].

There exists a colour version of the LIP model [12].

The LIP framework has been successfully applied to numerous problems for industry, medical applications, digital photography, etc. It gives access to new notions of contrast and metrics which take into account the variation of light, for example the Asplund's metric for functions.

2.2   Asplund's Metric for Functions Using the LIP Multiplicative Law

Let us remind the novel notion of Asplund's metric defined in [13,14] for functions in place of sets. It consists of using the logarithmic homothetic $\lambda \triangle f$.

Let $\mathcal{T}^* =]0, M[$ and the space of positive images be $\mathcal{I}^* = \mathcal{T}^{*D}$.

Definition 1. *Asplund's metric*. *Given two images f, $g \in \mathcal{I}^*$, g is chosen as the probing function for example, and we define the two numbers: $\lambda = \inf \{\alpha, f \leq \alpha \triangle g\}$ and $\mu = \sup \{\beta, \beta \triangle g \leq f\}$. The corresponding "functional Asplund's metric" $d_{As}^{\triangle}$ (with the LIP multiplication) is:*

$$d_{As}^{\triangle}(f,g) = \ln(\lambda/\mu) \tag{3}$$

From a mathematical point of view [11], $d_{As}^{\triangle}$ is a metric if the images $f, g \in \mathcal{I}^*$ are replaced by their equivalence classes $f^{\triangle} = \{g/\exists k > 0, k \triangle g = f\}$ and $g^{\triangle}$.

The relation $(\exists k > 0, k \triangle g = f)$ is clearly an equivalence relation written fRg, because it satisfies the three properties: (i) reflexivity $\forall f \in \mathcal{I}, fRf$, (ii) symmetry $\forall (f,g) \in \mathcal{I}^2$, $fRg \Leftrightarrow gRf$ and (iii) transitivity $\forall (f,g,h) \in \mathcal{I}^3$, $(fRg \text{ and } gRh) \Rightarrow fRh$. Let us now give a rigorous definition of the multiplicative Asplund's metric using the space of equivalence classes $\mathcal{I}^{\triangle}$

$$\forall (f^{\triangle}, g^{\triangle}) \in (\mathcal{I}^{\triangle})^2, \quad d_{As}^{\triangle}(f^{\triangle}, g^{\triangle}) = d_{As}^{\triangle}(f_1, g_1) \tag{4}$$

$d_{As}^{\triangle}(f_1, g_1)$ is defined by Eq. 3 between two elements f_1 and g_1 of the equivalence classes $f^{\triangle}$ and $g^{\triangle}$.

The demonstration of the metric properties are in the appendix (Sect. A).

Several examples have shown the interest of using Asplund's metric for pattern matching between a template function $t : D_t \to \mathcal{T}$ and the function f. For each point x of D, the distance $d_{As}^{\triangle}(f_{|D_t(x)}, t)$ is computed in the neighbourhood $D_t(x)$ centred in x, with $f_{|D_t(x)}$ being the restriction of f to $D_t(x)$. Therefore, one can define a map of Asplund's distances [23].

Definition 2. *Map of Asplund's distances*. *Given a grey-level image $f \in \mathcal{I}^*$ and a probe $t \in (\mathcal{T}^*)^{D_t}$, $t > 0$, their map of Asplund's distances is:*

$$As_t^{\triangle} f : \begin{cases} \mathcal{I}^* \times (\mathcal{T}^*)^{D_t} \to & (\mathbb{R}^+)^D \\ (f,t) & \to As_t^{\triangle} f(x) = d_{As}^{\triangle}(f_{|D_t(x)}, t) \end{cases} \tag{5}$$

$D_t(x)$ *is the neighbourhood associated to D_t centred in $x \in D$.*

One can notice that the template t is acting like a structuring element.

2.3 Short Reminder on Mathematical Morphology

In this subsection we give a reminder of the basis notions used in Mathematical Morphology (MM) [19]. MM is defined in complete lattices [3, 5, 9, 27].

Definition 3. *Complete lattice*. *Given a set $\mathcal{L}$ and a binary relation $\leq$ defining a partial order on $\mathcal{L}$, we say that $(\mathcal{L}, \leq)$ is a partially ordered set or poset. $\mathcal{L}$ is a complete lattice if any non empty subset $\mathcal{X}$ of $\mathcal{L}$ has a supremum (a least upper bound) and an infimum (a greatest lower bound). The infimum and the supremum will be denoted, respectively, by $\wedge \mathcal{X}$ and $\vee \mathcal{X}$. Two elements of the complete lattice $\mathcal{L}$ are important: the least element O and the greatest element I.*

The set of images from D to $[0, M]$, $\overline{\mathcal{I}} = [0, M]^D$, is a complete lattice with the partial order relation $\leq$, by inheritance of the complete lattice structure of $[0, M]$. The least and greatest elements are the constant functions f_0 and f_M whose values are equal respectively to 0 and M for all elements of D. The supremum and infimum are respectively, for any $\mathcal{X} \subset \overline{\mathcal{I}}$

$$
\begin{aligned}
(\wedge_{\overline{\mathcal{I}}} \mathcal{X})(x) &= \wedge_{[0,M]} \{f(x) : f \in \mathcal{X}, \, x \in D\} \\
(\vee_{\overline{\mathcal{I}}} \mathcal{X})(x) &= \vee_{[0,M]} \{f(x) : f \in \mathcal{X}, \, x \in D\}.
\end{aligned}
\tag{6}
$$

The set of functions $\overline{\mathbb{R}}^D$ is also a complete lattice with $\overline{\mathbb{R}} = \mathbb{R} \cup \{-\infty, +\infty\}$ with the usual order $\leq$, like the set of all subsets of D, written $\mathcal{P}(D)$, with the set inclusion $\subset$.

Definition 4. Erosion, dilation, anti-erosion, anti-dilation [3]. *Given $\mathscr{L}_1$ and $\mathscr{L}_2$ two complete lattices, a mapping $\psi \in \mathscr{L}_2^{\mathscr{L}_1}$ is*

1. *an erosion iff* $\quad \forall \mathcal{X} \subset \mathscr{L}_1, \, \psi(\wedge \mathcal{X}) = \wedge \psi(\mathcal{X})$, *then we write* $\varepsilon = \psi$;
2. *a dilation iff* $\quad \forall \mathcal{X} \subset \mathscr{L}_1, \, \psi(\vee \mathcal{X}) = \vee \psi(\mathcal{X})$, *then we write* $\delta = \psi$;
3. *an anti-erosion iff* $\quad \forall \mathcal{X} \subset \mathscr{L}_1, \, \psi(\wedge \mathcal{X}) = \vee \psi(\mathcal{X})$, *then we write* $\varepsilon^a = \psi$;
4. *an anti-dilation iff* $\quad \forall \mathcal{X} \subset \mathscr{L}_1, \, \psi(\vee \mathcal{X}) = \wedge \psi(\mathcal{X})$, *then we write* $\delta^a = \psi$.

As the definitions of these mappings apply even to the empty subset of $\mathscr{L}_1$, we have: $\varepsilon(I) = I$, $\delta(O) = O$, $\varepsilon^a(I) = O$ *and* $\delta^a(O) = I$.

Erosions and dilations are increasing mappings: $\forall \mathcal{X}, \mathcal{Y} \subset \mathscr{L}_1, \, \mathcal{X} \leq \mathcal{Y} \Rightarrow \psi(\mathcal{X}) \leq \psi(\mathcal{Y})$ while anti-erosions and anti-dilations are decreasing mappings: $\forall \mathcal{X}, \mathcal{Y} \subset \mathscr{L}_1, \, \mathcal{X} \leq \mathcal{Y} \Rightarrow \psi(\mathcal{Y}) \leq \psi(\mathcal{X})$.

Definition 5. Structuring element [9,26,29]. *Let us define a pulse function $i_{x,t} \in \overline{\mathcal{I}}$ of level t at the point x:*

$$
i_{x,t}(x) = t; \qquad i_{x,t}(y) = 0 \; if \; x \neq y.
\tag{7}
$$

The function f can be decomposed into the supremum of its pulses $f = \vee \{i_{x,f(x)}, x \in D\}$. It is easy to define dilations and erosions which are not translation-invariant (in the domain D). Let W be a map $\overline{\mathcal{I}} \to \overline{\mathcal{I}}$ associating to each pulse function $i_{x,t} \in \overline{\mathcal{I}}$ a (functional) "window" $W(i_{x,t})$. Then the operator $\delta_W : \overline{\mathcal{I}} \to \overline{\mathcal{I}}$ defined by:

$$
\delta_W(f) = \vee \{W(i_{x,f(x)}), x \in D\}
\tag{8}
$$

is a dilation. When all "windows" $W(i_{x,f(x)})$ are translation invariant (in D), they take the form $W(i_{x,f(x)}) = B(x)$ with $B(x) = B_x$ being a structuring element (or structuring function).

In this case the previously defined dilation δ and erosion ε, in the same lattice $(\overline{\mathcal{I}}, \leq)$, can be simplified:

$$
\begin{aligned}
(\delta_B(f))(x) &= \vee \{f(x - h) + B(h), h \in D_B\} = (f \oplus B)(x) \\
(\varepsilon_B(f))(x) &= \wedge \{f(x + h) - B(h), h \in D_B\} = (f \ominus B)(x)
\end{aligned}
\tag{9}
$$

$D_B \subset D$ is the definition domain of the structuring function $B : D_B \to \overline{\mathcal{T}}$. The symbols $\oplus$ and $\ominus$ represent the extension to functions [27] of Minkowski operations between sets [8,20]. Notice: in the case of a flat structuring element with its values equal to zero (i.e. $\forall x \in D_B$, $B(x) = 0$), we have $\delta_B(f)(x) = \vee \{f(x-h), h \in D_B\} = \delta_{D_B}(f)(x)$ and $\varepsilon_B(f)(x) = \wedge \{f(x+h), h \in D_B\} = \varepsilon_{D_B}(f)(x)$.

3   Map of Aplund's Distances and Mathematical Morphology

We now link the map of Asplund's distances with Mathematical Morphology.

Given $\overline{\mathbb{R}}^+ = [0, +\infty]$ a complete lattice with the natural order $\leq$, the map of the least upper bounds λ_B between the probe $B \in (\mathcal{T}^*)^{D_B}$ and the function $f \in \overline{\mathcal{I}}$ is defined as:

$$\lambda_B f : \begin{cases} \overline{\mathcal{I}} \times (\mathcal{T}^*)^{D_B} \to & \left(\overline{\mathbb{R}}^+\right)^D \\ (f, B) & \to \lambda_B f(x) = \wedge \{\alpha(x), f(x+h) \leq \alpha(x) \triangle B(h), h \in D_B\}. \end{cases} \tag{10}$$

The map of the greatest lower bounds $\mu^{\triangle}$ between the probe $B \in (\mathcal{T}^*)^{D_B}$ and the function $f \in \overline{\mathcal{I}}$ is defined as:

$$\mu_B f : \begin{cases} \overline{\mathcal{I}} \times (\mathcal{T}^*)^{D_B} \to & \left(\overline{\mathbb{R}}^+\right)^D \\ (f, B) & \to \mu_B f(x) = \vee \{\beta(x), \beta(x) \triangle B(h) \leq f(x+h), h \in D_B\}. \end{cases} \tag{11}$$

The two mappings λ_B and μ_B are defined between two complete lattices $\mathscr{L}_1 = (\overline{\mathcal{I}}, \leq)$ and $\mathscr{L}_2 = \left(\left(\overline{\mathbb{R}}^+\right)^D, \leq\right)$ with the natural order $\leq$. Therefore, the least element of $(\mathscr{L}_1, \leq)$ corresponds to the constant function equal to zero, $O = f_0$ and the greatest element is the constant function equal to M, $I = f_M$.

Using the Eqs. 10 and 11 the map of Asplund's distances (Eq. 5) can be simplified:

$$As_B^{\triangle} f = \ln\left(\frac{\lambda_B f}{\mu_B f}\right), \text{ with } f > 0. \tag{12}$$

In addition, $\forall x \in D$, $\forall h \in D_B$, $\forall \alpha \in \mathbb{R}^+$, we have:

$$\alpha(x) \triangle B(h) \geq f(x+h) \Leftrightarrow M - M(1 - B(h)/M)^{\alpha(x)} \geq f(x+h), \text{ (from Eq. 2)}$$

$$\Leftrightarrow \alpha(x) \geq \frac{\ln\left(1 - \frac{f(x+h)}{M}\right)}{\ln\left(1 - \frac{B(h)}{M}\right)}, \text{ because } \left(1 - \frac{B(h)}{M}\right) \in]0, 1[. \tag{13}$$

We assume that $\widetilde{f} = \ln(1 - f/M)$. Using Eq. 13, Eq. 10 becomes:

$$\lambda_B f = \wedge \left\{\alpha(x), \alpha(x) \geq \frac{\widetilde{f}(x+h)}{\widetilde{B}(h)}, h \in D_B\right\} = \vee \left\{\frac{\widetilde{f}(x+h)}{\widetilde{B}(h)}, h \in D_B\right\}. \tag{14}$$

In a similar way:

$$\mu_B f = \vee \left\{ \beta(x), \beta(x) \le \frac{\widetilde{f}(x+h)}{\widetilde{B}(h)}, h \in D_B \right\} = \wedge \left\{ \frac{\widetilde{f}(x+h)}{\widetilde{B}(h)}, h \in D_B \right\}. \quad (15)$$

3.1 Case of a Flat Structuring Element

In the case of a flat structuring element $B = B_0 \in T^*$ ($\forall x \in D_B$, $B(x) = B_0$), the Eqs. 14 and 15 can be simplified.

$$
\begin{aligned}
\lambda_{B_0} f &= \tfrac{1}{\widetilde{B_0}} \wedge \left\{ \widetilde{f}(x+h), h \in D_B \right\}, \quad \text{because } \widetilde{B}_0 < 0 \\
&= \tfrac{1}{\widetilde{B_0}} \ln \left(1 - \frac{\vee\{f(x-h),-h\in \check{D}_B\}}{M} \right) \\
&= \tfrac{1}{\widetilde{B_0}} \ln \left(1 - \frac{\delta_{\check{D}_B} f}{M} \right)
\end{aligned}
\quad (16)
$$

Notice: the infimum $\wedge$ is changed into an supremum $\vee$ because the function $\widetilde{f} : x \rightarrow \ln(1 - x/M)$ is a continuous decreasing mapping. The reflected (or transposed) domain $\check{D}_B$ is $\check{D}_B = \{-h, h \in D_B\}$ and the reflected structuring function $\check{B}$ is defined by the reflection of its definition domain $\forall x \in \check{D}_B$, $\check{B}(x) = B(-x)$ [28]. Similarly:

$$
\begin{aligned}
\mu_{B_0} f &= \tfrac{1}{\widetilde{B_0}} \vee \left\{ \widetilde{f}(x+h), h \in D_B \right\}, \quad \text{because } \widetilde{B}_0 < 0 \\
&= \tfrac{1}{\widetilde{B_0}} \ln \left(1 - \frac{\wedge\{f(x+h),h\in D_B\}}{M} \right) \\
&= \tfrac{1}{\widetilde{B_0}} \ln \left(1 - \frac{\varepsilon_{D_B} f}{M} \right).
\end{aligned}
\quad (17)
$$

With the Eqs. 16 and 17, the map of Asplund's distances (Eq. 12) becomes:

$$
As^{\triangle}_{B_0} f = \ln \left[\frac{\ln \left(1 - \frac{\delta_{\check{D}_B} f}{M} \right)}{\ln \left(1 - \frac{\varepsilon_{D_B} f}{M} \right)} \right], \quad \text{with } f > 0.
\quad (18)
$$

This important result shows that, with a flat probe, the map of Aplünd's distances can be computed using logarithms and operations of morphological erosion and dilation of an image. From an implementation point of view, the programming of the map of Aplünd's distances becomes easier, because the majority of image processing libraries contains morphological operations.

Notice: by replacing the dilation and erosion by rank-filters [26] one can compute the map of Asplund's distances with a tolerance [14,23].

3.2 General Case: A Structuring Function

Using a general structuring function in the Eqs. 14 and 15, the map of Asplund's distances is expressed as:

$$As_B^{\triangle} f = \ln\left(\frac{\lambda_B f}{\mu_B f}\right) = \ln\left(\frac{\vee\left\{\frac{\tilde{f}(x+h)}{\tilde{B}(h)}, h \in D_B\right\}}{\wedge\left\{\frac{\tilde{f}(x+h)}{\tilde{B}(h)}, h \in D_B\right\}}\right), \quad \text{with } f > 0. \quad (19)$$

Let us study the properties of mappings λ_B, $\mu_B \in \mathscr{L}_2^{\mathscr{L}_1}$, $\forall f, g \in \overline{\mathcal{I}}$

$$\lambda_B(f \vee g) = \vee\left\{\frac{\widetilde{f \vee g}(x+h)}{\tilde{B}(h)}, h \in D_B\right\}$$

$$= \vee\left\{\frac{\tilde{f}(x+h) \wedge \tilde{g}(x+h)}{\tilde{B}(h)}, h \in D_B\right\}, \text{ because } \tilde{f} \text{ is decreasing}$$

$$= \vee\left\{\frac{\tilde{f}(x+h)}{\tilde{B}(h)} \vee \frac{\tilde{g}(x+h)}{\tilde{B}(h)}, h \in D_B\right\}, \text{ because } \tilde{B}(h) < 0 \quad (20)$$

$$= \left[\vee\left\{\frac{\tilde{f}(x+h)}{\tilde{B}(h)}, h \in D_B\right\}\right] \vee \left[\vee\left\{\frac{\tilde{g}(x+h)}{\tilde{B}(h)}, h \in D_B\right\}\right]$$

$$= \lambda_B(f) \vee \lambda_B(g).$$

According to Definition 4, 2, λ_B is a dilation. In addition,

$$\lambda_B(O) = \lambda_B(f_0) = \wedge\left\{\alpha(x), \alpha(x) \geq \frac{\tilde{0}(x+h)}{\tilde{B}(h)}, h \in D_B\right\} = 0 = O. \quad (21)$$

Similarly, we have:

$$\mu_B(f \wedge g) = \wedge\left\{\frac{\widetilde{f \wedge g}(x+h)}{\tilde{B}(h)}, h \in D_B\right\}$$

$$= \wedge\left\{\frac{\tilde{f}(x+h) \vee \tilde{g}(x+h)}{\tilde{B}(h)}, h \in D_B\right\}, \text{ because } \tilde{f} \text{ is decreasing}$$

$$= \left[\wedge\left\{\frac{\tilde{f}(x+h)}{\tilde{B}(h)}, h \in D_B\right\}\right] \wedge \left[\wedge\left\{\frac{\tilde{g}(x+h)}{\tilde{B}(h)}, h \in D_B\right\}\right], \text{ because } \tilde{B}(h) < 0 \quad (22)$$

$$= \mu_B(f) \wedge \mu_B(g).$$

According to Definition 4, 1, μ_B is an erosion. In addition,

$$\mu_B(I) = \mu_B(f_M) = \vee\left\{\beta(x), \beta(x) \leq \frac{\widetilde{M}(x+h)}{\tilde{B}(h)}, h \in D_B\right\} = +\infty = I. \quad (23)$$

Therefore, the map of Asplund's distances is the logarithm of the ratio between a dilation and an erosion of the function f by the structuring function B. The map of the least upper bounds λ_B is an anti-dilation and the map of the greatest lower bounds μ_B is an anti-erosion. The two maps are defined from the mapping $(\mathscr{L}_1 = \overline{\mathcal{I}}, \leq)$ and the mapping $(\mathscr{L}_2 = (\overline{\mathbb{R}}^+)^D, \leq)$ with their respective natural orders.

4   Illustration

In Fig. 1(a), we extract a tile (i.e. the probe or the structuring function) in an image f and we look for the similar ones in a darken image, f^d, by means of a

LIP multiplication of 0.3. Physically, it corresponds to an object with a stronger light absorption. Importantly, the probe has a non convex domain shape and is not flat. We compute the map of Asplund's distances between the probe B and the image f^d with a tolerance, $As^{\triangle}_{B,p}f^d$, as introduced in [14,23]. This metric, robust to noise, is computed by discarding $p = 30\%$ of the points which are the closest to the least upper bounds and to the greatest lower bounds. The tiles are located at the local minima of the distance map which are extracted by a threshold of 0.7 (Fig. 1(b)). The tiles similar to the probe, according to the Asplund's distance, have been correctly detected (Fig. 1(c)). Notice that the domain of the probe is slightly smaller than the domain of the tiles.

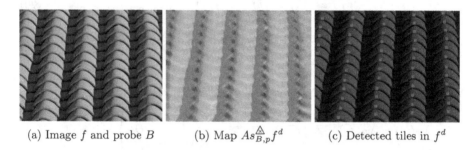

(a) Image f and probe B (b) Map $As^{\triangle}_{B,p}f^d$ (c) Detected tiles in f^d

Fig. 1. Detection of tiles using the map of Asplund's distances $As^{\triangle}_{B,p}f^d$ with a tolerance $p = 30\%$. (a) The probe B (in green) is extracted in the image f. (b) The minima (in blue) of the map of distances, $As^{\triangle}_{B,p}f^d$, between the probe and the darken image f^d are extracted with a threshold of 0.7. (c) Location of the detected tiles (red dots) in the darken image f^d. (Color figure online)

5   Conclusion

In the current paper, we have shown that the map of Asplund's distances between a probe and a function using the LIP multiplication is linked with morphological operations. The probe corresponds to a structuring function and the map of Asplund's distances is the logarithm of the ratio between a dilation and an erosion of the function by the structuring function into the lattice of positive functions $((\overline{\mathbb{R}}^+)^D, \leq)$. The dilation is the map of the least upper bounds $\lambda_B f$, between the function f and the probe B, while the erosion is the map of the greatest lower bounds $\mu_B f$.

The dilation and the erosion are mappings between the complete lattices of the images $(\overline{\mathcal{I}}, \leq)$ and the lattice $((\overline{\mathbb{R}}^+)^D, \leq)$ with the natural order. When using a flat structuring element, the expression of the map of Asplund's distances can be simplified with a dilation and an erosion of the image into the same lattice of the images $(\overline{\mathcal{I}}, \leq)$. An example of pattern matching has been presented with a non-flat structuring function.

The obtained results set the pattern matching approach by Asplund's distances in the well established framework of Mathematical Morphology. The current reasoning can be extended to the double-sided probing by Asplund's distances for colour and multivariate images using the LIP multiplicative or the LIP additive framework [11,23,24]. This will be presented in a coming paper.

A Appendix

Le us demonstrate that the Aplünd's metric $d_{As}^{\triangle}$ is a metric in the space of equivalence classes $\mathcal{I}^{\triangle}$. In order to be a metric on $(\mathcal{I}^{\triangle} \times \mathcal{I}^{\triangle}) \to \mathbb{R}^+$, $d_{As}^{\triangle}$ must satisfy the four following properties:

1. (positivity): $\forall f^{\triangle} \neq g^{\triangle} \in \mathcal{I}^{\triangle}$, $\forall x \in D$, $\lambda \triangle g^{\triangle}(x) > \mu \triangle g^{\triangle}(x)$ (Def. 1),
   because $g^{\triangle} > 0$
   $\Rightarrow \lambda > \mu$ because $\mathcal{I}^{\triangle}$ is an ordered set with the order $\leq$
   $\Rightarrow \forall f^{\triangle} \neq g^{\triangle} \in \mathcal{I}^{\triangle}$, $d_{As}^{\triangle}(f^{\triangle}, g^{\triangle}) > 0$.
2. (Axiom of separation):

$$\left.\begin{array}{l} d_{As}^{\triangle}(f^{\triangle}, g^{\triangle}) = 0 \Rightarrow \lambda = \mu \\ (\text{def.1}) \Rightarrow \lambda \triangle g^{\triangle} \geq f^{\triangle} \geq \mu \triangle g^{\triangle} \end{array}\right\} \Rightarrow \lambda \triangle g^{\triangle} = f^{\triangle} \Rightarrow f^{\triangle} = g^{\triangle} \text{ in } \mathcal{I}^{\triangle}(\text{A.1})$$

Reciprocally:

$$\left.\begin{array}{l} \forall f^{\triangle}, g^{\triangle} \in \mathcal{I}^{\triangle}, f^{\triangle} = g^{\triangle} \Rightarrow \lambda \triangle g^{\triangle} = f^{\triangle} \\ (\text{def. 1}) \lambda \triangle g^{\triangle} \geq f^{\triangle} \geq \mu \triangle g^{\triangle} \end{array}\right\} \Rightarrow \lambda \triangle g^{\triangle} = f^{\triangle} = \mu \triangle g^{\triangle}$$
(A.2)

$$\Rightarrow \lambda = \mu \Rightarrow d_{As}^{\triangle}(f^{\triangle}, g^{\triangle}) = 0$$

Equations A.1 and A.2 $\Rightarrow \left\{\forall f^{\triangle}, g^{\triangle} \in \mathcal{I}^{\triangle}, \ d_{As}^{\triangle}(f^{\triangle}, >) = 0 \Leftrightarrow f^{\triangle} = g^{\triangle}\right\}$.

3. (Triangle inequality): Let us define: $d_{As}^{\triangle}(f^{\triangle}, g^{\triangle}) = \ln(\lambda_1/\mu_1)$, $d_{As}^{\triangle}(g^{\triangle}, h^{\triangle}) = \ln(\lambda_2/\mu_2)$ and $d_{As}^{\triangle}(f^{\triangle}, h^{\triangle}) = \ln(\lambda_3/\mu_3)$. We have

$$d_{As}^{\triangle}(f^{\triangle}, g^{\triangle}) + d_{As}^{\triangle}(g^{\triangle}, h^{\triangle}) = \ln\left(\frac{\lambda_1\lambda_2}{\mu_1\mu_2}\right) \qquad (\text{A.3})$$

$$\text{Def. 1} \Rightarrow \left\{\begin{array}{l} \lambda_1 = \inf\left\{k_1 : \forall x, k_1 \triangle g^{\triangle}(x) \geq f^{\triangle}(x)\right\} \\ \lambda_2 = \inf\left\{k_2 : \forall x, k_2 \triangle h_j^{\triangle}(x) \geq g^{\triangle}(x)\right\} \\ \lambda_3 = \inf\left\{k_3 : \forall x, k_3 \triangle h_j^{\triangle}(x) \geq f^{\triangle}(x)\right\} \end{array}\right.$$

$$\Rightarrow \lambda_1\lambda_2 \leq \inf_{k_1}\left\{\inf_{k_2}\left\{\forall x, k_1 \triangle (k_2 \triangle h_j^{\triangle}(x)) \geq k_1 \triangle g^{\triangle}(x)\right\} \geq f^{\triangle}(x)\right\}$$

$$\leq \inf\left\{k' : \forall x, k' \triangle h_j^{\triangle}(x) \geq f^{\triangle}(x)\right\}, \text{ with } k' = k_1 \times k_2 \qquad (A.4)$$

$$\Rightarrow \lambda_1\lambda_2 \leq \lambda_3, \text{ with } \lambda_1, \lambda_2, \lambda_3 > 0$$

In the same way:

$$\mu_1\mu_2 \geq \mu_3, \text{ with } \mu_1, \mu_2, \mu_3 > 0 \qquad (A.5)$$

Equations A.3, A.4 and A.5 $\Rightarrow \frac{\lambda_1\lambda_2}{\mu_1\mu_2} \geq \frac{\lambda_3}{\mu_3}$

$\Rightarrow \forall f^{\triangle}, g^{\triangle}, h^{\triangle} \in \mathcal{I}^{\triangle}, d_{As}^{\triangle}(f^{\triangle}, h^{\triangle}) \leq d_{As}^{\triangle}(f^{\triangle}, g^{\triangle}) + d_{As}^{\triangle}(g^{\triangle}, h^{\triangle}).$

4. (Axiom of symmetry): Let us define: $d_{As}^{\triangle}(f^{\triangle}, g^{\triangle}) = \ln(\lambda_1/\mu_1)$, $d_{As}^{\triangle}(g^{\triangle}, f^{\triangle}) = \ln(\lambda_2/\mu_2)$.

Def. 1 $\Rightarrow \lambda_1 = \inf\left\{k : \forall x, g^{\triangle} \geq \frac{1}{k} \triangle f^{\triangle}\right\}$, because $k > 0$

$\Rightarrow \frac{1}{\lambda_1} = \sup\left\{k' : \forall x, g^{\triangle} \geq k' \triangle f^{\triangle}\right\} \Rightarrow \frac{1}{\lambda_1} = \mu_2.$

In the same way, we have $\frac{1}{\mu_1} = \lambda_2.$

Therefore,

$\forall f^{\triangle}, g^{\triangle} \in \mathcal{I}^{\triangle}, d_{As}^{\triangle}(f^{\triangle}, g^{\triangle}) = \ln(\lambda_1/\mu_1) = \ln(\lambda_2/\mu_2) = d_{As}^{\triangle}(g^{\triangle}, f^{\triangle}).$

References

1. Asplund, E.: Comparison between plane symmetric convex bodies and parallelograms. Math. Scand. **8**, 171–180 (1960)
2. Banon, G.J.F., Faria, S.D.: Morphological approach for template matching. In: Proceedings X Brazilian Symposium on Computer Graphics and Image Processing, pp. 171–178, October 1997
3. Banon, G.J.F., Barrera, J.: Decomposition of mappings between complete lattices by mathematical morphology, part I. General lattices. Signal Process. **30**(3), 299–327 (1993). http://www.sciencedirect.com/science/article/pii/0165168493900153
4. Barat, C., Ducottet, C., Jourlin, M.: Virtual double-sided image probing: a unifying framework for non-linear grayscale pattern matching. Pattern Recogn. **43**(10), 3433–3447 (2010). http://www.sciencedirect.com/science/article/pii/S003132031 0001962
5. Birkhoff, G.: Lattice Theory, American Mathematical Society Colloquium Publications, vol. 25, 3rd edn. American Mathematical Society, Providence (1967)
6. Brailean, J., Sullivan, B., Chen, C., Giger, M.: Evaluating the EM algorithm for image processing using a human visual fidelity criterion. In: International Conference on Acoustics, Speech, and Signal Processing, ICASSP 1991, vol. 4, pp. 2957–2960, April 1991
7. Grünbaum, B.: Measures of symmetry for convex sets. In: Proceedings of Symposia in Pure Mathematics, vol. 7, pp. 233–270 (1963)
8. Hadwiger, H.: Vorlesungen Über Inhalt, Oberfläche und Isoperimetrie. Grundlehren der Mathematischen Wissenschaften. Springer, Heidelberg (1957)
9. Heijmans, H., Ronse, C.: The algebraic basis of mathematical morphology I. Dilations and erosions. Comput. Vis. Graph. Image Process. **50**(3), 245–295 (1990). http://www.sciencedirect.com/science/article/pii/0734189X9090148O

10. Jourlin, M.: Chapter one - gray-level LIP model. Notations, recalls, and first applications. In: Jourlin, M. (ed.) Logarithmic Image Processing: Theory and Applications. Advances in Imaging and Electron Physics, vol. 195, pp. 1–26. Elsevier, Amsterdam (2016). http://www.sciencedirect.com/science/article/pii/S107656701 6300313

11. Jourlin, M.: Chapter three - metrics based on logarithmic laws. In: Jourlin, M. (ed.) Logarithmic Image Processing: Theory and Applications. Advances in Imaging and Electron Physics, vol. 195, pp. 61–113. Elsevier, Amsterdam (2016). http://www.sciencedirect.com/science/article/pii/S1076567016300337

12. Jourlin, M., Breugnot, J., Itthirad, F., Bouabdellah, M., Closs, B.: Chapter 2 - logarithmic image processing for color images. In: Hawkes, P.W. (ed.) Advances in Imaging and Electron Physics, vol. 168, pp. 65–107. Elsevier, Amsterdam (2011)

13. Jourlin, M., Carré, M., Breugnot, J., Bouabdellah, M.: Chapter 7 - logarithmic image processing: additive contrast, multiplicative contrast, and associated metrics. In: Hawkes, P.W. (ed.) Advances in Imaging and Electron Physics, vol. 171, pp. 357–406. Elsevier, Amsterdam (2012)

14. Jourlin, M., Couka, E., Abdallah, B., Corvo, J., Breugnot, J.: Asplünd's metric defined in the logarithmic image processing (LIP) framework: a new way to perform double-sided image probing for non-linear grayscale pattern matching. Pattern Recogn. **47**(9), 2908–2924 (2014)

15. Jourlin, M., Pinoli, J.: A model for logarithmic image processing. J. Microsc. **149**(1), 21–35 (1988)

16. Jourlin, M., Pinoli, J.: Logarithmic image processing: the mathematical and physical framework for the representation and processing of transmitted images. In: Hawkes, P.W. (ed.) Advances in Imaging and Electron Physics, vol. 115, pp. 129–196. Elsevier, Amsterdam (2001)

17. Jourlin, M., Pinoli, J.C.: Image dynamic range enhancement and stabilization in the context of the logarithmic image processing model. Signal Process. **41**(2), 225–237 (1995). http://www.sciencedirect.com/science/article/pii/0165168494001026

18. Khosravi, M., Schafer, R.W.: Template matching based on a grayscale hit-or-miss transform. IEEE Trans. Image Process. **5**(6), 1060–1066 (1996)

19. Matheron, G.: Eléments pour une théorie des milieux poreux. Masson, Paris (1967)

20. Minkowski, H.: Volumen und oberfläche. Math. Ann. **57**, 447–495 (1903). http://eudml.org/doc/158108

21. Noyel, G., Angulo, J., Jeulin, D.: Morphological segmentation of hyperspectral images. Image Anal. Stereol. **26**(3), 101–109 (2007)

22. Noyel, G., Angulo, J., Jeulin, D., Balvay, D., Cuenod, C.A.: Multivariate mathematical morphology for DCE-MRI image analysis in angiogenesis studies. Image Anal. Stereol. **34**(1), 1–25 (2014)

23. Noyel, G., Jourlin, M.: Asplünd's metric defined in the logarithmic image processing (LIP) framework for colour and multivariate images. In: IEEE International Conference on Image Processing (ICIP), pp. 3921–3925, September 2015

24. Noyel, G., Jourlin, M.: Spatio-colour Asplünd 's metric and logarithmic image processing for colour images (LIPC). In: CIARP2016 - XXI IberoAmerican Congress on Pattern Recognition. International Association for Pattern Recognition (IAPR), Lima, Peru, November 2016. https://hal.archives-ouvertes.fr/hal-01316581

25. Odone, F., Trucco, E., Verri, A.: General purpose matching of grey level arbitrary images. In: Arcelli, C., Cordella, L.P., di Baja, G.S. (eds.) IWVF 2001. LNCS, vol. 2059, pp. 573–582. Springer, Heidelberg (2001). doi:10.1007/3-540-45129-3_53

26. Serra, J.: Image Analysis and Mathematical Morphology: Theoretical Advances, vol. 2. Academic Press, London (1988)
27. Serra, J., Cressie, N.: Image Analysis and Mathematical Morphology, vol. 1. Academic Press, London (1982)
28. Soille, P.: Morphological Image Analysis: Principles and Applications, 2nd edn. Springer, New York (2003)
29. Verdú-Monedero, R., Angulo, J., Serra, J.: Anisotropic morphological filters with spatially-variant structuring elements based on image-dependent gradient fields. IEEE Trans. Image Process. **20**(1), 200–212 (2011)

Biomedical, Material Science and
Physical Applications

An Affinity Score for Grains Merging and Touching Grains Separation

Théodore Chabardès[(✉)], Petr Dokládal, Matthieu Faessel, and Michel Bilodeau

CMM, Center for Mathematical Morphology, PSL Research University - MINES
ParisTech, 35 rue Saint Honoré, Fontainebleau, France
{theodore.chabardes,petr.dokladal,matthieu.faessel,
michel.bilodeau}@mines-paristech.fr

Abstract. The physical properties of granular materials on a macroscopic scale derive from their microstructures. The segmentation of CT-images of this type of material is the first step towards simulation and modeling but it is not a trivial task. Non-spherical, elongated or non-convex objects fail to be separated with classical methods. Moreover, grains are commonly fragmented due to external conditions: aging, storage conditions, or even user-induced mechanical deformations. Grains are crushed into multiple fragments of different shape and volume; those fragments drift from one another in the binder phase. This paper focuses on reconstruction of grains from these fragments using scores that match the local thickness and the regularity of the interface between two objects from a given primary segmentation of the material. An affinity graph is built from those scores and optimized for a given application using a user-generated ground truth on a 2D slice of the tridimensional structures. A minimum spanning tree is generated, and a hierarchical cut is performed. This process allows to reassemble drifted fragments into whole grains and to solve the touching grains problem in tridimensional acquisitions.

1 Introduction

X-ray microtomography is now widely used by professionals and non-experts to characterize the structures of materials of interest. However, assessing precisely the structural information by conventional means is difficult due to the size of the acquisitions. Human interactions are unpractical for this amount of data. There is a need for automatic methods of analysis that can harvest the ever-increasing computing power.

This study focuses on composite materials made from a rubbery binder and brittle grains of various shapes and sizes mixed into a homogeneous mixture. Damage occurs when a mechanical shock is applied, causing fragmentation of these grains. The damaged structure affects the behavior of the material. Evaluating the extent of fragmentation may be useful for estimating effects on the reaction of the mixture. Highly clustered fragments and complex shapes require specific methods to identify all connections. The drift of fragments in the binder

© Springer International Publishing AG 2017
J. Angulo et al. (Eds.): ISMM 2017, LNCS 10225, pp. 423–434, 2017.
DOI: 10.1007/978-3-319-57240-6_34

Fig. 1. The fragment 2 is closer to 1 even though it should be associated to 3.

phase can not be solved using a method only based on the barycenter distance, as e.g. in [10], see Figs. 1 and 6.

In the first part, we present how to evaluate the shape of the interface between two objects of a given segmentation. Pixels equidistant to two objects border to border form separation segments. Each pixel of those segments is valued by the distance to border. Then, those segments are associated statistics as mean, standard deviation etc. of this valuation. Some already relate to affinities; others can be transformed to affinities using a Gaussian kernel. In a second part, the grains are reassembled from the derived fragments using the proposed affinity score in combination with a hierarchical clustering, and a set of predetermined measurements done on the unprocessed acquisition.

1.1   Materials and 3D Images

The studied materials are composite materials with an elastic binder degraded by a mechanical impact of a falling mass. Those samples were acquired by X-ray microtomography with a Skyscan 1172 high-resolution micro-CT system. Images of the work of Gillibert [6] were reused here for comparison purposes. The introduced method was also tested on our acquisitions. The following acquisitions illustrated in Fig. 2 are studied:

- MAT1 is a $1014 * 1155 * 250$ voxels image ($3650.4 * 4158 * 900$ μm^3). A mechanical impact is exerced from a 2 kg mass falling 15 cm. This material has been studied in [7].
- MAT2 is a $1000 * 1000 * 1000$ voxels image ($2097 * 2097 * 2097$ μm^3). A mechanic impact is exerced from a 2 kg mass falling 30 cm. This is a new material.

1.2   Initial Segmentation

The work presented in this paper requires a primary segmentation on the acquisition. We suppose each fragment for every grain isolated, which effectively induces an over-segmentation, where grains are fragmented. Then, we successively merge pairs of fragments, whose affinity is the highest. Several approaches can be considered to create this over-segmentation. Gillibert and al. used a multiscale stochastic watershed in [6]. The estimation of the granulometry makes it possible to select appropriate scales to perform several stochastic watersheds and combine them to obtain a correct segmentation for each size of grains. However, this approach is tedious and costly.

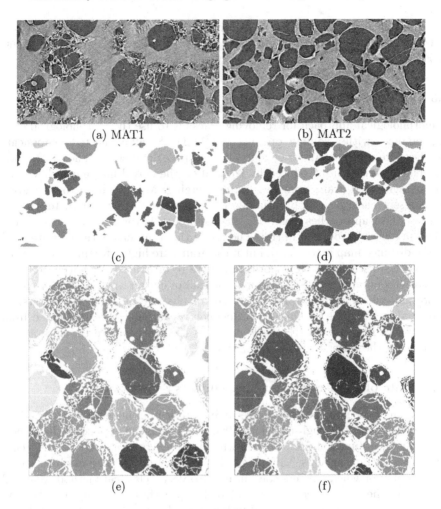

(a) MAT1   (b) MAT2

(c)   (d)

(e)   (f)

Fig. 2. (a, b) are 2D slices of 3D X-ray microtomographic images of fragmented granular material. (c, d) are the corresponding over-segmentation produced by h-minima markers. (e, f) are results obtained in [7]. (e) Watershed segmentation of the closed image after a h-minima filter. (f) Watershed segmentation using markers computed from the K-means. In (e) and (f), we can observe several grains over-segmented or under-segmented.

A less refined approach can be used, as over-segmented fragments will be merged during the clustering process: the resulting affinity will be high in this case. In this study, the image is first binarized using a method close to the Otsu threshold. The watershed is then computed from the inverse of the distance map, and the set of markers is generated by using the h-minima filter.

Other approaches could also be considered, such as using the distance map on the binarized image or a probability density map to generate random markers for the watershed transformation. The only requirement is to find a right amount of over-segmentation. Every fragment can be split into a small number of objects in the over-segmentation, but no object must gather more than one fragment. In Fig. 2, two segmentations have been provided for the studied materials.

1.3 State of the Art of Fragments Merging

A morphological approach for removing cracks is to use the morphological closing combined with a volumic opening, see [11,17]. Small connected components are therefore removed, and fragments close in space are merged. The constrained watershed transform introduced by [2] used on the closed image can differentiate a fragmented grain from two grains merging. Markers have to be chosen to represent each grain appropriately and are an important parameter in this method. Two approaches exist. The first approach is topological and uses the h-minima filter introduced in [18]. The use of the h-minima with the watershed on the distance map is standard, but if the grains are highly fragmented and the fragments are scattered, the algorithm fails to reconstruct grains accurately.

The second approach is based on a method of cluster analysis, the K-means clustering, which aims to partition a set of observations into K clusters, as described in [10]. The number of clusters can be automatically calculated from a covariance measure. In [7], the K-means is used to generate the appropriate markers for the watershed transformation.

1.4 State of the Art of Touching Objects Separation

Separation of touching objects is a recurring problem in image processing. Classical techniques like Hough-transform [9] or the watershed applied to the inverse of the distance function perform well when the objects of interest possess regular shapes such as spheres. Unfortunately, when shape and size of objects vary considerably or when clusters contain many objects, classical methods may fail to produce the desired separation. Different approaches exist [15]: morphology-based procedures [1], contour-based techniques [14], graph theoretic approaches, parametric fitting algorithms and level set approach [3,16]. Morphological multiscale decomposition can decompose clusters into size-specific scales, carrying markers for each disjoint region [8]. Methods exist for shape specific objects. In [19], a method based on a modified version of the pseudo-Euclidean distance transformation is able to split merged ellipses. In [4], a gap-filling method is proposed for elliptic shapes.

2 The Morphological Affinity Score for Adjacent Objects

2.1 Definitions

We start from an initial segmentation as described in Sect. 1.2. Let $f : \mathscr{D} \to \{0,1\}$ a binary image, where $\mathscr{D}$ is the definition domain.

Binary Objects. We define the set $\mathscr{X}$ of N objects as $\mathscr{X} = \{X_i\}_{i<N}$, so that $X_i \subset \mathscr{D}$ and $X_i \cap X_j = \emptyset$ for $i \neq j$. Each X_i represents a grain or a fragment. The complement $D \setminus \mathscr{X}$ is the binder.

Distance Mapping. We define the distance mapping $d : \mathscr{D} \to \mathbb{N}$ labeling each pixel with the distance to the nearest object $X_i \in \mathscr{X}$ according to a chosen metric.

SKIZ by Distance Function. A zone of influence of a object X_i of $\mathscr{X}$ is the subset of $\mathscr{D}$ that are closer to X_i than to any other object, and we note it $zi(i)$. A SKIZ is defined as the boundary of all zones of influence. We compute the SKIZ by using the watershed transform. We note this set $\mathrm{skiz}(\mathscr{X})$ and we assume it thin and connected using some connectivity C (26-connectivity in this paper).

Triple Points. The set $\mathscr{T}$ of triple points is the set of pixels that are equidistant to three or more distinct objects. We have extended the set $\mathscr{T}$ by including pixels on the border of f that are equidistant to two objects. The triple points are the extremities of a given segment of the SKIZ.

Interfaces. We define interfaces between X_i, X_j as part of the segments along the SKIZ. We note I the set of interfaces, and I_{ij} the interface between two distinct objects X_i and X_j. The distance decreases when going from triple points inwards, as shown in Fig. 4. However, this distance behaves differently when observed closer to the objects of interest. We define an incremental process to thin the skiz into a non-decremental segment, using C defined above:

$$I^0 = \mathrm{skiz}(\mathscr{X}), \ T^0 = \mathscr{T}$$
$$T^{i+1} = \{a \mid \exists b \mathsf{C} a, b \in T^i, \ d(b) > d(a)\}$$
$$I^{i+1} = I^i \setminus T^{i+1}$$

Finally, we define $I_{ij} = I^\infty \cap zi(i) \cap zi(j)$. Those interfaces are the support for building our new affinity scores. The distance mapping considered along the pixels of an interface gives various information regarding the shape of the interface between two objects. The Fig. 3(d) illustrates the interfaces I_{ij} between various objects given some initial segmentation Fig. 3(b).

2.2 Weights

In the following part, we propose several measures to weigh the interfaces. All proposed weights are $w : I \to \mathbb{R}$. However, it is important to notice that the range of w differs regarding to the considered weight.

The Barycenter Distance. The simplest measure already used in [7] is the Euclidean distance between barycenters of two distinct objects: $w_{bar}(I_{ij}) = \|\overline{X_i} - \overline{X_j}\|_{L^2}$, where $\overline{X}$ denotes the barycenter of the object labeled X in $\mathscr{X}$.

Sum on the Thinned Interface: $w_d(I_{ij}) = \sum_{x \in I_{ij}} d(x)$ with d the distance mapping previously introduced.

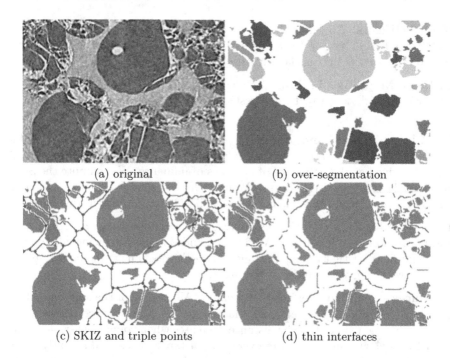

(a) original   (b) over-segmentation

(c) SKIZ and triple points   (d) thin interfaces

Fig. 3. The process to generate interfaces is illustrated. In (a) a granular material. (b) an example of segmentation that was used to produce the SKIZ. In (c), The SKIZ is drawn in red, the objects in gray, and triple points in blue. In (d) are shown interfaces generated using the Manhattan distance. (Color figure online)

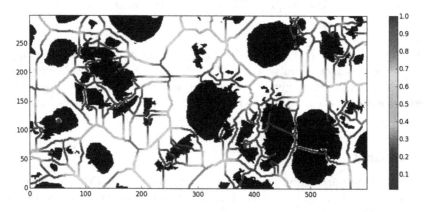

Fig. 4. The distance along the skiz increases towards the triple points.

Area of the Thinned Interface: $w_a(I_{ij}) = |I_{ij}|$ the area of each interface (length in 2D).

Mean: $w_\mu(I_{ij}) = w_d(I_{ij})/w_l(I_{ij})$ the mean of the distance.

Variance: $w_{\sigma^2}(I_{ij}) = \sum_{x \in I_{ij}} d(x)^2 - w_\mu^2(I_{ij})$ the variance of the distance.

2.3 Affinities

We note A, a positive and symetric matrix, with $a_{ij} \in (0,1)$ the affinity score of X_i and X_j, where a value close to 1 corresponds to two closely related objects and a value close to 0 to distinct objects. All previous weights can be transformed to an affinity. We use a Gaussian kernel for all weights that relate to a distance, where σ_w is the scale of the kernel, as follows:

$$A = G(w, \sigma_w) = \frac{1}{\sqrt{2\pi}\sigma_w} e^{-\frac{w^2}{2\sigma_w^2}}$$

Using the previous equations, we have A_{bar}, A_d, A_a, A_μ, A_{σ^2}.

Length-to-Surface Ratio. Other affinities can be obtained from the previous weights. Below, we use an affinity based on the surface of the objects. We define $s(X)$ the surface of the object X as $s(X) =| \{x, x \in G(f), x \in X\} |$, where $G(f)$ is a morphological gradient of the image f. We now define the new affinity A_s as follows:

$$(a_s)_{ij} = \frac{w_l(I_{ij})}{min(s(X_i), s(X_j))}$$

Note that $w_l(I_{ij})$ can be bigger than the surface of an object, and the resulting affinity has to be bounded to $(0,1)$.

All those affinities can be used as such with clustering techniques. However, those affinities complement each other and we can combine K affinities as follows:

$$A = \sum_{k<K} A_k \lambda_k \tag{1}$$

3 Fragments Merging

Using a linear combination of the affinities defined in the previous section, we merge fragments in a composite material fragmented by mechanical pressure. A maximum spanning tree is built. This maximum spanning tree can be used to form a hierarchy between fragments represented as a dendrogram. The estimation of grains size is used to cut this hierarchy and produce the final result. We have optimized the linear combination by minimizing a score between the resulting merging and a ground truth, as described in Sect. 3.2.

3.1 Minimum Spanning Tree and Hierarchical Cut

We can use the linear combination of the introduced affinities with several existing clustering methods, such as K-means, spectral clustering, DBSCAN and other [20]. The inconvenience is when the number of fragments increases computational issues appear (running time, instability of convergence, memory consumption). For given materials, we have chosen a hierarchical clustering and we perform a hierarchical cut using an estimated size of grain extracted from a covariance analysis.

The previous affinity matrix is an adjacency graph. The first step of the proposed clustering method is to produce a hierarchy using the minimum spanning tree of the graph $1 - A$ to remove edges of low affinity. We produce a dendrogram from the minimum spanning tree as illustrated in the Fig. 5. The strategies employed to build the hierarchy can be of any of the commonly used types: agglomerative or divisive.

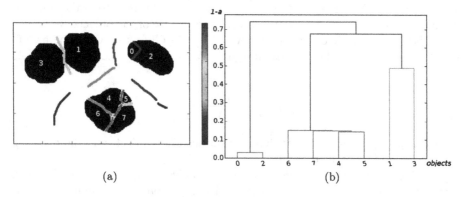

(a)   (b)

Fig. 5. A simple case of dendrogram built from a maximum spanning tree is illustrated. In (a), objects are shown in black, and interfaces are valued using a colormap scheme. The affinity used to generate the valued edges is a linear combination of A_μ, A_{σ^2} and A_s. The resulting dendrogram is shown in (b), where 3 clusters are visually identified, and the y-axis is $1 - A$. We can see that objects 1 and 3 have a lower affinity than 0 and 2, and are visually two distinct grains that are connected. (Color figure online)

Cutting the hierarchy depends on the material that is considered. MAT1 is a homogeneous composite with one size of grains. Therefore, the criterion that we used to cut the hierarchy is volumic. The covariance analysis of the raw data can give us an estimate of the mean size of the grains, see [5]. We have associated with each cluster the sum of the volumes of the individual objects composing the former. Cutting the hierarchy can be achieved by the pseudo-code (1).

Due to the previous thinning of the SKIZ, the initial graph might not be connected. The missing edges are diverging all along the SKIZ and therefore of low affinities. Moreover, filtering the graph by removing edges that feature low affinity values can be done to process graphs of lower density. Each connected component can be processed afterward independently.

Algorithm 1. Top-down cut of the hierarchy. $w(i)$ is the volume associated to cluster i. L is the volumic limit estimated by the covariance.

```
1: i ← root of the dendrogram
2: CUT(i)
3: procedure CUT(i)
4:     if i is a leaf OR w(i) < L then
5:         return i
6:     else
7:         return CUT(left child of i), CUT(right child of i)
```

3.2 Optimizing the Affinity

The choice of the affinity is crucial to obtain the desired result. One can see that a single affinity that was presented earlier is not sufficient to reassemble the fragments, see Fig. 1. The K-means algorithm is based on the barycenter distance and is not able to reassemble fragments that have drifted far away from their original positions. The size distribution of fragments is wide, the drift between similar fragments can be too large to be caught by the naive border-to-border distance as illustrated in Fig. 6. We assume that a combination of affinities is the key to obtain better results.

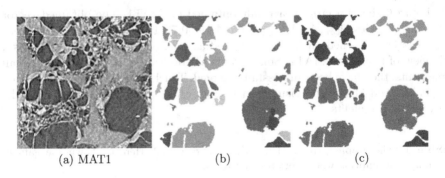

(a) MAT1   (b)   (c)

Fig. 6. (a) Shows two fragments that heavely drift from one another on the bottom left. (b) Shows the over-segmentation. (c) The result of the fragment merging. The round fragment in (b) in the bottom right corner colored in red and in blue, is oversegmented. In (c), the two objects were merged into one orange grain.) (Color figure online)

We choose a linear combination of the four affinities $A_{\mathrm{bar}}, A_\mu, A_{\sigma^2}$ and A_s. To optimize the affinity, we optimize the segmentation using a user-provided ground truth (GT) on one or more slices of the 3D image.

We construct a bivariate histogram $\mathscr{H}$ from the labels of GT and from the labels obtained by hierarchical clustering extracted from the same slices. $\mathscr{H} = [h_{ij}]$ is an $M \times N$ sparse matrix where M is the number of detected objects and N the number of objects in the GT. From $\mathscr{H}$, we compute the F_1 score of the segmentation in the following way:

A sum of a line $i = const, P(i) = \sum_j h_{ij}$, is the detection rate of label i. The index $l = \arg\max_j h_{ij}$, is the label corresponding to i in GT. We now compute the true positive $TP(i) = h_{il}$, the false positive $FP(i) = P(i) - TP(i)$.

The sensitivity and the positive predictive value for every i is: $Sens(i) = \frac{TP(i)}{P(i)}$ and $PPV(i) = \frac{TP(i)}{TP(i)+FP(i)}$.

A F_1 score is computed for every detected object, with H the harmonic mean: $F_1(i) = H(Sens(i), PPV(i))$. The overall F_1 score of the result is the harmonic mean of all scores: $F_1 = H(F_1(i)), 1 \le i \le M$. The vector (λ_k) in Eq. 1 is then determined by minimizing $1 - F_1$. The minimization method is a modification of Powell's method [12,13]. It performs sequential one-dimensional minimizations along each vector of the directions set.

3.3 Results

Table 1 shows the size of the generated graph for MAT1 and MAT2. One benefit of this method is to be able to work on a compressed representation of the studied materials from the raw tridimensional data. Further compression is achieved by using a ϵ-filtering of the weighted graph. Here, edges lower than mean$(a) - 2 * std(a)$ have been removed. An appropriate filtering will generate several connected components, which can each be processed independently and in parallel.

Figure 6 shows a crop of a result obtained on MAT1. The obtained vector (λ_k) for affinities $A_{bar}, A_\rho, A_{sigma^2}$ and A_s is $[0.73, 1.85, 0.30, 1.33]$. Notice that (λ_k) has been learned on MAT1 and used as such on MAT2. Figure 7 shows a 3D view of results on MAT1. Some grains are broken down into multiple small fragments, they have been successfully merged. Touching grains have been solved for MAT2 and MAT1 and visually satisfying results have been achieving for the merging of fragments.

Table 1. The graphs generated for MAT1 and MAT2. Here is shown the gain in memory consumption when working on graphs.

Materials	Number of voxels	Number of nodes	Number of edges	After ϵ-filtering
MAT1	$2.9 * 10^8$	720	5106	2456
MAT2	$1.0 * 10^9$	4660	28341	12846

4   Conclusion

We have introduced a method to calculate affinities between objects in tridimensional data such as CT images. Those affinities have been applied to the problem of fragments merging using a machine learning algorithm. Oversegmented fragments were also merged, as shown in Fig. 6, while touching grains remain separated. The touching objects problem was therefore also solved with

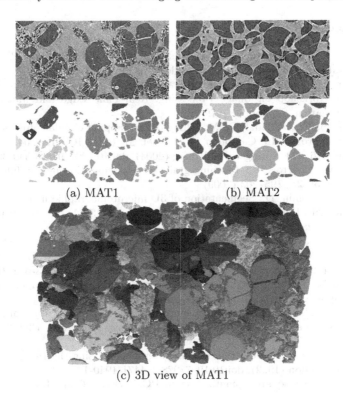

(a) MAT1   (b) MAT2

(c) 3D view of MAT1

Fig. 7. Results of the fragments merging method. (a) Slice of reassembled MAT1. (b) Slice of reassembled MAT2. In (a), grains are heavily fragmented such as the gray grain in the bottom-right corner, and heavily clustered such as green and dark green grains in the top-right corner. In (b), less fragmentation is observed but there is more connections between grains. In (c), reassembled grains from MAT1 are shown in 3D, using a similar color scheme. (Color figure online)

the same approach. Visually satisfying results were obtained for both, see Fig. 7. Those affinities are able to characterize frontiers between objects disregarding their shapes and sizes. Many perspectives arise from this preliminary work. Further works should first concern the influence of the primary segmentation and the metric employed to build the interfaces. More advanced clustering methods can be used in combination with the proposed affinities.

Acknowledgment. This work has received funding from DGA and is a collaboration between Transvalor, DGA, CEA Gramat. The acquisitions are carried out at the CEA Gramat, a french public laboratory affiliated to the Commission of Atomic Energy and Alternative Energies.

References

1. Beucher, S.: Segmentation d'Images et Morphologie Mathématique. Ph.D. thesis, Ecole Nationale Supérieure des Mines de Paris, June 1990
2. Beucher, S., Lantuejoul, C.: Use of watersheds in contour detection. In: International Workshop on Image Processing: Real-Time Edge and Motion Detection/Estimation, September 1979
3. Dejnozkova, E., Dokladal, P.: Modelling of overlapping circular objects based on level set approach. In: Campilho, A., Kamel, M. (eds.) ICIAR 2004. LNCS, vol. 3211, pp. 416–423. Springer, Heidelberg (2004). doi:10.1007/978-3-540-30125-7_52
4. Faessel, M., Courtois, F.: Touching grain kernels separation by gap-filling. Image Anal. Stereol. **28**, 195–203 (2009)
5. Faessel, M., Jeulin, D.: Segmentation of 3D microtomographic images of granular materials with the stochastic watershed. J. Microsc. **239**(1), 17–31 (2010). doi:10.1111/j.1365-2818.2009.03349.x
6. Gillibert, L., Jeulin, D.: Stochastic multiscale segmentation constrained by image content. In: Soille, P., Pesaresi, M., Ouzounis, G.K. (eds.) ISMM 2011. LNCS, vol. 6671, pp. 132–142. Springer, Heidelberg (2011). doi:10.1007/978-3-642-21569-8_12
7. Gillibert, L., Jeulin, D.: 3D reconstruction and analysis of the fragmented grains in a composite material. Image Anal. Stereol. **32**(2), 107–115 (2013)
8. Heijmans, H.J.A.M., van den Boomgaard, R.: Algebraic framework for linear and morphological scale-spaces. J. Vis. Commun. Image Represent. **13**(1–2), 269–301 (2002)
9. Leavers, V.F.: Shape Detection in Computer Vision Using the Hough Transform. Springer, London (1992). doi:10.1007/978-1-4471-1940-1
10. Lloyd, S.: Least squares quantization in PCM. IEEE Trans. Inf. Theory **28**(2), 129–137 (1982)
11. MacQueen, J.: Some methods for classification and analysis of multivariate observations. In: Cam, L.M.L., Neyman, J. (eds.) Proceedings of the 5th Berkeley Symposium on Mathematical Statistics and Probability, vol. 1, pp. 281–297 (1967)
12. Powel, M.J.D.: An efficient method for finding the minimum of a function of several variables without calculating derivatives. Comput. J. **7**(2), 155–162 (1964)
13. Press, W., Flannery, B.P., Teukolsky, S.A., Vetterling, W.T.: Numerical Recipes: The Art of Scientific Computing. Cambridge University Press, New York (1988)
14. Rosin, P.L.: Shape partitioning by convexity. IEEE Trans. Syst. Man Cybern. Part A Syst. Hum. **30**(2), 202–210 (2000). doi:10.1109/3468.833102
15. Schmitt, O., Hasse, M.: Morphological multiscale decomposition of connected regions with emphasis on cell clusters. Comput. Vis. Image Underst. **113**(2), 188–201 (2009)
16. Schüpp, S., Elmoataz, A., Fadili, M.-J., Bloyet, D.: Fast statistical level sets image segmentation for biomedical applications. In: Kerckhove, M. (ed.) Scale-Space 2001. LNCS, vol. 2106, pp. 380–388. Springer, Heidelberg (2001). doi:10.1007/3-540-47778-0_36
17. Serra, J.: Image Analysis and Mathematical Morphology. Academic Press, London (1983)
18. Soille, P.: Morphological Image Analysis: Principles and Applications. Springer, Heidelberg (2004). doi:10.1007/978-3-662-05088-0
19. Talbot, H., Appleton, B.C.: Elliptical distance transforms and object splitting. In: International Symposium on Mathematical, Morphology, pp. 229–240, April 2002
20. Xu, R., Wunsch, D.: Survey of clustering algorithms. IEEE Trans. Neural Netw. **16**(3), 645–678 (2005). doi:10.1109/TNN.2005.845141

Morphological Characterization of Graphene Plans Stacking

Albane Borocco[1(✉)], Clémentine Fellah[2,3], James Braun[2],
Marie-Hélène Berger[3], and Petr Dokládal[1]

[1] CMM - Centre for Mathematical Morphology, MINES ParisTech,
PSL Research University, 35, rue St. Honoré, 77305 Fontainebleau, France
{albane.borocco,petr.dokladal}@mines-paristech.fr
[2] CEA, DEN, SRMA, LTMEX, 91191 Gif-sur-Yvette, France
{clementine.fellah,james.braun}@cea.fr
[3] CMAT - Centre for Materials, MINES ParisTech,
PSL Research University, Évry, France
{clementine.fellah,marie-helene.berger}@mines-paristech.fr

Abstract. The graphene is a material obtained when carbon atoms form large planar molecules. Well organized, large graphene molecules stacked ontop each other convey to graphene particularly interesting properties useful in nuclear industry.

Understanding how the organization on the molecular scale influences the mechanical properties of the material is a key element in the material manufacturing process. In this scope, features like local orientation and length have already been largely explored in the literature.

This paper brings a new feature evaluating the number of plans stacked ontop each other and the length of this stacking. It allows obtaining other features such as the overall rate of organization or locality and preferential orientation. These informations, synthesized in the form of histograms provides a key information in the processus the material design.

Experimental results obtained on images taken by an electronic scanning microscope are presented to illustrate the proposed method.

Keywords: Mathematical morphology · Orientation space · Filtering · Graphene

1 Introduction

Ceramic matrix composite materials are candidates for thermostructural applications in extreme environments such as in turbojet engines or core materials for future nuclear reactors [1,7]. Foremost among them stand SiC/SiC composites which are composed of a SiC matrix by high purity silicon carbide (SiC) fibers reinforced. This combination of two refractory, brittle components gives rise to a damage tolerant material, thanks to a pyrocarbon (PyC) interphase deposited between the fibers and the matrix to control the fiber/matrix bonding strength

© Springer International Publishing AG 2017
J. Angulo et al. (Eds.): ISMM 2017, LNCS 10225, pp. 435–446, 2017.
DOI: 10.1007/978-3-319-57240-6_35

[10]. This interphase and their related interfaces allow crack deflection and crack bridging by the fibers during mechanical loading which is the key to achieve a pseudo-ductile macroscopic behavior. Optimization of their properties through the process must be performed and are from far a big concern. A thorough investigation of the carbon rich fiber surface structure and of the nature of the fiber/PyC interphase bonding is essential to control and improve the fiber/matrix coupling in SiC/SiC composites [2].

A variety of techniques could be used to characterize the carbon nanostructure of the SiC/SiC composite interface including Raman spectroscopy, X-ray Diffraction (XRD) and Transmission Electron Microscopy (TEM). The dark field and high-resolution modes in transmission electron microscopy allow imaging directly 002 lattice fringes of graphene layers and to identify their morphological, textural and structural parameters. But most of the information obtained from TEM images have been qualitative or semi-quantitative [13]. The direct interpretation of the TEM features is not easy considering contrast variations, layers twisting and overlapping and local changes in the organization of carbon nanostructure.

Image analysis allows eliminating problems of representativeness for the observations of graphene layers nanostructure and more quantitative data can be obtained [14,16,19]. Various parameters of the spatial arrangement of carbon lamellae can be measured such as lattice fringe length, size and shape, curvature and tortuosity of plans. A lot of studies are already published in the literature studying on the nanostructure of fibers, soot, carbon black, pyrolytic carbon, fullerenes and coal chars by image processing [4,5,8,12,17,20]. In this study, the proposed technique based on mathematical morphology involves the characterization of quantitative indices for describing the carbon interface of SiC/SiC composite. This new method leads rapidly to the detection of graphene layers stacking thanks to the development of a more important robustness to noise and to quantify especially their local and space orientations using HRTEM images. The final goal is to establish the local interaction mechanisms of carbon phases at the interface and the mechanisms that control mechanical behavior of SiC/SiC composites. Thereafter it will be possible to define optimized materials with these information for high-temperature applications.

2   Detection of Graphene Plans and Their Local Orientation

This section explains how to detect graphene plans and their orientation by modeling each plan by a centered, threadlike curve of same length and orientation.

2.1   Thresholding

The High-resolution transmission electron microscopy (HRTEM) allows to image individual graphene molecules (in bright) and the gaps in between (that appear darker). The molecules can be dissociated from the background using a

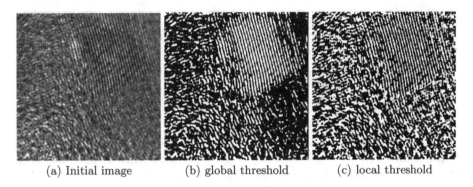

(a) Initial image   (b) global threshold   (c) local threshold

Fig. 1. Application of Otsu's threshold. (a) Initial image. (b) Global Otsu thresholding. (c) Sliding-window Otsu thresholding.

thresholding, see Fig. 1(a). The Otsu's method [11] is a thresholding algorithm that iterates over all possible threshold values and calculates the spread of the pixel levels each side to find the optimum that minimizes the sum of the foreground and background spreads, defined as a weighted sum of variances of the two classes:

$$t = arg(\min_t(\sigma_W^2(t))), \quad \text{with} \quad \sigma_W^2(t) = \omega_1(t)\sigma_1^2(t) + \omega_2(t)\sigma_2^2(t) \qquad (1)$$

with σ_i^2 being the variance of the class i and ω_i the weight given by its likelihood.

Since the image may contain globally darker and brighter regions, a global threshold is likely to fail, Fig. 1(b), the thresholding value should be calculated locally. For each pixel, an optimal threshold t is determined in a local neighborhood defined by a sliding window. Its size is fixed so as to contain several graphene plans and gaps, but small enough not to contain both a bright and a dark region. We have fixed the widow size to five times the plan width. Figure 1(c) illustrates the result of a local Otsu thresholding. It allows the extraction of the graphene plans despite local intensity variation. Let BW denote the binary image representing the graphene plans in the original grayscale image I.

$$BW(x,y) = \begin{cases} 1, & if \ I(x,y) > t(x,y) \\ 0, & otherwise \end{cases} \qquad (2)$$

2.2   Skeletonization

In order to model each plan by a centered threadlike curve with the same length, the skeleton is extracted from the binary image. This operation uses an algorithm [6] that removes the points of the outline of the shape while preserving its topological characteristics. Let Sk denote from now the skeleton.

$$Sk : I \rightarrow skeleton(BW) \qquad (3)$$

The skeleton is then pruned to remove spurious edges. The pruning is done using the Algorithm 1 that allows to remove spikes shorter than a certain size.

Algorithm 1. Skeleton filtering algorithm

Input: Sk - noisy skeleton, **Parameter:** n - pruning length
Output: Sk' - filtered skeleton

1: $X^0 \leftarrow Sk$
2: **for** i=1..n **do**
3:      $X^i \leftarrow X^{i-1} \setminus EP(X^{i-1})$
4: $Sk' \leftarrow R(X^n, X^0 \setminus TP(X^0)) \cup TP(X^0)$
5: **return** Sk'

6: **function** $TP(X)$ ▷ *Function Triple Points of X*
7:      $TP(X) \leftarrow \{p, p \in X, (X * N_8)(p) > 3\}$

8: **function** $EP(X)$ ▷ *Function End Points of X*
9:      $EP(X) \leftarrow \{p, p \in X, (X * N_8)(p) = 2\}$

10: **function** R(X,Y)   ▷ *Morph. reconstr. by dilation of X under Y*
11:      $X^0 \leftarrow X$
12:      **for** i=1..∞ **do**
13:          $X^i \leftarrow \delta(X^{i-1}) \cap Y$
14:      $R(X,Y) \leftarrow X^\infty$

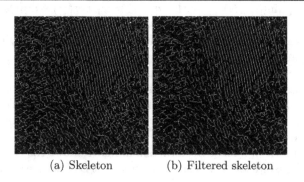

(a) Skeleton   (b) Filtered skeleton

Fig. 2. A skeleton before (a) and after (b) the pruning algorithm.

The functions EP and TP respectively detect end points and triple points of the skeleton by counting the number of neighbors in the N_8 neighborhood using a convolution, as denoted by $*N_8$.

Using this algorithm, spikes shorter than n are removed, isolated structures shorter than n are preserved and the connectivity is not modified. A convenient size for n is roughly equivalent to the interreticular distance. Figure 2 illustrates the impact of pruning on the skeleton.

2.3 Local Orientation

In order to determine the local orientation of each graphene plan, we first compute the local orientation of each pixel. This calculation is based on the

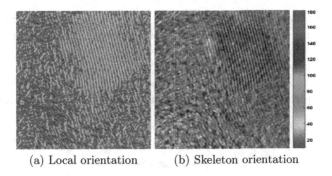

(a) Local orientation   (b) Skeleton orientation

Fig. 3. Application of linear opening to determine the local orientation. (a) Local orientation in false colors. (b) Combination of the local orientation and the skeleton superimposed on the initial image (Color figure online)

morphological opening of the grayscale image I by B_θ – a segment rotated by θ. The local orientation of a pixel is the one that maximizes the opening value.

$$\left| \begin{array}{ll} \theta: & D \to]0, \pi] \\ & (x, y) \mapsto \arg(\bigvee_{\theta \in]0, \pi]} (\gamma_{B_\theta}(I)(x, y))) \end{array} \right. \tag{4}$$

In the case of multiple maximum values, the local orientation is the first maximum found; the process to cope with that loss of information is explained in Sect. 3. The local orientation of each graphene plan is obtained by combining the local orientation of the gray scale image and the skeleton of the binary image. Figure 3 illustrates (in false colours) the results of local orientation on the gray-scaled image (a) and on the skeleton (b).

3 Orientation Space

Orientation space has been introduced by Chen and Hsu [3] and popularized later by others authors [9,18]. The orientation space is obtained by adding the orientation axis to the image support: it is the result of a transformation consisting in applying rotated copies of an orientation selective filter.

$$I(x, y) \to I^{[\theta]}(x, y, \theta) \tag{5}$$

with I the initial image, (x, y) the coordinates in the initial image, θ the local orientation, with $\theta \in]0, \pi]$ and $I^{[\theta]}$ the orientation-space representation of I.

It presents several advantages for the evaluation of orientation:

- If two differently-oriented structures intersect in the image, they will belong to different orientation plans, which will ease their separation.
- The noise spreads over the whole orientation space whereas the signal is located in specific position, which improves the signal-to-noise ratio.

The orientation space allows segmentation of overlapping objects where the connectivity is not sufficient to set them apart. Indeed, local orientation is a way to differentiate touching objects: two distincts segments can't both intersect and have the same orientation.

In order to separate the graphene plans according to their orientation, the skeleton of the picture is put in the orientation space, using the local orientation:

$$\left| \begin{array}{l} Sk^{[\theta]} : D \times \,]0 : \pi] \rightarrow \{0, 1\} \\ (x, y, \theta) \mapsto \begin{cases} 1, & \text{if } Sk(x, y) = 1 \text{ and } \theta(x, y) = \theta \\ 0, & \text{otherwise} \end{cases} \end{array} \right. \tag{6}$$

A one dimensional structure ξ in the image will be represented as a one dimensional structure in the orientation space. Moreover a smooth ξ will be connected in this space. Conversely, a continuous piece-wise smooth curve will be disconnected in non-differentiable points. Obviously, in order to identify graphene plans, each one should be a unique connected component even though its skeleton is not perfectly smooth. However, a perfectly smooth curve is difficult to obtain due to essentially two reasons: (1) we are in a discrete $\mathbb{Z}^2$ grid, (2) the local orientation is sampled (taken from a discrete set of angles in Eq. 4).

To cope with that issue, a convenient solution is to use a second-generation connectivity [15]. Two structures X and Y will be considered connected if they are connected after applying an operator ψ:

$$XC^{\psi}Y \Leftrightarrow \psi(X)C\psi(Y) \tag{7}$$

The operator ψ is chosen depending on the way structures are disconnected in the orientation space. We consider two types of imperfection:

1. A noise-corrupted, discontinuous, straight graphene plan is represented in the orientation space as composed of aligned but disconnected fragments. One can reconnect these fragments (and only these fragments) using a morphological closing by a segment oriented in the same direction. Since the orientation of the plans varies this closing also has to vary under translation. However, using the orientation space the spatially varying closing decomposes in a collection of translation-invariant closings. Each subspace $\theta = const.$ contains graphene plans oriented in θ. Hence, for every $\theta = const.$ the closing is a usual translation-invariant closing by a line segment oriented in θ.

   To ease the description of the structuring element (here and below in the text) we will use a rotated coordinate system. We will use two unit vectors $u_{\theta\|}$ and $u_{\theta\perp}$ oriented alongside and perpedicularly to the angle θ of rotation of the system that can be expressed using the initial coordinate system:

$$\begin{pmatrix} u_{\theta\|} \\ u_{\theta\perp} \\ \theta \end{pmatrix}_{\theta} = \begin{pmatrix} \cos(\theta) & -\sin(\theta) & 0 \\ \sin(\theta) & \cos(\theta) & 0 \\ 0 & 0 & 1 \end{pmatrix} \begin{pmatrix} x \\ y \\ \theta \end{pmatrix} \tag{8}$$

The coordinates in this new system will be expressed using the notation: $[., ., .]_{\theta}$. Using these coordinates, $[a, b, \theta]_{\theta}$ defines a rectangle $a \times b$ rotated by

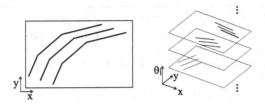

Fig. 4. Curved graphene plans (left) spread over different θ_i-subspaces of the orientation space (right).

θ and located at an altitude θ. Using this new coordinate system, the operator used in the noise-tolerant connectivity (Eq. 7) is a spatially varying closing by a segment:

$$\varphi_{B_\theta} \text{ with } B_\theta = [d, 0, \theta]_\theta \tag{9}$$

where B_θ is a segment of length d which is the maximum distance to consider two aligned segments as belonging to the same graphene plan, and θ gives the translation of the segment along the θ axis in the orientation space.

2. Since a discrete skeleton is not always smooth, its segments will distribute over different plans in the orientation space. Moreover, with the discrete angular dimension, there might be an abrupt change in local orientation. However, even though these continuous, smooth but tortuous structures may appear in the orientation space as disconnected they need to be identified as one connected component. The operator that allows these curves to be considered as connected is a closing by a vertical segment in the orientation space.

$$\varphi_A \text{ with } A = [0, 0, h] \tag{10}$$

The height h of this segment defines the maximum angle of two parts of the curve to consider them connected. When the angle exceeds $hd\theta$, where $d\theta$ is the angular sampling step used in Eq. 4, the two parts are considered disconnected.

4   Stacking Detection

The principal objective of this study is to identify and quantify stackings of plans. Two plans are stacked if: (i) they have the same orientation, and (ii) the distance between them is the inter-reticular distance, l. A stacking is a sequence of two-by-two stacked plans. In the orientation space, structures that are in the same plan and connected after a closing by a segment perpendicular to their orientation of length l are stacked.

The Algorithm 2 counts the number of stacked plans. The algorithm must handle the difficulty that even though curved structures spread in the orientation space over several θ_i-subspaces (see Fig. 4) their length shouldn't be

---

Algorithm 2. Quantification of stackings of graphene plans

Input: $Sk^{[\theta]}$ - the oriented-space representation of the skeleton;
Parameters: l - the interstitial distance
Output: H - bivaluate histogram count of the cumulated length

1: **for** $\theta_i = 1..180$ **do**
2:     $Sk^{\theta_i} \leftarrow R(Sk^{[\theta]} \cap (x, y, \theta_i), Sk^{[\theta]})$ ▷ select and reconstruct plans oriented along θ_i
3:     $Sk^{[\theta]} \leftarrow Sk^{[\theta]} \setminus Sk^{\theta_i}$ ▷ remove detected structures to avoid further processing
4:     $Sk^0 \leftarrow \varphi_{C_{\theta_i+90}}(Sk_{\theta_i})$ with $C_{\theta_i+90} = [0, l, \theta_i]_\theta$ ▷ connect stacked plans
5:     n=0   ▷ counter of stacked plans
6:     **while** $Sum(Sk^n) > 0$ **do**
7:         $Sk^{n+1} \leftarrow \gamma_{C'_{\theta+90}}(Sk^n)$ with $C'_{\theta+90} = [0, nl, \theta_i]_\theta$
8:         $H(\theta_i, n) \leftarrow Sum(R_{B_{\theta_i}}(Sk^n - Sk^{n+1}, Sk^n))$ with B_{θ_i} a segment tilted by θ_i
9:         $n \leftarrow n + 1$

---

underevaluated. This is overcome by the use of the second-generation connectivity allowing to detect a curve as one connected component.

The line 2 detects plans containing at least a portion oriented along θ_i and reconstructs it in its entirety by morphological reconstruction by dilation. $R(x, y)$ is a morphological reconstruction by dilation of x under y. See Sect. 5 below for the structuring element used in the reconstruction. The line 4 detects the stackings: plans connected after a closing by $C_{\theta+90}$ belong to the same stack. The lines 6 to 9 measure the cumulated length (using the Sum on the skeleton) using granulometry by opening by reconstruction. After, the stackings are quantified using a bivaluate distribution of the cumulated length over orientation and number of stacked plans.

5   Robustness to Noise

Even organized materials contain imperfections consisting in discontinuous and imperfectly-parallel plans. Such regions wouldn't be detected as stackings in a

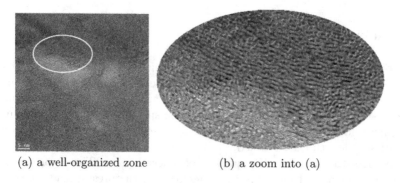

(a) a well-organized zone   (b) a zoom into (a)

Fig. 5. Zone of interest: a collection of disjoint and imperfectly parallel plans considered as a set of small stacks by the algorithm previously described

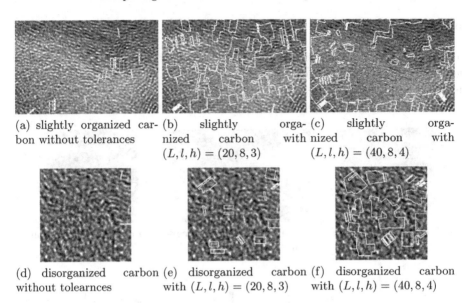

(a) slightly organized carbon without tolerances

(b) slightly organized carbon with $(L, l, h) = (20, 8, 3)$

(c) slightly organized carbon with $(L, l, h) = (40, 8, 4)$

(d) disorganized carbon without tolearnces

(e) disorganized carbon with $(L, l, h) = (20, 8, 3)$

(f) disorganized carbon with $(L, l, h) = (40, 8, 4)$

Fig. 6. Impact of tolerances on detection of stacks of at least four plans, in organized and disorganized carbon, the detected stacks are highlighted in boxes.

whole, but as fragmented stacking that would yield biased statistics. Figure 5 illustrates the situation. To cope with that issue, we set up tolerances to small discontinuities using second-generation connectivity. As in the previous case, the chosen operator is a closing, but the size of the structuring element has to be adapted to close only admissible imperfections.

The plans are disjoint and not perfectly parallel, so the structuring element needs a length L along the graphene plans orientation and a height h along the θ axis. Moreover, in order to detect stackings as collections of structures separated from an inter-reticular distance, the structuring element needs a thickness of the size of the inter-reticular distance l along the axis perpendicular structures' orientation. The operator used for the tolerances is a spatially varying morphological closing

$$\varphi_{T_\theta} \text{ with } T_\theta = [L, l, \theta]_\theta \times [0, 0, h] \tag{11}$$

with an element T_θ a rectangular parallelepiped $L \times l \times h$ rotated by θ (obtained as a product of a rotated flat rectangle and a vertically oriented segment). The use of φ_{T_θ} before the stacking detection allows the connection of every graphene plans that belong to the same stack. Each stack is then characterized using the previous method.

Figure 6 illustrates the impact of tolerances on stacking detection. Wider tolerances allow the detection larger stacking zones by gathering disjoint and imperfectly parallel plans. However, wide tolerances also imply the possibility of false detection of stacking in regions where graphene plans are disorganized. A trade-off must be found between false and non-detection of stackings.

6   Results

The results are presented in the form of a bivariate histogram, presenting the length or the area of the structure given its orientation and the number of stacked plans.

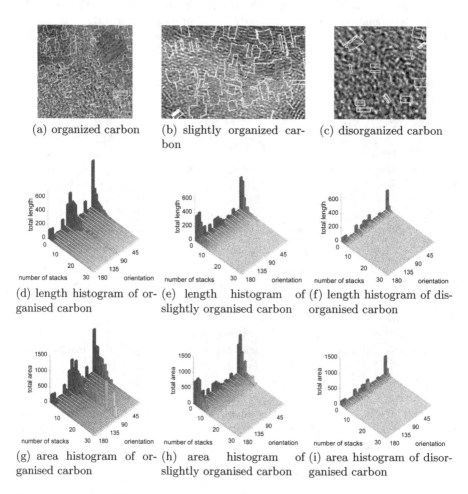

(a) organized carbon

(b) slightly organized carbon

(c) disorganized carbon

(d) length histogram of organised carbon

(e) length histogram of slightly organised carbon

(f) length histogram of disorganised carbon

(g) area histogram of organised carbon

(h) area histogram of slightly organised carbon

(i) area histogram of disorganised carbon

Fig. 7. Examples of (a) organized, (b) slightly organized and (c) disorganized carbon, detected stacks are highlighted in boxes. Below, the estimated distribution of the length (d–f) and area (g–i) of the stacks depending on the orientation and the number of stacked plans

Figure 7 illustrates the results for organized, slightly organized and disorganized samples. The organized carbon Fig. 7(a) shows one large stacking oriented around 90° in the upper-right zone of the image, and several smaller zones oriented around 0° in the left part of the image. These correspond to the two principal lobes around 90° and 0° in the histogram Fig. 7(d) and (g). Regarding the lobe around 90° one can observe that there are up to 30 stacked graphene plans (non-zero values up to 30 of number of stacks in the histogram). The *total length* and the *total area* histogram axis represents respectively the unnormalized length of the plans involved in a stacking and the area of those stackings. The slightly organized carbon, Fig. 7(b), contains a number of small stackings, most oriented roughly horizonally, represented by the lobe around 0° and 180° in Fig. 7(e). The disorganized carbon Fig. 7(c) contains almost no graphene stackings but a handful of tiny zones. Its histogram Fig. 7(f) is much poorer compared to that of the organized and slightly organized samples.

7   Conclusion

This paper presents an original method to identify graphene plans stackings in carbon. For every stacking, it (i) determines its orientation, (ii) counts the number of stacked plans and (iii) estimates the cumulative length of all stacked plans.

These measures can be synthesized in a statistical quantification representing the "organization" rate of a given sample in the form of a bivariate histogram. This is useful for statistical analysis and comparison of different samples. For example, one can evaluate the ratio of the organized area over the total area of the sample. This can be conveniently expressed as a cumulative distribution of organized area ratio per number of stacked graphene molecules.

This method is fast, simple and intuitive. An adjustable tolerance parameter avoids fractioning due to imperfectly aligned, or discontinuous plans. In this way the stackings identified by this method are closer to what one would intuitively delimit manually as a stacking despite present imperfections.

Experimental results obtained on images taken by an HRTEM microscope are provided to illustrate the validity of the proposed method.

References

1. Bansal, N.P., Lamon, J.: Ceramic Matrix Composites: Materials: Modeling and Technology. Wiley, New York (2014)
2. Buet, E., Sauder, C., Poissonnet, S., Brender, P., Gadiou, R., Vix-Guterl, C.: Influence of chemical and physical properties of the last generation of silicon carbide fibres on the mechanical behaviour of SiC/SiC composite. J. Eur. Ceram. Soc. **32**(3), 547–557 (2012)
3. Chen, Y.S., Hsu, W.H.: An interpretive model of line continuation in human visual perception. Patern Recogn. **22**(5), 619–639 (1989)
4. Da Costa, J.P., Weisbecker, P., Farbos, B., Leyssale, J.-M., Vignoles, G.L., Germain, C.: Investigating carbon materials nanostructure using image orientation statistics. Carbon **84**, 160–173 (2015)

5. Farbos, B., Weisbecker, P., Fischer, H.E., Da Costa, J.-P., Lalanne, M., Chollon, G., Germain, C., Vignoles, G.L., Leyssale, J.-M.: Nanoscale structure and texture of highly anisotropic pyrocarbons revisited with transmission electron microscopy, image processing, neutron diffraction and atomistic modeling. Carbon **80**, 472–489 (2014)

6. Guo, Z., Hall, R.W.: Parallel thinning with two-subiteration algorithms. Commun. ACM **32**(3), 359–373 (1989)

7. Katoh, Y., Snead, L.L., Henager, C.H., Hasegawa, A., Kohyama, A., Riccardi, R., Hegeman, H.: Current status and critical issues for development of SiC composites for fusion applications. J. Nucl. Mater. **367**, 659–671 (2007)

8. Leyssale, J.-M., Da Costa, J.-P., Germain, C., Weisbecker, P., Vignoles, G.L.: Structural features of pyrocarbon atomistic models constructed from transmission electron microscopy images. Carbon **50**(12), 4388–4400 (2012)

9. Luengo-Oroz, M.A.: Spatially-variant structuring elements inspired by the neuro-geometry of the visual cortex. In: Soille, P., Pesaresi, M., Ouzounis, G.K. (eds.) ISMM 2011. LNCS, vol. 6671, pp. 236–247. Springer, Heidelberg (2011). doi:10.1007/978-3-642-21569-8_21

10. Naslain, R.R.: The design of the fibre-matrix interfacial zone in ceramic matrix composites. Compos. Part A: Appl. Sci. Manuf. **29**(9), 1145–1155 (1998)

11. Otsu, N.: A threshold selection method from gray-level histograms. IEEE Trans. Syst. Man Cybern. **9**, 62–66 (1979)

12. Pré, P., Huchet, G., Jeulin, D., Rouzaud, J.-N., Sennour, M., Thorel, A.: A new approach to characterize the nanostructure of activated carbons from mathematical morphology applied to high resolution transmission electron microscopy images. Carbon **52**, 239–258 (2013)

13. Reznik, B., Hüttinger, K.J.: On the terminology for pyrolytic carbon. Carbon **40**(4), 621–624 (2002)

14. Rouzaud, J.-N., Clinard, C.: Quantitative high-resolution transmission electron microscopy: a promising tool for carbon materials characterization. Fuel Process. Technol. **77**, 229–235 (2002)

15. Serra, J.: Connectivity on complete lattices. J. Math. Imaging Vis. **9**(3), 231–251 (1998)

16. Shim, H.-S., Hurt, R.H., Yang, N.Y.C.: A methodology for analysis of 002 lattice fringe images and its application to combustion-derived carbons. Carbon **38**(1), 29–45 (2000)

17. Toth, P., Palotas, A.B., Eddings, E.G., Whitaker, R.T., Lighty, J.S.: A novel framework for the quantitative analysis of high resolution transmission electron micrographs of soot I. Improved measurement of interlayer spacing. Combust. Flame **160**(5), 909–919 (2013)

18. Van Ginkel, M., Van Vliet, L.J., Verbeek, P.W.: Applications of image analysis in orientation space. In: 4th Quinquennial Review 1996 – Dutch Society for Pattern Recognition and Image Processing. NVPHBV (2001)

19. Vander Wal, R.L., Tomasek, A.J., Pamphlet, M.I., Taylor, C.D., Thompson, W.K.: Analysis of HRTEM images for carbon nanostructure quantification. J. Nanopart. Res. **6**(6), 555–568 (2004)

20. Yehliu, K., Vander Wal, R.L., Boehman, A.L.: Development of an HRTEM image analysis method to quantify carbon nanostructure. Combust. Flame **158**(9), 1837–1851 (2011)

Segmentation of Collagen Fiber Bundles in 3D by Waterfall on Orientations

Michael Godehardt[1]([⊠]), Katja Schladitz[1], Sascha Dietrich[2], Renate Meyndt[2], and Haiko Schulz[2]

[1] Fraunhofer-Institut für Techno- und Wirtschaftsmathematik, Kaiserslautern, Germany
michael.godehardt@itwm.fraunhofer.de
[2] Forschungsinstitut für Leder und Kunststoffbahnen (FILK), Freiberg, Germany

Abstract. The micro-structure of bovine leather samples is imaged three dimensionally using micro-computed tomography. We report on the first algorithm for automatic segmentation of "typical" elements of this multiscale structure based on the reconstructed 3D images. In spite of the scales being hardly separable, a coarse segmentation and a finer substructure can be derived in a consistent way. For preprocessing, an adapted morphological shock filter is suggested. The segmentation algorithm for the coarse fiber bundles exploits the watershed transform and the waterfall paradigm on orientation. The fine substructure is reconstructed from core parts within the bounds given by the coarse bundles.

Keywords: Micro computed tomography · Multiscale structure · Bovine leather · Local orientation · Hessian matrix

1 Introduction

Genuine leather material is well established for premium applications in upholstery and automotive interiors. Knowledge regarding microstructure's influence on the resulting leather properties is as old as humankind. However, quantification e.g. of the impact of the collagen-based leather structure's anisotropy on the leather's physical properties is both, of great interest and still challenging.

The macroscopic material properties of leather are highly variable, not the least due to (micro)structural variation [1] depending on species, race, gender, age, husbandry conditions as well as individual body part and tanning process [2]. Microscopic studies of the microstructure of leather revealed significant structural differences [3]. The 3D microstructure of leather has been studied however only recently. Non destructive testing methods like ultrasound imaging, small angle X-ray scattering, and computed tomography (CT) have been applied to capture structural features of the collagen fiber bundles see [4–6]. 3D leather images have been first and so far only analyzed quantitatively in [5]. Two samples of vegetable tanned bovine leather were compared based on a binarization of the μCT images into collagen fiber bundle structure (foreground) and pore space (background)

© Springer International Publishing AG 2017
J. Angulo et al. (Eds.): ISMM 2017, LNCS 10225, pp. 447–454, 2017.
DOI: 10.1007/978-3-319-57240-6_36

by simple thresholding. Structure density, homogeneity, and anisotropy were analyzed in terms of area fraction and mean intercept lengths. These first 3D analyses proved that prominent features of the leather's microstructure can be captured and different samples can be quantitatively compared.

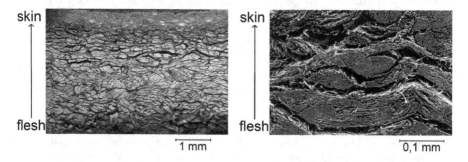

Fig. 1. Exemplary SEM images of a bovine axila leather sample revealing the bundle structure as well as the fiber and fibril sub-structure.

A deeper understanding of the specific leather structure, in particular of how the fiber bundles are interwoven and connected, requires however further segmentation. Moreover, our long-term goal of finite element simulations of mechanical properties requires splitting the interwoven bundle structure into elements of simple shapes and revealing their connectivity relations. This task is very demanding due to the multiscale nature of the leather's microstructure consisting of bundles formed by fibers that in turn consist of collagen fibrils, see e.g. [1,2]. These scales are not clearly separated, see Fig. 1. Moreover, being a natural material, the structure varies strongly. So far, segmentation of one individual fiber bundle with its substructure has been achieved by manually marking contours in virtual 2D sections of the CT image and tracing them through the whole stack [5]. This approach is however too tedious to obtain a number of bundles high enough for statistical analyses.

Here, we report on the first attempt to automatically segment typical structural elements of the leather's microstructure. A new soft shock filter is used for preprocessing. Subsequently the watershed transform yields a strongly oversegmented starting point for hierarchical coarsening following the waterfall paradigm [7] thriving on differences of features of the regions. As the decisive feature, we use local orientation information.

2   Samples and Image Data

Throughout this paper, we restrict ourselves to the two purely vegetable tanned samples of bovine leather from [5], see Fig. 2. The particular tanning process ensures that no leftovers of heavy metals destroy the weak X-ray absorption contrast of the collagen. The samples were scanned with a laboratory CT device featuring a Perkin-Elmer flat bed detector, at 75 kV/3.8W with pixel size 3.2 μm.

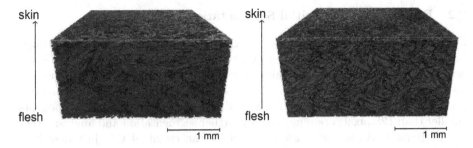

Fig. 2. Volume renderings ($500 \times 1\,000 \times 1\,000$ pixels) of the reconstructed μCT images of the two analyzed leather samples. Axila left, back right.

3 Segmentation Algorithm

3.1 Morphological Soft Shock Filter

The strong gray value fluctuations within the bundles caused by the substructure require extensive preprocessing. In order to remove these while preserving the fiber bundle edges, a locally adaptive filtering is needed. The morphological shock filter emphasizes noise and the substructure too much. We therefore suggest a "soft" shock filter. Let $\mathbb{L}^3 = a\mathbb{Z}^3$, $a > 0$ be a 3D orthogonal lattice and $f : \mathbb{L}^3 \mapsto \mathbb{R}$ the original image. Now recall the morphological midrange filter on gray value images: $\mathrm{mid}_s(f)(x) = 1/2(\epsilon_s(f)(x) + \delta_s(f)(x))$, where s denotes a structuring element and erosion ϵ and dilation δ are taken w.r.t. s centered in the current pixel x. Consider now the proportion $f/\mathrm{mid}_s(f)$. Values larger than 1 indicate, that the original f is closer to the dilation $\delta_s(f)$, values smaller than 1 that the original f is closer to the erosion $\epsilon_s(f)$.

The standard morphological shock filter toggles between erosion and dilation depending on which the original pixel value is closer to. The new filter ensures a smoother transition. Define the smooth operators δ^* and ϵ^* as weighted averages of erosion and dilation:

$$\delta_s^*(f) = \delta_s(f) \min(e^{(1-f/\mathrm{mid}_s(f))}, 1) + \epsilon_s(f) \max(1 - e^{(1-f/\mathrm{mid}_s(f))}, 0)$$
$$\epsilon_s^*(f) = \epsilon_s(f) \min(e^{(1-\mathrm{mid}_s(f)/f)}, 1) + \delta_s(f) \max(1 - e^{(1-\mathrm{mid}_s(f)/f)}, 0),$$

where the dilation dominates δ^* while the erosion dominates ϵ^*. Finally the "soft" shock filter σ^* is the filter toggling f between δ^* and ϵ^*. More precisely,

$$\sigma^*(f, SE_s)(x) = \begin{cases} \epsilon_s^*(f)(x), & \text{if } f(x) - \epsilon_s^*(f)(x) < \delta_s^*(f)(x) - f(x) \\ \delta_s^*(f)(x), & \text{if } \delta_s^*(f)(x) - f(x) < f(x) - \epsilon_s^*(f)(x) \\ f(x), & \text{else} \end{cases}.$$

The effect of the filter σ^* applied iteratively with growing structuring element on the leather image data is illustrated using 2D slices in Fig. 3 below.

3.2 Binarization and Initial Segmentation

Before individual fiber bundles can be segmented, the image has to be separated into foreground (leather structure) and background (pore system, air). This is achieved using the global gray value threshold according to Otsu's method [8] on the morphologically opened soft shock filtered original image. Remaining pores completely enclosed by leather as well as foreground connected components smaller than 125 pixels are removed. All further segmentation and analysis steps are performed within the mask provided by the result of this just described binarization.

The filtered gray value images are initially segmented by the classical watershed transform on the local orientation gradient image (Fig. 3(d)). The latter is obtained by first computing in each pixel the Hessian matrix of 2nd order gray value derivatives, subsequently deriving the local normal direction as the one corresponding to the largest eigenvalue of the Hessian, and finally calculating the local difference of normal directions in the great-circle metric on the sphere. Applying the watershed transform on the result yields a vast over-segmentation of the fiber structure, see Fig. 3(e).

3.3 Segmentation of the Fiber Bundle Structure

In this section, we unite regions generated by the initial steps using the waterfall algorithm [7], a paradigm for hierarchical coarsening of the segmentation based on removing watersheds dividing "similar" regions. This approach was chosen despite its known drawbacks [9], as it yields the needed natural combination of proximity in both the image as well as feature space.

For the first coarsening step, to each region of the initial segmentation from Sect. 3.2, the median gray value of this region in the soft shock filtered image is assigned, Fig. 3(f). Subsequently, the regions have grown large enough to switch to the main axis of inertia as criterion for the following coarsening steps. The computation of the main axis is stabilized by weighting the included pixels by their distance to the background. For the results, see Figs. 3(g)–(i).

The waterfall algorithm creates a hierarchy of structure elements. However, using the orientation information as described, results in prematurely merging regions of similar orientation (nearly parallel regions, regions branching under a small angle) without preserving the correct contact behavior. As this well-known drawback of the waterfall algorithm [9] is hard to overcome, the fine level bundle structure is reconstructed by the core part reconstruction algorithm from [10,11], applied to the binarization of the coarse level segmentation result and using the medial axis as marker image.

Roughly, the Euclidean distance transform $EDT(x)$ and the local radius $R(x)$ are computed in each foreground pixel x. A threshold p_{min} on the probability image $P = \min(EDT/R, 1)$ yields the fiber cores. Select a pixel x_0 from the marker image foreground. If $P(x_0) > p_{min}$, choose x_0 as initial coordinate for a fiber chain. Then extend the fiber chain towards the 26-neighbor y with maximum probability value until $P(y) \leq p_{min}$. Once the condition fails, treat the

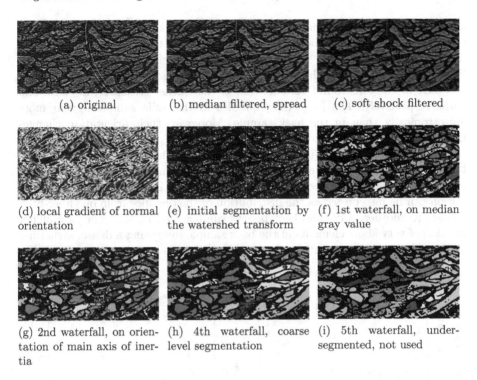

(a) original

(b) median filtered, spread

(c) soft shock filtered

(d) local gradient of normal orientation

(e) initial segmentation by the watershed transform

(f) 1st waterfall, on median gray value

(g) 2nd waterfall, on orientation of main axis of inertia

(h) 4th waterfall, coarse level segmentation

(i) 5th waterfall, under-segmented, not used

Fig. 3. Slices through the axila data set, steps of the preprocessing, the initial segmentation, and the subsequent waterfall coarsening.

opposite direction likewise. Finally, the fiber is reconstructed using the derived pixel coordinates and the radii from R. This procedure is repeated till all marker points are covered by a reconstructed fiber.

The core reconstruction results in a substructure of the coarse fiber bundles that is consistent in the following sense: Fine level elements tessellate the coarse level elements. Fine level elements start and end at extreme points, branching points, strongly curved areas or areas of thickness change of the corresponding coarse level element. See Fig. 4 for an example.

4   Application

For the 3D images of the microstructure of the axila and the back leather samples described in Sect. 2, the just derived algorithm yields 2 208 fiber bundles and 8 089 sub-bundles and 2 515 fiber bundles and 11 089 sub-bundles, respectively.

Exemplary, we report analysis results on orientation of the coarse bundle structure and on size of the substructure. For the first, the local orientation of the fiber bundle is estimated in each pixel using the eigenvalues of the Hessian matrix [12,13]. From these local orientations, the local 2nd order orientation tensor is derived. Averaging over all pixels within the fiber bundle and eigenvalue

analysis of the mean orientation tensor finally yields the mean orientation of
the bundle [14]. Figure 4 shows the mean orientation tensor averaged along the
sample thickness direction (x-direction) over all pixels within the 10% largest
fiber bundles. Note that these 10% largest fiber bundles represent about 49%
and 54% of the foreground volume, that is the leather structure, for back and
axila, respectively. The large fiber bundles in the axila sample appear more
anisotropically than in the back sample. Moreover, their orientation changes
more strongly from flesh to grain (skin) side.

The substructure size is measured by the length of the minimal volume
bounding boxes of the fine scale elements approximated using the algorithm
of [15] on the vertices of the convex hull as described in [16]. This length distribution can be interpreted as the distribution of the distance from one fiber-fiber
contact to another. Axila and back behave similar, see Fig. 4, except for a higher
fraction of very short elements in the back sample indicating a denser structure.
This result is consistent both with visual impression and other analysis results.

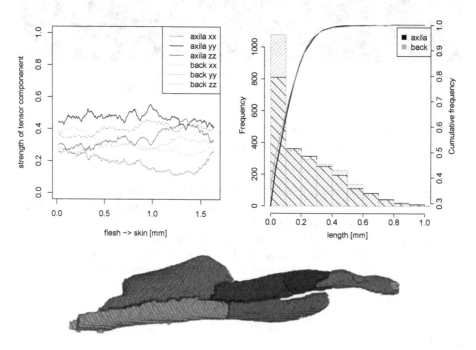

Fig. 4. Diagonal components of the mean orientation tensor along the thickness direction for the 10% largest coarse fiber bundles, length distribution of the fine scale elements, and volume rendering of one selected bundle with its substructure.

5   Discussion

This paper describes the first algorithm for automatic segmentation of representative elements of leather's microstructure in 3D. Application to two μCT data

sets of bovine leather yields fiber bundle samples on two consistent hierarchy levels. These samples are sufficiently large for statistical analyses and allow to quantitatively describe the connectivity of the structure.

Given that there is no ground truth available, the algorithm can not be proved to be correct. Consistency with visual impression and previous findings was however checked in each step.

The bundle structure on the finer scale is very susceptible to slight changes in processing steps and parameters. Robustness w.r.t. imaging conditions was tested using images of the same samples taken at ID 19 of the European Synchrotron Radiation Facility. These trials failed although SRμCT increases the gray value contrast, lowers the noise level, and enables exploitation of the inline X-ray phase contrast [17]. Indeed, the edges of the fiber bundles are much clearer defined, see Fig. 5. However, less substructure within the bundles remains and thus the orientation information does not suffice to steer the watershed and the waterfall transforms.

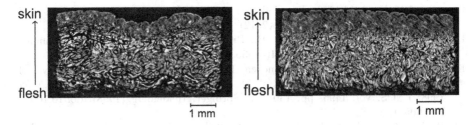

Fig. 5. Slices through the reconstructed SRμCT image of the axila (left) and the back (right) sample. Visualized are $2\,539 \times 1\,211$ and $2\,308 \times 1\,080$ pixels.

Further effort is needed to develop a robust segmentation method. Moreover, the substructure should result directly from the hierarchical segmentation. P-algorithms [9] have to be tested to this end.

Acknowledgment. This research was funded through IGF-project 18102 BG of the Research Association Leather by the AiF on behalf of the Bundesministerium für Wirtschaft und Technologie based on a resolution of Deutscher Bundestag.

References

1. Haines, B.: Fiber structure and physical properties of leather. J. Am. Leather Chem. Assoc. **69**, 96–111 (1974)
2. Attenburrow, G.: The rheology of leather a review. J. Soc. Leather Technol. Chem. **77**, 107–114 (1993)
3. Qu, J.B., Zhang, C.B., Feng, J.Y., Gao, F.T.: Natural and synthetic leather: a microstructural comparison. J. Soc. Leather Technol. Chem. **92**, 8–13 (2008)

4. Basil-Jones, M.M., Edmonds, R.L., Allsop, T.F., Cooper, S.M., Holmes, G., Norris, G.E., Cookson, D.J., Kirby, N., Haverkamp, R.G.: Leather structure determination by small-angle x-ray scattering (saxs): cross sections of ovine and bovine leather. J. Agric. Food Chem. **58**(9), 5286–5291 (2010)

5. Bittrich, E., Schladitz, K., Meyndt, R., Schulz, H., Godehardt, M.: Micro-computed tomography studies for three-dimensional leather structure analysis. J. Am. Leather Chem. Assoc. **109**(11), 367–371 (2014)

6. Wells, H.C., Holmes, G., Haverkamp, R.G.: Looseness in bovine leather: microstructural characterization. J. Sci. Food Agric. **96**(8), 2731–2736 (2016)

7. Beucher, S.: Watershed, hierarchical segmentation and waterfall algorithm. In: Serra, J., Soille, P. (eds.) Mathematical Morphology and Its Applications to Image Processing. Computational Imaging and Vision, vol. 2. Kluwer, Dordrecht (1994)

8. Otsu, N.: A threshold selection method from gray level histograms. IEEE Trans. Syst. Man Cybern. **9**, 62–66 (1979)

9. Beucher, S.: Towards a unification of waterfalls, standard and p algorithms. Technical report, Centre for Mathematical Morphology, MINES ParisTech (2012)

10. Altendorf, H., Jeulin, D.: Fiber separation from local orientation and probability maps. In: Abstract book of the 9th ISMM, Groningen (2009)

11. Altendorf, H.: Analysis and modeling of random fiber networks. Ph.D. thesis, TU Kaiserslautern, Mines ParisTech (2012)

12. Eberly, D., Gardner, R., Morse, B., Pizer, S., Scharlach, C.: Ridges for image analysis. J. Math. Imaging Vis. **4**(4), 353–373 (1994)

13. Frangi, A.F., Niessen, W.J., Vincken, K.L., Viergever, M.A.: Multiscale vessel enhancement filtering. In: Wells, W.M., Colchester, A., Delp, S. (eds.) MICCAI 1998. LNCS, vol. 1496, pp. 130–137. Springer, Heidelberg (1998). doi:10.1007/BFb0056195

14. Wirjadi, O., Schladitz, K., Easwaran, P., Ohser, J.: Estimating fibre direction distributions of reinforced composites from tomographic images. Image Anal. Stereology **35**(3), 167–179 (2016)

15. Barequet, G., Har-Peled, S.: Efficiently approximating the minimum-volume bounding box of a point set in three dimensions. J. Algorithms **38**(1), 91–109 (2001)

16. Vecchio, I., Schladitz, K., Godehardt, M., Heneka, M.: 3D geometric characterization of particles applied to technical cleanliness. Image Anal. Stereology **31**(3), 163–174 (2012)

17. Paganin, D., Mayo, S.C., Gureyev, T.E., Miller, P.R., Wilkins, S.W.: Simultaneous phase and amplitude extraction from a single defocused image of a homogeneous object. J. Microsc. **206**(1), 33–40 (2002)

Brain Lesion Detection in 3D PET Images Using Max-Trees and a New Spatial Context Criterion

Hélène Urien[1]($\boxtimes$), Irène Buvat[2], Nicolas Rougon[3], Michaël Soussan[2], and Isabelle Bloch[1]

[1] LTCI, Télécom ParisTech, Université Paris-Saclay, 75013 Paris, France
helene.urien@telecom-paristech.fr
[2] IMIV, Inserm, CEA, Université Paris-Sud, CNRS,
Université Paris-Saclay, 91400 Orsay, France
[3] MAP5, CNRS, Télécom SudParis, Université Paris-Saclay, 91011 Evry, France

Abstract. In this work, we propose a new criterion based on spatial context to select relevant nodes in a max-tree representation of an image, dedicated to the detection of 3D brain tumors for ^{18}F-FDG PET images. This criterion prevents the detected lesions from merging with surrounding physiological radiotracer uptake. A complete detection method based on this criterion is proposed, and was evaluated on five patients with brain metastases and tuberculosis, and quantitatively assessed using the true positive rates and positive predictive values. The experimental results show that the method detects all the lesions in the PET images.

Keywords: Max-tree representation · Spatial context · Brain tumors · Positron Emission Tomography · Detection

1 Introduction

Automatic tumor detection in Positron Emission Tomography (PET) imaging, usually performed as a first step before segmentation, is a difficult task due to the coexistence of physiological and pathological radiotracer uptake, both resulting in a high signal intensity. For example, in ^{18}F-FDG PET imaging, the distinction between brain metastases and the whole physiological brain uptake is not obvious, especially for small lesions. In clinical routine, the detection problem is overcome by manually defining a volume surrounding the tumor. The segmentation is then performed within this volume of interest using various strategies [1].

In a multimodal segmentation process, combining PET with an anatomical modality, such as Magnetic Resonance Imaging (MRI), tumor detection can also turn to be a critical initialization step to assess the location and number of lesions, and so influences the final result. A typical initialization method consists in thresholding the PET signal intensity, which can be further refined using mathematical morphology [2]. However, using such a threshold as a detection step has two main limitations. First, it is not adapted to patients having several lesions of different metabolisms, and can lead to an under- or over-detection

© Springer International Publishing AG 2017
J. Angulo et al. (Eds.): ISMM 2017, LNCS 10225, pp. 455–466, 2017.
DOI: 10.1007/978-3-319-57240-6_37

according the threshold value. Secondly, the use of a PET threshold does not prevent the detected tumor from merging with adjacent structures of physiological uptake, such as basal nuclei or other brain regions when using ^{18}F-FDG as a radiotracer. Thus, our goal was to design a detection method for brain PET images, providing lesion markers for a subsequent PET-MR segmentation, and detecting all the pathological areas of increased uptake, while preventing them from merging with regions with physiologically increased uptake.

In this context, we propose a method to detect brain tumors on PET images, embedding spatial context information about the tumor in a hierarchical approach. We designed a new criterion to select relevant nodes in a max-tree representation, based on contextual information modeling reasonable hypotheses about the appearance of tumors in PET images. We first describe the tumor detection method, which prevents the tumor from merging with nearby physiological uptake. Then, we show and discuss the results, which will further be used as an initialization step to our previous segmentation method performed on MRI and guided by PET information [3].

2   Proposed Detection Method

The max-tree, as used in [4], is a hierarchical representation of an image based on the study of its intensity thresholds. It can be built according to various methods (see e.g. [5,6] for a comparison). Regions of interest are then selected according to a given criterion. Since PET image threshold is a common method for tumor segmentation [1], and since the max-tree representation highlights bright areas in an image, and so potential tumors, it has already been used on PET images for tumor segmentation, for example using a shape criterion [7]. In addition, other criteria have been evaluated in hierarchical approaches, such as spatial context [8]. In this paper, we propose a new spatial context criterion applied on a max-tree representation of the image to detect potential tumors in PET images. The detection results are then refined using topological and symmetry information.

2.1   3D Hierarchical Tumor Detection Using Spatial Context Information

In our method, the max-tree representation of the PET image is computed on the SUV, and leads to a hierarchical representation of its flat zones (identified by a letter in Fig. 1a), that are homogeneous regions of unique intensity value (identified by a number in Fig. 1a). The SUV (Standardized Uptake Value) is a standardization of the PET image widely used for quantification purposes [9], which measures the tumor metabolism and depends of parameters proper to the exam (duration and radiotracer) and the patient (weight). Formally, let Ω be the spatial domain (here $\mathbb{Z}^3$ for discrete formulation). An image is defined as a function I from Ω into $\mathbb{N}$ of $\mathbb{R}^+$. The max-tree of image I is a hierarchical representation of the connected components of all its upper level sets L_n^+, namely

$L_n^+(I) = \{x \in \Omega \mid I(x) \geq n\}$, given $n \in \mathbb{N}$, within a defined neighborhood. The tree is composed of nodes, associated to the previous connected components, and edges, embedding the inclusion relation between them. A node N is said to be a descendant of a node M if a path in the tree allows linking them and if N is at a higher position in the tree than M (for example nodes F, E, D and C are descendants of node B in Fig. 1a). Inversely, M is said to be an ancestor of node N. In this paper, a node N is said to be a direct descendant of a node M if N is a descendant of M that is directly connected to it in the tree (for example nodes D and E are the direct descendants of node C in Fig. 1a).

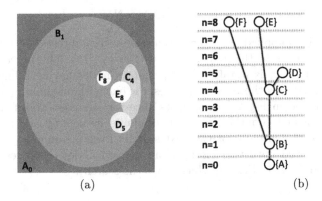

(a)   (b)

Fig. 1. A synthetic image (a) and its max-tree (b). A flat zone R_n in (a) is represented by a node R at level n in (b).

To simplify the notations, we will use the notation N both for a node of the max-tree and the corresponding region in Ω. The set of all the nodes is denoted by $MT(I)$. Let $DD(N)$ be the set of direct descendant nodes of N, $DD(N) = \{N^i, i \in I(N)\}$, with $I(N)$ the index set of the direct descendants of N. We also define $D(N)$ and $A(N)$, respectively the set of descendant and ancestor nodes of N. Let $\widehat{N}$ be the region made of the node N and all its descendant nodes $D(N)$: $\widehat{N} = N \cup D(N)$. Finally, $|N|$ denotes the number of voxels of node N, $i.e$ its volume in Ω.

Once the tree is created, relevant nodes are selected according to a given criterion. We expect tumors to correspond to some of the leaf nodes. However, the distinction between nearby physiological and pathological radiotracer uptake is difficult. We propose to first apply a criterion χ_t preventing a tumor node from merging with a node including physiological uptake. We base this criterion on two hypotheses. First, an acceptable merge should result in a node having several direct descendant nodes, and a large difference in volume compared to at least one of them. We thus design the following criterion χ_t^1, taking values in $[0, 1]$:

$$\chi_t^1(N) = \max_{N^i \in DD(N)} \frac{\left|\widehat{N}\right| - \left|\widehat{N^i}\right|}{\left|\widehat{N}\right|} \tag{1}$$

Moreover, the merge does not concern too voluminous nodes, associated with almost all both physiological and pathological radiotracer uptake. The resulting criterion χ_t^2 is then designed as follows:

$$\chi_t^2(N) = e^{-\left(\frac{|\tilde{N}|}{K_t|I_b|}\right)^2} \tag{2}$$

where K_t is a positive parameter and I_b the binary mask of the brain, created by thresholding the PET image converted into SUV. The final criterion χ_t combines the two previous criteria in a conjunctive way:

$$\chi_t(N) = \chi_t^1(N)\chi_t^2(N) \tag{3}$$

The criterion χ_t is applied only to nodes having several direct descendant nodes, called merged nodes in this paper (B and C in Fig. 1a). The process for node selection based on χ_t is iterative, and promotes nodes having a high χ_t value (Algorithm 1). Among these merged nodes, the one with maximal χ_t value (node C depicted in red in Fig. 2a) and all its ancestors (crossed nodes A and B in Fig. 2a) are removed from the tree. Its descendants are kept (E and D in Fig. 2a), but not taken into account for the next search of the maximum χ_t value. This process is repeated until there is no more merged nodes to process (the only remaining node is F in Fig. 2a, which stops the search).

Algorithm 1. Node selection using χ_t

Input: I, $k = 0$, $MT^k = MT$
Output: MT (filtered max-tree according to χ_t)
   while $\exists N \in MT^k \,|\, |DD(N)| \geq 2$ **do**
     $k = k + 1$
     $N^k = \underset{N||DD(N)|\geq 2}{\arg\max} \; \chi_t(N)$
     $MT^k = MT^{k-1}\backslash \{N^k \cup A(N)\}$
   end while
   $MT = MT^k$

After applying the χ_t criterion, we define a context criterion χ_c for tumor detection. The spatial context volume is defined by using the distance transform of $\hat{N}$, associating to each voxel x of the PET volume the value $D_{\hat{N}}(x) = \min_{y\in\hat{N}} d(x,y)$, where y is a voxel included in $\hat{N}$, and d is the distance between two voxels (the Euclidian distance was used in this work). The context volume $C_{\hat{N}}$ is obtained by thresholding this function, which can be formally written as $C_{\hat{N}} = \{x \,|\, 0 < D_{\hat{N}}(x) \leq s\}$, where s is a positive value. Finally, the PET image is thresholded within the spatial context volume using Otsu's method [10], to exclude white matter.

The χ_c criterion design follows three hypotheses. First, the intensity of the tumor in the PET image should be higher than the one of its surroundings. This is embedded in the criterion χ_c^1, which should be greater than one for nodes associated to tumors:

$$\chi_c^1(N) = \frac{\mu(I_{\mathrm{PET}}, \widehat{N})}{\mu(I_{\mathrm{PET}}, C_{\widehat{N}})} \tag{4}$$

where $\mu(I_{\mathrm{PET}}, V)$ is the mean intensity value of the PET image I_{PET} inside volume V, $C_{\widehat{N}}$ the context volume.

Then, the tumor should not be too voluminous. In fact, some voluminous nodes, associated to almost all pathological and physiological radiotracer uptake, can have a χ_c^1 value greater than one. The criterion χ_c^2 prevents from this situation:

$$\chi_c^2(N) = e^{-\left(\frac{|\widehat{N}|}{K_c^1 |I_b|}\right)^2} \tag{5}$$

where K_c^1 is a positive parameter.

Finally, the tumor should not be too small. The criterion χ_c^3 embedds this last hypothesis:

$$\chi_c^3(N) = \frac{K_c^2}{|\widehat{N}|} \tag{6}$$

where K_c^2 is a positive parameter.

The context criterion χ_c is then created by combining the three previous criteria, for each node $N \in MT$:

$$\chi_c(N) = \chi_c^1(N)\chi_c^2(N) - \chi_c^3(N) \tag{7}$$

The product of $\chi_c^1(N)$ and $\chi_c^2(N)$ is less than 1 for a voluminous node N, even if $\chi_c^2(N) \geq 1$. Substracting $\chi_c^3(N)$, increasing if $|\widehat{N}|$ decreases, from $\chi_c^1(N)\chi_c^2(N)$ allows having a final $\chi_c(N)$ value less than 1 for a node N of too small volume. Thus, only the nodes with a χ_c value greater than 1 are taken into account in the node selection process leading to the binary image of the detected tumors T (Algorithm 2).

The nodes maximizing the χ_c value are then selected. Among the nodes kept after the χ_t process (D, E and F in Fig. 2b), only the nodes with a value greater than 1 are considered. Among these nodes, the one of maximum χ_c value is selected, and its descendants and ancestors are removed. This process is repeated until there is no node of χ_c value greater than 1 to process. Each selected node is filled with its descendant nodes, which gives the signal intensity-based detection, shown in red in Fig. 2c.

Thus, the aim of our method is twofold: removing nodes embedding both pathological and physiological uptake, and then selecting only the tumors among the remaining nodes.

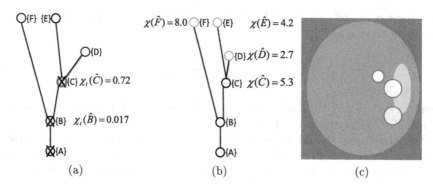

Fig. 2. Node selection process. (a) Assignment of a χ_t value to each merged node. The node of maximal χ_t value is depicted in red. The nodes and their ancestors coming from a too large node merging are not taken into account for further process (crossed nodes). (b) Assignment of a χ_c value to nodes accepted after applying the criterion χ_t (in green). (c) Contours of the resulting segmentation (in red). (Color figure online)

Algorithm 2. Node selection using χ_c

Input: I, $k = 0$, $MT^k = \{N \mid N \in MT \wedge \chi_c(N) \geq 1\}$, $T^k = \emptyset$
Output: T (selected nodes)
 while $MT^k \neq \emptyset$ **do**
  $k = k + 1$
  $N^k = \arg\max_{N \in MT^k} \chi_c(N)$
  $MT^k = MT^{k-1} \backslash \{N^k \cup A(N) \cup D(N)\}$
  $T^k = T^{k-1} \cup \{N^k \cup D(N)\}$
 end while
 $T = T^k$

2.2 Refinement Using Topological and Symmetry Information

Once the detection is obtained using the max-tree approach, we study each of the detected lesions to remove false positives due to physiological radiotracer uptake, using topological and symmetry hypotheses. First, we keep the detected lesions having only one connected component (8-connectivity) by slice. Then, based on the hypothesis that the PET signal is higher in the tumor than its symmetrical region, we compute for the remaining lesions the ratio between the mean intensity of the PET within the lesion and the mean intensity of the PET in the symmetrical region of the lesion with respect to the inter-hemispheric plane, which is automatically identified using the method from [11]. The symmetrical region is thresholded via Otsu's method [10], to exclude white matter. The set of obtained ratio values that are greater than 1 is divided into two classes (using k-means algorithm), and only the one corresponding to the highest values is kept.

3   Experimental Results

3.1   Patient Data

The proposed method was evaluated on images from 5 patients having brain lesions (tuberculosis for patient P2, and metastasis for the others), who underwent a whole body PET-MR scan. The exam was performed with a PET/MR

Table 1. Quantitative comparison between automatic lesion detections with or without post-processing (PP). The True Positive Rate (TPR) and Positive Predictive Value (PPV) are computed for each patient, before and after (TPR-PP and PPV-PP) post-processing.

Patient	P1	P2	P3	P4	P5	$\mu \pm \sigma$
TPR	1.00	1.00	1.00	1.00	1.00	1.00 ± 0.00
TPR-PP	1.00	1.00	1.00	1.00	1.00	1.00 ± 0.00
PPV	0.09	0.12	0.11	0.08	0.09	0.09 ± 0.02
PPV-PP	1.00	0.40	0.33	0.50	0.50	0.55 ± 0.23

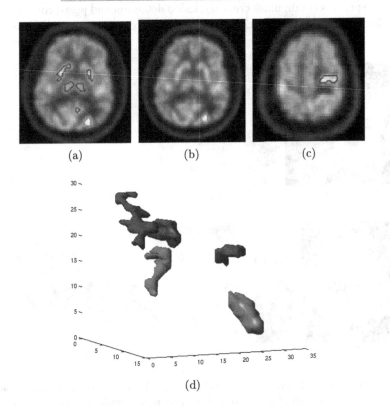

Fig. 3. Evaluation of the PP for P2: comparison between the detected lesions, true (in green) and false (in red) positives. Results shown on PET slices before (a) and after ((b) and (c)) PP, and in 3D after applying PP (d). (Color figure online)

scanner (GE SIGNA) after an initial injection of 320 MBq ^{18}F-FDG and about 100 min delay between injection and PET/MR acquisition. The voxel size in the PET images is $3.12 \times 3.12 \times 3.78\,\mathrm{mm}^3$.

3.2 Hyperparameter Setting

The 3D 6-connectivity was used to define the connected flat zones in the SUV image and create the max-tree, and to differentiate the detected lesions while applying the post-processing. The parameter in the criterion χ_t was set to $K_t = 0.5$, while those of the spatial context criterion χ_c were set to $K_c^1 = 0.5$ and $K_c^2 = 4$. The spatial context parameter was set to $s = 1$ voxel. These hyperparameters were set experimentally to reduce the number of detected lesions for the patients having the smallest tumors (P1 and P3), and then applied to all the dataset.

3.3 Results

The ground truth PET brain lesions were those visually detected by a medical expert using only the PET data. For each patient, the computational time for the whole process (brain mask creation, lesion detection and post-processing) was about 3 min limiting the detection to a brain mask containing about 53000 voxels.

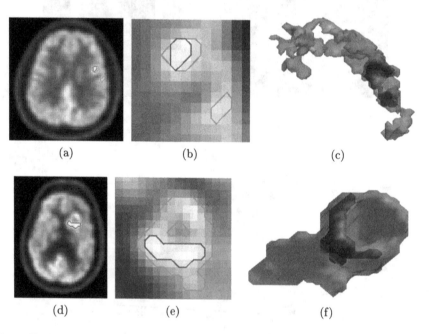

(a)   (b)   (c)

(d)   (e)   (f)

Fig. 4. Evaluation of the criterion χ_t for P2 (first row) and P5 (second row): comparison between the true positive lesions detected with (in red) or without (in green) previously applying the criterion χ_t. Results shown on a PET slice ((a) and (d)) zoomed in (b) and (e), and in 3D ((c) and (f)). (Color figure online)

The detection performance was characterized using the True Positive Rate (TPR), defined as the ratio between the number of true tumors among the detected lesions and the real number of tumors, and the Positive Predictive Value (PPV), defined as the ratio between the number of true tumors among the detected lesions and the total number of detected lesions. The true tumors were defined as the detected tumors having a non empty intersection with the ground truth. As shown by the TPR in Table 1, the algorithm detected all the lesions visible in the PET images. As shown by the PPV in Table 1, the algorithm also detected false positives, but drastically less after using topological and symmetry information in a post-processing (PP) step.

The usefulness of the PP was also assessed visually. As shown in Fig. 3b, the PP eliminates false positives due to the physiological uptake of symmetrical structures such as basal nuclei. The remaining false positives are due to other asymetrical physiological uptake in the brain (Fig. 3c).

Moreover, applying the criterion χ_t before the spatial context criterion χ_c prevents the true detected lesions from merging with basal nuclei (Fig. 4c) or other brain physiological uptake regions (Fig. 4f). However, as previously shown

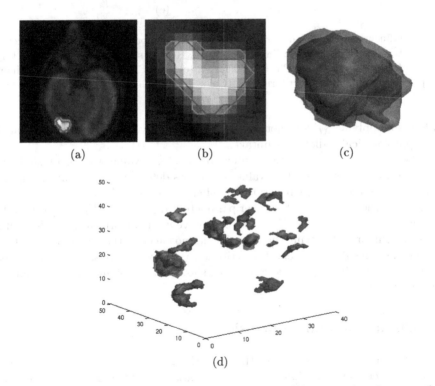

Fig. 5. Influence of the spatial context parameter s for P4: comparison between the detection setting $s = 1$ (in green) and $s = 5$ (in red). Results shown on a PET slice (a), zoomed in (b), and in 3D for only the true positive (c) or all the detected lesions before PP (d). (Color figure online)

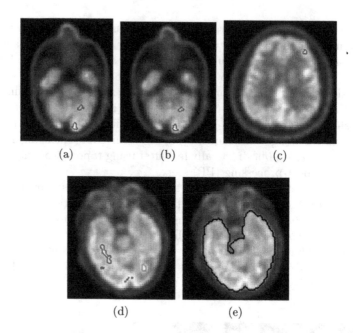

(a)   (b)   (c)

(d)   (e)

Fig. 6. Influence of the parameters K_c^1, K_c^2 and K_t: comparison between the detected lesions, true (in green) or false (in red) positives. Results shown on PET slices setting $K_c^2 = 4$ (a) or $K_c^2 = 5$ (b) for P3, $K_t^1 = 1$ for P2 (c), $K_c^1 = 1$ for P1 with (d) or without (e) previously applying the criterion χ_t. (Color figure online)

in Fig. 3b, applying χ_t does not reduce the detected volume for tumors isolated from physiological radiotracer uptake.

The influence of the spatial context parameter s value was also tested. As shown in Fig. 5 increasing the s value still allows detecting the actual lesion, but reduces its volume and increases the number of false positives.

Finally, the influence of the other parameters K_c^1, K_c^2 and K_t were tested. As shown in Fig. 6, increasing K_c^2 or K_t can prevent the algorithm from detecting too small tumors. Modifying the K_c^1 value has no effect on the detection (Fig. 6d). However, the result is different if the criterion χ_t is not previously applied (Fig. 6e where the whole brain is detected), which shows that applying χ_t also penalizes too large nodes.

4   Discussion and Conclusion

In this paper, we proposed a method based on a 3D max-tree representation and a new selection criterion based on the spatial context to detect brain lesions on 3D PET images. Our algorithm identifies tumor locations, preventing the detected lesions from merging with spatially close physiological uptake regions, at the price of a final reduced tumor volume. Thus, it is intended to be a detection method that can serve as a preliminary step for a segmentation method, such

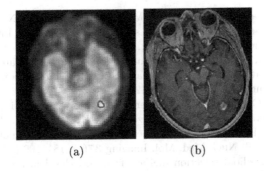

(a)   (b)

Fig. 7. Results of a MR segmentation method initialized with the PET for P1. (a) Superimposing of the PET image on the contours of the detected lesion (in red). (b) Superimposing of the MR image on the contours of the segmented lesion (in red). (Color figure online)

as the one described in [3] to segment brain tumors on MR images in a variational approach using information from the PET (Fig. 7). However, the proposed method still detects asymetrical areas of physiological increased uptake, which could be further removed in our approach using complementary information from MRI volumes. Moreover, using MRI allows detecting other lesions, not visible in the ^{18}F-FDG PET, which is not the standard imaging procedure for metastasis diagnosis. Thus, our detection algorithm depends on the quality of the information given by the PET image. Finally, our method can also be extended to other brain lesions as shown in [12].

Acknowledgments. This work was supported by the "Lidex-PIM" project funded by the IDEX Paris-Saclay, ANR-11-IDEX-0003-02.

References

1. Foster, B., Bagci, U., Mansoor, A., Xu, Z., Mollura, D.J.: A review on segmentation of positron emission tomography images. Comput. Biol. Med. **50**, 76–96 (2014)
2. Bagci, U., Udupa, J.K., Mendhiratta, N., Foster, B., Xu, Z., Yao, J., Chen, X., Mollura, D.J.: Joint segmentation of anatomical and functional images: applications in quantification of lesions from PET, PET-CT and MRI-PET-CT images. Med. Image Anal. **17**, 929–945 (2013)
3. Urien, H., Buvat, I., Rougon, R., Boughdad, S., Bloch, I.: PET-driven multi-phase segmentation on meningiomas in MRI. In: IEEE International Symposium on Biomedical Imaging (ISBI), pp. 407–410 (2016)
4. Salembier, P., Oliveras, A., Garrido, L.: Antiextensive connected operators for image and sequence processing. IEEE Trans. Image Process. **7**(4), 555–570 (1998)
5. Carlinet, E., Géraud, T.: A comparative review of component tree computation algorithms. IEEE Trans. Image Process. **23**(9), 3885–3895 (2014)
6. Salembier, P., Wilkinson, M.H.F.: Connected operators. IEEE Sig. Process. Mag. **26**(6), 136–157 (2009)

7. Grossiord, E., Talbot, H., Passat, N., Meignan, M., Tervé, P., Najman, L.: Hierarchies and shape-space for PET image segmentation. In: IEEE International Symposium on Biomedical Imaging (ISBI), pp. 1118–1121 (2015)

8. Xu, Y., Géraud, T., Najman, L.: Context-based energy estimator: application to object segmentation on the tree of shapes. In: IEEE International Conference on Image Processing, pp. 1577–1580 (2012)

9. Boellaard, R., ODoherty, M.J., Weber, W.A., Mottaghy, F.M., Lonsdale, M.N., Stroobants, S.G., Oyen, W.J.G., Kotzerke, J., Hoekstra, O.S., Pruim, J., et al.: FDG PET and PET/CT: EANM procedure guidelines for tumour PET imaging: version 1.0. Eur. J. Nucl. Med. Mol. Imaging **37**(1), 181–200 (2010)

10. Otsu, N.: A threshold selection method from gray-level histograms. Automatica **11**, 285–296 (1975)

11. Tuzikov, A.V., Colliot, O., Bloch, I.: Evaluation of the symmetry image plane in 3D MR brain images. Pattern Recogn. Lett. **24**(14), 2219–2233 (2003)

12. Urien, H., Buvat, I., Rougon, N., Bloch, I.: A 3D hierarchical multimodal detection and segmentation method for multiple sclerosis lesions in MRI. In: MICCAI Multiple Sclerosis SEGmentation (MSSEG) Challenge, pp. 69–74 (2016)

Morphological Texture Description from Multispectral Skin Images in Cosmetology

Joris Corvo[1]([✉]), Jesus Angulo[2], Josselin Breugnot[1], Sylvie Bordes[1],
and Brigitte Closs[1]

[1] SILAB, BP 213, 19108 Brive Cedex, France
{j.corvo,j.breugnot,s.bordes,b.closs}@silab.fr
[2] Center for Mathematical Morphology, Mines-ParisTech,
PSL Research University, 77300 Fontainebleau, France
jesus.angulo@mines-paristech.fr

Abstract. In this paper, we propose methods to extract texture features from multispectral skin images. We first describe the acquisition protocol and corrections we applied on multispectral skin images. In the framework of a cosmetology application, a skin morphological texture evaluation is then proposed using either multivariate approach on multispectral dataset or marginal on a dataset whose dimensionality has been reduced by a multivariate analysis based on PCA.

Keywords: Multispectral imaging · Mathematical morphology · Texture analysis · Cosmetology

1 Introduction

Most common image texture features have been primarily designed for 2D or 3D scalar images. Those features focus on variations in the spatial domain living up information contained in spectral domain. However, in some classification studies [10,25] the definition of color or multispectral texture features led to better results than classical scalar texture features. This study tackles the issue of multispectral texture extraction in the context of skin description and classification.

In literature, three main approaches are used to obtain texture features from multi-chromatic images [12,25]:

- In the *parallel approach*, texture and spectral data are considered separately. In other words, spatial information is extracted on a greyscale version of initial images, whereas spectrum (or color) information is measured globally. This approach is often used in image retrieval [24].
- Frequently used in material analysis [19,30], the *sequential approach* consists in labeling the image spectra before extracting texture features. Usual scalar texture algorithms are processed on the labeled imaged since each label value can be considered as a scalar. The major disadvantage of this method is its lack of reproducibility induced by the preliminary step of image labeling.

J. Angulo et al. (Eds.): ISMM 2017, LNCS 10225, pp. 467–477, 2017.
DOI: 10.1007/978-3-319-57240-6_38

- The *integrative approach* regroups two alternatives for multivariate image texture analysis. The first one is the *marginal integrative approach*. It proposes to compute usual scalar texture features on each image component separately [10,27]. On the contrary, the second alternative is called *multivariate integrative approach* and relies on different algorithms allowing to extract texture features using all image components simultaneously [1,15,16,25,26,31,34].

In this paper, we adopt the integrative approach since we intend to define multispectral texture descriptors which are not content-dependent. In particular, we focus on building up multispectral texture features corresponding to the multivariate integrative approach, i.e., texture features for which every single value is obtained using all available wavelengths of a given multispectral image.

The new multispectral texture descriptors are applied to a database of human skin multispectral images acquired in-vivo within the context of a cosmetological study. Texture data are employed to analyze and predict the degradation of a foundation make-up. In the next section, we briefly introduce the database acquisition protocol and the different preprocessing steps applied to each image. Section 2 presents an original multivariate analysis called *Inverted and Permuted Principal Component Analysis* (IP-PCA) [9], used to reduce image dimensionality by projecting images components into consistent orthogonal spaces of representation. All multispectral texture parameters proposed in this paper are detailed in Sect. 4. Finally, in Sect. 5 we use data from the cosmetological study to illustrate and assess the multispectral texture efficiency.

2   Multispectral Image Acquisition and Preprocessing

2.1   Acquisition Device

Our base of skin images was obtained using a multispectral camera called Asclepios System. This camera designed by the Le2i Laboratory was originally dedicated to dermatological issues [17,18,35]. Inside the Asclepios System, light emitted by a constant light source passes through a wheel of 10 spectral filters and is guided to a hand-held acquisition device directly placed on the volunteers skin. The device contains a grey-level CCD camera synchronized to the filter wheel so that it acquires reflection images whose illuminant only depends on single spectral filter of the wheel. Functional schema of Fig. 1 gives an overview of the system different components.

Each multispectral acquisition is taken in about 2.5 s and is composed of 10 gray level spectral images between 430 nm and 85 nm. A raw acquisition is illustrated in Fig. 2. Spectral images are sized 1312×1082 pixels for approximately $35 \, mm \times 28 \, mm$ surface of skin (2.67×10^{-2} mm par pixel). This definition has proved to be sufficient to observe the impact of make-up on skin texture.

2.2   Cosmetology Protocol for Database Generation

The study of foundation make up effect and degradation on multispectral images skin was led on a panel of 30 volunteers. All of them applied the same amount

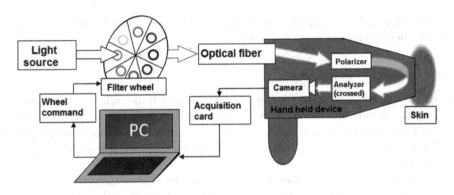

Fig. 1. Functional schema of the Asclepios System.

Fig. 2. Example of skin image raw acquisition from Asclepios System.

of foundation make up on their faces. Multispectral acquisitions were taken on their cheek. Only left side of their faces was made up so that the other side could serve as normalization value. In order to track in time the make-up degradation, 4 acquisitions series were realized on each volunteer. The two firsts corresponding to the skin before and right after application, whereas the two lasts were taken respectively 3 and 6 h after application. An example of image time series is given in Fig. 3. We note that for the purpose of visualization, a RGB color rendering from the multispectral is used. This rendering is based on pixel-wise linear regressions that associate RGB values to each pixel expressed in the Asclepios

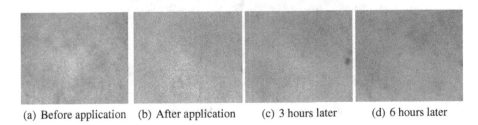

(a) Before application (b) After application (c) 3 hours later (d) 6 hours later

Fig. 3. Color-rendering of Asclepios system acquisitions of the same volunteer on the left cheek at four different times. (Color figure online)

spectral space. Those regressions are estimated using several Asclepios multi-spectral acquisitions of a Macbeth colorchart.

2.3 Image Prepocessing

Images acquired through the Asclepios system suffer from various perturbations leading to degraded images. In order to increase the quality of computed texture features, we seek to improve collected data by correcting and standardizing the multispectral raw images. A chain of four different preprocessing steps is applied to the data set in a specific order.

First of all, image normalization is performed by subtracting from each spectral image the average spectral image of the same wavelength. This results in eliminating the illumination gradient and some constant artifacts, e.g., Fig. 4(b).

Then a Vector Median Filtering (VMF) [2] is implemented to filter out noise of multispectral images without introducing artificial spectral values, e.g., Fig. 4(c).

In practice, we can observe small displacements between spectral images of the same acquisition. Those might be caused by movements of experimenter hand, volunteer facial expression changes or vibrations of the filter wheel during the 2.5 s acquisition. We compensate these shifts by computing a non-rigid registration vector field between all pairs of successive spectral images. Vector fields are then composed and applied in a way that all spectral images are registered to the same reference spectral image, e.g., Fig. 4(d).

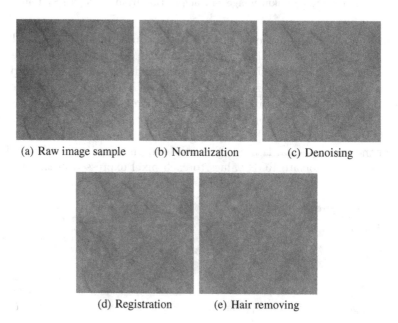

(a) Raw image sample   (b) Normalization   (c) Denoising

(d) Registration   (e) Hair removing

Fig. 4. Color-rendering of a multispectral skin image after successive preprocessing steps. (Color figure online)

The last preprocessing step is dedicated to deal with skin hairiness as it is not a component we want to take into account for the foundation make-up study. We designed a Digital Hair Removing (DHR) algorithm whose first part is to detect pixels belonging to hairiness using a linear opening and a binary thresholding. Second part is an inpainting method (Fast Marching inpainting method described in [6]) to fill in pixels classified as hairiness with values from their neighborhoods, e.g., Fig. 4(e).

3   Dimensionality Reduction by IP-PCA

In our set of acquisitions, we noticed important redundancy level between spectral images which is a recurrent issue of multispectral imaging. Principal Component Analysis (PCA) is classically used to reduce image dimensionality since this analysis method is canonical [3,5,20,28,35] and was proved to be the most effective in the case of real application datasets [33]. Given a multispectral image, PCA computes a reduced space of representation composed of the image eigenvectors. It allows us to transform a 10 components multispectral image into 4 eigenimages containing at least 95% of the original image variance.

However, PCA is only suitable for single group study, in our case, for a single multispectral image. Applying PCA to each multispectral skin image separately will generate several reduced spaces non homogeneous bases to one another because of possible inversions and permutations of eigenvectors and consequently, eigenimages which are not consistent between different acquisitions. Thus, we practice Inverted and Permuted Principal Component Analysis (IP-PCA) [9] to generate group-wise reduced spaces corrected by means of inversions and permutations as shown in Fig. 5. The algorithm detailed in [8,9], was proved to be useful on our dataset [8].

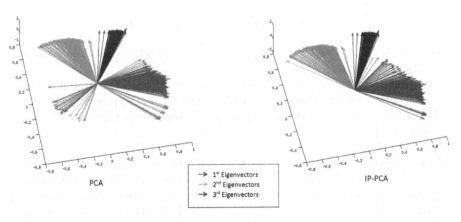

Fig. 5. Spatial representation (for the first three coordinates) of the eigenvectors of rank $\{1, 2, 3\}$ for the entire images base in the case of PCA (left side) and IP-PCA (right side). Each base of 3 eigenvectors corresponds to one of the multispectral images.

Now that redundancy is decreased, we can focus on extracting significant texture parameters on resulting eigenimages.

4   Texture Extraction from Multispectral Images

Starting from our multispectral dataset, the goal is to quantify the presence of foundation make-up. This section explains the strategy of texture extraction directly on multispectral data (integrative approach). This objective can be driven following either a marginal approach, i.e., component by component, or a multivariate one using all components simultaneously.

4.1   Marginal Approach from Eigenimages

Marginal approach has the advantage of using classical grey level algorithms on each image component. In our case, marginal texture features are extracted separately on the 4 first IP-PCA eigenimages, which are linear combinations of initial spectral images. In this sense, even marginally computed texture data take all spectral images into account. Thus, texture descriptors obtained that way can be qualified as multispectral texture descriptors.

Several texture parameters are considered in order to enrich the skin surface analysis. In the following definitions, we consider a multispectral image $F(x) = \{f_1(x), \ldots, f_D(x)\}$ composed of D components defined for each pixel x.

- *Statistical moments*
  Statistical moments until the 4th order (i,e., the mean, variance, asymmetry coefficient and kurtosis) are computed from the gray level distribution.
- *Variogram*
  The geometric variogram function [22] is a morphological texture description defined for a given angle θ and a distance h by:

$$\gamma_{f,\theta}(h) = \frac{1}{2} Var\left(f(x) - f(x + h_\theta)\right), \tag{1}$$

  where $x + h_\theta$ is the translated pixel from x at the amplitude h in the direction θ.
  The variogram itself is not used as a texture parameter. Instead some characteristic values measured on it, for instance, the range, the practical range, the slope [7]. In practice, we studied 6 parameters from variograms in 2 directions $\theta \in \{0°, 90°\}$.
- *Granulometry*
  A morphological granulometry is the study on an image size distribution [23,29]. Pattern spectrum (PS) is a granulometric curve obtained by measuring the difference between successive size growing morphological openings or closings. The *granulometric PS* function (PS^+) and the *anti-granulometric PS* function (PS^-) are just defined by the two following equations:

$$PS^+(f, n) = \frac{m\left(\gamma_n(f)\right) - m\left(\gamma_{n+1}(f)\right)}{m(f)}, \tag{2}$$

and

$$PS^-(f, -n) = \frac{m\left(\varphi_n(f)\right) - m\left(\varphi_{n-1}(f)\right)}{m(f)}, \tag{3}$$

where m is the Lebesgue integral of a gray level image, i.e., the sum of its pixels value, $\gamma_n(f)$ is the morphological opening of f by the nth element of a size increasing family of structuring elements and $\varphi_n(f)$ is the dual morphological closing.

Finally, the mean, variance, asymmetry coefficient and kurtosis of the PS^+ and PS^- functions are computed as texture parameters.

– *Haralick's parameters*
  Gray level co-occurrence matrices into 4 directions ($0°$, $45°$, $90°$, $135°$) are also computed. Among the 14 Haralick's parameters [14] we extracted the 6 parameters considered to be the less correlated [4], i.e., the energy, contrast, entropy, correlation, homogeneity and variance.

4.2 Multivariate Approach from Multispectral Images

Multivariate approach takes into account all wavelengths at the same time and is practiced on original images space (instead of IP-PCA reduced spaces). A spectral distance is necessary to build texture descriptors. We selected the Spectral Angle Mapper (SAM) and the Euclidian distance because they were proved to be the most effective with our dataset [8]. For two spectra $F(x) = \{f_1(x), \ldots, f_D(x)\}$ and $F(y) = \{f_1(y), \ldots, f_D(y)\}$, the distance $SAM(F(x), F(y))$ is given by:

$$SAM(F(x), F(y)) = cos^{-1}\left(\frac{\sum_{d=1}^{D} f_d(x)f_d(y)}{\|F(x)\|.\|F(y)\|}\right).$$

Once the spectral distance is defined, we can extend some marginal texture definitions to the multivariate case as follows

– *Multispectral variogram*
  A multispectral variogram function can be obtained by generalizing the solution proposed in [21], which consists in replacing in (1) the variance operator by a spectral distance:

$$\gamma_{F,\theta}^{Dist}(h) = \frac{1}{2}\sum_x Dist^2\left(F(x), F(x+h)\right), \tag{4}$$

where $Dist$ is a spectral distance (SAM or Euclidian in our case), cumulated on all the pixels of the image. Likewise, the same parameters are measured on the multispectral variogram curve.

– *Multispectral granulometry*
  In order to set up a multispectral pattern spectrum function, we first rely on the color top-hat [13], to propose a multispectral circular top-hat associated to a spectral distance:

$$\rho_B(F(x)) = \sup_{y \in B_x} \{-\xi_B(y)\}, \tag{5}$$

where B_x is the structuring element centered on x and

$$\xi_B(y) = \inf_{z \in B_y} \{-Dist(F(y), F(z))\}. \tag{6}$$

Then, the proposed multispectral circular Pattern Spectrum PS corresponds to (2) where $\gamma_n(f)$ is replaced by $\rho_n(F)$. Contrary to an usual top-hat, the ρ has no interpretation in the sense of positive peak extraction or image order. Hence, there is no need to define an anti-granulometric function.

- *Texture differential image*
  To extract statistical moments and Haralick's parameters in the context of a multivariate integrative approach, we transform the multispectral initial images into texture differential images in which each pixel (of D components) is replaced by the mean distance between the pixel spectrum and spectra of the pixel neighborhood. Haralick's parameters and statistical moments are extracted afterwards on the differential image.

5   Applications to a Cosmetology Study

The protocol (see Sect. 2) to study foundation make-up presence on skin with the Asclepios System mentioned that 4 acquisitions were taken on each volunteer corresponding to 4 states of the applied product (before and after foundation make-up application, 3 h and 6 h after application). Our goal is to predict the acquisition time of each image using its texture descriptors as predictive variables. We expect this multiclass prediction case to be able to illustrate the relevance of texture data depending on whether they are extracted marginally or in a multivariate way.

5.1   Prediction Process

Instead of using Asclepios System entire images, texture descriptors are extracted on 25 smaller samples of initial surface. This method is called bootstraping and allows us to artificially grow the number of experiments from 720 to 180000. Samples dimension (256×256 pixels) is chosen larger than 4 times the size of estimated biggest skin structures.

   In order to process the 4 times prediction, Support Vector Machines (SVM) [32] with a Radial Basis Function (RBF) kernel and a constant σ automaticaly defined (by the hill climbing algorithm) are used. SVM were selected among several predictors (Random Forest, Naives Bayes Classifier, linear regression,...) because they provided the best results. Their extension to multiclass prediction is handled by the One-vs-All model. Precision score i.e. rate of correct class attribution is the considered performance criterium. It is estimated using a cross validation method compatible with the bootstrapping: the 30 volunteers are splitted into 5 groups of 6, then data issued from a group of 6 volunteers (i.e. 20% of the available data) are iteratively used as test samples whereas data from the other groups serve for learning.

Since marginal approach produces more texture features than multivariate approach, we limit the number of multivariate features using a backward feature elimination routine (belonging to the wrapper methods) [11]. Likewise, when blending both marginal and multivariate texture features to proceed the time prediction, the same feature elimination process is used.

5.2   Prediction Results

Prediction accuracy obtained with different texture parameters and approaches of extraction (marginal or multivariate) is given in Table 1. We can notice that the marginal integrative approach seems more effective. However, mixing both multivariate and marginal approaches provides the best precision rates. Thus, multivariate parameters are also interesting in our task of prediction.

Table 1. Precision scores of acquisition time prediction obtained with different texture features computed with marginal (Marg.) and multivariate (Multi.) approaches.

Parameters	Marg.	Multi.	Marg.+Multi.
Statistical moments (8 parameters)	58.4%	42.7%	67.1%
Variogram (18 parmeters)	61.9%	50.1%	61.9%
Granulometry (32 parameters)	61.7%	61.4%	69.2%
Haralick (12 parameters)	61.9%	56.7%	76.0%
All parameters (70 parameters)	82.2%	72.1%	84.8%

6   Conclusion

This paper presented new multispectral texture features based on mathematical morphology as well as two alternatives of multispectral images texture analysis. The first is to reduce images dimensionality before measuring standard parameters on each eigenimage separately, the other is to process multivariate texture extraction directly on all the image wavelengths.

After a dedicated preprocessing chain, data from our study of foundation make-up on skin allow us to demonstrate the efficiency of both approaches and simultaneously the validity of the new multivariate texture features.

The next step is to confirm our methods efficiency on another dataset from a different context.

References

1. Arvis, V., Debain, C., Berducat, M., Benassi, A.: Generalization of the cooccurrence matrix for colour images: application to colour texture classification. Image Anal. Stereology **23**(1), 63–72 (2011)
2. Astola, J., Haavisto, P., Neuvo, Y.: Vector median filters. Proc. IEEE **78**(4), 678–689 (1990)
3. Avena, G., Ricotta, C., Volpe, F.: The influence of principal component analysis on the spatial structure of a multispectral dataset. Int. J. Remote Sens. **20**(17), 3367–3376 (1999)
4. Baraldi, A., Parmiggiani, F.: An investigation of the textural characteristics associated with gray level cooccurrence matrix statistical parameters. IEEE Trans. Geosci. Remote Sens. **33**(2), 293–304 (1995)
5. Baronti, S., Casini, A., Lotti, F., Porcinai, S.: Principal component analysis of visible and near-infrared multispectral images of works of art. Chemometr. Intell. Lab. Syst. **39**(1), 103–114 (1997)
6. Bertalmio, M., Sapiro, G., Caselles, V., Ballester, C.: Image inpainting. In: Proceedings of the 27th Annual Conference on Computer Graphics and Interactive Techniques, pp. 417–424. ACM Press/Addison-Wesley Publishing Co. (2000)
7. Bohling, G.: Introduction to geostatistics and variogram analysis. Kansas Geol. Surv. **1**, 1–20 (2005)
8. Corvo, J.: Characterization of cosmetologic data from multispectral skin images. Ph.D. thesis, Ecole Nationale Supérieure des Mines de Paris (2016)
9. Corvo, J., Angulo, J., Breugnot, J., Borbes, S., Closs, B.: Common reduced spaces of representation applied to multispectral texture analysis in cosmetology. In: SPIE BiOS, p. 970104. International Society for Optics and Photonics (2016)
10. Drimbarean, A., Whelan, P.F.: Experiments in colour texture analysis. Pattern Recogn. Lett. **22**(10), 1161–1167 (2001)
11. Guyon, I., Elisseeff, A.: An introduction to variable and feature selection. J. Mach. Learn. Res. **3**, 1157–1182 (2003)
12. Hanbury, A., Kandaswamy, U., Adjeroh, D.A.: Illumination-invariant morphological texture classification. In: Ronse, C., Najman, L., Decencière, E. (eds.) Mathematical Morphology: 40 Years On, pp. 377–386. Springer, Heidelberg (2005)
13. Hanbury, A.G., Serra, J.: Morphological operators on the unit circle. IEEE Trans. Image Process. **10**(12), 1842–1850 (2001)
14. Haralick, R.M., Shanmugam, K., Dinstein, I.H.: Textural features for image classification. IEEE Trans. Syst. Man Cybern. **6**, 610–621 (1973)
15. Healey, G., Wang, L.: Illumination-invariant recognition of texture in color images. JOSA A **12**(9), 1877–1883 (1995)
16. Jain, A., Healey, G.: A multiscale representation including opponent color features for texture recognition. IEEE Trans. Image Process. **7**(1), 124–128 (1998)
17. Jolivot, R.: Développement d'un outil d'imagerie dédié à l'acquisition, l'analyse et a la caractérisation multispectrale des lésions dermatologiques. Ph.D. thesis, Le2i laboratory, Universite de Bourgogne (2011)
18. Jolivot, R., Benezeth, Y., Marzani, F.: Skin parameter map retrieval from a dedicated multispectral imaging system applied to dermatology/cosmetology. Int. J. Biomed. Imaging **2013**, 26–41 (2013)
19. Kukkonen, S., Kälviäinen, H., Parkkinen, J.: Color features for quality control in ceramic tile industry. Opt. Eng. **40**(2), 170–177 (2001)

20. Lanir, J., Maltz, M., Rotman, S.R.: Comparing multispectral image fusion methods for a target detection task. Opt. Eng. **46**(6), 066402 (2007)
21. Li, P., Cheng, T., Guo, J.: Multivariate image texture by multivariate variogram for multispectral image classification. Photogram. Eng. Remote Sens. **75**(2), 147–157 (2009)
22. Matheron, G.: Principles of geostatistics. Econ. Geol. **58**(8), 1246–1266 (1963)
23. Matheron, G.: Eléments pour une théorie des milieux poreux. Masson (1967)
24. Messer, K., Kittler, J.: A region-based image database system using colour and texture. Pattern Recogn. Lett. **20**(11), 1323–1330 (1999)
25. Palm, C.: Color texture classification by integrative co-occurrence matrices. Pattern Recogn. **37**(5), 965–976 (2004)
26. Palm, C., Keysers, D., Lehmann, T., Spitzer, K.: Gabor filtering of complex hue/saturation images for color texture classification. In: International Conference on Computer Vision, vol. 2, pp. 45–49 (2000)
27. Paola, J., Schowengerdt, R.: A review and analysis of backpropagation neural networks for classification of remotely-sensed multi-spectral imagery. Int. J. Remote Sens. **16**(16), 3033–3058 (1995)
28. Paquit, V.C., Tobin, K.W., Price, J.R., Mériaudeau, F.: 3d and multispectral imaging for subcutaneous veins detection. Opt. Express **17**(14), 11360–11365 (2009)
29. Serra, J.: Image Analysis and Mathematical Morphology V.1. Academic Press, Cambridge (1982)
30. Song, K.Y., Kittler, J., Petrou, M.: Defect detection in random colour textures. Image Vis. Comput. **14**(9), 667–683 (1996)
31. Suen, P.H., Healey, G.: Modeling and classifying color textures using random fields in a random environment. Pattern Recogn. **32**(6), 1009–1017 (1999)
32. Suykens, J.A., Vandewalle, J.: Least squares support vector machine classifiers. Neural Process. Lett. **9**(3), 293–300 (1999)
33. Van Der Maaten, L., Postma, E., Van den Herik, J.: Dimensionality reduction: a comparative. J. Mach. Learn. Res. **10**, 66–71 (2009)
34. Van de Wouwer, G., Scheunders, P., Livens, S., Van Dyck, D.: Wavelet correlation signatures for color texture characterization. Pattern Recogn. **32**(3), 443–451 (1999)
35. Yamaguchi, M., Mitsui, M., Murakami, Y., Fukuda, H., Ohyama, N., Kubota, Y.: Multispectral color imaging for dermatology: application in inflammatory and immunologic diseases. In: Color and Imaging Conference, vol. 2005, pp. 52–58. Society for Imaging Science and Technology (2005)

Application of Size Distribution Analysis to Wrinkle Evaluation of Textile Materials

Chie Muraki Asano[1], Akira Asano[2(✉)], and Takako Fujimoto[3]

[1] Faculty of Human Life and Environmental Sciences,
Nagoya Women's University, Nagoya, Japan
7mikeneko@gmail.com
[2] Faculty of Informatics, Kansai University, Takatsuki, Osaka, Japan
a.asano@kansai-u.ac.jp
[3] Faculty of Education, Hokkaido University of Education, Sapporo, Japan

Abstract. An evaluation method of wrinkle shapes on fabrics using simple morphological operations is proposed. It calculates the size density function of the object indicating the region surrounded by a folded fabric in the experimental condition of the standard wrinkle/crease angle test. The characteristics of the size density function and the parameters of the linear and cubic function fitted to the size density function indicate the fabric shape characteristics, which correspond to visual edge sharpness and roundedness of pleat lines, as well as their mechanical properties.

Keywords: Size distribution analysis · Textile · Wrinkle evaluation

1 Introduction

Size distribution analysis has been one of the most fundamental applications of mathematical morphology since the early days of its development. Size distribution analysis is widely applied to evaluating granularity of textures, roughness of material surfaces, etc.

We propose an application of size distribution analysis to textile sciences. We employ it for evaluating shapes of wrinkles on fabrics. Our objective is describing quantitatively visual characteristics of wrinkles, which are directly connected to visual aesthetic sense of textile materials. The wrinkle is semipermanent deformation of textile materials caused by mechanical stress, and it has been characterized by measurements of rheological properties such as bending stiffness and hysteresis of the fabrics or textile materials and statistical regression methods in textile sciences.

The characterization of textile products based on the material mechanical properties is, however, not always directly related to the visual aesthetic sense, i.e. evaluation of finally appearing shape characteristics, because of the complexity of these structures. Additionally, the conventional method has also difficulties when it is applied to newly developed materials.

We propose a method of quantitatively describing shape characteristics of wrinkles, as an extended version of the preliminary proposal in a conference

© Springer International Publishing AG 2017
J. Angulo et al. (Eds.): ISMM 2017, LNCS 10225, pp. 478–485, 2017.
DOI: 10.1007/978-3-319-57240-6_39

on human affective sciences [1], and contribute to textile sciences by providing characteristics connected to visual aesthetic sense. The wrinkle shapes are determined by several construction levels; Fiber structures, yarn structures, which are twisted fiber bundles, and two-dimensional structures of textile materials that are woven, knitted, or non-woven. According to this fact, we propose a method describing the shape of wrinkle or crease by categorizing the size density function of the shapes based on the fitted polynomial functions, and find the relationship between the parameters of the polynomial functions and visual characteristics.

Firstly, we derive size density functions of artificial wrinkle shapes, including a sharp triangular shape, which is the model of wrinkle/crease of paper-like materials, and a rounded shape, which is the model of wrinkles caused by all of the mechanical and structural properties of textile materials; Fiber, yarn, and two-dimensional structures. We obtain the parameters of the polynomial functions fitted to the size density functions, and observe the relationship between the parameters and shapes. We then apply the method to the images of the pleats on actual textile materials, and show the parameters and their relationships to visual aesthetic sense.

2   Wrinkle Evaluation of Textile Materials

Evaluation of shapes of wrinkles, creases and pleats has attracted garment producers and consumers since the 1960's in the field of textile sciences. It was at that time that the standards for the measurement of wrinkle/crease property were established worldwide [2], and since then various apparatuses for this purpose have been designed and developed. Those instruments employ the same simple principle of measurement. It is called wrinkle/crease angle test, described as follows:

A specimen of fabric in rectangle shape of 10 cm × 20 cm is prepared. A fold of fabric in half is pressed with a fixed weight load between two plates, which are tiles of 10 cm × 10 cm, so as to make coincide the folding line. After releasing the pressure, an angle of two wings of the specimen, centralizing the pleat line, is measured.

The above method observes a fold/crease of one fabric under a fixed weight load for a fixed time. The method has been widely used since it is directly related to mechanical properties of fabrics. The method of folding one fabric observes the angle of the two wings of the specimen. The angle, or expansion of the two wings, indicates the resilience of the fabric. However, there are more characteristics to be observed in this experiment. The edge sharpness of the pleat line influences the visual impression of clothes, and the curve of the wings indicates the elasticity of the fabric.

The method proposed in this paper aims to evaluate these characteristics of wrinkle using image processing techniques. Some methods of evaluating wrinkle roughness on the surface of the whole fabric using image processing were proposed, since wrinkles on the whole fabric have an influence to visual impressions on cloths [3,4]. Our method, on the other hand, evaluate them in the standard experimental condition of the wrinkle/crease angle test by employing simple morphological operations.

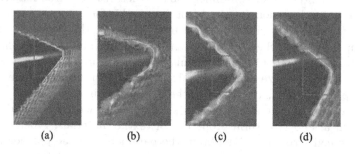

    (a)   (b)   (c)   (d)

Fig. 1. Examples of folded fabrics in the experimental condition. Squared regions are extracted for the proposed method. Traces along the wings are also indicated.

3   Method

The procedure of the proposed method is summarized as follows:

1. The shape of the folded material is traced, and a binary image expressing the region surrounded by the fabric is generated.
2. The size density function of the binary image is calculated. The disk-shape structuring element is employed. The size n is defined as a radius of each disk in pixel. The structuring element in the discrete space is defined as the set of pixels that are located no greater than n away from the origin in Euclidean distance.
3. The size density function is approximated by removing the noisy components caused by jagged edge of the region. Let $f(n)$ be the original size density function and $g(n)$ be the approximated size density function. We set $g(0) = 0$, and repeat the following procedure for $n = 1, 2, \ldots, N$:

$$g(n) = \begin{cases} f(n) & \text{if } f(n) \geq \max[g(i)] \text{ for } i = 1, 2 \ldots n - 1, \\ 0 & \text{otherwise.} \end{cases} \quad (1)$$

   The maximum size N is the smallest number satisfying $f(n') = 0$ for all $n' > N$.

   The approximation extracts peaks that are monotonically increasing. It is performed to extract characteristic peaks in the size density function using the property of the images that the two wings of the specimen expand monotonically.
4. The element at the maximum size, $g(N)$, is removed, since it indicates the area of the residual part outside the region surrounded by the wings.
5. A polynomial function is fitted to the approximated size density function $g(n)$, and the parameters are estimated. These parameters indicate the shape of the wrinkle and suggest the property of the target material.

4   Experimental Results and Discussion

4.1   Experiments on Artificial Shapes

We firstly applied our method to binary images of three artificially generated shapes, resembling shapes of folded fabrics. They are accounted to be models of typical types of actual fabrics. Application of the method to actual fabrics classifies them to one of the models and presents the most effective construction level of the fabric to the generation of the wrinkle, as shown in the next subsection. The size of images is 100×100 pixels.

Figure 2 shows the artificial shape AS1, its original size density function, its approximated size density function, and the linear function fitted to the approximated size density function. The value of the size density function is magnified 10^4 times. The values of the size density function at small sizes has meaningful information on the wrinkle, and the values at these sizes are very small. The magnification is applied in order to avoid that the parameters of the polynomial functions fitted to the size density function of meaningful sizes becomes too small numbers. The estimated formula of the fitting and the coefficient of determination, denoted by R^2, are also indicated. Figures 3 and 4 are the results for the cases of the other artificial shapes, denoted by AS2 and AS3, respectively. The original size density functions are omitted, and the approximated functions only are shown.

The shape AS1 is a model of fabrics that have significant semipermanent deformation and low resilience. The pleat line is very sharp and the wings are

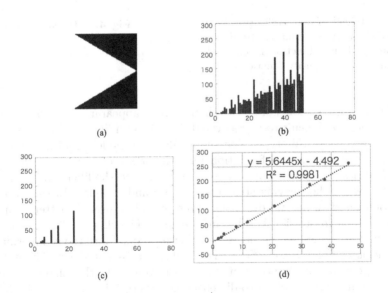

Fig. 2. Artificial shape AS1. (a) Artificial shape. (b) Original size density function. (c) Approximated size density function. (d) Linear function fitted to the approximated size density function. The estimated formula and the coefficient of determination (R^2) are indicated.

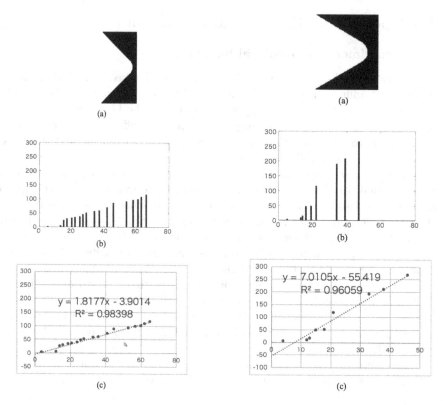

Fig. 3. Artificial shape AS2. (a) Artificial shape. (b) Approximated size density function. (c) Linear function fitted to the approximated size density function.

Fig. 4. Artificial shape AS3.

straight. This is the model of fabrics showing poor appearance because the wrinkles on them are difficult to be recovered. Plain-woven linen and thin paper are typical examples of this type. The shape AS2 is a model of fabrics that have partially semipermanent deformation but still preserve elastic resilience. This type of fabrics is composed of short or thin fibers and the fiber bundle is easily deformed, while the weaving structure is dense and complex. In this case, the evaluation based on the two-dimensional rheological property is better than AS1, however, visual impression is sometimes poorer. The shape AS3 is a model of fabrics that are composed of longer and thicker fibers. Since the semipermanent deformation of fibers is weak, the edge of the pleat is not sharp. The wrinkles appearing on fabrics composed of thick fibers resembles this shape, since the weight per unit area of the two-dimensional structure is large and the fabrics can form rounded shapes easily. The appearance of wrinkles on this type of fabrics is not so poor, and the wrinkles can be recovered by annealing processes.

The minimum sizes appearing in the elements, except isolated small elements, of the approximated size density functions of AS1, AS2, and AS3 indicate the edge sharpness of the pleat lines. They correspond to the deformability of these fabrics by mechanical stress. The coefficient of determination, R^2, of the linear fitting is smaller in AS3 than in AS1 and AS2. The lower fitness indicates the rounded shape caused by the two-dimensional structures of longer and thicker fibers. The effect of the structures can be examined by fitting higher-order polynomial functions, as explained in the next subsection.

4.2 Experiments on Actual Fabrics

Our method was applied to actual fabrics shown in Fig. 1(a)–(d), denoted by AF1-AF4, respectively. The size of the binarized images is 128×128 pixels. The results are shown in Figs. 5, 6, 7 and 8.

Figure 5 shows that the fitness, evaluated by R^2, of the linear function is high and the minimum size is large, as well as AS2, in the case of AF1. It indicates that the visual impression caused by yarn and two-dimensional structures is similar to AS2, and that caused by fiber structures is similar to AS3. In the

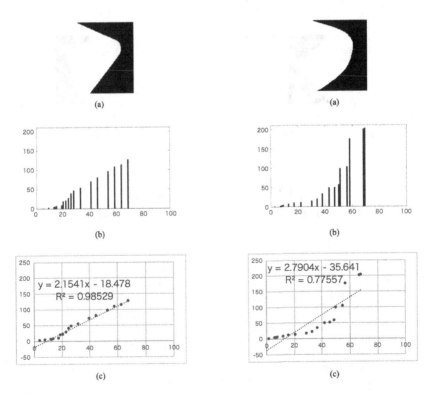

Fig. 5. Actual fabric AF1, extracted from Fig. 1(a).

Fig. 6. Actual fabric AF2, extracted from Fig. 1(b).

case of AF2, as shown in Fig. 6, the fitness R^2 is small and the minimum size is large. It indicates that this sample has higher resilience and higher ability of the recovery of wrinkles than the model AS3. The sample AF3 shows high fitness and the minimum size is small, as shown in Fig. 7. It indicates that the sample is close to AS1 and AS2 in fiber structures. It has a small coefficient on the linear term, and it indicates that the visual impression caused by yarn and two-dimensional structures is similar to AS2. Finally, the sample AF4 shows low fitness and the minimum size is large, as shown in Fig. 8. It indicates that the sample is close to AS3 in general, however, it is closer to AS2 in two-dimensional structures since it has a small coefficient on the linear term.

Figure 9(a) shows the fitting of the cubic function to the approximated size density function of AF2. In this case R^2 is significantly higher than in the case of the linear fitting. The shapes showing the characteristic of two-dimensional structures are expressed by cubic curves. It indicates that the sample AF2 has higher elasticity than the others. The fitness of the linear function in the case of AF4, as shown in Fig. 8(d), is also rather lower. The fitting of the cubic function to the case of AF4 is shown in Fig. 9(b). The fitness becomes higher, and it indicates that the shape caused by the two-dimensional structure in AF4 shows a curve caused by its elasticity.

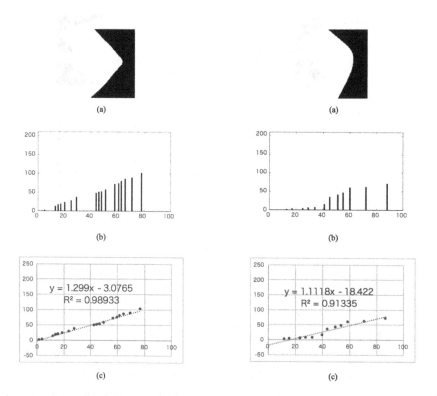

Fig. 7. Actual fabric AF3, extracted from Fig. 1(c).

Fig. 8. Actual fabric AF4, extracted from Fig. 1(d).

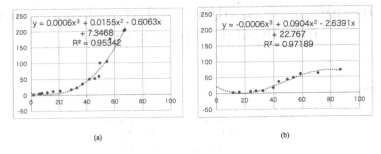

Fig. 9. Fitting by cubic function. (a) Actual fabric AF2 and (b) AF4.

5   Conclusions and Future Remarks

We have proposed a method of evaluating the shape of wrinkle/crease on fabrics in the experimental condition using the morphological size density function. The evaluated characteristics of the shapes correspond to fabrics' properties, which are edge sharpness on the pleat line, as well as resilience and elasticity of fabrics.

We employed the fixed disk-shape structuring elements for calculating the size density function. If some flexibility is introduced to the shapes of the structuring elements, it is expected that richer and more precise information is obtained. We have proposed methods of finding shapes of structuring elements fitting image objects using an optimization technique [5], and the methods can be utilized for this.

Acknowledgments. This research was partially financially supported by JSPS KAK-ENHI Grant No. 15K00706 and Kansai University Researcher Grant 2016.

References

1. Muraki Asano, C., Asano, A., Ouchi, R., Fujimoto, T.: A new evaluation method for fabric wrinkles using the morphological technique. In: Proceedings of the 5th Kansei Engineering & Emotion Research International Conference, no. 123 (2014)
2. Industrial standards for wrinkle evaluations: IS4681, BS3086, ASTM D 1295 and JIS L 1059-1
3. Su, J., Xu, B.: Fabric wrinkle evaluation using laser triangulation and neural network classifier. Opt. Eng. **38**(10), 1688–1693 (1999)
4. Matsudaira, M., Han, J., Yang, M.: Objective evaluation method for appearance of fabric wrinkling replica by image processing system. J. Text. Eng. **48**(1), 11–16 (2002)
5. Yang, L., Li, L., Muraki Asano, C., Asano, A.: Primitive and grain estimation using flexible magnification for a morphological texture model. In: Soille, P., Pesaresi, M., Ouzounis, G.K. (eds.) ISMM 2011. LNCS, vol. 6671, pp. 190–199. Springer, Heidelberg (2011). doi:10.1007/978-3-642-21569-8_17

Morphological Analysis of Brownian Motion for Physical Measurements

Élodie Puybareau[1,2], Hugues Talbot[2(✉)], Noha Gaber[3], and Tarik Bourouina[2]

[1] EPITA Research and Development,
14-16 rue Voltaire, 94276 Le Kremlin-Bicetre, France
[2] Université Paris-Est/ESIEE,
2 Boulevard Blaise-Pascal, 93162 Noisy-le-Grand, France
hugues.talbot@esiee.fr
[3] National Oceanic and Atmospheric Administration, Washington, D.C., USA

Abstract. Brownian motion is a well-known, apparently chaotic motion affecting microscopic objects in fluid media. The mathematical and physical basis of Brownian motion have been well studied but not often exploited. In this article we propose a particle tracking methodology based on mathematical morphology, suitable for Brownian motion analysis, which can provide difficult physical measurements such as the local temperature and viscosity. We illustrate our methodology on simulation and real data, showing that interesting phenomena and good precision can be achieved.

Keywords: Random walk · Particle suspension · Particle tracking · Segmentation

1 Introduction

Brownian motion was first described by Robert Brown in 1828 [1]. He observed the jittery motion of microscopic particles suspended in water, but could not explain it at the time. Einstein, in one of his 1905 *annus mirabilis* articles [2] was able to propose a consistent theory of Brownian motion based on random walks. He explained that the observed motion was the result of random collisions between water molecules and the observed particles. Due to the discrete nature of these collisions, at any point in time the forces acting on a particle are constantly changing, resulting in an unpredictable trajectory. Jean Perrin carefully designed experiments that proved that the Einstein theory was correct, the first convincing observational evidence of the existence of molecules [3]. The Einstein theory together with the Perrin experiments allowed estimations of the size of atoms and how many atoms there are in a mole (the Avogadro number). This earned Jean Perrin the Nobel prize in physics in 1926.

Einstein had sought to estimate the distance a Brownian particle travels in a given time. Because the number of collisions between molecules and particles is enormous, classical mechanics could not be used. Instead Einstein called upon

J. Angulo et al. (Eds.): ISMM 2017, LNCS 10225, pp. 486–497, 2017.
DOI: 10.1007/978-3-319-57240-6_40

the notion of random walks. Using the ensemble motion of a large number of particles, Einstein was able to show that the density of particles obeys a diffusion equation.

Random walks are an interesting and ubiquitous model for a number of stochastic processes. Besides physics, they are useful in finance [4], ecology [5], biology [6], chemistry [7], and of course imaging [8]. They form the basis of stochastic optimisation, particularly MCMC methods [9,10]. In imaging, these approaches led to Markov Random Fields methods [11].

Random walk also allow the construction of complex models [12], both unpredictable at the short timescale and highly regular in the long run. Indeed, many stochastic processes converge to the Brownian motion model, which is one of the simplest yet most irregular stochastic evolution process.

Random walks have been very widely studied. In this article, we are not so much interested in their phenomena per se, as we are in the image analysis of Brownian motion. Indeed, we show that the image analysis of Brownian motion can lead to useful physical measurements. As far as we know, Brownian motion has been used to estimate temperature before [13] but not through automated image analysis methods.

In the remainder of this article, we define the Brownian motion model and its more precise links to random walks and diffusion phenomena in Sect. 2. We describe our sequence processing pipeline in Sect. 3. We then formalize the expected results of physical Brownian motion via particle tracking in Sect. 4. Finally we show results on simulations and actual data.

2 Brownian Motion, Random Walks and Diffusion

What is now widely referred to as "Brownian motion" is no longer the physical phenomenon that Robert Brown observed in 1828. Instead, it is a process model, also called the Wiener process.

Definition 1. *Let W_t be a random process with the following properties:*

1. $W_0 = 0$ *a.s;*
2. *W has independent increments, meaning if $0 \leq s_1 < t_1 \leq s_2 < t_2$ then $W_{t_1} - W_{s_1}$ and $W_{t_2} - W_{s_2}$ are independent random variables;*
3. *W has Gaussian increment, meaning $W_{t+u} - W_t \sim \mathcal{N}(0, u)$, i.e. $W_{t+u} - W_t$ is Normally distributed with variance u;*
4. *W is continuous a.s.*

This process is non-smooth at all scale (no matter how small $t_1 - s_1$ is), and so largely theoretical, but can be well approximated by random walks.

Definition 2. *Let $\xi_1, \xi_2, \ldots$ be i.i.d random variables with mean 0 and variance 1, Let W_t^r be the following random step function:*

$$W_n^r(t) = \frac{1}{\sqrt{n}} \sum_{1 \leq k \leq \lfloor nt \rfloor} \xi_k$$

This function is called a random walk. We now have the following theorem:

Theorem 1. *As $n \to \infty$, $W_n^r(t)$ converges to the Wiener process W_t.*

Proof. The proof is given in [14], but the idea is intuitive. Indeed, $W_n^r(0) = 0$; moreover, since the ξ_i are independent, $W_n^r(t)$ has independent increments; for large n, $W_n^r(t) - W_n^r(s)$ is close to $\mathcal{N}(0, t-s)$ by the central limit theorem (CLT). The continuity argument is more difficult. Since W_n^r is a step function, it is continuous a.e, but this is of little utility in the limit. However as $n \to +\infty$, the jumps tend to zero even though jumps may be arbitrary large initially. This guarantees the almost sure continuity.

In practical terms, this means that a sum of i.i.d Normal random variables converges to the Brownian motion process, and so is easy to simulate. In fact, because so many distributions converge to the Normal distribution by the CLT, we can use simpler distributions for the steps. For instance, binary steps where $\xi_i \in \{+1/-1\}$ with equal probabilities instead of Normal steps are fairly standard.

This process extends to any dimension by choosing vector steps with i.i.d. components (binary or Normal). In the binary case, this corresponds to random walks on an infinite regular grid. There are a number of interesting facts regarding random walks and Brownian motion, in particular:

$$E[W_t] = 0 \tag{1}$$
$$V[W_t] = t \tag{2}$$

A proof in the discrete case is given below. Pólya showed [15] that random walks return to the origin with probability 1 in 1-D and 2-D but not for dimensions greater than 2.

2.1   Average and Variance of the Position

In the simplest case, we consider a random walker making steps of length l in one dimension. This means that at each timestep s_i of duration τ, this random walker can move one position to the right or left with equal probability.

$$s_i = \begin{cases} -l \text{ with probability } 1/2 \\ +l \text{ with probability } 1/2 \end{cases} \tag{3}$$

After N steps (at time $N\tau$), the position of the walker is

$$x(N) = \sum_{i=1}^{N} s_i. \tag{4}$$

with going left or right equiprobable. Starting from 0, the average position is

$$E[x(N)] = 0 \tag{5}$$

i.e. the average position is always at the origin. However the variance of the position changes with time. We write

$$x^2(N) = \left(\sum_{i=1}^{N} s_i\right)^2 = \sum_{i=1}^{N} s_i^2 + \sum_{i=1}^{N} \sum_{j=1,j\neq i}^{N} s_i s_j \qquad (6)$$

However, the quantity $s_i s_j$ for a pair $i, j, i \neq j$ is

$$s_i s_j = \begin{cases} -l^2 \text{ with probability } 1/2 \\ +l^2 \text{ with probability } 1/2 \end{cases}, \qquad (7)$$

so on average it will be zero. On the other hand s_i^2 will always be l^2, therefore

$$E[x^2(N)] = l^2 N. \qquad (8)$$

after N steps. Since the average of x is zero, this is also the variance. We see that it increases linearly with time.

2.2   Relation to Diffusion

We now see how random walks behave as the timestep τ tends to zero [16]. Let $P(i, N)$ denote the probability that a walker is at position i after N timesteps. Due to the equal probability for a walker to move left or right, we have the recursive equation:

$$P(i, N) = \frac{1}{2} P(i+1, N-1) + \frac{1}{2} P(i-1, N-1) \qquad (9)$$

We write $x = il$ and $t = \tau N$, since these probabilities are scale-independent we find

$$P(x, \tau) = \frac{1}{2} P(x+l, t-\tau) + \frac{1}{2} P(x-l, t-\tau) \qquad (10)$$

Subtracting $P(x, t-\tau)$ and dividing by τ, we have

$$\frac{P(x,t) - P(x,t-\tau)}{\tau} = \frac{l^2}{2\tau} \frac{P(x+l,t-\tau) + P(x-l,t-\tau) - 2P(x,t-\tau)}{l^2} \qquad (11)$$

The left-hand side is a first order finite difference approximation of $\frac{\partial P}{\partial t}$ and the right-hand side is a first order approximation of $\frac{\partial^2 P}{\partial x^2}$. As τ and l tend to zero but $l^2/2\tau$ remains constant, we have

$$\frac{\partial P}{\partial t} = \frac{l^2}{2\tau} \frac{\partial^2 P}{\partial x^2}, \qquad (12)$$

which is the well-known one-dimensional diffusion equation, a continuous process. A similar derivation can be achieved in arbitrary dimension.

3   Processing of Brownian Motion Sequences

In this section, we describe our sequence processing pipeline. The data we wish to process comes from a bespoke microfluidic device produced at ESIEE Paris [17]. The objective of this device is to allow the study of optical trapping at the micro scale. Because of the small size of the device and of the particles, existing particle tracking system do not work sufficiently well [18].

Pre-processing. The sequences were recorded under a Leica inverted optical microscope, observing a capillary tube embedded in an optical trapping device. The tube contains the particles and is surrounded by the microfluidic system itself. Only the area containing the particles is of interest to us, so we first cropped the sequences around the capillary and automatically corrected the luminosity variations that occur during acquisition, by reference to the average image over the entire sequence. We then removed the non-moving components (fluidics system, capillary) by subtracting this average image. We named this pre-processed sequence $\mathcal{S}_0$. A sample frame from this output is shown on Fig. 1.

(a)   (b)

Fig. 1. A sample frame from the initial sequence, and the output of the pre-processing for that frame.

Particle Segmentation. We simplified $\mathcal{S}_0$ using an area black top-hat on the h-maxima on each frame of the sequence:

$$\forall I \in \mathcal{S}_0, \quad I_{\mathrm{hm}} = \varphi_R(I - h, I) - I \tag{13}$$

$$I_{\mathrm{bta}} = I_{\mathrm{hm}} - \gamma_\alpha(I_{\mathrm{hm}}) \tag{14}$$

These are classical mathematical morphology operators [19]. φ_R is the closing by reconstruction [20]; γ_α is the area opening of parameter α [21], and h is the height parameter. These are efficiently implemented using the component tree [22,23].

This pipeline retains the particles and erases the background. It depends on few parameters. The area parameter α is twice the known area of the particles and h was hand-tuned at 15. We then smoothed the result using a 2D+t median filter on this sequence considered as a 3D image, of size $3 \times 3 \times 10$. We thresholded and denoised this sequence by erasing the very small remaining components via a small 3×3 opening. The result is named $\mathcal{S}_1^{3D}$ (Fig. 2).

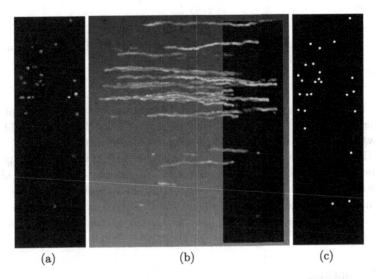

(a)   (b)   (c)

Fig. 2. In (a) the result of the black top-hat. In (b) the 2D+t traces. In (c), the segmented particles.

Trajectory Separation. We used a morphological erosion from a Euclidean distance map on $\mathcal{S}_1^{3D}$ to filter out small components and separate some traces. From this result, we estimated a discrete trajectory from the thresholded center on each slice. We then dilated these detected centroids with a disk in order to obtain a smooth and regular mask of the trajectories. We then computed the 3D skeleton of this binary mask [24]. Because the particles float in a 3D medium, they can appear to overlap and so their trajectory can merge. We detect crossing points in our skeleton that would cause non-unique labels on each slice. We then removed the triple points before labeling the trajectory.

Now that our trajectories are unique and identified, we dilated them with a disk with a greater radius than that of the particles to obtain a labeled mask.

Because the particles are round, a more precise way to find their centroids is to compute a weighted average of the coordinates of the pixels belonging to the trajectory in the mask according to the grey-level intensities of the initial sequence. We hence obtain sub-pixel accurate coordinates of the particles on each frame associated with a unique label ensuring the unicity of the tracking. We now study how to use these detected trajectories.

4   Particle Tracking

We now wish to exploit the potential of Brownian motion for physical measurements

In (12), the quantity $l^2/2\tau$ is called the diffusion coefficient D.

$$D \equiv \frac{l^2}{2\tau} \tag{15}$$

Einstein [2] has shown that in a purely viscous fluid, with no external force influence

$$D \propto \frac{T}{\eta}, \tag{16}$$

where T is the absolute temperature and η the viscosity. In general these physical quantities are difficult to measure at the microscopic level. The study of diffusion via Brownian motion analysis is a potentially powerful method.

For this we assume neutrally buoyant, non-interacting test particles floating in a fluid and observed under a microscope. We assume that we are able to measure the position of these particles with arbitrary spatial and time resolution. This is not an outrageous demand, since Brownian motion is scale-independent, as we have seen. It means that we can trade some spatial and time resolution with each other.

4.1   Simulations

To assess the potential of Brownian motion to measure physical quantities, we simulate a 2D field with temperature varying in space and time, but with constant viscosity, where test particles are present.

Each particle is assumed to evolve independently of each other (no shocks or other interaction) irrespective of the density. To simulate temperature change, we vary the diffusion parameter D of Eq. (12) subject to arbitrary, but controlled change. This changes the expected spatial step according to (15). To measure the temperature, we allow ourselves to only use the parameters of the Brownian motion of each particle.

4.2   Mean Square Displacement

Because a single spatial step of an arbitrary particle can only be expected to resemble a random deviate from a Normal distribution of variance proportional to l^2, it cannot be used in isolation. Instead, we can estimate D by an averaging process. For this we define the mean square displacement (MSD) as follows:

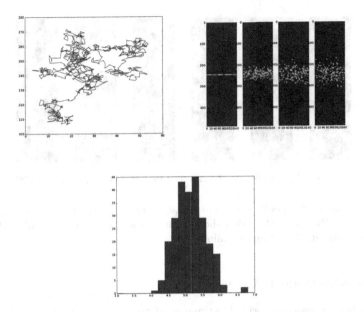

Fig. 3. (a) A simulated random walk. (b) A simulation of a diffusive process by random walk at time 0, 50, 150 and 300. The color of each particle represents the estimation of the diffusion coefficient. (c) The dispersion of the diffusion coefficient estimation at time 100 over all particles. (Color figure online)

$$\text{MSD}(\tau) = E[\Delta r^2(\tau)] = \frac{1}{m} \sum_{t=1}^{m} |r(t+\tau) - r(t)|^2, \tag{17}$$

where $r(t)$ is the measured position of the studied particle at time t, and τ the timestep. Since $\text{MSD}(\tau) = l^2/2n$, from (15), we should expect $\text{MSD}(\tau)/2\tau$ to be constant and a reasonable approximation of D. To provide a stable estimate of the diffusion coefficient from a particle, we must consider a varying τ. In the discrete case, we can consider averaging the sequence $\text{MSD}(n\tau)/n\tau$.

$$D \approx \frac{1}{2n} \sum_{i=1}^{n} \frac{\text{MSD}(n\tau)}{n\tau}. \tag{18}$$

Averaging over several particles provides an even stabler estimation. On Fig. 3, we show the output of a simulation of diffusing particles starting from a single line over time with a constant D. The color of each particle represents the estimate of the diffusion coefficient given by the motion analysis of that particle. For this estimation, we used $m = 150$ in (17) and $n = 5$ in (18). As seen in Fig. 3, the estimation of this coefficient has a large variance but is still be usable. Indeed the simulated D was 5.0 and the median estimated D was 5.05.

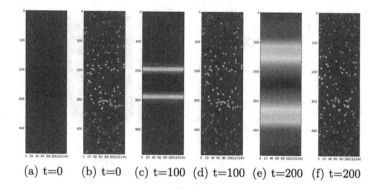

(a) t=0   (b) t=0   (c) t=100   (d) t=100   (e) t=200   (f) t=200

Fig. 4. Simulation of a cooling medium (a,c,e) sampled by the random walk analysis of 200 particles (b,d,f). Color code indicates the temperature, proportional to the local diffusion coefficient. (Color figure online)

4.3 Time-Dependent Analysis

To test whether motion-derived estimates of the diffusion coefficient are sufficiently precise, we simulate a time-dependent process illustrated on Fig. 4. In this test, a hot rectangular area is allowed to cool over time by conduction in a cooler medium. 200 particles are uniformly randomly placed and tracked over

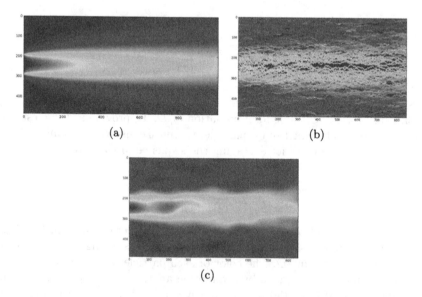

Fig. 5. (a): 1D+t ground-truth representation of the diffusion coefficient in the cooling medium of Fig. 4. (b): Superposition of the trace estimation for the diffusion coefficient D. (c) 2D Kriging estimation of the same diffusion coefficient. The color code is the same in all three images. (Color figure online)

time. The estimated MSD is computed and associated with each particle and is color-coded with the same scheme. Since the problem is really only 1D+t, as there is no variation along the horizontal x axis, we can project all the estimated MSD onto the vertical y axis, as shown on Fig. 5(a). The MSD estimation is quite sparse, so it is beneficial to interpolate it. To achieve this, we used 2D universal kriging [25,26]. Our input data are all the traces points with their estimated diffusion coefficients.

In this experiment, D is estimated over 150 causal trace steps, so to avoid border effects, the last 150 steps are not estimated, shortening the sequence by that amount. As we can see in Fig. 5(c), the estimation is now of reasonable quality and appears bias-free.

5   Results on Real Data

Starting from our estimated traces from Sect. 3, we first verify their random walk qualities, then estimate the diffusion coefficient image in the same way as in the simulation.

On Fig. 6, we show the trace of one of the particle over time, and we estimate the associated MSD from (17). We note the subpixel accuracy of the trace, and the linear aspect of MSD(τ) with respect to τ. All sufficiently long trace (time length > 150 step) were found to exhibit similar characteristics. From these we estimate the MSD at every point of all the segmented traces, and interpolated the data as above. This is illustrated on Fig. 7.

On this experiment, the particles were held in an optical trap before $t = 0$ and then released at that time. We expect the particles to diffuse and heat as they move into the medium, i.e. we expect the traces to expand and become redder in the false-color rendition of the result, which is what we are indeed observing.

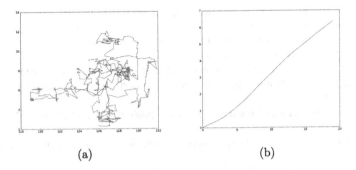

(a)   (b)

Fig. 6. (a) The trace of a real particle. Note the subpixel accuracy of the trace and similarity to Fig. 3(a) ; (b) the MSD(τ) vs. τ associated with this trace, which is linear as expected.

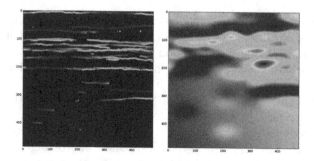

Fig. 7. (a) Real, segmented traces with associated estimated MSD as in Fig. 5. (b) Interpolated MSD.

6   Conclusion

In this article we have shown that subpixel-accurate trace analysis of the Brownian motion of microscopic particles is possible, even in challenging situations, and that it can provide estimates of the local diffusion coefficient, which is proportional to the temperature divided by the viscosity of the medium. We have provided a full pipeline, validated on simulated data and tested on real data. Future work will validate the physical measurements in more controlled acquisitions where a stable temperature gradient can be established. Also, the current diffusion coefficient estimation includes an integrating step over a long period (150 time steps in our study) for accuracy. This integration blurs the estimation in time and in the direction of the particle travel. We plan to correct this effect by considering it as an inverse problem.

References

1. Brown, R.: A brief account of microscopical observations made in the months of June, July and August, 1827, on the particles contained in the pollen of plants; and on the general existence of active molecules in organic and inorganic bodies. Philos. Mag. **4**(21), 161–173 (1827)
2. Einstein, A.: Über die von der molekularkinetischen theorie der wärme geforderte bewegung von in ruhenden flüssigkeiten suspendierten teilchen. Ann. Phys. **322**(8), 549–560 (1905)
3. Perrin, J.: Mouvement brownien et réalité moléculaire. Ann. Chim. Phys. **18**(8), 5–114 (1909)
4. Bachelier, L.: Théorie de la spéculation. Ann. Sci. l'École Normale Supér. **3**(17), 21–86 (1900)
5. Skellam, J.G.: Random dispersal in theoretical populations. Biometrika **38**(1/2), 196–218 (1951)
6. Colding, E., et al.: Random walk models in biology. J. R. Soc. Interface **5**, 813–834 (2008)
7. De Gennes, P.G.: Scaling Concepts in Polymer Physics. Cornell University Press, Ithaca (1979)

8. Grady, L.: Random walks for image segmentation. IEEE Trans. Pattern Anal. Mach. Intell. **28**(11), 1768–1783 (2006)
9. Metropolis, N., Rosenbluth, A.W., Rosenbluth, M.N., Teller, A.H., Teller, E.: Equation of state calculations by fast computing machines. J. Chem. Phys. **21**(6), 1087–1092 (1953)
10. Hastings, W.: Monte Carlo sampling methods using Markov chains and their applications. Biometrika **57**(1), 97–109 (1970)
11. Geman, S., Geman, D.: Stochastic relaxation, Gibbs distributions, and the Bayesian restoration of images. IEEE Trans. Pattern Anal. Mach. Intell. **6**, 721–741 (1984)
12. Bertsimas, D., Vempala, S.: Solving convex programs by random walks. J. ACM (JACM) **51**(4), 540–556 (2004)
13. Park, J., Choi, C., Kihm, K.: Temperature measurement for a nanoparticle suspension by detecting the Brownian motion using optical serial sectioning microscopy (OSSM). Meas. Sci. Technol. **16**(7), 1418 (2005)
14. Donsker, M.: An invariance principle for certain probability limit theorems. Mem. Am. Math. Soc. **6** (1951)
15. Pólya, G.: Über eine aufgabe betreffend die irrfahrt im strassennetz. Math. Ann. **84**, 149–160 (1921)
16. Nordlund, K.: Basics of Monte Carlo simulations. http://www.acclab.helsinki.fi/~knordlun/mc/mc5nc.pdf
17. Gaber, N., Malak, M., Marty, F., Angelescu, D.E., Richalot, E., Bourouina, T.: Optical trapping and binding of particles in an optofluidic stable fabry-pérot resonator with single-sided injection. Lab Chip **14**(13), 2259–2265 (2014)
18. Allan, D., et al.: Trackpy: fast, flexible particle-tracking toolkit. http://soft-matter.github.io/trackpy
19. Najman, L., Talbot, H. (eds.): Mathematical Morphology: from Theory to Applications. ISTE-Wiley, London, September 2010. ISBN 978-1848212152
20. Vincent, L.: Morphological grayscale reconstruction in image analysis: applications and efficient algorithms. IEEE Trans. Image Process. **2**(2), 176–201 (1993)
21. Vincent, L.: Grayscale area openings and closings, their efficient implementation and applications. In: Proceedings of the Conference on Mathematical Morphology and Its Applications to Signal Processing, Barcelona, Spain, pp. 22–27, May 1993
22. Meijster, A., Wilkinson, H.: A comparison of algorithms for connected set openings and closings. IEEE Trans. Pattern Anal. Mach. Intell. **24**(4), 484–494 (2002)
23. Géraud, T., Talbot, H., Vandroogenbroeck, M.: Algorithms for mathematical morphology. In: [19] Chap. 12, pp. 323–354. ISBN 978-1848212152
24. Bertrand, G., Couprie, M.: Transformations topologiques discretes. In: Coeurjolly, D., Montanvert, A., Chassery, J. (eds.) Géométrie discrète et images numériques, pp. 187–209. Hermès, Mumbai (2007)
25. Matheron, G.: The Theory of Regionalized Variables and Its Applications, vol. 5. École national supérieure des mines, Paris (1971)
26. Olea, R.A.: Optimal contour mapping using universal kriging. J. Geophys. Res. **79**(5), 695–702 (1974)

Author Index